Thermodynamics: From Concepts to Applications

Thermodynamics: From Concepts to Applications

Edited by **Barney Tyler**

NY RESEARCH PRESS

New York

Published by NY Research Press,
23 West, 55th Street, Suite 816,
New York, NY 10019, USA
www.nyresearchpress.com

Thermodynamics: From Concepts to Applications
Edited by Barney Tyler

International Standard Book Number: 978-1-63238-525-3 (Hardback)

Printed in China.

Contents

Preface

This book has been a concerted effort by a group of academicians, researchers and scientists, who have contributed their research works for the realization of the book. This book has materialized in the wake of emerging advancements and innovations in this field. Therefore, the need of the hour was to compile all the required researches and disseminate the knowledge to a broad spectrum of people comprising of students, researchers and specialists of the field.

Thermodynamics is the branch of physics that is concerned with the study of temperature and heat in relation to work and energy. It mainly focuses on macroscopic explanations of processes and bodies. Entropy, internal energy and pressure are the commonly known variables that are dealt with in this discipline. It is applied to a wide variety of areas such as chemical engineering, mechanical engineering and physical chemistry. This book provides comprehensive insights into the field of thermodynamics. The various studies that are constantly contributing towards advancing technologies and evolution of this field are examined in detail. The aim of this text is to present researches that have transformed this discipline and aided its advancement. Scientists and students actively engaged in this area of study will find this book full of crucial and unexplored concepts.

At the end of the preface, I would like to thank the authors for their brilliant chapters and the publisher for guiding us all-through the making of the book till its final stage. Also, I would like to thank my family for providing the support and encouragement throughout my academic career and research projects.

Editor

1

Studies on Thermophysical Properties of CaO and MgO by γ-Ray Attenuation

A. S. Madhusudhan Rao[1] and K. Narender[2]

[1] *Department of Physics, Varadha Reddy College of Engineering, Warangal 506009, India*
[2] *Kakatiya University, Warangal 506009, India*

Correspondence should be addressed to A. S. Madhusudhan Rao; madhusudhanammiraju@yahoo.com

Academic Editor: Mohammad Al-Nimr

The study on temperature dependent γ-ray attenuation and thermophysical properties of CaO and MgO has been carried out in the temperature range 300 K–1250 K using different energies of γ-beam, namely, Am (0.0595 MeV), Cs (0.66 MeV), and Co (1.173 MeV and 1.332 MeV) on γ-ray densitometer fabricated in our laboratory. The linear attenuation coefficients (μ_l) for the pellets of CaO and MgO as a function of temperature have been determined using γ-beam of different energies. The coefficients of temperature dependence of density have been reported. The variation of density and linear thermal expansion of CaO and MgO in the temperature range of 300 K–1250 K has been studied and compared with the results available in the literature. The temperature dependence of linear attenuation coefficients, density, and thermal expansion has been represented by second degree polynomial. Volume thermal expansion coefficients have been reported.

1. Introduction

Density and thermal expansion are fundamental thermophysical properties of solids. The study of temperature dependence of these properties is very important in understanding the temperature variation of other properties like elastic constants, refractive indices, dielectric constants, thermal conductivity, diffusion coefficients, and other heat transfer dimensionless numbers. Thermal expansion of solids is of technical importance as it determines the thermal stability and thermal shock resistance of the material. In general the thermal expansion characteristics decide the choice of material for the construction of metrological instruments and the choice of container material in nuclear fuel technology. A number of methods have evolved for the determination of density and thermal expansion of solids at high temperature like Archimedean method [1–3], pycnometry [4–8], dilatometry [9–12], electromagnetic levitation [13], method of maximal pressure in gas bubble [14–18], method of sessile drop [19], hydrostatic weighing [20, 21], high temperature electrostatic levitation [22], and gamma ray densitometry [23–34]. Using γ-ray attenuation technique Drotning [23] measured thermal expansion of solid materials at high temperatures. He studied thermal expansion of aluminum and type 303 stainless steel at high temperatures and such studies have been extended by him to study the thermal expansion of metals and glasses in the condensed state [24]. The γ-radiation attenuation technique for the determination of thermophysical properties in the condensed state has several advantages over other methods at high temperatures. This is possible because the γ-ray is not in any kind of physical or thermal contact with the material and hence the thermal losses are also reduced and in addition eliminate sample and probe compatibility problem.

We extended, for the first time, the γ-ray attenuation technique, to carry out the studies on temperature dependence of γ-ray attenuation and thermophysical properties of CaO and MgO. In the present communication, we report the temperature dependence of linear attenuation coefficient for different energies of γ-beam [Am (0.0595 MeV), Cs (0.66 MeV), Co (1.173 MeV and 1.332 MeV)], density, and thermal expansion of CaO and MgO in the temperature range 300 K–1250 K. In order to carry out this work, we have fabricated in our laboratory a γ-ray densitometer and

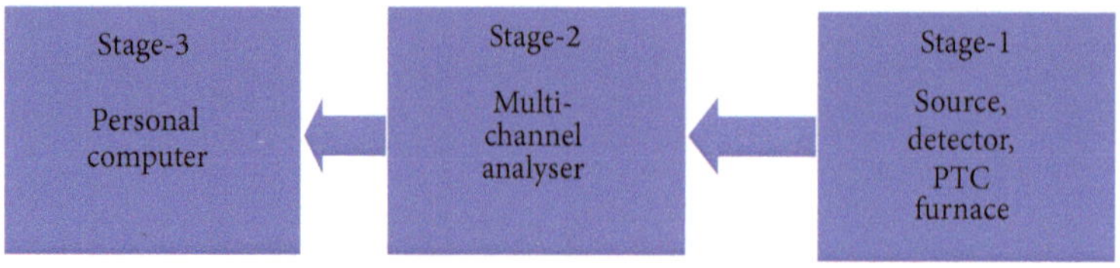

Figure 1: Block diagram of γ-ray densitometer.

a programmable temperature controlled furnace (PTC) which can reach high temperatures. The data obtained in the present work for coefficient of linear thermal expansion of CaO and MgO as a function of temperature have been compared with experimental and theoretical data in the literature. Authors have carried out such studies on different materials like wrought aluminum alloys [29] and alkali halides [30] using the same experimental setup.

2. Theory

The technique of γ-ray attenuation method is based on the following fundamental equation:

$$I = I_0 \exp\left[-\mu\rho l\right], \tag{1}$$

where I_0 is the intensity of γ-ray before passing through the sample, I is the intensity of γ-ray after passing through the sample, μ is the mass attenuation coefficient of the sample, ρ is the density of the sample, and l is the thickness of the sample. It is clear from (1) that any change in the temperature of the solid is accompanied by change in its density causing a change in the measured intensity. The density and thermal expansion of the materials studied in the present work have been determined following the method suggested by Drotning [23]. The relation between coefficient of volumetric thermal expansion (α_ρ) and coefficient of linear thermal expansion (α_1) is given by

$$\alpha_\rho \equiv -3\alpha_1\left(1 - 2\alpha_l \Delta T\right), \tag{2}$$

where α_ρ and α_l are mean values over a temperature interval:

$$\Delta T = T_2 - T_1,$$
$$\text{such that } \alpha_l = \frac{(l_2 - l_1)}{(\Delta T)\, l_1}, \qquad \alpha_\rho = \frac{(\rho_2 - \rho_1)}{(\Delta T)\, \rho_1}, \tag{3}$$

where $\rho_1 = \rho(T_1)$, $l_1 = l(T_1)$, and so forth.

Rewrite (2) as

$$(\Delta T)^2 \alpha_1 \alpha_\rho = z - (\Delta T)\,\alpha_1 - (\Delta T)\,\alpha_\rho, \tag{4}$$

where z is defined by

$$z = \ln \frac{\left[I(T_1)\, I_0(T_2) / I(T_2)\, I_0(T_1)\right]}{(\mu\rho_1 l_1)} = \left(\frac{\rho_2 l_2}{\rho_1 l_1}\right) - 1. \tag{5}$$

Substituting for α_ρ from (2) gives

$$-3(\Delta T)^2 \alpha_l^2 \left(1 - 2\alpha_l \Delta T\right) = z - (\Delta T)\,\alpha_l + 3(\Delta T)\,\alpha_l\left(1 - 2\alpha_l \Delta T\right), \tag{6}$$

which can be rewritten as

$$6x^3 + 3x^2 - 2x - z = 0, \tag{7}$$

where

$$x \equiv (\Delta T)\,\alpha_l. \tag{8}$$

The intensities of γ-radiation with sample I and without sample I_0 are recorded at every temperature. At room temperature T_1, thickness of the sample l_1 is measured and using (1) μ is determined. Further measurements of I and I_0 at different temperatures enable the determination of z by (5) and hence x can be found from the solution of (7). From the value of x, mean linear thermal expansion (α_l) can be determined as a function of temperature.

3. Experimental

The experimental setup used for determination of density of materials utilizing γ-ray attenuation technique is called a γ-ray densitometer. A programmable temperature controlled furnace with sample inside the air tight quartz tube is introduced in the γ-radiation path allowing the beam to pass through the sample and to the detector without any interruption. The temperature of the sample is varied to study the attenuation at various temperatures. The block diagram of γ-ray densitometer used in the present work is shown in Figure 1.

The samples studied in the present work were in the form of pellets. The weight of the powder is 20.0 gm and the thickness of pellet is 1.40 cm with a die set by applying hydraulic press. CaO and MgO pellets were sintered at a temperature of 600 K for densification. The pellet was then firmly mounted on the round sample holder made of flat stainless steel strip inserted into an air tight quartz tube. The sample holder along with the sample was slid through a cork into an air tight quartz tube and was fixed firmly. A diffusion pump was then connected to the sample holder tube for evacuation. For inert atmosphere, argon gas was introduced into the quartz tube through the sample holder tube. Then the quartz tube assembly along with the sample was slid into the programmable temperature controlled (PTC) furnace and fixed at appropriate position ensuring a perfect alignment of sample with collimation on either side. A programmable temperature controlled furnace with sample inside the air tight quartz tube is introduced in the γ-radiation path allowing the beam to pass through the sample and to the detector without any interruption. The temperature of the sample is varied to study the attenuation at various temperatures. The PTC furnace was programmed in such a way that the furnace temperature is increased by 50 K in every step starting from room temperature and stabilizes there for 5 minutes. Temperature equilibrium is achieved within 5 minutes and temperature accuracy is ±1% of set point temperature after stabilization. Heating rate between each step of 50 K is 5 K/min. At each temperature the γ-ray counts of photon energies [0.0595 MeV, 0.662 MeV, and 1.173 MeV and 1.332 MeV] emitted by 10 mCi ^{241}Am, 30 mci ^{137}Cs, and 11.73 μCi ^{60}Co radioactive point sources, respectively, with sample (I) and without sample (I_o) were detected and recorded using a multichannel analyzer. The recording of γ-ray counts was done for a period of 20 minutes at each programmed temperature. After a time of 20 minutes of isothermal holding, the difference of programmed and sample temperature (error) was ±1 K of set point. Measurement of γ-ray attenuation counts at every step of temperature was repeated before and after the sample was introduced and the average value was considered in all our calculations. The cooling rate varied between 10 K/min from 1300 K to 800 K, 6 K/min from 800 K to 500 K, 4 K/min from 500 K to 400 K, and 2 K/min thereafter up to 300 K. The γ-ray counts were recorded while heating and cooling the sample. This procedure was repeated until the desired temperature range was covered in each case. The gamma radiation detector used in our study is a sodium iodide-thallium activated detector. The 0.0762 m diameter and 0.0762 m thick crystal are integrally coupled to a 0.0762 m diameter photomultiplier tube (PMT). The PMT has a 14-pin base and can be mounted on two types of PMT preamplifier units. The one used in our study is a coaxial in-line preamplifier. The detector has a resolution of 8.5% for 0.662 MeV of ^{137}Cs.

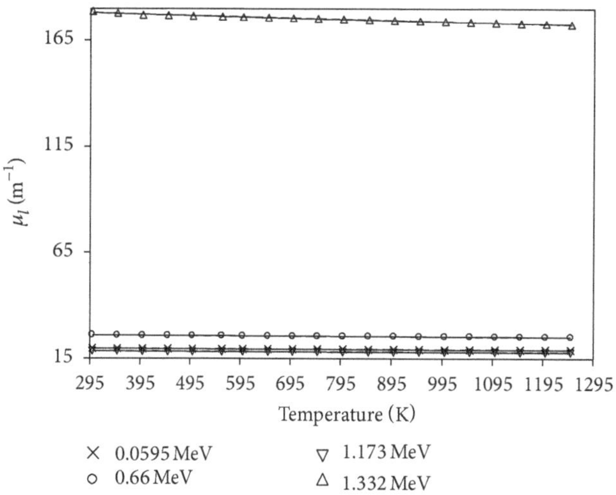

FIGURE 2: Variation of linear attenuation coefficient of CaO with temperature for different γ-sources.

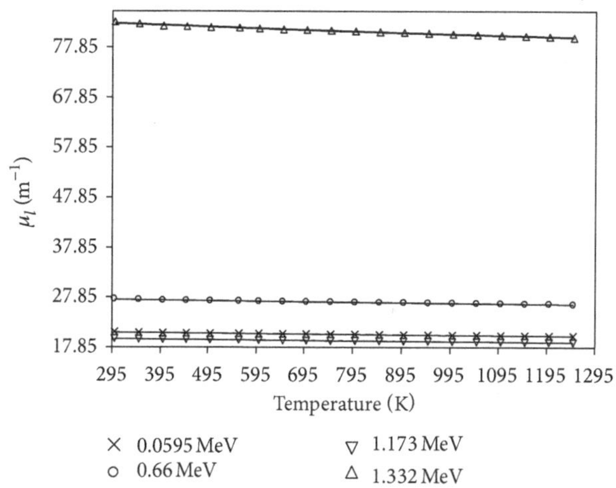

FIGURE 3: Variation of linear attenuation coefficient of MgO with temperature for different γ-sources.

4. Results and Discussion

The experimentally determined values of mass attenuation coefficients (μ) of CaO and MgO for γ-beam of different energies are compared with the theoretical values obtained from National Institute of Standards and Technology (NIST-X-COM) in Table 1. The results obtained for the temperature dependence of the linear attenuation coefficients (μL) of CaO and MgO for γ-beam of different energies are summarized in Table 2. The results obtained for the temperature dependence of the density (ρ) and the coefficient of linear thermal expansion (α) of CaO and MgO are presented in Table 3. The measurements have been carried out in solid phase only. The experimental data obtained in the present work for the density, the linear attenuation coefficient, and coefficient of linear thermal expansion have been fit to a least squares quadratic polynomial of the following form:

$$\rho(T) = a + bT + cT^2. \quad (9)$$

Since the measurements have been made in the limited temperature range the coefficient of volumetric thermal expansion (CVTE) was calculated using the following equation:

$$\beta = \left(\frac{1}{\rho}\right)\left(\frac{d\rho}{dT}\right), \quad (10)$$

where $(d\rho/dT)$ is the first derivative of density with respect to the absolute temperature which is determined from (9). The variation of linear attenuation coefficients of CaO and MgO with temperature for γ-beam of different energies is shown in Figures 2 and 3, respectively. The variation of density and linear thermal expansion with temperature of CaO and MgO has been shown in Figures 4 and 5, respectively.

4.1. CaO. The temperature dependence of linear attenuation coefficient of CaO is a negative linear function of temperature. The temperature dependence of linear attenuation coefficient for different energies of γ-beam [Am (0.0595 MeV),

TABLE 1: Comparison of mass attenuation coefficients (μ) of CaO and MgO of γ-rays with different energies.

(μ) 10^{-3} m^2kg^{-1}	Cs (0.66 MeV)		Am (0.0595 MeV)		Co (1.173 MeV)		Co (1.332 MeV)	
	NIST X-COM	EXPTL	NIST X-COM	EXPTL	NIST X-COM	EXPTL	NIST X-COM	EXPTL
CaO	7.78	7.76	53.39	53.35	5.89	5.90	5.52	5.52
MgO	7.69	7.67	23.30	23.15	5.84	5.85	5.48	5.47

TABLE 2: Variation of linear attenuation coefficient of CaO and MgO with temperature for different γ energies.

T K	CaO (μ_l m^{-1})				MgO (μ_l m^{-1})			
	Cs (0.66) MeV	Am (0.0595) MeV	Co (1.173) MeV	Co (1.33) MeV	Cs (0.66) MeV	Am (0.0595) MeV	Co (1.173) MeV	Co (1.33) MeV
300	26.01	178.72	19.73	18.51	27.46	82.88	20.91	19.61
350	25.89	177.87	19.64	18.42	27.33	82.48	20.81	19.51
400	25.76	177.02	19.54	18.32	27.2	82.09	20.71	19.41
450	25.73	176.8	19.52	18.30	27.16	81.95	20.68	19.38
500	25.69	176.54	19.49	18.27	27.11	81.81	20.64	19.35
559	25.66	176.27	19.46	18.24	27.07	81.7	20.61	19.32
600	25.62	176.00	19.43	18.21	27.03	81.56	20.58	19.29
650	25.58	175.79	19.41	18.19	26.99	81.42	20.55	19.26
700	25.54	175.52	19.38	18.16	26.94	81.28	20.51	19.22
750	25.51	175.25	19.35	18.14	26.89	81.16	20.48	19.19
800	25.47	174.99	19.32	18.11	26.84	81.03	20.44	19.16
850	25.43	174.72	19.29	18.08	26.80	80.89	20.41	19.13
900	25.39	174.45	19.26	18.05	26.76	80.75	20.37	19.09
950	25.35	174.19	19.24	18.03	26.72	80.61	20.34	19.06
1000	25.31	173.92	19.21	18.01	26.67	80.47	20.31	19.03
1050	25.27	173.65	19.18	17.98	26.62	80.33	20.27	19.00
1100	25.23	173.39	19.14	17.94	26.57	80.19	20.23	18.96
1150	25.19	173.12	19.11	17.91	26.52	80.05	20.19	18.93
1200	25.15	172.80	19.08	17.88	26.47	79.89	20.15	18.89
1250	25.11	172.53	19.05	17.85	26.42	79.75	20.12	18.86

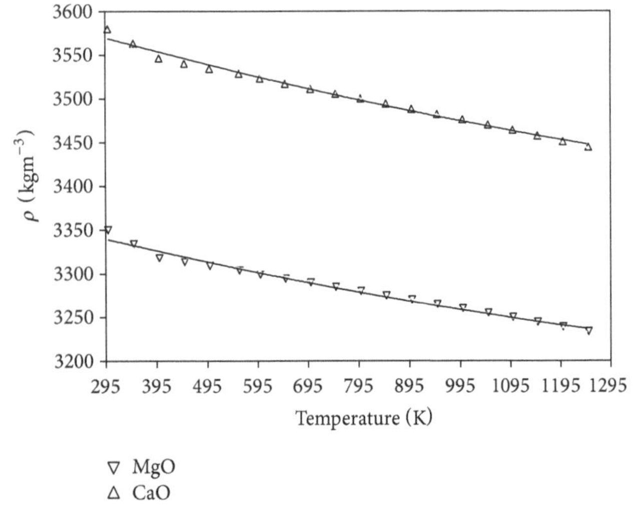

FIGURE 4: Variation of density of CaO and MgO with temperature.

Cs (0.66 MeV), and Co (1.173 MeV and 1.332 MeV)] has been represented by second degree polynomial, respectively:

$$\mu_l(T) = (180.49) + \left(-8.32 \times 10^{-3}\right)T + \left(1.66 \times 10^{-6}\right)T^2,$$
$$\mu_l(T) = (27.59) + \left(-1.29 \times 10^{-3}\right)T + \left(2.60 \times 10^{-7}\right)T^2,$$
$$\mu_l(T) = (19.92) + \left(-8.90 \times 10^{-4}\right)T + \left(1.67 \times 10^{-7}\right)T^2,$$
$$\mu_l(T) = (18.70) + \left(-9.00 \times 10^{-4}\right)T + \left(1.93 \times 10^{-7}\right)T^2. \quad (11)$$

The density of CaO decreases from a value of 3350 kgm^{-3} at 300 K to a value of 3234 kgm^{-3} at 1250 K with a decrease of

TABLE 3: Variation of density and coefficient of linear thermal expansion of CaO and MgO with temperature.

T K	CaO		MgO	
	ρ kgm^{-3}	α (10^{-6}) K^{-1}	ρ kgm^{-3}	α (10^{-6}) K^{-1}
300	3350	11.91	3580	10.42
350	3334	12.01	3563	11.22
400	3318	12.11	3546	12.02
450	3314	12.31	3540	12.47
500	3309	12.51	3534	12.91
559	3304	12.65	3529	13.17
600	3299	12.78	3523	13.42
650	3295	12.93	3517	13.66
700	3290	13.07	3511	13.90
750	3285	13.19	3506	14.10
800	3280	13.31	3500	14.30
850	3275	13.41	3494	14.43
900	3270	13.50	3488	14.55
950	3265	13.62	3482	14.68
1000	3260	13.73	3476	14.81
1050	3255	13.85	3470	14.96
1100	3250	13.97	3464	15.11
1150	3245	14.07	3458	15.28
1200	3239	14.17	3451	15.45
1250	3234	14.29	3445	15.60

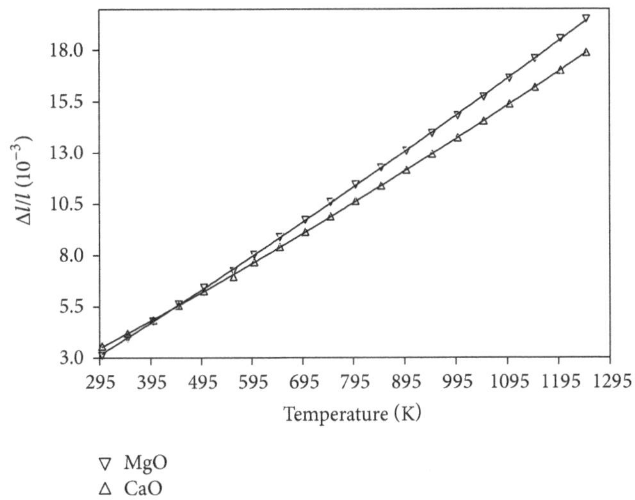

FIGURE 5: Variation of thermal expansion of CaO and MgO with temperature.

about 1.16%. The temperature dependence of density is a negative linear function of temperature. For CaO the temperature dependence of density is represented by quadratic equation:

$$\rho(T) = (3382.77) + \left(-1.55 \times 10^{-1}\right) T + \left(3.04 \times 10^{-5}\right) T^2. \quad (12)$$

The coefficient of temperature dependence of density is $-0.122\,\text{kgm}^{-3}\,\text{K}^{-1}$ and the coefficient of volume thermal expansion is $4.0 \times 10^{-5}\,\text{K}^{-1}$ in the temperature range 300 K–1250 K. The thermal expansion increases linearly with temperature and the results on thermal expansion in the temperature range from 300 K to 1250 K have been analyzed by least squares method and are represented by the following polynomial equation:

$$\frac{\Delta l}{l(T)} = \left(-1.35 \times 10^{-3}\right) + \left(1.47 \times 10^{-5}\right) T + \left(1.54 \times 10^{-9}\right) T^2. \quad (13)$$

The coefficient of linear thermal expansion (α) of CaO in the temperature range 300 K–1250 K has been represented by second degree polynomial and has been shown in Figure 6 along with data obtained by other methods for comparison:

$$\alpha(T) = \left(10.84 \times 10^{-6}\right) + \left(3.63 \times 10^{-9}\right) T + \left(-7.18 \times 10^{-13}\right) T^2. \quad (14)$$

4.2. MgO. The temperature dependence of linear attenuation coefficient of MgO is a negative linear function of temperature. The temperature dependence of linear attenuation coefficient for different energies of γ-beam [Am (0.0595 MeV),

Cs (0.66 MeV), and Co (1.173 MeV and 1.332 MeV)] has been represented by second degree polynomial, respectively:

$$\mu_l(T) = (83.78) + \left(-4.10 \times 10^{-3}\right)T + \left(7.46 \times 10^{-7}\right)T^2,$$
$$\mu_l(T) = (29.12) + \left(-1.39 \times 10^{-3}\right)T + \left(2.39 \times 10^{-7}\right)T^2,$$
$$\mu_l(T) = (21.13) + \left(-1.01 \times 10^{-3}\right)T + \left(1.68 \times 10^{-7}\right)T^2,$$
$$\mu_l(T) = (19.83) + \left(-9.99 \times 10^{-4}\right)T + \left(1.89 \times 10^{-7}\right)T^2. \quad (15)$$

The density of MgO decreases from a value of 3580 $\mathrm{kgm^{-3}}$ at 300 K to a value of 3445 $\mathrm{kgm^{-3}}$ at 1250 K with a decrease of about 1.35%. The temperature dependence of density is a negative linear function of temperature. For MgO the temperature dependence of density is represented by the following quadratic equation:

$$\rho(T) = (3619.02) + \left(-1.77 \times 10^{-1}\right)T + \left(3.19 \times 10^{-5}\right)T^2. \quad (16)$$

The coefficient of temperature dependence of density is $-0.142\,\mathrm{kgm^{-3}\,K^{-1}}$ and the coefficient of volume thermal expansion is $4.0 \times 10^{-5}\,\mathrm{K^{-1}}$. The thermal expansion increases linearly with temperature and the results on thermal expansion in the temperature range from 300 K to 1250 K have been analyzed by least squares method and are represented by the following equation:

$$\frac{\Delta l}{l(T)} = \left(-2.42 \times 10^{-4}\right) + \left(1.20 \times 10^{-5}\right)T + \left(2.01 \times 10^{-9}\right)T^2. \quad (17)$$

The coefficient of linear thermal expansion (α) of MgO in the temperature range 300 K–1250 K has been represented by second degree polynomial and has been shown in Figure 7 along with data obtained by other methods for comparison:

$$\alpha(T) = \left(7.72 \times 10^{-6}\right) + \left(0.01 \times 10^{-6}\right)T + \left(-4.89 \times 10^{-12}\right)T^2. \quad (18)$$

The decrease in density in CaO and MgO with temperature can be attributed to the increase in the equilibrium concentration of thermally generated Schottky defects. Another source of generation of defects in CaO and MgO is the irradiation with γ-rays as the technique used in the present study is γ-attenuation technique. However, the values of density seem to have not been affected much by irradiation of γ-rays since the values of coefficient of linear thermal expansion at different temperatures obtained in the present work agree well with data reported in the literature from other methods [35–38] as shown in Figures 6 and 7, respectively. However, the results on variation of density and linear attenuation coefficient of the samples with temperature are not available from other methods for comparison.

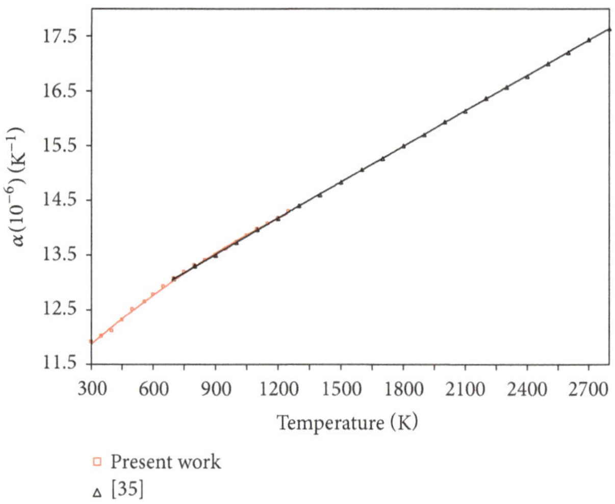

FIGURE 6: Comparison of coefficient of thermal expansion of CaO with data in the literature.

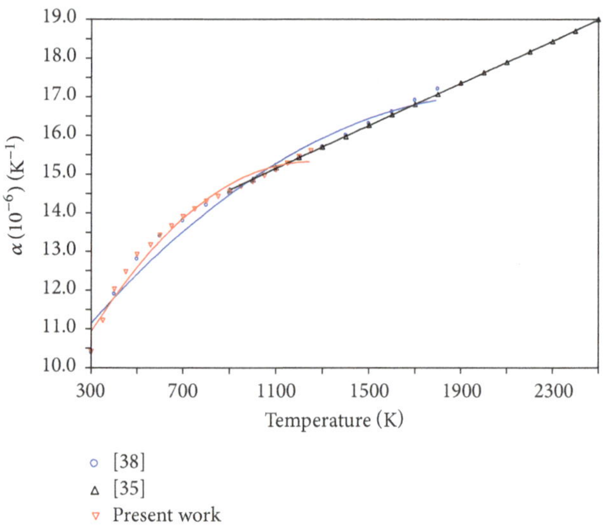

FIGURE 7: Comparison of coefficient of thermal expansion of MgO with data in the literature.

The uncertainty in the measured physical parameters depends on uncertainty in the furnace temperature and measurement of the mass attenuation coefficient, which has been estimated from errors in intensities I_0, I and thickness (l) using the following relation [39–41]:

$$\Delta(\mu_m) = \frac{1}{\rho l}\left[\left(\frac{\Delta I_0}{I}\right)^2 + \left(\frac{\Delta I}{I}\right)^2 + \left(\ln\frac{I_0}{I}\right)^2 + \left(\frac{\Delta l}{l}\right)^2\right]^{1/2}, \quad (19)$$

where ΔI_0, ΔI, and Δl are the errors in the intensities I_0, I and thickness l, respectively. In this experiment, the intensities I_0 and I have been recorded for the same time and under the same experimental conditions. Estimated error in these measurements was around 2%.

5. Conclusions

The pellets were prepared with fine powder of CaO and MgO with a diameter of 20 mm with varying thicknesses with a die set by applying hydraulic press. The γ-ray attenuation measurements of CaO and MgO have been made using γ-beam of different energy sources on a γ-ray densitometer. The results on the variation of density and linear thermal expansion with temperature of both the pellets are found to be equal for all the γ-energy sources. The results on the variation of linear attenuation coefficient, density, and linear thermal expansion with temperature of these pellets have been reported and these variations have been represented by quadratic equations. The results on these pellets by using γ-ray attenuation technique are being reported for the first time.

Conflict of Interests

The authors hereby declare that there is no issue of any type of conflict of interests in any manner.

Acknowledgment

The authors thank University Grants Commission (UGC), New Delhi, for the financial assistance through Special Assistance Programme (SAP) no. F.530/8/DRS/2009 (SAP-1).

References

[1] B. B. Alchagirov and A. M. Chochaeva, "Temperature dependence of the density of liquid tin," *High Temperature*, vol. 38, no. 1, pp. 44–48, 2000.

[2] L. Wang, Q. Wang, A. Xian, and K. Lu, "Precise measurement of the densities of liquid Bi, Sn, Pb and Sb," *Journal of Physics: Condensed Matter*, vol. 15, no. 6, pp. 777–783, 2003.

[3] X. Chen, Q. Wang, and K. Lu, "Temperature and time dependence of the density of molten indium antimonide measured by an improved archimedean method," *Journal of Physics: Condensed Matter*, vol. 11, no. 50, pp. 10335–10341, 1999.

[4] B. B. Alchagirov, A. G. Mozgovoy, T. M. Taova, and T. A. Sizhahzev, *AdvancedMaterials*, vol. 6, no. 35, 2005.

[5] B. B. Alchagirov, T. M. Shamparov, and A. G. Mozgovoi, "Experimental investigation of the density of molten lead-bismuth eutectic," *High Temperature*, vol. 41, no. 2, pp. 210–215, 2003.

[6] A. F. Crawley, "Densities and viscosities of some liquid alloys of zinc and cadmium," *Metallurgical Transactions B*, vol. 3, no. 4, pp. 971–975, 1972.

[7] K. Mukai, F. Xiao, K. Nogi, and Z. Li, "Measurement of the density of Ni–Cr alloy by a modified pycnometric method," *Materials Transactions*, vol. 45, no. 7, pp. 2357–2363, 2004.

[8] Y. Sato, T. Nishizuka, K. Hara, T. Yamamura, and Y. Waseda, "Density measurement of molten silicon by a pycnometric method," *International Journal of Thermophysics*, vol. 21, no. 6, pp. 1463–1471, 2000.

[9] U. Jauch, G. Hasse, and B. Schulz, "Part 11: thermophysical properties of Li(17)Pb(83) eutectic alloy," in *Thermophysical Properties in the System Li-Pb*, vol. 25, Kernforschungszentrum Karlsruhe, Karlsruhe, Germany, 1986.

[10] L. Wang and Q. Mei, "Density measurement of liquid metals using dilatometer," *Journal of Materials Science & Technology*, vol. 22, no. 4, pp. 569–571, 2006.

[11] G. K. White and J. G. Collins, "The thermal expansion of alkali halides at low temperatures. II. Sodium, rubidium and caesium halides," *Proceedings of the Royal Society A: Mathematical, Physical & Engineering Sciences*, vol. 333, pp. 237–259, 1973.

[12] "Pb-free solders and other materials," *Journal of Electronic Materials*, 2011.

[13] J. Brillo, I. Egry, and I. Ho, "Density and thermal expansion of liquid Ag-Cu and Ag-Au alloys," *International Journal of Thermophysics*, vol. 27, no. 2, pp. 494–506, 2006.

[14] S. A. Been, H. S. Edwards, C. E. Tecter, and V. P. Calkins, "ORNL: fairchild and airplane corporation," NEPA Report 1585, 1950.

[15] Fairchild Engine and Airplane Corporation, Oak Ridge, Tenn, USA.

[16] H. Ruppersberg and W. Speicher, "Density and compressibility of liquid Li–Pb alloys," *Zeitschrift Naturforschung Teil A*, vol. 31, pp. 47–52, 1976.

[17] J. Saar and H. Ruppersberg, "Calculation of $C_p(T)$ for liquid Li/Pb alloys from experimental $\rho(T)$ and $(\delta p/\delta T)$s data," *Journal of Physics F: Metal Physics*, vol. 17, no. 2, pp. 305–314, 1987.

[18] F. T. Firdu and P. Taskinen, Aalto University Publications in Material Science and Engineering, Espoo TKK-MT-215, 2010.

[19] I. V. Kazakova, S. A. Lyamkyn, and B. M. Lepinskikh, *The Journal of Physical Chemistry*, vol. 58, p. 1534, 1984.

[20] N. A. Nikol'skyi, N. A. Kalakutskaya, I. M. Pehelkin, T. V. Klassen, and V. A. Vel'mishcheva, "Teploenergetika (Power Engineering)," vol. 2, no. 92, 1959.

[21] N. A. Nikol'skyi, N. A. Kalakutskaya, I. M. Pehelkin, T. V. Klassen, and V. A. Vel'mishcheva, "Voprocsy teploobmena," Problems of Heat Transfer.

[22] S. K. Chung, D. B. Thiessen, and W.-K. Rhim, "A noncontact measurement technique for the density and thermal expansion coefficient of solid and liquid materials," *Review of Scientific Instruments*, vol. 67, no. 9, pp. 3175–3181, 1996.

[23] W. D. Drotning, "Thermal expansion of solids at high temperatures by the gamma attenuation technique," *Review of Scientific Instruments*, vol. 50, no. 12, Article ID 121567, pp. 1567–1570, 1979.

[24] W. D. Drotning, "Thermal expansion of the group IIb liquid metals zinc, cadmium and mercury," *Journal of The Less-Common Metals*, vol. 96, pp. 223–227, 1984.

[25] W. D. Drotning, "Thermal expansion of iron, cobalt, nickel, and copper at temperatures up to 600 K above melting," *High Temperatures-High Pressures*, vol. 13, no. 4, pp. 441–458, 1981.

[26] G. M. Kalinin et al., "Study of Lil7–Pb83 eutecticproperties," *USSR Contribution to ITER*. In press.

[27] R. A. Khairulin, A. S. Kosheleva, and S. V. Stankus, "Thermal properties of liquid alloys of magnesium-lead system," *Thermophysics and Aeromechanics*, vol. 14, no. 1, pp. 75–80, 2007.

[28] R. A. Khairulin, S. V. Stankus, R. N. Abdullaev, Y. A. Plevachuk, and K. Y. Shunyaev, "The density and the binary diffusion coefficients of silver-tin melts," *Thermophysics and Aeromechanics*, vol. 17, no. 3, pp. 391–396, 2010.

[29] K. Narender, A. S. M. Rao, K. G. K. Rao, and N. G. Krishna, "Thermo physical properties of wrought aluminum alloys 6061, 2219 and 2014 by gamma ray attenuation method," *Thermochimica Acta*, vol. 569, pp. 90–96, 2013.

[30] A. S. M. Rao, K. Narender, K. G. K. Rao, and N. G. Krishna, "Thermophysical properties of rubidium and lithium halides by gamma ray attenuation technique," *High Temperature*. In press.

[31] S. V. Stankus, R. A. Khairulin, A. G. Mozgovoy, V. V. Roshchupkin, and M. A. Pokrasin, "The density and thermal expansion of eutectic alloys of lead with bismuth and lithium in condensed state," *Journal of Physics: Conference Series*, vol. 98, no. 6, Article ID 062017, 2008.

[32] S. V. Stankus and P. V. Tyagel'skii, "Density of high-purity dysprosium in the solid and liquid states," *High Temperature*, vol. 38, no. 4, pp. 555–559, 2000.

[33] S. V. Stankus, R. A. Khairulin, A. G. Mozgovoi, V. V. Roshchupkin, and M. A. Pokrasin, "An experimental investigation of the density of bismuth in the condensed state in a wide temperature range," *High Temperature*, vol. 43, no. 3, pp. 368–378, 2005.

[34] S. V. Stankus, R. A. Khairulin, and A. G. Mozgovoi, "Experimental study of density and thermal expansion of the advanced materials and heat transfer agents for liquid metal systems of thermonuclear reactor: lithium," *High Temperature*, vol. 49, no. 2, pp. 187–192, 2011.

[35] S. K. Srivastava, P. Sinha, and M. Panwar, "Thermal expansivity and isothermal bulk modulus of ionic materials at high temperatures," *Indian Journal of Pure & Applied Physics*, vol. 47, no. 3, pp. 175–179, 2009.

[36] K. Y. Singh and B. R. K. Gupta, "A simple approach to analyse the thermal expansion in minerals under the effect of high temperature," *Physica B: Condensed Matter*, vol. 334, no. 3-4, pp. 266–271, 2003.

[37] B. P. Singh, H. Chandra, R. Shyam, and A. Singh, "Analysis of volume expansion data for periclase, lime, corundum and spinel at high temperatures," *Bulletin of Materials Science*, vol. 35, no. 4, pp. 631–637, 2012.

[38] R. E. Taylor, *Thermal Expansion of Solids*, ASM International, Materials Park, Ohio, USA, 1998.

[39] I. Han and L. Demir, "Studies on effective atomic numbers, electron densities and mass attenuation coefficients in Au alloys," *Journal of X-Ray Science and Technology*, vol. 18, no. 1, pp. 39–46, 2010.

[40] D. Demir, A. Turşucu, and T. Öznülüer, "Studies on mass attenuation coefficient, effective atomic number and electron density of some vitamins," *Radiation and Environmental Biophysics*, vol. 51, no. 4, pp. 469–475, 2012.

[41] N. Kucuk, Z. Tumsavas, and M. Cakir, "Determining photon energy absorption parameters for different soil samples," *Journal of Radiation Research*, vol. 54, no. 3, pp. 578–586, 2013.

2

Thermodynamic Properties of Vapors from Speed of Sound

Muhamed Bijedić and Sabina Begić

Faculty of Technology, University of Tuzla, 75000 Tuzla, Bosnia and Herzegovina

Correspondence should be addressed to Muhamed Bijedić; muhamed.bijedic@untz.ba

Academic Editor: Angelo Lucia

A numerical procedure for deriving the thermodynamic properties (Z, c_v, and c_p) of the vapor phase in the subcritical temperature range from the speed of sound is presented. The set of differential equations connecting these properties with the speed of sound is solved as the initial-value problem in T-ϕ domain ($\phi = \rho/\rho_{\text{sat}}$). The initial values of Z and $(\partial Z/\partial T)_\rho$ are specified along the isotherm T_0 with the highest temperature, at a several values of ϕ [0.1, 1.0]. The values of Z are generated by the reference equation of state, while the values of $(\partial Z/\partial T)_\rho$ are derived from the speed of sound, by solving another set of differential equations in T-ρ domain in the transcritical temperature range. This set of equations is solved as the initial-boundary-value problem. The initial values of Z and c_v are specified along the isochore in the limit of the ideal gas, at several isotherms distributed according to the Chebyshev points of the second kind. The boundary values of Z are specified along the same isotherm T_0 and along another isotherm with a higher temperature, at several values of ρ. The procedure is tested on Ar, N_2, CH_4, and CO_2, with the mean AADs for Z, c_v, and c_p at 0.0003%, 0.0046%, and 0.0061%, respectively (0.0007%, 0.0130%, and 0.0189% along the saturation line).

1. Introduction

The speed of sound is the property of a fluid which is measured with an exceptional accuracy (Ewing and Goodwin [1], Estrada-Alexanders and Trusler [2], Costa Gomes and Trusler [3], Trusler and Zarari [4], and Estrada-Alexanders and Trusler [5]). While the speed of sound itself is a piece of useful information about a fluid, other thermodynamic properties like the density and the heat capacity, which may be derived from it, make the speed of sound even more useful. However, the speed of sound in a fluid is connected to these properties through the set of nonlinear partial differential equations of the second order. The general solution to this set of equations has not been found yet, except for the gaseous phase under low pressures (Trusler et al. [6] and Estela-Uribe and Trusler [7]). At moderate and higher pressures a procedure based on the numerical integration may be used for obtaining the particular solution (Estrada-Alexanders et al. [8] and Bijedić and Neimarlija [9–11]). The main disadvantage of this approach is the need for the initial and/or the boundary values of the quantities being calculated (the Dirichlet conditions) and their derivatives with respect to an independent variable (the Neumann conditions).

In the previous paper (Bijedić and Neimarlija [11]) it was shown that the compressibility factor and the heat capacity of a gas may be derived from the speed of sound in the supercritical temperature range only with the initial/boundary conditions of the Dirichlet type. In this paper, that procedure is applied to the transcritical temperature range in order to generate the initial conditions of the Neumann type, which are then used for deriving the compressibility factor and the heat capacity of a vapor in the subcritical temperature range. In this way, the main disadvantage of the approach based on the numerical integration is minimized, because only a few data points of the compressibility factor from other sources are needed to cover the whole gaseous phase except the region around the critical point. The boundary values of the compressibility factor are specified along two isotherms, several degrees below the critical point. The maximum density is chosen so as to correspond to that of the saturated vapor at the initial temperature. In this way, the saturation line becomes a boundary of the domain of integration, and the thermodynamic properties of the saturated vapor are derived as well. While the derived properties of a gas in the transcritical temperature range have a very good agreement with the corresponding reference data they are not discussed in this paper (see Table 7).

2. Theory

The thermodynamic speed of sound (the mechanical disturbance of small amplitude and low frequency) is connected to other macroscopic thermodynamic properties through the following fundamental relation (Trusler [12]):

$$u^2 = \left(\frac{\partial p}{\partial \rho}\right)_s, \tag{1}$$

where u is the speed of sound, p is the pressure, ρ is the density, and s is the specific entropy. However, since s is not a measurable property it is convenient to avoid it. This can be accomplished with the help of the following thermodynamic relation (Moran and Shapiro [13]):

$$\left(\frac{\partial p}{\partial \rho}\right)_s = \frac{c_p}{c_v}\left(\frac{\partial p}{\partial \rho}\right)_T, \tag{2}$$

where T is the thermodynamic temperature, c_p is the specific heat capacity at constant pressure, and c_v is the specific heat capacity at constant volume. Now, (1) takes the following form:

$$u^2 = \frac{c_p}{c_v}\left(\frac{\partial p}{\partial \rho}\right)_T. \tag{3}$$

The isobaric heat capacity may be eliminated from (3) on account of the following thermodynamic relation (Moran and Shapiro [13]):

$$c_p - c_v = -T\left(\frac{\partial v}{\partial T}\right)_p^2\left(\frac{\partial p}{\partial v}\right)_T, \tag{4}$$

where v is the specific volume. If the specific volume is replaced with the density, (4) becomes

$$c_p - c_v = \frac{T}{\rho^2}\left(\frac{\partial \rho}{\partial T}\right)_p^2\left(\frac{\partial p}{\partial \rho}\right)_T. \tag{5}$$

According to the chain rule from the calculus, one may write

$$\left(\frac{\partial p}{\partial \rho}\right)_T\left(\frac{\partial \rho}{\partial T}\right)_p\left(\frac{\partial T}{\partial p}\right)_\rho = -1. \tag{6}$$

From (6) it follows that

$$\left(\frac{\partial \rho}{\partial T}\right)_p = -\left(\frac{\partial \rho}{\partial p}\right)_T\left(\frac{\partial p}{\partial T}\right)_\rho. \tag{7}$$

If the right hand side of (7) is introduced into (5) one obtains

$$c_p = c_v + \frac{T}{\rho^2}\left(\frac{\partial \rho}{\partial p}\right)_T\left(\frac{\partial p}{\partial T}\right)_\rho^2. \tag{8}$$

Introducing the right hand side of (8) into (3), one finally obtains

$$u^2 = \left(\frac{\partial p}{\partial \rho}\right)_T + \frac{T}{\rho^2 c_v}\left(\frac{\partial p}{\partial T}\right)_\rho^2. \tag{9}$$

In (9), not only the density but also the isochoric heat capacity is unknown. Therefore, another equation, which connects the heat capacity with the thermal properties, is necessary. For this purpose, the following thermodynamic relation is suitable:

$$\left(\frac{\partial c_v}{\partial \rho}\right)_T = -\frac{T}{\rho^2}\left(\frac{\partial^2 p}{\partial T^2}\right)_\rho. \tag{10}$$

If the density is replaced with a less varying compressibility factor

$$Z = \frac{Mp}{\rho RT}, \tag{11}$$

where Z is the compressibility factor, M is the molar mass, and R is the universal gas constant, (9) and (10), after rearrangement, take the following forms, respectively,

$$\left(\frac{\partial Z}{\partial \rho}\right)_T = \frac{1}{\rho}\left\{\frac{Mu^2}{RT} - Z - \frac{R}{Mc_v}\left[Z + T\left(\frac{\partial Z}{\partial T}\right)_\rho\right]^2\right\} \tag{12}$$

$$\left(\frac{\partial c_v}{\partial \rho}\right)_T = -\frac{R}{M\rho}\left[2T\left(\frac{\partial Z}{\partial T}\right)_\rho + T^2\left(\frac{\partial^2 Z}{\partial T^2}\right)_\rho\right]. \tag{13}$$

Equations (12) and (13) may be solved simultaneously for Z and c_v in the range of p and T in which accurate values of the speed of sound are available (Bijedić and Neimarlija [11]). This set of equations is designed to be solved in rectangular T-ρ domain.

If the subcritical vapor phase is considered, it is convenient to introduce new independent variable instead of the density. For this purpose, the ratio of the density and its maximum value at the observed temperature (at the saturation) is suitable, that is,

$$\phi = \frac{\rho}{\rho_s}, \tag{14}$$

where ϕ is the new independent variable and ρ_s is the density of the saturated vapor at the observed temperature. In such a way new rectangular T-ϕ domain is obtained, which allows one to cover the temperature range from the critical point down to the triple point and the density range from the ideal gas limit up to the saturation line.

According to the calculus, one may write

$$\left(\frac{\partial Z}{\partial T}\right)_\phi = \left(\frac{\partial Z}{\partial T}\right)_\rho + \left(\frac{\partial Z}{\partial \rho}\right)_T\left(\frac{\partial \rho}{\partial T}\right)_\phi. \tag{15}$$

If new variable a is defined as

$$a = \left(\frac{\partial Z}{\partial T}\right)_\rho, \tag{16}$$

TABLE 1: ρ-T-p ranges covered for the subcritical domain.

Substance	ρ [kg × m^{-3}] min	ρ [kg × m^{-3}] max	T [K] min	T [K] max	p [MPa] min	p [MPa] max
Ar	0	178.85835	90	140	0	3.16822
N_2	0	87.29420	70	115	0	1.93704
CH_4	0	61.37508	100	180	0	3.28518
CO_2	0	171.96269	220	290	0	5.31772

TABLE 2: ρ-T-p ranges covered for the transcritical domain.

Substance	ρ [kg × m^{-3}] min	ρ [kg × m^{-3}] max	T [K] min	T [K] max	p [MPa] min	p [MPa] max
Ar	0	178.85835	140	280	0	9.75263
N_2	0	87.29420	115	255	0	6.39904
CH_4	0	61.37508	180	320	0	9.14395
CO_2	0	171.96269	290	430	0	11.84715

(13) now looks like

$$\left(\frac{\partial c_v}{\partial \rho}\right)_T = -\frac{R}{M\rho}\left[2Ta + T^2\left(\frac{\partial a}{\partial T}\right)_\rho\right]. \quad (17)$$

According to the calculus, one may also write

$$\left(\frac{\partial a}{\partial T}\right)_\phi = \left(\frac{\partial a}{\partial T}\right)_\rho + \left(\frac{\partial a}{\partial \rho}\right)_T\left(\frac{\partial \rho}{\partial T}\right)_\phi. \quad (18)$$

If (17) and (18) are combined one obtains

$$\left(\frac{\partial a}{\partial T}\right)_\phi = -\frac{1}{T}\left[\frac{M\rho}{RT}\left(\frac{\partial c_v}{\partial \rho}\right)_T + 2a\right] + \left(\frac{\partial a}{\partial \rho}\right)_T\left(\frac{\partial \rho}{\partial T}\right)_\phi. \quad (19)$$

Now, (12), (15), and (19) may be solved simultaneously for Z and c_v in the range of p and T in which accurate values of the speed of sound are available.

When the integration is complete, c_p is calculated from the following relation:

$$c_p = c_v\frac{Mu^2}{ZRT}\left[1 - \frac{p}{Z}\left(\frac{\partial Z}{\partial p}\right)_T\right] \quad (20)$$

which is obtained from (3) when the density is replaced with the compressibility factor.

3. Results and Discussion

The speed of sound is usually available in the form $u = u(T, p)$ and not in the form $u = u(T, \rho)$. However, the numerical integration is performed with T and ρ/ρ_s as the independent variables, rather than T and p/p_s, because more accurate results are obtained (Bijedić and Neimarlija [10]). For that reason, the lines of constant ϕ are generated in advance by the Peng-Robinson equation of state (Peng and Robinson [14]), except the line $\phi = 1.0$ which is generated by the corresponding reference equation of state. Since the compressibility factor is calculated before the speed of sound is required, in each integration step, the latter is interpolated along the isotherms with respect to the pressure ($p = Z\rho RT/M$). Then, $M \times N$ partial derivatives $(\partial\rho/\partial T)_\phi$ are estimated from the interpolating polynomial in the Lagrange form. The total number of the isotherms, M, is 7 to 10, and the total number of the lines with constant ϕ, N, is 10. The values of $(\partial\rho/\partial T)_\phi$ off the isotherms are interpolated along the lines of constant ϕ with respect to the temperature.

The calculation procedure is initialized by specifying values of Z and a along the isotherm T_0 with the highest temperature (see Table 1), at 10 equidistant values of ϕ [0.1, 1.0]. The values of Z are generated by the corresponding reference equation of state, while values of a are derived from the speed of sound, by solving the set of (12) and (13) in T-ρ domain in the transcritical temperature range (see Tables 2 and 3). Then, the partial derivatives $(\partial Z/\partial \rho)_T$ and $(\partial a/\partial \rho)_T$ are estimated along T_0 from the interpolating polynomial in

TABLE 3: The calculated and the reference $(\partial Z/\partial T)_\rho$ at T_0.

Substance	p [MPa]	$(\partial Z/\partial T)_\rho \times 10^3$ [K^{-1}] Calculated	$(\partial Z/\partial T)_\rho \times 10^3$ [K^{-1}] Reference
Ar, $T_0 = 140$ K	0.8107563	0.9873	0.9873
	1.4993690	1.9681	1.9681
	2.0729260	2.9453	2.9455
	2.5387200	3.9234	3.9240
	2.9042320	4.9117	4.9134
	3.1682270	5.8890	5.8933
N_2, $T_0 = 115$ K	0.4788111	1.1037	1.1037
	0.8932118	2.2079	2.2081
	1.2462020	3.3152	3.3158
	1.5408300	4.4349	4.4364
	1.7801230	5.5829	5.5865
	1.9370410	6.5622	6.5692
CH_4, $T_0 = 180$ K	0.8598106	0.8467	0.8467
	1.5796330	1.6841	1.6841
	2.1691930	2.5078	2.5079
	2.6385480	3.3268	3.3273
	2.9977680	4.1575	4.1589
	3.2851810	5.1380	5.1412
CO_2, $T_0 = 290$ K	1.5005140	0.6973	0.6973
	2.7299300	1.3783	1.3783
	3.7114140	2.0445	2.0445
	4.4685400	2.7004	2.7005
	5.0247630	3.3553	3.3555
	5.3177280	3.8461	3.8472

TABLE 4: The maximum RD in the subcritical domain.

Substance	RD [%] Z	RD [%] c_v	RD [%] c_p
Ar	+0.0014	+0.0012	+0.0016
	−0.0001	−0.0532	−0.0801
N_2	+0.0025	+0.0135	+0.0142
	−0.0000	−0.0875	−0.1085
CH_4	+0.0021	+0.0352	+0.0361
	−0.0004	−0.0482	−0.0787
CO_2	+0.0018	+0.0073	+0.0074
	−0.0003	−0.0211	−0.0562

Table 5: AAD in the subcritical domain.

Substance	AAD [%]		
	Z	c_v	c_p
Ar	0.0001	0.0025	0.0037
N_2	0.0003	0.0048	0.0061
CH_4	0.0003	0.0062	0.0078
CO_2	0.0005	0.0048	0.0069
Mean	**0.0003**	**0.0046**	**0.0061**

the Lagrange form. Next, c_v is calculated from (12) and the partial derivatives $(\partial c_v/\partial \rho)_T$ are estimated along T_0 also from the interpolating polynomial in the Lagrange form. Now, values of Z and a are calculated from (15) and (19), along the isotherm $T_1 = T_0 - \Delta T$, by the Runge-Kutta-Verner fifth-order and six-order method with the adaptive step-size ΔT (Hull et al. [15]). The whole procedure is repeated until the lowest temperature of the range is reached (see Table 1).

The results are assessed by means of the relative deviation (RD) and the absolute average deviation (AAD), which are computed from the following relations:

$$\mathrm{RD} = \frac{X_{\mathrm{cal}} - X_{\mathrm{ref}}}{X_{\mathrm{ref}}} \times 100, [\%]$$
$$\mathrm{AAD} = \frac{100}{(M-1)N} \sum_{i=2}^{M} \sum_{j=1}^{N} \frac{|X_{\mathrm{cal}} - X_{\mathrm{ref}}|}{X_{\mathrm{ref}}}, [\%], \tag{21}$$

where X_{cal} is the calculated value and X_{ref} is the reference value of Z, c_v, and c_p (Tegeler et al. [16], Span et al. [17], Setzmann and Wagner [18], and Span and Wagner [19]).

As one could anticipate, the worst results are obtained along the saturation line ($\phi = 1.0$), because it exhibits the highest nonlinearity of Z with respect to T, and because all the partial derivatives with respect to the density are the worst exactly along this boundary line (Runge's phenomenon). For that reason, the maximum relative deviations given in Table 4 actually refer to the saturation line. While these results are worse than the rest, they are still in the limits of the experimental uncertainties of the direct measurements (see Table 8). However, the mean AADs of the results are 2 orders of magnitude (1-2 along the saturation line) better than the experimental uncertainties of the corresponding direct measurements (see Tables 5 and 6).

4. Conclusions

The compressibility factor and the heat capacity (the isochoric and the isobaric) of a vapor in the subcritical temperature range (including the saturation line) may be derived from the speed of sound with the AADs two orders of magnitude better than the experimental uncertainties of the corresponding direct measurements. This can be accomplished on account of just a few data points of the compressibility factor obtained from the direct measurement of the density. Moreover, the temperature derivatives of the compressibility factor, needed for initialization of the numerical integration, may also be derived from the speed of sound. Practically, the whole gaseous phase from the triple point to far above the critical point and from the ideal gas limit to the saturation line may be covered with just a few thermal data points from other source.

Table 6: AAD along the saturation line.

Substance	AAD [%]		
	Z	c_v	c_p
Ar	0.0004	0.0093	0.0142
N_2	0.0009	0.0180	0.0235
CH_4	0.0009	0.0155	0.0220
CO_2	0.0004	0.0093	0.0160
Mean	**0.0007**	**0.0130**	**0.0189**

Table 7: AAD in the transcritical domain.

Substance	AAD [%]		
	Z	c_v	c_p
Ar	0.0031	0.0573	0.0680
N_2	0.0039	0.0671	0.0672
CH_4	0.0016	0.0435	0.0957
CO_2	0.0009	0.0222	0.0943
Mean	**0.0024**	**0.0475**	**0.0813**

Table 8: The uncertainty of the reference data.

Substance	Uncertainty [%]		
	Z	c_v	c_p
Ar	±0.02	±0.30	±0.30
N_2	±0.02	±0.30	±0.30
CH_4	±0.03	±1.00	±1.00
CO_2	±0.03	±1.00	±1.00

Appendix

See Tables 1, 2, 3, 4, 5, 6, 7, and 8.

Conflict of Interests

The authors declare that there is no conflict of interests regarding the publication of this paper.

References

[1] M. B. Ewing and A. R. H. Goodwin, "An apparatus based on a spherical resonator for measuring the speed of sound in gases at high pressures. Results for argon at temperatures between 255 K and 300 K and at pressures up to 7 MPa," *The Journal of Chemical Thermodynamics*, vol. 24, no. 5, pp. 531–547, 1992.

[2] A. F. Estrada-Alexanders and J. P. M. Trusler, "The speed of sound in gaseous argon at temperatures between 110 K and 450 K and at pressures up to 19 MPa," *The Journal of Chemical Thermodynamics*, vol. 27, no. 10, pp. 1075–1089, 1995.

[3] M. F. Costa Gomes and J. P. M. Trusler, "The speed of sound in nitrogen at temperatures between $T = 250$ K and $T =$

350 K and at pressures up to 30 MPa," *Journal of Chemical Thermodynamics*, vol. 30, no. 5, pp. 527–534, 1998.

[4] J. P. M. Trusler and M. Zarari, "The speed of sound and derived thermodynamic properties of methane at temperatures between 275 K and 375 K and pressures up to 10 MPa," *The Journal of Chemical Thermodynamics*, vol. 24, no. 9, pp. 973–991, 1992.

[5] A. F. Estrada-Alexanders and J. P. M. Trusler, "Speed of sound in carbon dioxide at temperatures between (220 and 450) K and pressures up to 14 MPa," *Journal of Chemical Thermodynamics*, vol. 30, no. 12, pp. 1589–1601, 1998.

[6] J. P. M. Trusler, W. A. Wakeham, and M. P. Zarari, "Model intermolecular potentials and virial coefficients determined from the speed of sound," *Molecular Physics*, vol. 90, no. 5, pp. 695–703, 1997.

[7] J. F. Estela-Uribe and J. P. M. Trusler, "Acoustic and volumetric virial coefficients of nitrogen," *International Journal of Thermophysics*, vol. 21, no. 5, pp. 1033–1044, 2000.

[8] A. F. Estrada-Alexanders, J. P. M. Trusler, and M. P. Zarari, "Determination of thermodynamic properties from the speed of sound," *International Journal of Thermophysics*, vol. 16, no. 3, pp. 663–673, 1995.

[9] M. Bijedić and N. Neimarlija, "Thermodynamic properties of gases from speed-of-sound measurements," *International Journal of Thermophysics*, vol. 28, no. 1, pp. 268–278, 2007.

[10] M. Bijedić and N. Neimarlija, "Thermodynamic properties of carbon dioxide derived from the speed of sound," *Journal of the Iranian Chemical Society*, vol. 5, no. 2, pp. 286–295, 2008.

[11] M. Bijedić and N. Neimarlija, "Speed of sound as a source of accurate thermodynamic properties of gases," *Latin American Applied Research*, vol. 43, no. 4, pp. 393–398, 2013.

[12] J. P. M. Trusler, *Physical Acoustics and Metrology of Fluids*, Adam Hilger, Bristol, UK, 1991.

[13] M. J. Moran and H. N. Shapiro, *Fundamentals of Engineering Thermodynamics*, John Wiley & Sons, Chichester, UK, 5th edition, 2006.

[14] D. Y. Peng and D. B. Robinson, "A new two-constant equation of state," *Industrial and Engineering Chemistry Fundamentals*, vol. 15, no. 1, pp. 59–64, 1976.

[15] T. E. Hull, W. H. Enright, and K. R. Jackson, "User's guide for DVERK—a subroutine for solving non-stiff ODEs," Tech. Rep., University of Toronto, Toronto, Canada, 1976.

[16] C. Tegeler, R. Span, and W. Wagner, "A new equation of state for argon covering the fluid region for temperatures from the melting line to 700 K at pressures up to 1000 MPa," *Journal of Physical and Chemical Reference Data*, vol. 28, no. 3, pp. 779–850, 1999.

[17] R. Span, E. W. Lemmon, R. T. Jacobsen, W. Wagner, and A. Yokozeki, "A reference equation of state for the thermodynamic properties of nitrogen for temperatures from 63.151 to 1000 K and pressures to 2200 MPa," *Journal of Physical and Chemical Reference Data*, vol. 29, no. 6, pp. 1361–1433, 2000.

[18] U. Setzmann and W. Wagner, "A new equation of state and tables of thermodynamic properties for methane covering the range from the melting line to 625 K at pressures up to 1000 MPa," *Journal of Physical and Chemical Reference Data*, vol. 20, no. 6, pp. 1061–1155, 1991.

[19] R. Span and W. Wagner, "A new equation of state for carbon dioxide covering the fluid region from the triple-point temperature to 1100 K at pressures up to 800 MPa," *Journal of Physical and Chemical Reference Data*, vol. 25, no. 6, pp. 1509–1596, 1996.

3

MHD Flow and Heat Transfer in a Channel Bounded by a Shrinking Sheet and a Porous Medium Bed: Homotopy Analysis Method

Dileep Singh Chauhan and Rashmi Agrawal

Department of Mathematics, University of Rajasthan, Jaipur 302004, India

Correspondence should be addressed to Dileep Singh Chauhan; dileepschauhan@gmail.com

Academic Editors: G. L. Aranovich and R. R. Burnette

MHD flow of viscous conducting fluid is considered between a shrinking sheet and a porous medium bed. Suction is applied at the upper shrinking sheet and its surface temperature is always maintained higher than the temperature of the lower porous bed surface. Similarity transformations and HAM are used to solve the governing equations for velocity and temperature fields. The effects of various pertinent parameters on the results are discussed graphically.

1. Introduction

The study of boundary layer viscous flow and heat transfer due to stretching/shrinking surfaces is very significant because of its several applications in engineering and industrial processes, such as extrusion of polymer sheets from a die, drawing of plastic films, polyester thin wall heat shrink tubing, shrink film, wire drawing, glass fiber, and paper production. It is very important to control drag and the rate of heat transfer from the sheet for a better quality of the final product. Crane [1] first examined such flow over a linearly stretching sheet and obtained an exact similarity solution for the steady boundary layer flow of a viscous fluid in a closed form. His work was extended under various situations and physical conditions, by several researchers, such as Rajagopal et al. [2], Dandapat and Gupta [3], Chiam [4], Cortell [5], Liu [6], Ariel et al. [7, 8], Abbas et al. [9], Abd El-Aziz [10], Chauhan and Rastogi [11], Chauhan and Agrawal [12], and Chauhan and Olkha [13].

Recently the study of boundary layer flow and heat transfer adjacent to a shrinking sheet is given attention because of its several industrial applications, particularly in packaging industry. Flow due to a shrinking sheet is relatively a new phenomenon, where the velocity on the sheet is directed towards a fixed point. Steady flow in such a case does not occur normally, since the vorticity generated due to the shrinking of the sheet is not confined in the boundary layer unless a proper suction is applied at the sheet. Wang [14] reported the flow development adjacent to a shrinking sheet. Miklavčič and Wang [15] studied the existence and uniqueness of steady flow due to shrinking sheet and reported that for specific suction values dual solutions exist. Wang [16] studied stagnation-point flow over a shrinking sheet. Nadeem and Awais [17] investigated fluid flow through porous medium due to unsteady shrinking sheet with variable viscosity. Fang et al. [18] obtained a new solution for the Blasius equation of a shrinking sheet problem. Muhaimin et al. [19] investigated effects of heat and mass transfer on MHD flow over a shrinking sheet with suction. Rahimpour et al. [20] obtained the analytical solution of axisymmetric stagnation flow towards a shrinking sheet. Using HAM, Hayat et al. [21] investigated MHD rotating flow over a shrinking surface. Sajid and Hayat [22] examined MHD viscous flow due to a shrinking sheet using homotopy analysis method. The boundary layer flow adjacent to a shrinking sheet with power-law surface velocity is investigated by Fang [23] with mass transfer. A closed form analytical solution was presented by Fang and Zhang [24] for MHD flow over a shrinking sheet. Fang et al. [25] investigated the unsteady flow over a shrinking sheet with mass suction.

Yao and Chen [26] presented the new analytical solution of Blasius equation for a shrinking sheet problem. Mohd and Hashim [27] studied MHD flow and heat transfer over a shrinking sheet embedded in a porous medium with suction. Cortell [28] examined MHD flow due to shrinking sheet for axisymmetric and two-dimensional cases with suction. Noor et al. [29] obtained a series solution for MHD flow due to shrinking sheet by ADM (Adomian decomposition method). Hayat et al. [30] developed an analytical solution for a rotating flow due to a shrinking sheet. Fang and Zhang [31] presented an exact analytical solution for thermal boundary layer over a shrinking sheet with mass transfer effects. Fang et al. [32] investigated analytically flow adjacent to a shrinking-sheet with a second-order slip flow model. Ishak et al. [33] investigated stagnation-point flow of a micropolar fluid over a shrinking sheet. The channel flows with stretching/shrinking walls are also very important problems in fluid dynamics. Brady and Acrivos [34], Borkakoti and Bharali [35], Chauhan and Jakhar [36], Hayat et al. [37], and Chauhan and Agrawal [38] investigated flow and heat transfer problem in channels with one wall as stretching or shrinking sheet.

The purpose of the present investigation is to discuss the analytically solutions of electrically conducting viscous fluid flow and heat transfer in a channel bounded by a porous shrinking sheet and a porous medium bed of very small permeability, in the presence of a magnetic field. The equations governing velocity and temperature fields are simplified by a suitable similarity transformation, and then the reduced nonliner boundary value problem is solved analytically using homotopy analysis method. The physical effects of pertinent parameters on the flow and heat transfer are depicted graphically and discussed. The convergence of the solutions has also discussed by the so-called h-curves.

2. Formulation of the Problem

An electrically conducting, incompressible, viscous fluid flow in a horizontal channel bounded by a shrinking sheet and a porous medium bed is considered. A constant suction w_0 is applied at the upper shrinking sheet of the channel. The lower porous bed of the channel is of very small permeability k_0. We model flow in porous bed by Darcy's law. So, the flow in the porous bed is assumed to be zero in the absence of any external pressure gradient. A slip boundary condition suggested by Saffman [39] is applied at the fluid-porous medium interface. The permeability of the porous bed affects the flow in the channel through this boundary condition. Thus the effect of the porous bed is to introduce slip at the lower boundary of the horizontal channel. A constant magnetic field of strength B_0 is applied to the normal direction of the horizontal channel. Magnetic Reynolds number is taken to be small; therefore, induced magnetic field is neglected. No external electric field is applied and the polarization effect of the ionized fluid is negligible. So the electric field E is assumed to be zero. A Cartesian coordinate system (x, y, z) is assumed with origin at the porous bed surface. The axis of x and axis of y are taken on the surface of the porous bed and the axis of z is normal to it (see schematic diagram, Figure 1).

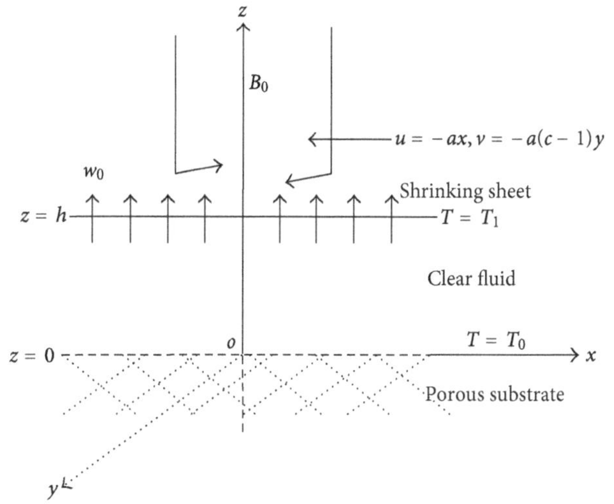

FIGURE 1: Schematic diagram.

(u, v, w) are considered velocity components in x, y, and z directions, respectively, and T is the temperature of the fluid in the channel. The width of the channel is assumed h, so that walls of the channel are at $z = 0$ and $z = h$.

The governing equations for the present problem of MHD flow and heat transfer are given by

$$\frac{\partial u}{\partial x} + \frac{\partial v}{\partial y} + \frac{\partial w}{\partial z} = 0, \tag{1}$$

$$u\frac{\partial u}{\partial x}+v\frac{\partial u}{\partial y}+w\frac{\partial u}{\partial z} = -\frac{1}{\rho}\frac{\partial P}{\partial x}+\nu\left(\frac{\partial^2 u}{\partial x^2}+\frac{\partial^2 u}{\partial y^2}+\frac{\partial^2 u}{\partial z^2}\right)-\frac{\sigma B_0^2}{\rho}u, \tag{2}$$

$$u\frac{\partial v}{\partial x}+v\frac{\partial v}{\partial y}+w\frac{\partial v}{\partial z} = -\frac{1}{\rho}\frac{\partial P}{\partial y}+\nu\left(\frac{\partial^2 v}{\partial x^2}+\frac{\partial^2 v}{\partial y^2}+\frac{\partial^2 v}{\partial z^2}\right)-\frac{\sigma B_0^2}{\rho}v, \tag{3}$$

$$u\frac{\partial w}{\partial x}+v\frac{\partial w}{\partial y}+w\frac{\partial w}{\partial z} = -\frac{1}{\rho}\frac{\partial P}{\partial z}+\nu\left(\frac{\partial^2 w}{\partial x^2}+\frac{\partial^2 w}{\partial y^2}+\frac{\partial^2 w}{\partial z^2}\right), \tag{4}$$

$$\rho C_p\left(u\frac{\partial T}{\partial x}+v\frac{\partial T}{\partial y}+w\frac{\partial T}{\partial z}\right) = k\left(\frac{\partial^2 T}{\partial x^2}+\frac{\partial^2 T}{\partial y^2}+\frac{\partial^2 T}{\partial z^2}\right). \tag{5}$$

The corresponding boundary conditions are given by

$$\begin{aligned} &\text{at } z = h, \quad u = -ax, \quad v = -a(c-1)y, \\ &\qquad w = w_0, \quad T = T_1, \\ &\text{at } z = 0, \quad \tau_{zx} = \frac{\mu\alpha}{\sqrt{k_0}}u, \quad \tau_{zy} = \frac{\mu\alpha}{\sqrt{k_0}}v, \\ &\qquad w = 0, \quad T = T_0, \end{aligned} \tag{6}$$

where P is the pressure; ρ is the density of the fluid; ν is the kinematic viscosity; σ is the electrical conductivity; C_p

is the specific heat at constant pressure; k is the thermal conductivity in the clear fluid medium; T_0 is the porous medium wall temperature; T_1 is the temperature of upper stretching wall; k_0 is the permeability of the porous medium; μ is the viscosity of the clear fluid; α is a nondimensional constant that depends only on the structure of the porous matrix.

Here, $a > 0$ is the shrinking constant. c is a constant, where $c = 1$ indicates one-direction shrinking and $c = 2$ corresponds to axisymmetric shrinking.

3. Method of Solution

We introduce the following similarity transformations:

$$u = axf'(\eta), \quad v = a(c-1)yf'(\eta), \quad w = -ahcf(\eta),$$

$$\eta = \frac{z}{h}, \quad \theta(\eta) = \frac{T - T_0}{T_1 - T_0}, \tag{7}$$

where $T_1 - T_0 = Dx$, D being a nonzero constant.

By using similarity transformations defined in (7), we find that the continuity equation (1) is automatically satisfied. Further, we see that (4) reduces to

$$w\frac{\partial w}{\partial z} = -\frac{1}{\rho}\frac{\partial P}{\partial z} + v\left(\frac{\partial^2 w}{\partial z^2}\right). \tag{8}$$

On integration, we get

$$\frac{w^2}{2} = -\frac{P}{\rho} + v\frac{dw}{dz} + \text{constant}. \tag{9}$$

Equations (2) and (3), using (9) and similarity transformations (7), are reduced to

$$f^{iv} - M^2 f'' - R\left((2-c)f'f'' - cff'''\right) = 0. \tag{10}$$

Similarly, using (7), we see that (5) is reduced to

$$\theta'' = R\Pr\left(f'\theta - cf\theta'\right). \tag{11}$$

The corresponding boundary conditions for the present problem are reduced to

$$\begin{aligned} &\text{at} \quad \eta = 1, \quad f' = -1, \quad f = -\frac{w_0}{\text{ach}} = -\lambda, \quad \theta = 1, \\ &\text{at} \quad \eta = 0, \quad f'' = \frac{\alpha}{K}f', \quad f = 0, \quad \theta = 0, \end{aligned} \tag{12}$$

where $M = \sqrt{(\sigma/\rho v)}B_0 h$, the Hartmann number; $R = ah^2/v$, the nondimensional shrinking parameter; $\Pr = \mu C_p/k$, the Prandtl number; $K = \sqrt{k_0}/h$, the permeability parameter.

4. Homotopy Analysis Method

Following Liao [40], we now solve the nonlinear system consisting of (10)–(12) by HAM. For HAM solutions, we choose the initial guesses $f_0(\eta)$ and $\theta_0(\eta)$ and the auxiliary linear operators L_f and L_θ in the following forms:

$$\begin{aligned} f_0(\eta) &= \frac{2K(1-3\lambda)}{4K+\alpha}\eta + \frac{\alpha(1-3\lambda)}{4K+\alpha}\eta^2 \\ &\quad + \frac{\alpha(2\lambda-1) - 2K(1-\lambda)}{4K+\alpha}\eta^3, \\ \theta_0(\eta) &= \eta, \\ L_f(f) &= \frac{d^4 f}{d\eta^4}, \\ L_\theta(\theta) &= \frac{d^2\theta}{d\eta^2}. \end{aligned} \tag{13}$$

Here, auxiliary linear operators have the property that

$$\begin{aligned} L_f(f_0) &= L_f\left(a_{0,0} + a_{0,1}\eta + a_{0,2}\eta^2 + a_{0,3}\eta^3\right) = 0, \\ L_\theta(\theta_0) &= L_\theta(b_{0,0} + b_{0,1}\eta) = 0. \end{aligned} \tag{14}$$

Let $p \in [0,1]$ denote the embedding parameter and h_1, h_2 indicate the nonzero auxiliary parameters. Then we construct the following homotopies:

$$\begin{aligned} H_f\left[\overline{f}(\eta;p);p\right] &= (1-p)L_f\left[\overline{f}(\eta;p) - f_0(\eta)\right] \\ &\quad - p\hbar_1 N_f\left[\overline{f}(\eta;p)\right], \\ H_\theta\left[\overline{\theta}(\eta;p);p\right] &= (1-p)L_\theta\left[\overline{\theta}(\eta;p) - \theta_0(\eta)\right] \\ &\quad - p\hbar_2 N_\theta\left[\overline{\theta}(\eta;p)\right], \end{aligned} \tag{15}$$

where

$$\begin{aligned} N_f\left[\overline{f}(\eta;p)\right] &= \frac{\partial^4\overline{f}(\eta;p)}{\partial\eta^4} - M^2\frac{\partial^2\overline{f}(\eta;p)}{\partial\eta^2} \\ &\quad - R\left[(2-c)\frac{\partial\overline{f}(\eta;p)}{\partial\eta}\frac{\partial^2\overline{f}(\eta;p)}{\partial\eta^2}\right. \\ &\quad \left. - c\,\overline{f}(\eta;p)\frac{\partial^3\overline{f}(\eta;p)}{\partial\eta^3}\right], \\ N_\theta\left[\overline{\theta}(\eta;p)\right] &= \frac{\partial^2\overline{\theta}(\eta;p)}{\partial\eta^2} - R\Pr \\ &\quad \times\left[\overline{\theta}(\eta;p)\frac{\partial\overline{f}(\eta;p)}{\partial\eta} - c\overline{f}(\eta;p)\frac{\partial\overline{\theta}(\eta;p)}{\partial\eta}\right]. \end{aligned} \tag{16}$$

When $p = 0$, we have

$$\begin{aligned} H_f\left[\overline{f}(\eta;0);0\right] &= L_f\left[\overline{f}(\eta;0) - f_0(\eta)\right], \\ H_\theta\left[\overline{\theta}(\eta;0);0\right] &= L_\theta\left[\overline{\theta}(\eta;0) - \theta_0(\eta)\right]. \end{aligned} \tag{17}$$

If $p = 1$, we have

$$H_f\left[\overline{f}(\eta;1);1\right] = -\hbar_1 N_f\left[\overline{f}(\eta;1)\right],$$
$$H_\theta\left[\overline{\theta}(\eta;1);1\right] = -\hbar_2 N_\theta\left[\overline{\theta}(\eta;1)\right]. \tag{18}$$

4.1. Zeroth Order Deformation Equations. Thus, if we take

$$H_f\left[\overline{f}(\eta;p);p\right] = 0,$$
$$H_\theta\left[\overline{\theta}(\eta;p);p\right] = 0, \tag{19}$$

we obtain the family of equations as follows:

$$(1-p)L_f\left[\overline{f}(\eta;p) - f_0(\eta)\right] = p\hbar_1 N_f\left[\overline{f}(\eta;p)\right],$$
$$(1-p)L_\theta\left[\overline{\theta}(\eta;p) - \theta_0(\eta)\right] = p\hbar_2 N_\theta\left[\overline{\theta}(\eta;p),\overline{f}(\eta;p)\right]. \tag{20}$$

These equations are subjected to the boundary conditions:

$$\overline{f'}(1;p) = -1, \qquad \overline{f}(1;p) = -\lambda, \qquad \overline{\theta}(1;p) = 1,$$
$$\overline{f''}(0;p) = \frac{\alpha}{K}\overline{f'}(0;p), \qquad \overline{f}(0;p) = 0, \qquad \overline{\theta}(0;p) = 0. \tag{21}$$

Obviously, when $p = 0$, we see that

$$\overline{f}(\eta;0) = f_0(\eta),$$
$$\overline{\theta}(\eta;0) = \theta_0(\eta), \tag{22}$$

and when $p = 1$, we have

$$\overline{f}(\eta;1) = f(\eta),$$
$$\overline{\theta}(\eta;1) = \theta(\eta). \tag{23}$$

Thus, as the embedding parameter p increases from 0 to 1, the mappings $\overline{f}(\eta;p)$ and $\overline{\theta}(\eta;p)$ vary smoothly from their known initial guesses, $f_0(\eta)$ and $\theta_0(\eta)$, to the required solutions, $f(\eta)$ and $\theta(\eta)$, respectively.

Expanding $\overline{f}(\eta;p)$ and $\overline{\theta}(\eta;p)$ by Taylor's series in the embedding parameter p and using (22), we can write

$$\overline{f}(\eta;p) = f_0(\eta) + \sum_{k=1}^{\infty} f_k(\eta)\,p^k, \tag{24}$$

where $f_k(\eta) = [(1/k!)(\partial^k \overline{f}(\eta;p)/\partial p^k)]_{p=0}$ and

$$\overline{\theta}(\eta;\ p) = \theta_0(\eta) + \sum_{k=1}^{\infty}\theta_k(\eta)\,p^k, \tag{25}$$

where $\theta_k(\eta) = [(1/k!)(\partial^k \overline{\theta}(\eta;p)/\partial p^k)]_{p=0}$.

Now assume that for properly chosen values of $\hbar_1$ and $\hbar_2$ the series (24) and the series (25) are convergent at $p = 1$. Thus we obtain

$$f(\eta) = f_0(\eta) + \sum_{k=1}^{\infty} f_k(\eta),$$
$$\theta(\eta) = \theta_0(\eta) + \sum_{k=1}^{\infty}\theta_k(\eta). \tag{26}$$

4.2. mth Order Deformation Equations. The governing equations of $f_m(\eta)$ and $\theta_m(\eta)$ can be deduced from the zeroth order deformation equations (20) by differentiating these m times with respect to the embedding parameter p, setting $p = 0$ and finally dividing them by $\underline{|m}$. Thus we have

$$L_f\left[f_m(\eta) - \chi_m f_{m-1}(\eta)\right] = \hbar_1 R_m^f(\eta),$$
$$L_\theta\left[\theta_m(\eta) - \chi_m \theta_{m-1}(\eta)\right] = \hbar_2 R_m^\theta(\eta), \tag{27}$$

subject to corresponding boundary conditions:

$$f'_m(1) = 0, \qquad f_m(1) = 0, \qquad \theta_m(1) = 0,$$
$$f''_m(0) = \frac{\alpha}{K} f'_m(0), \qquad f_m(0) = 0, \qquad \theta_m(0) = 0, \tag{28}$$

where

$$R_m^f(\eta) = \frac{1}{\underline{|m-1}} \left.\frac{\partial^{m-1} N_f\left[\overline{f}(\eta;p)\right]}{\partial p^{m-1}}\right|_{p=0}$$
$$= f_{m-1}^{iv} - M^2 f''_{m-1} + \mathrm{Re}\sum_{r=0}^{m-1}\left((2-c)f'_{m-1-r}f''_r - c f_{m-1-r}f'''_r\right),$$
$$R_m^\theta(\eta) = \frac{1}{\underline{|m-1}} \left.\frac{\partial^{m-1} N_\theta\left[\overline{\theta}(\eta;p),\overline{f}(\eta;p)\right]}{\partial p^{m-1}}\right|_{p=0}$$
$$= \theta''_{m-1} - \mathrm{RPr}\sum_{r=0}^{m-1}\left(f'_{m-1-r}\theta_r - c f_{m-1-r}\theta'_r\right). \tag{29}$$

Here, χ_m is defined as

$$\chi_m = \begin{cases} 0, & \text{for } m \le 1, \\ 1, & \text{for } m > 1. \end{cases} \tag{30}$$

The mth order ($m \ge 1$) deformation equations (27) are linear, and using the initial guesses $f_0(\eta)$, $\theta_0(\eta)$, we can obtain $f_1(\eta)$, $f_2(\eta)$, $f_3(\eta)$, and so forth and $\theta_1(\eta)$, $\theta_2(\eta)$, $\theta_3(\eta)$, and so forth using boundary conditions (28). With the help of the symbolic computation software MATHEMATICA, the solution series for (27) and (28) can be expressed in the following form:

$$f(\eta) = \sum_{m=0}^{\infty} f_m(\eta) = \lim_{M\to\infty}\left[\sum_{n=1}^{4M+3}\sum_{m=n-1}^{4M+2} a_{m,n}\eta^n\right],$$
$$\theta(\eta) = \sum_{m=0}^{\infty} \theta_m(\eta) = \lim_{M\to\infty}\left[\sum_{n=1}^{4M+1}\sum_{m=n-1}^{4M} b_{m,n}\eta^n\right]. \quad (31)$$

The coefficient $a_{m,n}$ and $b_{m,n}$ can be obtained using the following recurrence relations:

$$a_{m,0} = 0,$$
$$a_{m,1} = \chi_m a_{m-1,1} + \frac{2K}{4K+\alpha}\sum_{k=4}^{4m+3}\frac{\Gamma_{m,k-4}}{k(k-1)(k-2)},$$
$$a_{m,2} = \chi_m a_{m-1,2} + \frac{\alpha}{4K+\alpha}\sum_{k=4}^{4m+3}\frac{\Gamma_{m,k-4}}{k(k-1)(k-2)},$$
$$a_{m,3} = \chi_m a_{m-1,3} - \sum_{k=4}^{4m+3}\left[\left\{\frac{\alpha+2K}{4K+\alpha}(k-3)+1\right\}\frac{\Gamma_{m,k-4}}{k(k-1)(k-2)(k-3)}\right],$$
$$a_{m,k} = \chi_m a_{m-1,k} + \frac{\Gamma_{m,k-4}}{k(k-1)(k-2)(k-3)},$$
$$4 \le k \le 4m+3, \quad (32)$$

where

$$\Gamma_{m,1}^{k} = \hbar_1\left(a4_{m-1,k} - \left(M^2\right)a2_{m-1,k} - R\left\{(2-c)\,\delta1_{m,k} - c\delta2_{m,k}\right\}\right), \quad 0 \le k \le 4m-1,$$
$$\delta1_{m,n} = \sum_{r=0}^{m-1}\sum_{k=\max\{0,n-4m+4r+1\}}^{\min\{n,4r+3\}} a1_{r,k}a2_{m-r-1,n-k},$$
$$0 \le n \le 4m-1,$$
$$\delta2_{m,n} = \sum_{r=0}^{m-1}\sum_{k=\max\{0,n-4m+4r+1\}}^{\min\{n,4r+3\}} a_{r,k}a3_{m-r-1,n-k},$$
$$0 \le n \le 4m-1,$$
$$a4_{m,k} = (k+1)\,a3_{m,k+1}, \quad 0 \le k \le 4m-1,$$
$$a3_{m,k} = (k+1)\,a2_{m,k+1}, \quad 0 \le k \le 4m,$$
$$a2_{m,k} = (k+1)\,a1_{m,k+1}, \quad 0 \le k \le 4m+1,$$
$$a1_{m,k} = (k+1)\,a_{m,k+1}, \quad 0 \le k \le 4m+2,$$

$$b_{m,0} = 0,$$
$$b_{m,1} = \chi_m b_{m-1,1} - \sum_{k=2}^{4m+1}\frac{\gamma_{m,k-2}}{k(k-1)},$$
$$b_{m,k} = \chi_m b_{m-1,k} + \frac{\gamma_{m,k-2}}{k(k-1)}, \quad 2 \le k \le 4m+1, \quad (33)$$

where

$$\gamma_{m,1}^{k} = \hbar_2\left(b2_{m-1,k} - R\Pr\left\{\Delta1_{m,k} - c\Delta2_{m,k}\right\}\right),$$
$$0 \le k \le 4m-1,$$
$$\Delta1_{m,n} = \sum_{r=0}^{m-1}\sum_{k=\max\{0,n-4m+4r+1\}}^{\min\{n,4r+3\}} a1_{r,k}b_{m-r-1,n-k},$$
$$0 \le n \le 4m-1, \quad (34)$$
$$\Delta2_{m,n} = \sum_{r=0}^{m-1}\sum_{k=\max\{0,n-4m+4r+1\}}^{\min\{n,4r+3\}} a_{r,k}b1_{m-r-1,n-k},$$
$$0 \le n \le 4m-1,$$
$$b2_{m,k} = (k+1)\,b1_{m,k+1}, \quad 0 \le k \le 4m-1,$$
$$b1_{m,k} = (k+1)\,b_{m,k+1}, \quad 0 \le k \le 4m.$$

5. Convergence of the HAM Solutions

Equations (31) are the series solutions of the present problem of MHD flow and heat transfer in a horizontal channel bounded by a shrinking sheet and a porous bed of very small permeability. As Liao [40] pointed out that the convergence region and convergence rate of the HAM solution series can be adjusted by means of the value of the auxiliary parameters $\hbar_1$ and $\hbar_2$ here we have taken $\hbar_1 = \hbar_2 = \hbar$. In order to find the convergence region for the series (31), we have plotted the so-called h-curves. We have plotted the curves of $f''(0)$ versus $\hbar$ and $\theta'(0)$ versus $\hbar$ in Figure 2, for 20th, 21st, and 22nd order approximation. It clearly indicates that the valid region of $\hbar$ is about $-1.8 \le \hbar \le -0.4$. In all numerical calculations for plotting graphs, the order of approximation is taken as $m = 22$ and $\hbar = -1.1$.

6. Discussion

In the present investigation, the effects of a porous bed and a shrinking wall on the MHD flow and heat transfer are examined in a horizontal channel. The governing nonlinear partial differential equations for flow and heat transfer are converted into ordinary differential equations using similarity transformations. Then series solutions are obtained for velocity and temperature fields using homotopy analysis method (HAM). The convergence region of the solution series by HAM is obtained by plotting the so-called h-curve graph in Figure 2. To ensure the steady flow near the

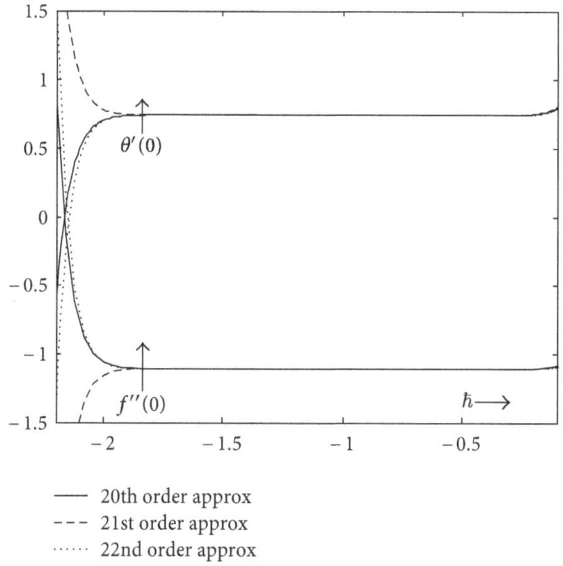

FIGURE 2: h-curve for $f''(0)$ and $\theta'(0)$ for $M = 1$, $c = 3$, $K = 0.001$, $\lambda = 0.5$, $\alpha = 0.1$, Re $= 2$, and Pr $= 2$.

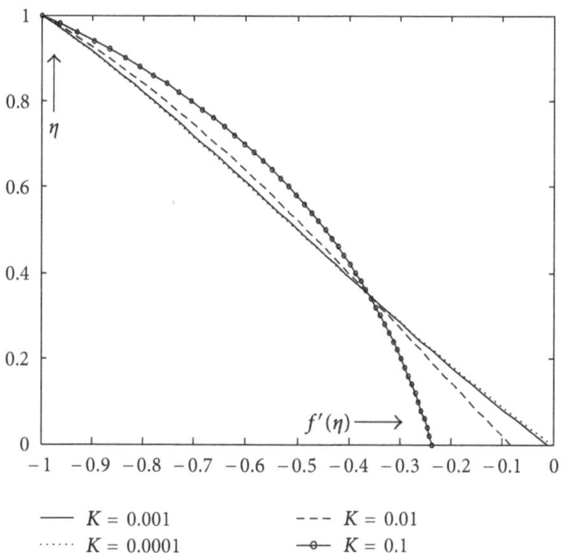

FIGURE 3: $f'(\eta)$ for $M = 1$, $c = 3$, $\alpha = 0.1$, $\lambda = 0.5$, and $R = 2$.

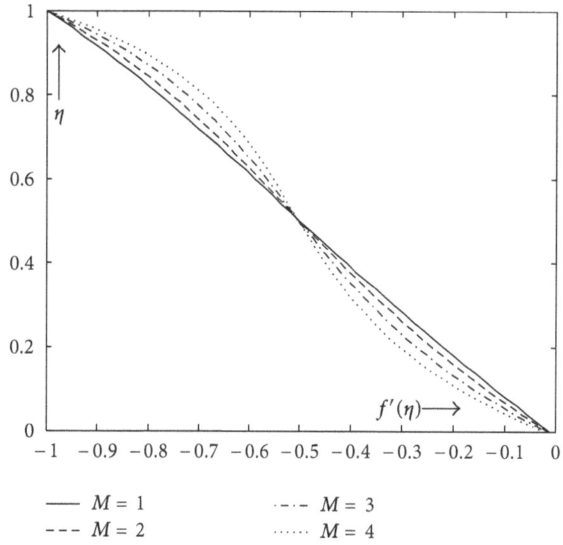

FIGURE 4: $f'(\eta)$ for $c = 3$, $K = 0.001$, $\lambda = 0.5$, $\alpha = 0.1$, and $R = 2$.

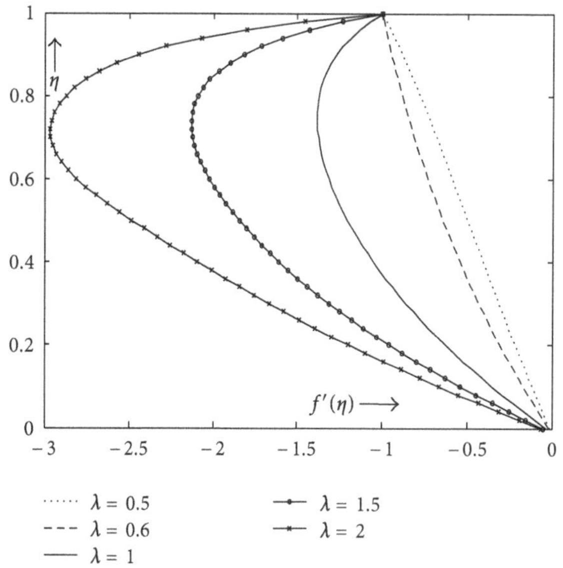

FIGURE 5: $f'(\eta)$ for $M = 1$, $c = 3$, $K = 0.001$, $\alpha = 0.1$, and $R = 2$.

shrinking sheet by confining the generated vorticity in the boundary layer, the opposite forces such as the wall mass suction and the magnetic field are taken as sufficiently high by assigning suitable values of M or λ.

Figures 3, 4, 5, 6, and 7 illustrate the variations of axial velocity for various values of the pertinent parameters. Figure 3 shows that magnitude of the axial velocity increases near the porous bed in the channel with the increase of the permeability parameter (K), while it decreases near the upper shrinking sheet. Permeability is the measure of ease to flow and it enhances fluid flow near the porous bed. The effect of the fluid saturated porous medium bed is to introduce a slip at one wall of the channel, and therefore the flow velocity at this wall is no longer equal to zero. This slip depends on the parameters of structure and permeability of the porous matrix and affects the flow inside the channel through boundary condition suggested by Saffman [39]. In Figure 4, we observe that the magnitude of the axial velocity decreases with the increase of the magnetic parameter (M) near the upper shrinking sheet since Lorentz force opposes the flow and leads to the deceleration of fluid flow near upper wall, while it increases near porous bed. Figure 5 illustrates that with increase in the value of suction parameter (λ), magnitude of the axial flow increases in the channel. Figure 6 shows that the effect of the shrinking parameter (R) is to reduce the flow near the upper sheet while the reverse effect is observed near the lower porous bed. Figure 7 depicts that the axial flow near the upper wall is greater when the sheet shrinks axisymmetrically than that when it shrinks only in

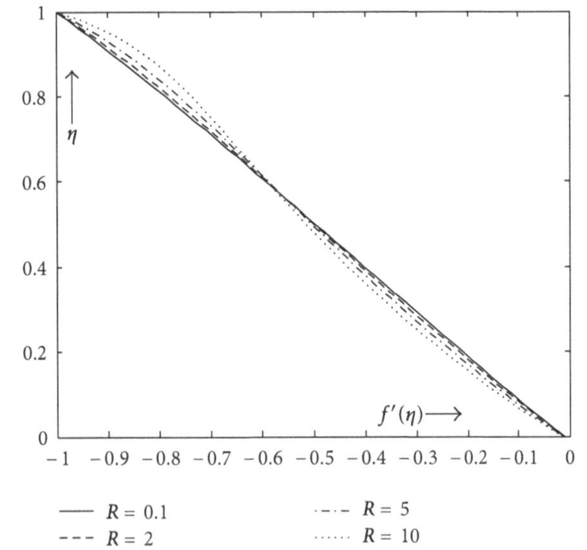

FIGURE 6: $f'(\eta)$ for $M = 1$, $c = 3$, $K = 0.001$, $\lambda = 0.5$, and $\alpha = 0.1$.

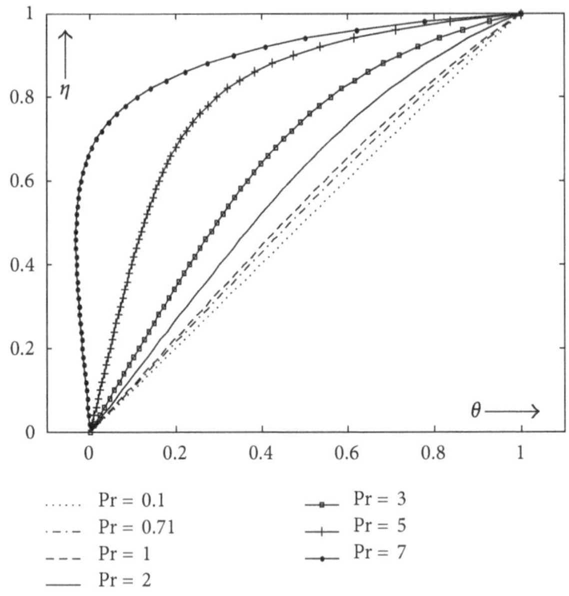

FIGURE 8: $\theta(\eta)$ for $M = 1$, $c = 3$, $K = 0.001$, $\lambda = 0.5$, $\alpha = 0.1$, and $R = 2$.

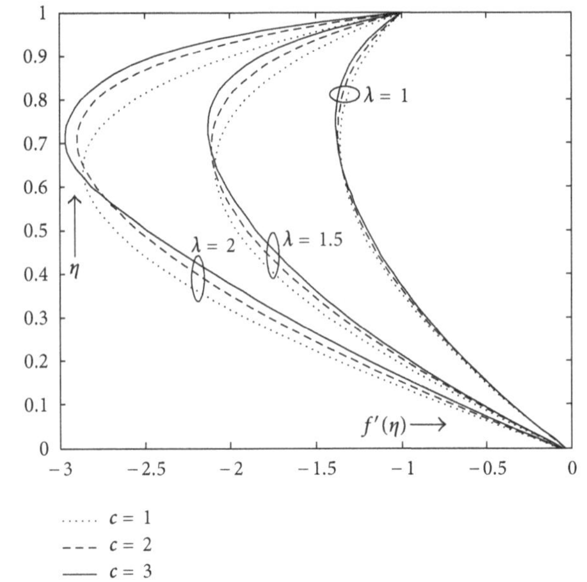

FIGURE 7: $f'(\eta)$ for $M = 1$, $K = 0.001$, $\alpha = 0.1$, and $R = 2$.

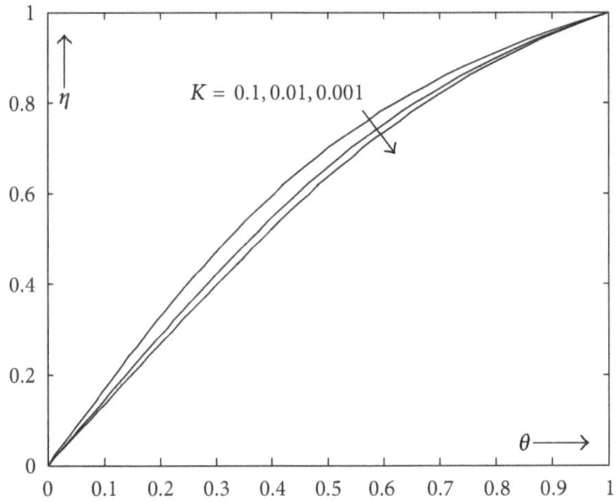

FIGURE 9: $\theta(\eta)$ for $M = 1$, $c = 3$, $\lambda = 0.5$, $\alpha = 0.1$, $R = 2$, and Pr = 2.

the x direction, while near the porous bed reverse effect is observed.

Figures 8, 9, 10, 11, 12, and 13 illustrate profiles of the nondimensional temperature against the coordinate η. The effect of Prandtl number (Pr) on the temperature profiles can be seen in Figure 8. Increase in the Pr values leads to the decrease of the temperature in the channel and the thickness of the thermal boundary layer becomes smaller. As Pr value increases, the thermal diffusivity decreases which reduces the energy transfer ability; hence the temperature decreases. Note that very small values of Pr correspond to liquid metals, Pr $\sim$ 1 corresponds to air and diatomic gases, and very large values of Pr correspond to highly viscous oils. At room temperature, Pr $\sim$ 7 corresponds to water. Figure 9 depicts temperature distribution in the channel for various values of the permeability parameter (K). It is found that the effect of the permeability of lower bed is to decrease the temperature throughout the channel. It is also observed in Figures 10, 11, and 12 that the temperature in the channel decreases with increase in the value of suction parameter (λ) or magnetic parameter (M) or shrinking parameter (R). Further, it is observed in Figure 13 that the temperature throughout the channel for the axisymmetric shrinking case ($c = 2$) is less than that for the case of one-direction shrinking ($c = 1$).

Many industrial processes, for example, in metallurgical and packaging units, involve the drawing and cooling of continuous strips, sheets, and so forth and there quality depends on the rate of cooling and shear stresses on them. Therefore the skin friction coefficient and the rate of heat transfer are the physical quantities of interest to engineers

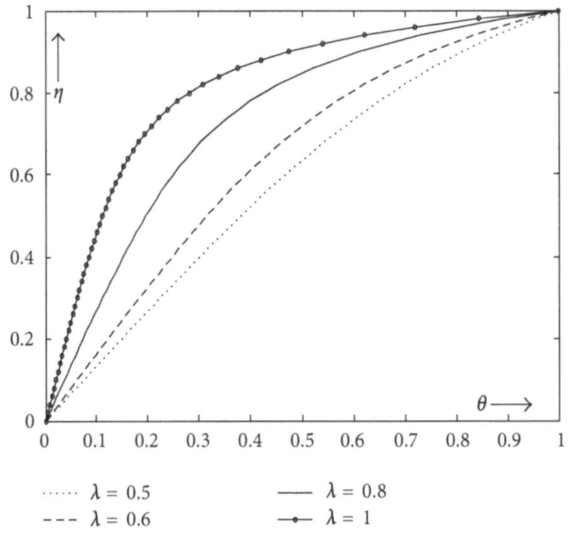

FIGURE 10: $\theta(\eta)$ for $M = 1$, $c = 3$, $K = 0.001$, $R = 2$, $\Pr = 2$, and $\alpha = 0.1$.

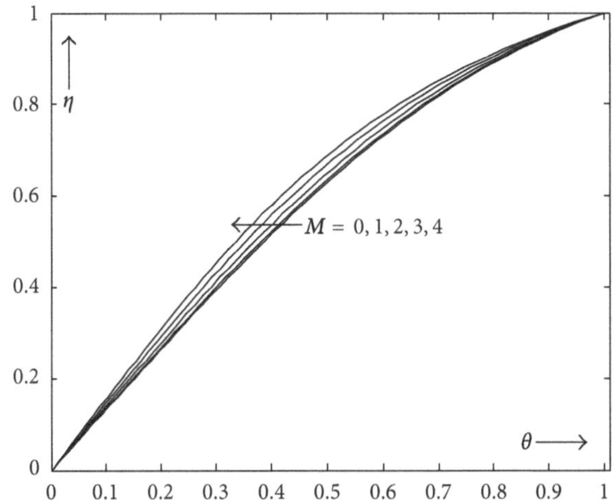

FIGURE 11: $\theta(\eta)$ for $\Pr = 2$, $c = 3$, $K = 0.001$, $\lambda = 0.5$, $\alpha = 0.1$, and $R = 2$.

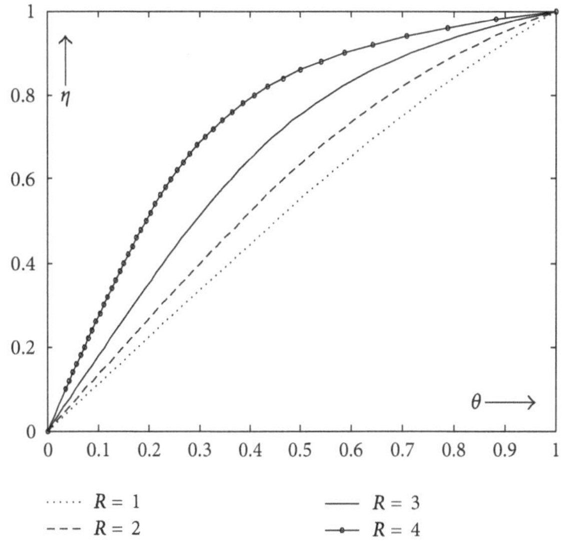

FIGURE 12: $\theta(\eta)$ for $M = 1$, $c = 3$, $\lambda = 0.5$, $\alpha = 0.1$, $K = 0.001$, and $\Pr = 2$.

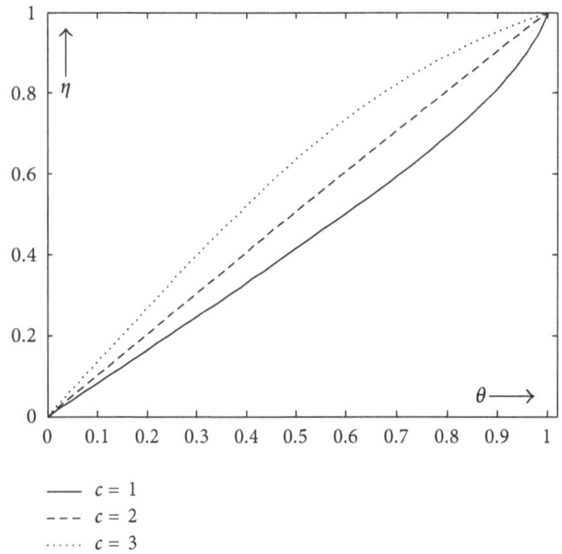

FIGURE 13: $\theta(\eta)$ for $M = 1$, $\lambda = 0.5$, $K = 0.001$, $R = 2$, $\Pr = 2$, and $\alpha = 0.1$.

and these are plotted in Figures 14 and 15 for various values of the pertinent parameters. It is found that the magnetic field increases the surface shear stress for axial flow, that is, the local skin friction at the shrinking sheet $\eta = 1$, because an increase in the strength of magnetic field leads to a thinner boundary layer, thereby causing an increase in the velocity gradient $f''(1)$ at the sheet. Shrinking parameter (R) or permeability parameter (K) also leads to an increase in the skin friction at the shrinking sheet, whereas the effect of the suction parameter (λ) is to reduce skin friction at the shrinking sheet. It is also noticed that the skin friction at shrinking sheet for axisymmetric shrinking is greater than that for the case of one-direction shrinking. It is observed in Figure 15 that the temperature gradient at the shrinking sheet increases as Prandtl number (Pr) increases, since a higher Pr fluid is more viscous and has a thinner thermal boundary layer which results at the sheet in a higher rate of heat transfer. It is also noticed that the temperature gradient at the shrinking sheet increases with the increase in the value of the shrinking parameter (R), or permeability (K), or suction parameter (λ), or magnetic parameter (M). It is also observed that the temperature gradient at the shrinking sheet is greater for the axisymmetric shrinking case than to one-direction shrinking.

7. Conclusion

The present paper studies the flow and heat transfer in a channel bounded by a porous medium bed and a shrinking sheet in the presence of a magnetic field. Using similarity

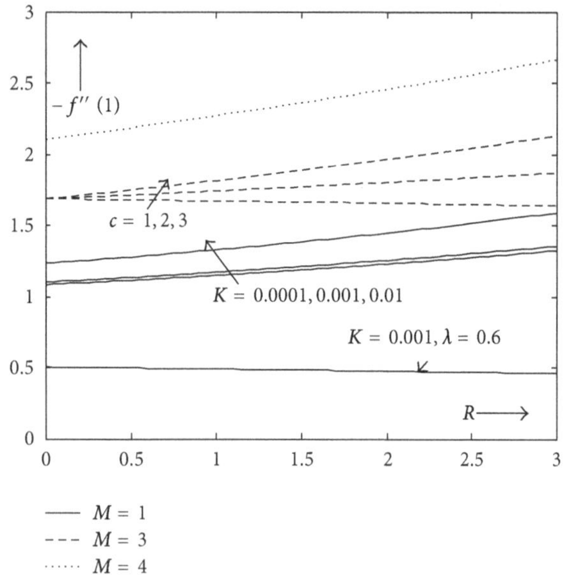

FIGURE 14: $-f''(1)$ versus R for $c = 3$, $K = 0.001$, $\lambda = 0.5$, and $\alpha = 0.1$.

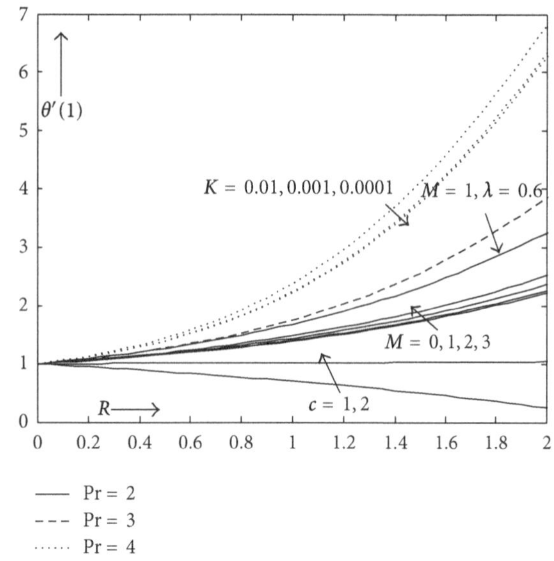

FIGURE 15: $\theta'(1)$ versus R for $M = 1$, $c = 3$, $K = 0.001$, $\lambda = 0.5$, and $\alpha = 0.1$.

transformations and homotopy analysis method, analytical series solutions are obtained and presented graphically to discuss features of these solutions for various values of the pertinent parameters. Convergence of the HAM series solution is ensured by taking suitable value of the parameter h. It is observed that the axial flow in the channel adjacent to upper wall is greater when it shrinks axisymmetrically to that when it shrinks in one direction, while reverse effect is observed at the other wall. Further, the effect of suction parameter is to increase the axial flow in the channel. Temperature in the channel decreases as the value of Prandtl number increases, and the effect of permeability parameter of the lower porous bed is to decrease temperature throughout the channel. It is also found that the temperature in the channel for the axisymmetric shrinking case is less than that for the case of one-direction shrinking. It is seen that the physical parameters such as suction and magnetic field control the rate of cooling and skin friction at the shrinking surface.

Acknowledgment

The support provided by the Council of Scientific and Industrial Research through Senior Research Fellowship to one of the authors, Rashmi Agrawal, is gratefully acknowledged.

References

[1] L. J. Crane, "Flow past a stretching plate," *Zeitschrift für Angewandte Mathematik und Physik*, vol. 21, no. 4, pp. 645–647, 1970.

[2] K. R. Rajagopal, T. Y. Na, and A. S. Gupta, "Flow of a viscoelastic fluid over a stretching sheet," *Rheologica Acta*, vol. 23, no. 2, pp. 213–215, 1984.

[3] B. S. Dandapat and A. S. Gupta, "Flow and heat transfer in a viscoelastic fluid over a stretching sheet," *International Journal of Non-Linear Mechanics*, vol. 24, no. 3, pp. 215–219, 1989.

[4] T. C. Chiam, "Hydromagnetic flow over a surface stretching with a power-law velocity," *International Journal of Engineering Science*, vol. 33, no. 3, pp. 429–435, 1995.

[5] R. Cortell, "Flow and heat transfer of a fluid through a porous medium over a stretching surface with internal heat generation/absorption and suction/blowing," *Fluid Dynamics Research*, vol. 37, no. 4, pp. 231–245, 2005.

[6] I. C. Liu, "Flow and heat transfer of an electrically conducting fluid of second grade in a porous medium over a stretching sheet subject to a transverse magnetic field," *International Journal of Non-Linear Mechanics*, vol. 40, no. 4, pp. 465–474, 2005.

[7] P. D. Ariel, T. Hayat, and S. Asghar, "The flow of an elastico-viscous fluid past a stretching sheet with partial slip," *Acta Mechanica*, vol. 187, no. 1–4, pp. 29–35, 2006.

[8] P. D. Ariel, T. Hayat, and S. Asghar, "Homotopy perturbation method and axisymmetric flow over a stretching sheet," *International Journal of Nonlinear Sciences and Numerical Simulation*, vol. 7, no. 4, pp. 399–406, 2006.

[9] Z. Abbas, Y. Wang, T. Hayat, and M. Oberlack, "Hydromagnetic flow in a viscoelastic fluid due to the oscillatory stretching surface," *International Journal of Non-Linear Mechanics*, vol. 43, no. 8, pp. 783–793, 2008.

[10] M. Abd El-Aziz, "Flow and heat transfer over an unsteady stretching surface with Hall effect," *Meccanica*, vol. 45, no. 1, pp. 97–109, 2010.

[11] D. S. Chauhan and P. Rastogi, "Heat transfer and entropy generation in MHD flow through porous medium past a stretching sheet," *International Journal of Energy and Technology*, vol. 3, pp. 1–13, 2011.

[12] D. S. Chauhan and R. Agrawal, "MHD flow through a porous medium adjacent to a stretching sheet: numerical and an approximate solution," *European Physical Journal Plus*, vol. 126, article 47, 2011.

[13] D. S. Chauhan and A. Olkha, "Slip flow and heat transfer of a second-grade fluid in a porous medium over a stretching sheet with power-law surface temperature or heat flux," *Chemical*

Engineering Communications, vol. 198, no. 9, pp. 1129–1145, 2011.

[14] C. Y. Wang, "Liquid film on an unsteady stretching sheet," *Quarterly of Applied Mathematics*, vol. 48, pp. 601–610, 1990.

[15] M. Miklavčič and C. Y. Wang, "Viscous flow due to a shrinking sheet," *Quarterly of Applied Mathematics*, vol. 64, no. 2, pp. 283–290, 2006.

[16] C. Y. Wang, "Stagnation flow towards a shrinking sheet," *International Journal of Non-Linear Mechanics*, vol. 43, no. 5, pp. 377–382, 2008.

[17] S. Nadeem and M. Awais, "Thin film flow of an unsteady shrinking sheet through porous medium with variable viscosity," *Physics Letters A*, vol. 372, no. 30, pp. 4965–4972, 2008.

[18] T. Fang, W. Liang, and C. F. F. Lee, "A new solution branch for the Blasius equation-A shrinking sheet problem," *Computers and Mathematics with Applications*, vol. 56, no. 12, pp. 3088–3095, 2008.

[19] Muhaimin, R. Kandasamy, and A. B. Khamis, "Effects of heat and mass transfer on nonlinear MHD boundary layer flow over a shrinking sheet in the presence of suction," *Applied Mathematics and Mechanics*, vol. 29, no. 10, pp. 1309–1317, 2008.

[20] M. Rahimpour, C. R. Mohebpour, A. Kimiaeifar, and G. H. Bagheri, "On the analytical solution of axisymmetric stagnation flow towards a shrinking sheet," *International Journal of Mechanical*, vol. 2, pp. 1–10, 2008.

[21] T. Hayat, T. Javed, and M. Sajid, "Analytic solution for MHD rotating flow of a second grade fluid over a shrinking surface," *Physics Letters A*, vol. 372, no. 18, pp. 3264–3273, 2008.

[22] M. Sajid and T. Hayat, "The application of homotopy analysis method for MHD viscous flow due to a shrinking sheet," *Chaos, Solitons and Fractals*, vol. 39, no. 3, pp. 1317–1323, 2009.

[23] T. Fang, "Boundary layer flow over a shrinking sheet with power-law velocity," *International Journal of Heat and Mass Transfer*, vol. 51, no. 25-26, pp. 5838–5843, 2008.

[24] T. Fang and J. Zhang, "Closed-form exact solutions of MHD viscous flow over a shrinking sheet," *Communications in Nonlinear Science and Numerical Simulation*, vol. 14, no. 7, pp. 2853–2857, 2009.

[25] T. Fang, J. Zhang, and S. Yao, "Viscous flow over an unsteady shrinking sheet with mass transfer," *Chinese Physics Letters*, vol. 26, Article ID 014703, 2009.

[26] B. Yao and J. Chen, "A new analytical solution branch for the Blasius equation with a shrinking sheet," *Applied Mathematics and Computation*, vol. 215, no. 3, pp. 1146–1153, 2009.

[27] N. F. Mohd and I. Hashim, "MHD flow and heat transfer adjacent to a permeable shrinking sheet embedded in a porous medium," *Sains Malaysiana*, vol. 38, no. 4, pp. 559–565, 2009.

[28] R. Cortell, "On a certain boundary value problem arising in shrinking sheet flows," *Applied Mathematics and Computation*, vol. 217, no. 8, pp. 4086–4093, 2010.

[29] N. F. M. Noor, S. Awang Kechil, and I. Hashim, "Simple non-perturbative solution for MHD viscous flow due to a shrinking sheet," *Communications in Nonlinear Science and Numerical Simulation*, vol. 15, no. 2, pp. 144–148, 2010.

[30] T. Hayat, S. Iram, T. Javed, and S. Asghar, "Shrinking flow of second grade fluid in a rotating frame: an analytic solution," *Communications in Nonlinear Science and Numerical Simulation*, vol. 15, no. 10, pp. 2932–2941, 2010.

[31] T. Fang and J. Zhang, "Thermal boundary layers over a shrinking sheet: an analytical solution," *Acta Mechanica*, vol. 209, no. 3-4, pp. 325–343, 2010.

[32] T. Fang, S. Yao, J. Zhang, and A. Aziz, "Viscous flow over a shrinking sheet with a second order slip flow model," *Communications in Nonlinear Science and Numerical Simulation*, vol. 15, no. 7, pp. 1831–1842, 2010.

[33] A. Ishak, Y. Y. Lok, and I. Pop, "Stagnation-point flow over a shrinking sheet in a micropolar fluid," *Chemical Engineering Communications*, vol. 197, no. 11, pp. 1417–1427, 2010.

[34] J. F. Brady and A. Acrivos, "Steady flow in a channel or tube with an accelerating surface velocity. An exact solution to the Navier-Stokes equations with reverse flow," *Journal of Fluid Mechanics*, vol. 112, pp. 127–150, 1981.

[35] A. K. Borkakoti and A. Bharali, "Hydromagnetic flow and heat transfer between two horizontal plates, the lower plate being a stretching sheet," *Quarterly of Applied Mathematics*, vol. 40, no. 4, pp. 461–467, 1983.

[36] D. S. Chauhan and P. K. Jakhar, "Non-Newtonian stretching coupled-flow in a channel bounded by a highly porous layer," *Modelling, Measurement and Control B*, vol. 71, no. 3-4, pp. 33–40, 2002.

[37] T. Hayat, Z. Abbas, T. Javed, and M. Sajid, "Three-dimensional rotating flow induced by a shrinking sheet for suction," *Chaos, Solitons and Fractals*, vol. 39, no. 4, pp. 1615–1626, 2009.

[38] D. S. Chauhan and R. Agrawal, "MHD flow and heat transfer in a channel bounded by a shrinking sheet and a plate with porous substrate," *Journal of Engineering Physics and Thermophysics*, vol. 84, pp. 1034–1046, 2011.

[39] P. G. Saffman, "On the boundary condition at the surface of a porous medium," *Studies in Applied Mathematics*, vol. 50, no. 2, pp. 93–101, 1971.

[40] S. J. Liao, *Beyond Perturbation: Introduction to the Homotopy Analysis Method*, Chapman and Hall, Boca Raton, Fla, USA, 2004.

4

Pedagogical Visualization of a Nonideal Carnot Engine

Jonas Johansson

Solid State Physics, Lund University, P.O. Box 118, 221 00 Lund, Sweden

Correspondence should be addressed to Jonas Johansson; jonas.johansson@ftf.lth.se

Academic Editor: Shripad T. Revankar

We have implemented a visualization tool for the demonstration of a nonideal Carnot engine, operating at finite time. The cycle time can be varied using a slide bar and the pressure-volume, temperature-entropy, power-time, and efficiency-time diagrams change interactively and are shown on one screen. We have evaluated the visualization tool among engineering students at university level during an introductory course on thermodynamics and we review and discuss the outcome of the evaluation.

1. Introduction

In 1975, Curzon and Ahlborn made the observation that most undergraduate textbooks on thermodynamics do not treat the time aspects of thermodynamic cycles; thus lacking the explanation of how power is generated from, for example, heat engines. In their investigation, Curzon and Ahlborn considered a Carnot cycle operating at finite time by modeling the time-dependent energy losses in the isotherms [1]. This approach enabled them to derive a general expression for the efficiency at maximum power, depending only on the temperatures of the reservoirs, just as the Carnot efficiency. Curzon and Ahlborn's motivation was purely pedagogical and at the time they were probably not aware that the same expression had been published already in 1957 by Novikov [2] (the original paper in Russian was published in 1957) and by Chambadal [3]. Nevertheless, this efficiency is most commonly referred to as the Curzon-Ahlborn efficiency and their work paved the way for finite time thermodynamics [4].

The purpose of this investigation is to construct an interactive visualization tool in the form of a computer program illustrating the time dependence of a nonideal Carnot engine and to evaluate this demonstration tool among university students during an introductory thermodynamics course. The reason for doing this is the lack of time and power aspects of thermodynamic cycles in undergraduate textbooks. The visualization tool may thus fill a gap in thermodynamics courses since it increases the awareness of these important engineering aspects. Illustrating the concepts by an interactive visualization may enable a more holistic understanding, which, in turn, will allow for deep learning.

2. Theoretical Model

We base our nonideal Carnot engine on Curzon-Ahlborn's model [1]; that is, the engine spends finite times in the isotherms so that the high temperature of the working medium (which we will implicitly consider to be an ideal gas), T_{hw}, will not reach the temperature of the hot reservoir, T_h. Correspondingly, the low temperature of the working medium, T_{cw}, will not reach the temperature of the cold reservoir, T_c, and we have $T_{hw} \leq T_h$ and $T_{cw} \geq T_c$. Apart from these heat losses, the engine is an ideal Carnot engine and it is schematically depicted in Figure 1.

The heat input, Q_h, and heat output, Q_c, are given by the times spent in the isotherms multiplied by the heat fluxes through the vessel containing the working medium, such that $Q_h = F_h t_h$ and $Q_c = F_c t_c$, where F_h and t_h (F_c and t_c) are the heat flux and the time spent in the hot (cold) isotherm. The heat fluxes are modeled as being proportional to the temperature differences between the working medium (an ideal gas) and the reservoirs, $F_h = \alpha(T_h - T_{hw})$ and $F_c = \beta(T_{cw} - T_c)$, where α and β are heat conduction coefficients, which depend on the material and the geometry of the vessel. If the material of the vessel is uniform, its thickness is constant, and the areas exposed to the reservoirs are the same for the hot and the cold reservoir, then $\alpha = \beta$.

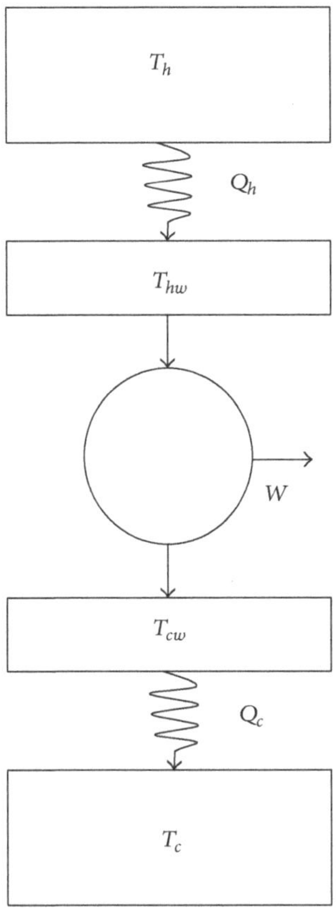

FIGURE 1: Schematic illustration of the Curzon-Ahlborn engine. The symbols are explained in the text. The zigzag arrows indicate the heat losses in the isotherms. The process operating between T_{hw} and T_{cw} is an ideal Carnot engine.

In our following analysis we take this to be the case and α is given by $\alpha = \kappa S/d$, where κ is the thermal conductivity of the vessel material, S is the area of the vessel that is exposed to any of the reservoirs, and d is the thickness of the vessel.

Due to the reversibility of the adiabatic stages, we note that it must hold that

$$\frac{Q_h}{T_{hw}} = \frac{Q_c}{T_{cw}} = \lambda, \tag{1}$$

where λ is a constant. Based on (1), the high and the low temperatures of the working medium can be expressed as

$$T_{hw} = \frac{\alpha t_h}{\alpha t_h + \lambda} T_h, \tag{2}$$

$$T_{cw} = \frac{\alpha t_c}{\alpha t_c - \lambda} T_c, \tag{3}$$

respectively. In the following analysis we let the engine spend equal amounts of time in the hot and the cold reservoir; that is, $t_h = t_c = t$. As $t \to \infty$, we see that $T_{hw} \to T_h$ and $T_{cw} \to T_c$. By taking this limit in (1), Q_h approaches its ideal value $nRT_h \ln(V_2/V_1)$ and we can identify $\lambda = nR\ln(V_2/V_1)$.

Power is defined as work per time and the output power of the engine can be calculated as work per cycle, $Q_h - Q_c$, divided by the cycle time. If we assume that the times spent in the adiabats are considerably shorter than the times spent in the isotherms, t for each isotherm, the cycle time is approximately given by $2t$ and the power can be expressed as

$$P = \frac{Q_h - Q_c}{2t} = \frac{\lambda\alpha}{2}\left(\frac{T_h}{\alpha t + \lambda} - \frac{T_c}{\alpha t - \lambda}\right). \tag{4}$$

The efficiency is given by

$$\eta = \frac{Q_h - Q_c}{Q_h} = 1 - \frac{T_{cw}}{T_{hw}} = 1 - \frac{\alpha t + \lambda}{\alpha t - \lambda}\frac{T_c}{T_h} \tag{5}$$

and we immediately see that η approaches the Carnot efficiency, $\eta_C = 1 - T_c/T_h$, for long cycle times ($t \to \infty$). By differentiating (4) with respect to t, we see that P has optima at the times given by

$$\frac{\alpha t - \lambda}{\alpha t + \lambda} = \pm\sqrt{\frac{T_c}{T_h}}. \tag{6}$$

The equation with negative right hand side in (6) can be discarded, since it is nonphysical. Its solution is a time shorter than the minimum time, which in turn is a solution to $P = 0$. By instead inserting the time corresponding to the positive right hand side in (6) into (5), we get the efficiency at maximum power, the so-called Curzon-Ahlborn efficiency,

$$\eta_{CA} = 1 - \sqrt{\frac{T_c}{T_h}}. \tag{7}$$

3. Visualization Tool

With our visualization tool we show the time dependence of four interesting properties of the nonideal Carnot engine simultaneously on one screen: the pressure-volume, the temperature-entropy, the power-time, and the efficiency-time diagrams. The reason that we restrict the tool to show precisely these four properties is that they are the most essential for the understanding of the nonideal, finite time Carnot engine. The pressure-volume and the temperature-entropy diagrams are well covered in introductory thermodynamics courses; therefore, it is instructive to see how these diagrams change as an effect of a finite cycle time. The power-time diagram shows the output power as a function of time. It should also be instructive to see that this curve has a maximum and also at what efficiency this maximum power is reached, which is shown in the efficiency-time diagram. In these two time-dependent diagrams it is also obvious that the engine cannot generate any power when operated at Carnot efficiency.

The time the engine spends in one of the isotherms can be interactively varied. We have implemented the tool as a cdf (computational document format) document [5]. This is an interactive document format that is programmed using the software Mathematica. cdf documents can be used and viewed with the free cdf player that can be downloaded at [5]. This format can be considered as a candidate for innovative,

Carnot cycle operating in finite time

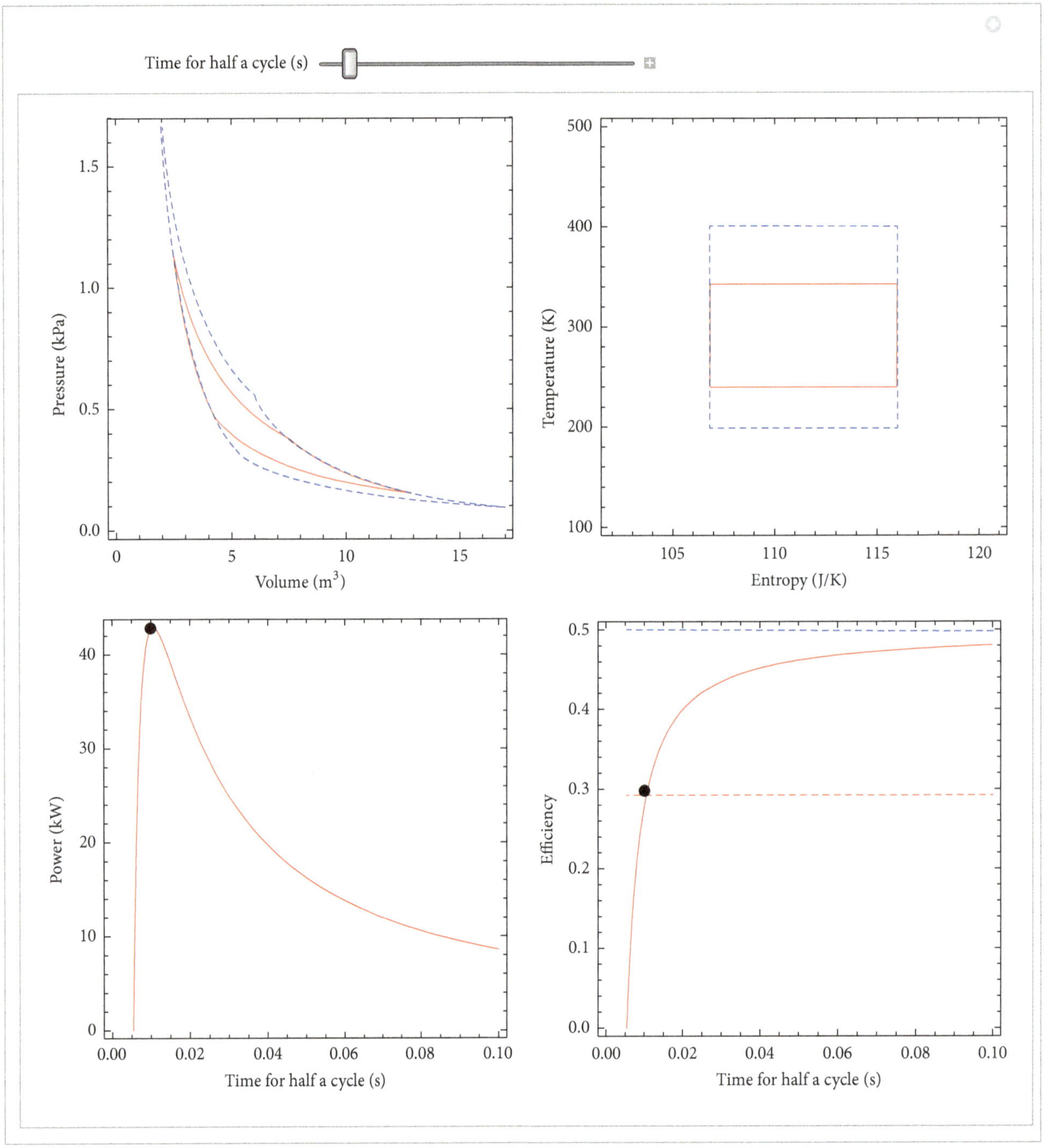

FIGURE 2: Snapshot of the visualization tool. The time spent in one of the isotherms (time for half a cycle) can be interactively changed. The blue dashed curves in the pressure-volume and temperature-entropy diagrams represent the curves for the ideal Carnot engine (infinite cycle time), while the red curves correspond to the current value of the time for half a cycle. The black dots on the power-time and efficiency-time curves display the current, time-dependent values of power and efficiency. The blue and red dashed lines in the efficiency-time diagram represent the Carnot and Curzon-Ahlborn efficiencies, respectively.

interactive, and electronic textbooks [6] as well as for online course materials.

In Figure 2 we display a snapshot of our visualization tool. In the upper left corner the p-V (pressure-volume) diagram is shown and in the upper right corner the T-S (temperature-entropy) diagram is shown. The dashed blue curves represent the ideal Carnot cycles, while the solid red curves represent the finite time Carnot cycles. The lower left graph shows the power as a function of time and the lower right graph shows the efficiency as a function of time (red solid curve). The black solid circles in the lower graphs indicate the instantaneous time, the blue dashed line in the lower right graph indicates the Carnot efficiency, and the red dashed line indicates the Curzon-Ahlborn efficiency. The time, or more precisely, the time spent in one of the isotherms can be adjusted with the slide bar, entitled "time for half a cycle," above the graphs. The red solid cycles in the upper diagrams and the black solid circles in the lower diagrams change interactively and continuously on the screen, as the slide bar is moved. At long times, the red cycles approach the blue dashed cycles in the p-V and T-S diagrams, while $P \to 0$ and $\eta \to \eta_C$. As the time is decreased, the area of the red solid cycles decreases and the power increases, while the efficiency decreases. After a while, the power passes through a maximum, corresponding to $\eta = \eta_{CA}$, and for even shorter times, the power decreases.

The curves making up the p-V cycle were calculated according to the ideal gas law, where the pressure as function of volume is given by $p = nRT/V$ for the isotherms and by $p = nRT_1V_1^{\gamma-1}/V^{\gamma}$ for the adiabats, where (T_1, V_1) is a reference state along the adiabat and γ is the adiabatic index. The entropies in the T-S cycle were modeled so that the entropy change during the isotherms is given by $S_2 - S_1 = nR\ln(V_2/V_1)$. For further details, see [7], where two versions of the cdf file are available.

The visualization tool has a very simple and user-friendly interface, since everything is contained on the same screen view; see Figure 2. Thus no switching between screens is necessary. Moreover, the only way a user can interact with the tool is to change the time, that is, to change the setting of the slide bar. If, however, an advanced user wants to add more functionality, this can easily be done, since the cdf file is publicly available for download [7].

4. Evaluation Method

We evaluated the visualization tool in the setting of an introductory course on thermodynamics for first year students at a university level engineering program. Prior to our demonstration of the visualization tool, the students had been taught about thermodynamic cycles, so this moment came timely in their course.

The evaluation was conducted on the entire class but set up during two consecutive, half-class problem solving occasions (12 students in group A and 16 students in group B). We started with a short (approximately 10 min) discussion about the Carnot cycle, how to calculate the work as the area enclosed in the p-V diagram, Carnot efficiency, and how power is defined and calculated as work divided by time. In addition, we showed the students data on temperatures and efficiencies of some real power plants (from Curzon and Ahlborn's paper [1]). After this, each student was given a piece of paper with the question (translated from Swedish):

(1) *Why do most power plants and engines have an efficiency that is so far from the Carnot efficiency?*

After all students had written their answers, the papers were collected and the visualization tool was demonstrated. Care was taken to go through the process slowly and systematically and point out the simultaneous changes in all four diagrams. Special emphasis was put on the interconnection time-power efficiency. That is, for long cycle time, the power approaches zero, while the efficiency approaches its theoretical maximum, the Carnot efficiency. For shorter cycle times, the power passes through a maximum, and the efficiency at maximum power is known as the Curzon-Ahlborn efficiency.

Following the demonstration, we again looked at the power plant data and we explained to the students that the observed efficiencies match very well the Curzon-Ahlborn efficiencies and that no power plant will operate at the Carnot efficiency since the output power in this case will approach zero. Instead, the Curzon-Ahlborn efficiency is in many cases a better measure of the working efficiency of power plants and engines. Effectively, this reasoning provided the answer to question 1 above. After this, each student was given the remaining evaluation questions (translated from Swedish):

(2) *Did you find the interactive demonstration helpful for your understanding? If so, in which way?*

(3) *Was there anything difficult to understand? In that case, what?*

(4) *Which alternative do you think you would prefer (mark your choice):*

   (a) *An instructor shows the demonstration and discusses*

   (b) *Work individually with the demonstration together with written facts and exercises*

   (c) *First a demonstration and then possibility to work individually*

   (d) *Other, namely...*

(5) *Would you like to see more interactive demonstrations of this kind in your education?*

(6) *If you answered yes to question 5, what do you think is important for the lecturer to keep in mind when showing interactive demonstrations or visualizations?*

In addition there was one note where we asked the students to send an email to us if they were interested to work with the visualization themselves. If so, we would reply and send them the visualization tool (the cdf) together with the download link for the free cdf reader [5].

5. Results and Discussion

Since the preparation of the two groups was identical and we did not find any systematic differences among the answers, we do not distinguish between the groups in our analysis but rather treat the two groups as one. Still, the number of students is quite small (28), which prevents any quantitative conclusions. Instead we discuss the results qualitatively.

The purpose of question 1 was to assess the prior knowledge of the students. While analyzing the answers, it soon became apparent that they could be divided into the following categories: heat losses—unspecified, heat losses—frictional, ideality aspects, timing aspects, and others. The number of answers belonging to each category is found in Table 1. In a few cases the answers contain parts from more than one category (e.g., mentioning both heat losses due to friction and timing aspects). In such cases, the answer will also contribute to more than one category.

Not very surprising, the category with most of the answers is heat losses—unspecified with 14 answers. It is interesting to note that five of the answers from the heat losses categories also include ideality aspects; see Table 1. Many of these students appear to believe that the reason that real power plants do not operate at Carnot efficiency is that the Carnot process is ideal and thus not possible to reach in practice, because of heat losses. This is in part true but it is clear from the answers that the students do not understand why there are heat losses. Their answers are also phrased in a way that one would think that they believe that the designer of the power plant strived for Carnot efficiency (instead of high power output). Besides, a few of the answers in the ideality aspects category show a vague understanding of the ideality concept. One such answer reads (translated from Swedish): *Because I think that the Carnot process refers to an ideal process, the reality is not always like that.*

Five of the answers in the heat loss categories also contain timing aspects. It is clear that these students have reached a higher level of understanding. They seem to understand that Carnot efficiency can in principle be reached, but that it would require a very long cycle time and a minimum of heat losses, which is not practically feasible. Only one student mentions power in the answer (translated from Swedish): *The isothermal reaction does not proceed slowly enough, which gives rise to energy losses. This decreases the efficiency. However, it cannot proceed infinitely slowly because of the power one would like to extract.*

It is interesting to note that one student in the timing aspects category gives a confused answer: *Because the time in the isotherms cannot be allowed to approach zero in reality.* This student knows that there is a timing issue but somehow misunderstands and believes that the time in the isotherms should approach zero instead of infinity in order to reach Carnot efficiency.

The two answers in the other category are not relevant. One of these specifically mentions the regenerator, which is not part of the Carnot but the Stirling engine. The other one reads: *To get a really high efficiency, a very low minimum temperature is required. This is difficult to achieve in a power plant and as we see on the chart, the low temperatures are high; that is,* T_{cold}/T_{warm} *is still a fairly high ratio.* This answers a different question, for example, What would you do to get a high Carnot efficiency?

The evaluation of the prior knowledge shows that some students believe that the Carnot process is an idealization that one cannot reach because of energy losses, while some students realize that there is a tradeoff between cycle time and efficiency. In any case, the prior knowledge of all the students seemed to be on the right level for further evaluation of the visualization to be meaningful.

All the responses to question 2 are positive with one exception. Some selected positive answers read (translated from Swedish and separated by semicolons): *Yes, it was interesting to see how the time directly influenced the efficiency of the cycle*; *Yes, a little bit about understanding the connection between efficiency, power, and so forth*; *Yes, it was good to see what happened with all the graphs as the time was changed*; *Yes, it was very good to see the different kinds of graphs at the same time*; *Yes, to a certain degree. It is easier to grasp when you see a visual demonstration like this*; *Definitely. To see how power and efficiency changed in relation to the ideal Carnot process with respect to time was very pedagogical.* The one negative answer here reads: *Not so much, take it slower and comment more on each separate graph.* From the collected answers to question 2, we conclude that the students in general appreciate the visualization since it offers a complementary way to understand the Carnot process.

While the answers to question 2 are very much aligned, the answers to question 3 are quite diverse. 10 of the 28 students give negative answers; that is, there was nothing that was difficult to understand. The other answers are so scattered, so no pattern is discernible. One student answers that *I don't think I fully understood the graphs* and another one writes: *The figure where the Carnot process changed its size.* At least two students seemed to need more time. They answer: *It is hard to get the time to think things through* and *The different graphs could be explained a little bit more.* This is in contrast to two other answers: *No, the demonstration was almost on a too low level* and *No, it was calm and methodical.*

The answers to question 4, which addresses how the students would like to use visualizations, are consistent. All students except one would prefer that an instructor demonstrates the visualization. After this demonstration half of the students would like to have a discussion, led by the instructor. The other half would like to have the possibility to work individually. Only one student would prefer the second option, that is, to work individually together with written facts and exercises.

The responses to question 5 are overwhelmingly positive. All students would, to a varying degree, like to see more of this kind of interactive demonstrations in their education. About half (12) of the students simply answer *Yes*, while some of the students specify or motivate their answer. Some

Table 1: Categorization of answers to question 1. In the middle column, the contributions of the individual students are shown. The students are labeled as A1–A12 and B1–B16.

Category	Students	Number of answers
Heat losses—unspecified	A1, A3, A9, A12 B2, B3, B5, B7, B8, B9, B11, B12, B15, B16	14
Heat losses—frictional	A2, A8, A10 B5	4
Ideality aspects	A3, A5, A6, A7 B3, B4, B6, B7, B9	9
Timing aspects	A4, A9, A10, A11, A12 B13, B14, B15, B16	9
Other	B1, B10	2

examples of this are as follows: *Yes, I like practical examples and demonstrations provided that they are at the right level* and *It would certainly be helpful at some moments*, as well as *Yes, especially on topics that are difficult to grasp intuitively the first time one encounters them*. Only one of the students is more reserved and answers *Maybe*.

The responses to question 6 are quite diversified and most students have clear ideas on what the instructor needs to think about. However, not all students agree in their answers. The pace of the presentation seems to be the factor that most students have opinions about and here we find opposing views. One student writes: *Fast, clear, consice* [*sic*], while another one answers: *Calm and methodical*. Yet another student notes that the pace should be adapted to the group of students. One student thinks that it is important that the understanding of the theory precedes the visualization, while another one wants the visualization to be on the same difficulty level as the one they are currently studying. Moreover, according to the students it is important that the visualization makes the understanding easier and does not complicate matter. There should be some degree of discussion and feedback. A holistic view and a sense of reality are also desirable. Only one of the students requested the visualization tool as cdf file for individual study.

Most students welcome more demonstrations of this kind in the education but they think the lecturer needs to be careful so that the demonstrations or visualizations do not complicate or confuse things. It is also important that the visualization is at the right level and that the pace of the demonstration is appropriate. However, among the students there is a lack of consensus about the right level and the appropriate pace. Some students want the visualization to be aligned with what they are currently studying, while others want the theory to precede the visualization. Concerning the pace, some students want a calm and methodical demonstration, while others want it to be fast and concise. Vavra et al. [8] have proposed a set of recommendations for the use of visualizations in science teaching and it is interesting to see how well aligned some of the student's answers are with these recommendations.

In one of the recommendations it is pointed out that "students require a repertoire of knowledge and skills to use visualization objects effectively" [8] and partly due to this there is a risk that only the best students benefit from the visualization, which has been pointed out by Gelaan [9]. Those students can use the visualization to test and enhance the understanding of concepts that they already have grasped, while the less able students are likely to misunderstand significant parts of the visualization [9]. Since the prior knowledge and understanding can vary a lot among students, this was probably the case in the current investigation, where the prior knowledge was diverse (see Table 1). Some of the answers to question 2 in the survey confirm this.

Even if it is unfeasible to require all students to have the same prior knowledge, the lecturer should at least make sure that all students have prior knowledge enough to make predictions of the outcome of the visualization, no matter if the predictions are correct or not. It has been shown that if students can make predictions, they are more likely to understand the demonstration in a correct way [10].

All students were positive about the visualization and would like to see more of this in their education. We are, however, cautious of drawing any conclusions about the changes in the students' understanding, as we have not quantified this. It has previously been reported that students find learning with visualization engaging and enjoyable [11]. We believe that this could facilitate a deep approach to learning [12].

It is hardly surprising that only one of the students wanted to work individually with the visualization. Many engineering students tend to choose the easiest path, in this case that an instructor demonstrates the visualization, even if this might not be optimal for their learning. Nevertheless, to actively work with the visualization is the most efficient way to use the visualization tool [13].

6. Conclusion

We have constructed an interactive visualization of a nonideal Carnot engine operating at finite time. When the cycle time is changed, the pressure-volume, temperature-entropy, power-time, and efficiency-time diagrams, which are displayed in parallel on one screen, change interactively. We have implemented the visualization in Wolfram's computational document format (cdf).

Our evaluation of the visualization tool in the setting of an introductory course on thermodynamics for engineering students who had just been taught about thermodynamic cycles showed that their prior knowledge about which aspects that make the Carnot cycle ideal is vague. According to the survey, all students found the interactive demonstration helpful in their understanding. A general remark we extracted from several of the students' answers is that the relationships between efficiency, power, and cycle time became clearer. Several students also conclude that seeing the different diagrams being varied simultaneously gives increased holistic understanding.

Since students are individuals and at more or less different levels of understanding, care must be taken when designing the visualization tool and planning the pace of the demonstration. From the lack of consensus concerning what the students think that the lecturer should keep in mind, we conclude that it will be extremely difficult to make a visualization that pleases and facilitates learning for all students in a class.

Conflict of Interests

The author declares that there is no conflict of interests in this paper.

Acknowledgments

This work was performed within the Nanometer Structure Consortium at Lund University (nmC@LU). The authors are grateful to E. Nilsson and M. Borgström for letting them demonstrate the visualization and conduct the survey in their classes. The authors thank T. Roxå and A. Diehl for their ideas on the construction of the survey and E. Nilsson and K. Deppert for their help to improve the paper.

References

[1] F. L. Curzon and B. Ahlborn, "Efficiency of a carnot engine at maximum power output," *American Journal of Physics*, vol. 43, no. 1, pp. 22–24, 1975.

[2] I. I. Novikov, "The efficiency of atomic power stations (a review)," *Journal of Nuclear Energy (1954)*, vol. 7, no. 1-2, pp. 125–128, 1958.

[3] P. Chambadal, *Les centrales nucléaires*, Armand Colin, Paris, France, 1957.

[4] C. van den Broeck, "Thermodynamic efficiency at maximum power," *Physical Review Letters*, vol. 95, no. 19, Article ID 190602, 2005.

[5] http://www.wolfram.com/cdf.

[6] W. L. Briggs and L. Cochran, *Calculus: Early Transcendentals*, Pearson, 2011.

[7] "Supplemental material for "Pedagogical visualization of a non-ideal Carnot engine"," https://sites.google.com/site/stenjonas/.

[8] K. L. Vavra, V. Janjic-Watrich, K. Loerke, L. M. Phillips, S. P. Norris, and J. Macnab, "Visualization in science education," *Alberta Science Education Journal*, vol. 41, no. 1, pp. 22–30, 2011.

[9] D. Gelaan, "Effectiveness of scientific visualizations in year 11 chemistry and physics education," in *Proceedings of the Australian Computers in Education Conference: It's Time (ACEC '12)*, Perth, Australia, 2012.

[10] K. Miller, N. Lasry, K. Chu, and E. Mazur, "Role of physics lecture demonstrations in conceptual learning," *Physical Review Special Topics—Physics Education Research*, vol. 9, no. 2, Article ID 020113, 2013.

[11] L. A. Annetta, J. Minogue, S. Y. Holmes, and M. Cheng, "Investigating the impact of video games on high school students'engagement and learning about genetics," *Computers and Education*, vol. 53, no. 1, pp. 74–85, 2009.

[12] F. Marton and R. Säljö, "On qualitative differences in learning 1. Outcome and process," *British Journal of Educational Psychology*, vol. 46, pp. 4–11, 1976.

[13] C. D. Hundhausen, S. A. Douglas, and J. T. Stasko, "A meta-study of algorithm visualization effectiveness," *Journal of Visual Languages and Computing*, vol. 13, no. 3, pp. 259–290, 2002.

Thermophysical Properties of Binary Mixtures of Dimethylsulfoxide with 1-Phenylethanone and 1,4-Dimethylbenzene at Various Temperatures

Harmandeep Singh Gill and V. K. Rattan

Dr. SSB University Institute of Chemical Engineering and Technology, Panjab University, Chandigarh 160014, India

Correspondence should be addressed to Harmandeep Singh Gill; harman_gill@outlook.com

Academic Editor: K. A. Antonopoulos

This research article reports the experimental results of the density, viscosity, refractive index, and speed of sound analysis of binary mixtures of dimethylsulfoxide (DMSO) + 1-phenylethanone (acetophenone) and + 1,4-dimethylbenzene (*para*-xylene) over the whole composition range at 313.15, 318.15, 323.15, and 328.15 K and at atmospheric pressure. The excess molar volumes (V^E), viscosity deviations ($\Delta\eta$), excess Gibbs energy of activation (G^E), deviations in isentropic compressibility (K_S^E), deviations in speed of sound (u^E), and deviations in the molar refraction (ΔR) were calculated from the experimental data. The computed quantities were fitted to the Redlich-Kister equation to derive the coefficients and estimate the standard error values. The viscosities have also been correlated with two, and three-parameter models, that is, Heric correlation, McAllister model, and Grunberg-Nissan correlation, respectively.

1. Introduction

This paper is a continuation of our ongoing research on the solution properties. Studies of the thermodynamic properties of binary mixtures play an important role in the fundamental understanding of different molecules and the interactions prevalent in them. In the present study, data on density, viscosity, refractive index, and speed of sound of binary mixtures of dimethylsulfoxide (DMSO) + 1-phenylethanone (acetophenone) and 1,4-Dimethylbenzene (*para*-xylene) at 313.15, 318.15, 323.15, and 328.15 K has been measured experimentally. From these results, the excess molar volumes, viscosity deviations, deviations in molar refraction, deviations in speed of sound, and isentropic compressibility have been derived. Dimethylsulfoxide is a versatile nonaqueous dipolar aprotic solvent having wide range of applications like in veterinary medicine, dermatology, microbiology, experimental immunology, and enzyme catalyzed reactions. It can easily pass through membranes, a quality which has been verified by numerous researchers. It has the ability to penetrate through living tissues without damaging them. Therefore an anesthetic or penicillin can be carried through the skin without using a needle which makes it paramount in medicinal field. Acetophenone is the simplest aromatic ketone organic compound. It can easily dissolve in water, but, since it is denser than water, it tends to sink. Its vapor is heavier than air and when inhaled in high concentrations it can be narcotic and also mild irritant to the eyes and skin. It is mostly used to create fragrances that smell like cherry, almond, strawberry, or other fruits. Acetophenone can also be found naturally occurring in fruits such as apple and banana. *Para*-xylene is an aromatic hydrocarbon based on benzene with two methyl substituents, opposite to each other. It is a colorless, flammable liquid and is insoluble in water. It is used as a thinner for paint and in paints and varnishes. The study of the thermodynamic properties of DMSO + 1-phenylethanone (acetophenone) and + 1,4-dimethylbenzene (*para*-xylene) mixtures is of interest mainly in industrial fields where solvent mixtures could be used as selective solvents for numerous reactions. In principle, interactions between the molecules can be established from the study of the deviations from ideal behavior of physical properties

TABLE 1: Physical properties of the components at 298.15 K.

Component	ρ (g·cm^{-3})		η (mPa·s)		n_D	
	Exptl.	Lit.	Exptl.	Lit.	Exptl.	Lit.
Acetophenone	1.0283	1.0263[17]	0.01681	0.01681[17]	1.5050	1.5005[17]
DMSO	1.0940	1.09537	1.9834	1.9910	1.4798	1.4775
p-Xylene	8.5661	8.5662[16]	0.609	0.611[16]	1.4933	1.4933[16]

such as molar volume and isentropic compressibility. The negative or positive deviations from the ideal value depend on the type and the extent of the interactions between the unlike molecules, as well as on the composition and the temperature. The variation of the isentropic compressibility is analogous of that of the excess molar volume, whereas the change of the deviation in speed of sound tends to become the inverse [1]. Physical and transport properties of liquid mixtures also affect most separation procedures, such as liquid-liquid extraction, gas absorption, and distillation [2]. The mixture DMSO-*p*-xylene has been earlier reported twice in literature at different temperatures [1, 3].

2. Experimental Section

2.1. Materials. The chemicals used are of analytical reagent grade. Dimethylsulfoxide (DMSO) is from Riedel, Germany; 1-phenylethanone (acetophenone) and 1,4-dimethylbenzene (*para*-xylene) are from S-D Fine Chemicals, Mumbai. The chemicals were purified using standard procedure [4] and were stored over molecular sieves. The purity of the chemicals was verified by comparing density, viscosity, and refractive index with the known values reported in the literature as shown in Table 1. All the compositions were prepared by using SARTORIUS balance. The possible uncertainty in the mole fraction is estimated to be less than $\pm 1 \times 10^{-4}$.

2.2. Viscosity. Kinematic viscosities were measured by using a calibrated modified Ubbelohde viscometer [5]. The calibration of viscometer was done at each temperature in order to determine the constants A and B of the following equation:

$$\nu = \frac{\eta}{\rho} = At + \frac{B}{t}. \quad (1)$$

The viscometer was kept vertically in a transparent-walled water bath with a thermal stability of ± 0.05 K for about 30 minutes to attain thermal equilibrium. Flow time was measured with an electronic stop watch with precision of ± 0.01 s. The corresponding uncertainty in the kinematic viscosity is $\pm 0.001 \times 10^{-6}\,\mathrm{m^2\,s^{-1}}$. The efflux time was repeated at least three times for each composition and the average of these readings was taken. The temperature of the bath was maintained constant with the help of a circulating type cryostat (type MK70, MLW, Germany). The dynamic viscosities were found out after the DSA analysis, that is, by dividing the above found kinematic viscosity by density. The uncertainty in the values of dynamic viscosity is within ± 0.003 mPa·s.

2.3. Density and Speed of Sound. Density and speed of sound were measured with the help of an ANTON PAAR density meter (DSA 5000). The accuracy in the measurement of density and speed of sound is $\pm 0.000005\,\mathrm{g\,cm^{-3}}$ and $\pm 0.5\,\mathrm{ms^{-1}}$, respectively. The density meter was calibrated by using triply distilled degassed water.

2.4. Refractive Index. Refractive indices were measured for sodium D-line by ABBE-3L refractometer having Bausch and Lomb lenses. The temperature was maintained constant with the help of water bath used for the viscosity measurement. A minimum of three independent readings were taken for each composition, and the average value was considered in all the calculations. Refractive index values are accurate up to ± 0.0001 units.

3. Experimental Results and Correlations

At least three independent readings of all the physical property measurements of density (ρ), viscosity (η), refractive index (n_D), and speed of sound (u) were taken for each composition and the averages of these experimental values are presented in Tables 2 and 3 for both systems. The experimentally determined values are used for the deviation calculations.

3.1. Excess Molar Volume. Density values are used to evaluate excess molar volume by the equation

$$V^E = \frac{x_1 M_1 + x_2 M_2}{\rho} - \frac{x_1 M_1}{\rho_1} - \frac{x_2 M_2}{\rho_2}, \quad (2)$$

where ρ_1, ρ_2 are the densities of pure components and ρ is the density of the mixture. M_1, M_2 are the molar mass of the two components and x_1, x_2 are the mole fraction of DMSO.

Excess Gibbs' free energy of activation has been also calculated using the viscosity and density of the mixture by the equation

$$\Delta G^E = RT\left[\ln(\eta V) - \sum_{i=1}^{2} x_i \ln(\eta_i V_i)\right], \quad (3)$$

where R is a universal gas constant, T is the temperature of the mixture, and η and η_i are the viscosities of the mixture and pure compound, respectively. V, V_i refer to the molar volume of the mixture and pure components, respectively.

3.2. Viscosity Calculations. The deviation in viscosity is obtained by the following equation:

$$\Delta\eta = \eta - \eta_1 x_1 - \eta_2 x_2, \quad (4)$$

TABLE 2: Refractive indices, n_D, density, ρ, speed of sound, u, and viscosity, η, for DMSO(1) + acetophenone(2) system at different temperatures.

x_1	n_D	ρ (g·cm^{-3})	u (m·s^{-1})	η (mPa·s)
		DMSO(1) + acetophenone(2)		
		313.15 K		
0.00000	1.5142	1.0106	1421.81	1.2920
0.17422	1.5090	1.0173	1424.53	1.3093
0.31320	1.5046	1.0238	1426.25	1.3150
0.43453	1.5001	1.0300	1427.15	1.3180
0.54292	1.4955	1.0361	1427.99	1.3609
0.63921	1.4915	1.0430	1428.88	1.4000
0.72696	1.4877	1.0502	1430.33	1.4295
0.80600	1.4841	1.0578	1432.57	1.4515
0.87681	1.4806	1.0651	1434.55	1.4721
0.94521	1.4771	1.0730	1436.77	1.4959
1.00000	1.4742	1.0802	1438.85	1.5169
		318.15 K		
0.00000	1.5102	1.0072	1403.51	1.1161
0.17422	1.5055	1.0144	1406.42	1.1362
0.31320	1.5011	1.0208	1408.41	1.1459
0.43453	1.4967	1.0271	1409.52	1.1633
0.54292	1.4921	1.0332	1410.57	1.2100
0.63921	1.4880	1.0400	1411.59	1.2550
0.72696	1.4843	1.0476	1413.17	1.2873
0.80600	1.4807	1.0549	1415.38	1.3150
0.87681	1.4772	1.0623	1417.55	1.3400
0.94521	1.4735	1.0700	1419.97	1.3707
1.00000	1.4703	1.0768	1422.30	1.3968
		323.15 K		
0.00000	1.5061	0.9987	1385.35	1.0645
0.17422	1.5019	1.0064	1388.33	1.0659
0.31320	1.4978	1.0134	1390.60	1.0735
0.43453	1.4933	1.0199	1391.91	1.0870
0.54292	1.4887	1.0261	1393.10	1.1275
0.63921	1.4847	1.0332	1394.32	1.1675
0.72696	1.4810	1.0408	1396.05	1.1969
0.80600	1.4775	1.0486	1398.30	1.2205
0.87681	1.4739	1.0560	1400.55	1.2425
0.94521	1.470	1.0636	1403.10	1.2675
1.00000	1.4665	1.0702	1405.68	1.2972
		328.15 K		
0.00000	1.5030	0.9856	1367.34	0.9873
0.17422	1.4991	0.9944	1370.47	0.9815
0.31320	1.4950	1.0024	1372.93	0.9875
0.43453	1.4904	1.0095	1374.35	1.0010
0.54292	1.4859	1.0168	1375.75	1.0405
0.63921	1.4817	1.0245	1377.15	1.0770
0.72696	1.4780	1.0331	1379.00	1.1065
0.80600	1.4744	1.0417	1381.30	1.1305
0.87681	1.4707	1.0498	1383.50	1.1505
0.94521	1.4668	1.0583	1386.10	1.1755
1.00000	1.4630	1.0655	1389.12	1.2086

TABLE 3: Refractive indices, n_D, density, ρ, speed of sound, u, and viscosity, η, for DMSO(1) + p-xylene(2) system at different temperatures.

x_1	n_D	ρ (g·cm^{-3})	u (m·s^{-1})	η (mPa·s)
		DMSO(1) + p-xylene(2)		
		313.15 K		
0.00000	1.4850	0.8445	1247.96	0.5175
0.15820	1.4843	0.8720	1264.82	0.5979
0.30056	1.4839	0.8999	1281.64	0.6777
0.42694	1.4828	0.9293	1297.89	0.7389
0.52448	1.4813	0.9577	1310.14	0.8106
0.62739	1.4790	0.9923	1325.23	0.8875
0.72506	1.4778	1.0233	1346.79	0.9889
0.80799	1.4765	1.0477	1373.38	1.0997
0.87462	1.4747	1.0557	1395.79	1.2275
0.94123	1.4727	1.0723	1418.20	1.3757
1.00000	1.4717	1.0802	1439.15	1.5116
		318.15 K		
0.00000	1.4831	0.8395	1227.99	0.4923
0.15820	1.4823	0.8665	1246.38	0.5675
0.30056	1.4816	0.8942	1263.94	0.6422
0.42694	1.4803	0.9233	1281.67	0.6971
0.52448	1.4789	0.9517	1295.39	0.7555
0.62739	1.4767	0.9870	1310.77	0.8179
0.72506	1.4754	1.0183	1331.90	0.9090
0.80799	1.4737	1.0417	1358.39	1.0220
0.87462	1.4719	1.0497	1380.68	1.1401
0.94123	1.4701	1.0668	1404.05	1.2705
1.00000	1.4690	1.0768	1423.09	1.3822
		323.15 K		
0.00000	1.4812	0.8348	1207.94	0.4603
0.15820	1.4802	0.8610	1226.95	0.5401
0.30056	1.4794	0.8882	1245.22	0.6030
0.42694	1.4777	0.9167	1263.54	0.6550
0.52448	1.4761	0.9443	1277.98	0.7107
0.62739	1.4744	0.9804	1292.79	0.7697
0.72506	1.4727	1.0110	1314.76	0.8557
0.80799	1.4710	1.0333	1340.33	0.9776
0.87462	1.4690	1.0410	1363.23	1.0895
0.94123	1.4675	1.0590	1385.97	1.1999
1.00000	1.4665	1.0702	1404.14	1.2972
		328.15 K		
0.00000	1.4793	0.8298	1188.15	0.4385
0.15820	1.4781	0.8553	1208.59	0.5150
0.30056	1.4771	0.8823	1227.53	0.5707
0.42694	1.4754	0.9106	1247.59	0.6092
0.52448	1.4736	0.9377	1262.23	0.6579
0.62739	1.4715	0.9733	1278.99	0.7175
0.72506	1.4699	1.0045	1300.23	0.7917
0.80799	1.4683	1.0261	1325.12	0.9158
0.87462	1.4663	1.0341	1348.21	1.0109
0.94123	1.4652	1.0530	1370.98	1.1152
1.00000	1.4640	1.0655	1388.57	1.1974

where η is the viscosity of mixture and η_1, η_2 refer to the viscosities of pure components.

McAllister [6] model

$$\ln \nu = x_1^3 \ln \nu_1 + x_2^3 \ln \nu_2 + 3x_1^2 x_2 \ln \eta_{12} + 3x_1 x_2^2 \ln \eta_{21} - \ln\left[x_1 + x_2 \frac{M_2}{M_1}\right] + 3x_1^2 x_2 \ln\left[\frac{2 + M_2/M_1}{3}\right] + 3x_1 x_2^2 \ln\left[\frac{1 + (2M_2/M_1)}{3}\right] + x_2^3 \ln\left[\frac{M_2}{M_1}\right]. \quad (5)$$

Herric [6] correlation

$$\ln \nu = x_1 \ln \nu_1 + x_2 \ln \nu_2 + x_1 x_2 \left[\alpha_{12} + \acute{\alpha}_{12}(x_1 - x_2)\right] - \ln M_{\text{mix}} + x_1 \ln M_1 + x_2 \ln M_2, \quad (6)$$

and Grunberg-Nissan [7] equation

$$\ln(\eta) = x_1 \ln(\eta_1) + x_2 \ln(\eta_2) + d_{12} x_1 x_2 \quad (7)$$

have been fitted to viscosity data and it was found that both have the same standard errors at each temperature.

3.3. Isentropic Compressibility. The experimental results for the speed of sound of binary mixtures are listed in Tables 2 and 3. The isentropic compressibility was evaluated by using $K_S = u^{-2}\rho^{-1}$ and the deviation in isentropic compressibility is calculated using the following equation:

$$K_S^E = K_S - K_S^{\text{id}}, \quad (8)$$

where K_S^{id} stands for isentropic compressibility for an ideal mixture calculated using Benson-Kiyohara model [8, 9]:

$$K_S^{\text{id}} = \sum_{i=1}^{2} \Phi_i \left[K_{S,i} + \frac{TV_i(\alpha_i^2)}{C_{p,i}}\right] - \left\{\frac{T\left(\sum_{i=1}^2 x_i V_i\right)\left(\sum_{i=1}^2 \Phi_i a_i\right)^2}{\sum_{i=1}^2 x_i C_{p,i}}\right\}, \quad (9)$$

where a_i and C_p are the thermal expansion coefficient and molar heat capacity of the ith components, respectively.

The deviation in speed of sound is given by

$$\Delta u = u - x_1 u_1 - x_2 u_2. \quad (10)$$

3.4. Molar Refraction. Refractive indices have been used for the calculation of molar refraction (R_m) that is obtained by using Lorentz-Lorenz equation [8].

Deviation in molar refraction (ΔR) is calculated by the following equation:

$$R = R_m - \sum \Phi_i R_i, \quad \Phi_i = \frac{x_i}{\sum x_j V_j}, \quad (11)$$

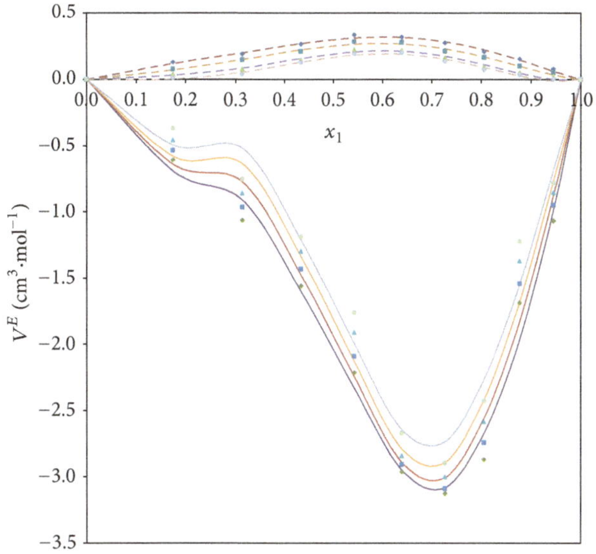

Figure 1: Experimental and calculated excess molar volume for (i) DMSO(1) + acetophenone(2) and (ii) DMSO(1) + p-Xylene(2) at 313.15 K, ✦, 318.15 K, ■, 323.15 K, ▲, 328.15 K, ●; symbols represent the experimental values; dotted lines represent DMSO-acetophenone mixture and solid lines represent DMSO-p-Xylene mixture, both optimised by Redlich-Kister parameters.

where n_D refers to the refractive index, R_m is molar refraction of the mixture, R_i is molar refraction of the ith component, and Φ is ideal state volume fraction.

All the deviations (V^E, ΔR, $\Delta\eta$, Δu, and K_S^E) have been fitted to Redlich-Kister polynomial regression of the type

$$\Delta Y = x_1 x_2 \sum_{i=1}^{m} A_i (1 - 2x_1)^{i-1} \quad (12)$$

to derive the constant A_i using the method of the least square.

Standard deviation for each case is calculated by

$$\sigma = \left[\frac{\sum(\Delta Y_{\text{exptl}} - \Delta Y_{\text{calcd}})^2}{m - n}\right]^{0.5}, \quad (13)$$

where m is the number of data points and n is the number of coefficients. Derived parameters of the Redlich-Kister equation (12) and standard deviations (13) are presented in Tables 4 and 5.

4. Discussions

The excess molar volume from 313.15 to 328.15 K versus the mole fraction of both mixtures with respect to DMSO is shown in Figure 1. The molar volume of the mixtures and the viscosity data have been used for the calculation of Gibbs' free energy presented in Figure 5. The V^E values decrease with increasing temperatures for the systems but are positive in case of DMSO-acetophenone mixture and negative for DMSO-p-xylene mixture. Treszczanowicz et al.

TABLE 4: Derived parameters of Redlich-Kister equation (12) and standard deviation (13) for various functions of the binary mixtures at different temperatures (DMSO-acetophenone).

T/K	A_0	A_1	A_2	A_3	σ
			V^E ($cm^3 \cdot mol^{-1}$)		
313.15	1.2005	0.741	−0.1784	−0.6838	0.01377
318.15	1.0176	0.7493	−0.5481	−0.7711	0.01379
323.15	0.7709	0.8700	−0.7689	−1.2620	0.01527
328.15	0.6714	0.9374	−0.9528	−1.3257	0.01381
			$\Delta\eta$ (mPa·s)		
313.15	−0.2371	0.2135	0.2487	−0.4972	0.00461
318.15	−0.2575	0.2259	0.1846	−0.4935	0.00323
323.15	−0.2722	0.2375	0.0974	−0.4693	0.00309
328.15	−0.2887	0.2407	0.0508	−0.4755	0.00290
			K_s^E (TPa^{-1})		
313.15	4.8515	15.4271	−1.3121	−12.5359	0.11076
318.15	3.5099	15.7048	−1.8315	−12.902	0.09236
323.15	2.0047	21.1502	−4.7008	−34.8976	0.44672
328.15	0.0146	15.1308	−2.0538	−11.8737	0.17247
			ΔR		
313.15	0.06817	0.039577	−0.06186	−0.09945	0.006271
318.15	0.102475	0.056959	−0.00441	−0.06411	0.003073852
323.15	0.147987	0.106458	0.080891	−0.10664	0.005976
328.15	0.175329	0.134545	0.138756	−0.14317	0.008927
			ΔG^E ($J \cdot mol^{-1}$)		
313.15	−66.3882	516.3512	492.0794	−981.8602	9.00998
318.15	−141.309	625.565	410.846	−1119.199	7.38255
323.15	−234.912	700.1368	230.6251	−1143.3850	7.33602
328.15	−321.946	780.1317	132.3463	−1256.91	7.21134

[10] and Roux and Desnoyers [11] suggested that V^E is the resultant contribution from several opposing effects. These effects can be primarily divided into three types, namely, chemical, physical, and structural. A physical contribution, that is, specific interactions between the real species present in the mixture, contribute in negative terms to V^E. The chemical or specific intermolecular interactions result in a volume decrease, and these include charge transfer type forces and other complex forming interactions. This effect also contributes in negative values to V^E. The structural contributions are mostly negative and can arise from several effects, especially from changes of free volume and interstitial accommodation. In other words, structural contributions arising from geometrical fitting of one component into the other due to the differences in the free volume and molar volume between components lead to a negative contribution to V^E. The viscosity and deviations are presented in Table 2 and plotted in Figure 2, respectively, for both systems. The viscosity deviations decrease with the increase in temperature for both systems. The negative $\Delta\eta$ values are generally observed for systems where dispersion or weak dipole-dipole forces are primarily responsible for interaction between the component molecules. The viscosity data is also fitted to the two, and the three-parameter models, that is, Herric correlation, the McAllister model, and Grunberg-Nissan correlation, and the evaluated parameters are presented in Tables 6 and 7. The deviations in molar refraction for both systems are shown in Figure 3. The ΔR values are positive for acetophenone system for the whole composition range which goes on increasing as the temperature of the solution increases. The ΔR values are negative for *para*-xylene system for the whole composition range which goes on decreasing as the temperature of the solution increases. In general, the negative values of ΔR suggest that we have weak interactions between the component molecules in the mixture. The results of excess isentropic compressibility (K_S^E) are also plotted in Figure 4. The deviations for DMSO-acetophenone system are initially negative and then become positive when mole fraction is around 0.5, whereas for DMSO-p-xylene system they are negative over the entire composition range. Deviation in Gibbs free energy for DMSO-acetophenone system follows an arbitrary path, going from negative to positive and vice versa twice, while for DMSO-*para*-xylene system the deviations are negative and increase with increasing temperature.

TABLE 5: Derived parameters of Redlich-Kister equation (12) and standard deviation (13) for various functions of the binary mixtures at different temperatures (DMSO-p-xylene).

T/K	A_0	A_1	A_2	A_3	σ
			V^E ($cm^3 \cdot mol^{-1}$)		
313.15	−8.5850	−15.1234	−6.084	12.8652	0.15687
318.15	−8.2010	−15.9715	−5.4345	15.6022	0.17024
323.15	−7.7110	−16.483	−4.7908	17.7396	0.19337
328.15	−7.1898	−16.3123	−4.0863	18.0512	0.20482
			$\Delta\eta$ (mPa·s)		
313.15	−0.9028	−0.7268	−0.2127	0.25	0.00553
318.15	−0.8069	−0.7852	−0.1061	0.5942	0.00475
323.15	−0.7589	−0.7364	0.093	0.6582	0.00619
328.15	−0.7126	−0.6943	0.2137	0.6476	0.0067
			K_s^E (TPa^{-1})		
313.15	−119.0039	−64.4234	−74.9361	−36.2431	1.03835
318.15	−137.2976	−76.3846	−70.5856	−21.5844	0.97745
323.15	−148.0003	−81.8649	−67.7606	−17.6011	0.84832
328.15	−166.5237	−92.1709	−61.6439	−2.9525	0.87227
			ΔR		
313.15	−3.25811	4.787247	−1.77213	−3.9986	0.226222
318.15	−3.20078	4.955419	−1.49922	−4.5735	0.239049
323.15	−3.14001	5.033874	−1.26969	−4.89488	0.245337
328.15	−3.0793	5.147192	−1.01287	−5.28097	0.254675
			ΔG^E ($J \cdot mol^{-1}$)		
313.15	−1022.2560	−1753.98	−140.348	1173.282	15.8755
318.15	−1013.8590	−2190.4	88.3584	2252.137	14.30563
323.15	−1011.6820	−2200.58	769.1765	2310.619	22.20041
328.15	−1113.8360	−2249.67	1294.863	2307.221	24.1139

TABLE 6: Interaction parameters for the McAllister model (5), Herric correlation (6), and Grunberg-Nissan correlation (7) for viscosity at different temperatures (DMSO-acetophenone).

	McAllister model		
T/K	η_{12}	η_{21}	$\sigma(\eta)$/mPa·s
313.15	1.3242	1.3473	0.00016
318.15	1.350624	1.334886	0.00025
323.15	1.377883	1.314174	0.00038
328.15	1.348994	1.240874	0.00042
	Herric correlation		
T/K	α_{12}	$\acute{\alpha}_{12}$	$\sigma(\eta)$/mPa·s
313.15	−0.02338	−0.01368	0.00016
318.15	−0.026	−0.01677	0.00025
323.15	−0.02874	−0.01993	0.00038
328.15	−0.03047	−0.02084	0.00042
	Grunberg-Nissan correlation		
T/K	d_{12}		$\sigma(\eta)$/mPa·s
313.15	−0.0956181		0.156414
318.15	−0.1201454		0.095161
323.15	−0.1618181		0.124392
328.15	−0.1988545		0.2077

Table 7: Interaction parameters for the McAllister model (5), Herric correlation (6), and Grunberg-Nissan correlation (7) for viscosity at different temperatures (DMSO-p-xylene).

	McAllister model		
T/K	η_{12}	η_{21}	$\sigma(\eta)$/mPa·s
313.15	1.04511700	0.88014850	0.00304
318.15	1.03668800	0.87514410	0.00310
323.15	1.02613100	0.87012170	0.00325
328.15	1.01597400	0.86517980	0.00327
	Herric correlation		
T/K	α_{12}	$\acute{\alpha}_{12}$	$\sigma(\eta)$/mPa·s
313.15	0.04364338	0.13563440	0.00304
318.15	0.03664824	0.13067020	0.00310
323.15	0.03026920	0.12422880	0.00325
328.15	0.02245963	0.11706220	0.00327
	Grunberg-Nissan correlation		
T/K	d_{12}		$\sigma(\eta)$/mPa·s
313.15	−0.408281		0.01591
318.15	−0.377472		0.025161
323.15	−0.320472		0.01239
328.15	−0.312945		0.02016

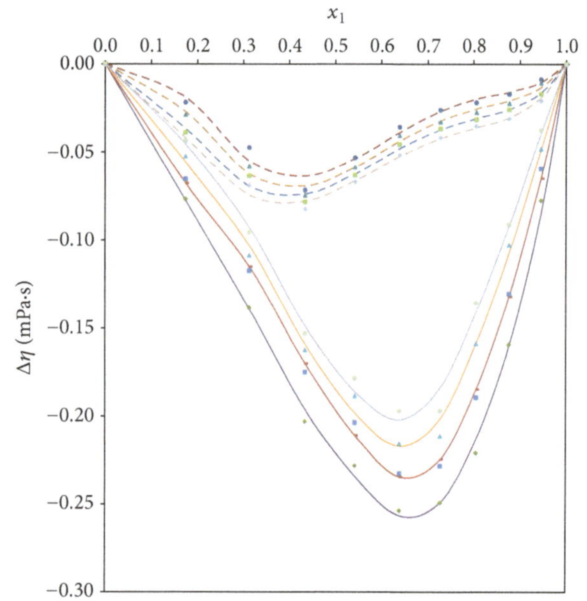

Figure 2: Experimental and calculated deviations in viscosity for (i) DMSO(1) + acetophenone(2) and (ii) DMSO(1) + p-xylene(2) at 313.15 K, ♦, 318.15 K, ■, 323.15 K, ▲, 328.15 K, ●; symbols represent the experimental values; dotted lines represent DMSO-acetophenone mixture and solid lines represent DMSO-p-xylene mixture, both optimised by Redlich-Kister parameters.

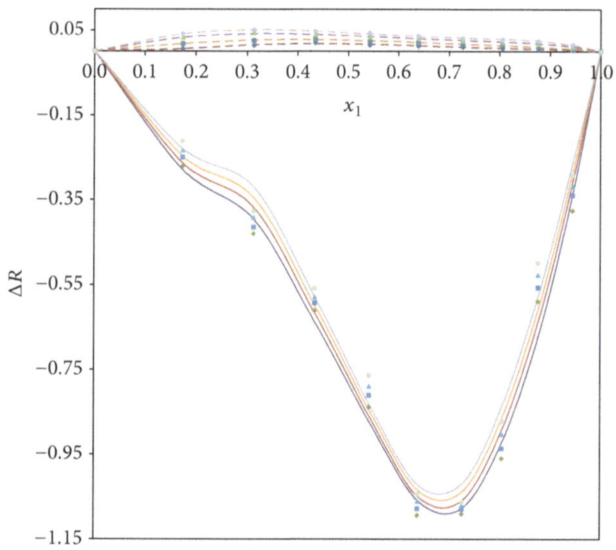

Figure 3: Experimental and calculated deviations in molar refraction for (i) DMSO(1) + acetophenone(2) and (ii) DMSO(1) + p-xylene(2) at 313.15 K, ♦, 318.15 K, ■, 323.15 K, ▲, 328.15 K, ●; and symbols represent the experimental values; dotted lines represent DMSO-acetophenone mixture and solid lines represent DMSO-p-xylene mixture, both optimised by Redlich-Kister parameters.

Symbols Used

- A_1, A_2, A_3, A_4: Parameters of Redlich-Kister equation
- A_{12}, A_{21}: Interaction coefficients of McAllister model
- $\alpha_{12}, \acute{\alpha}_{12}$: Coefficients of Herric's correlation
- ν: Kinematic viscosity (m^2s^{-1})
- ρ: Density ($g\,cm^{-3}$)
- σ: Standard deviation
- α: Thermal expansion coefficient (K^{-1})
- η: Dynamic viscosity (mPa·s)
- ΔG^E: Excess Gibbs free energy ($Jmol^{-1}$)
- V^E: Excess molar volume (m^3mol^{-1})
- ΔK_S^E: Excess isentropic compressibility (TPa^{-1})

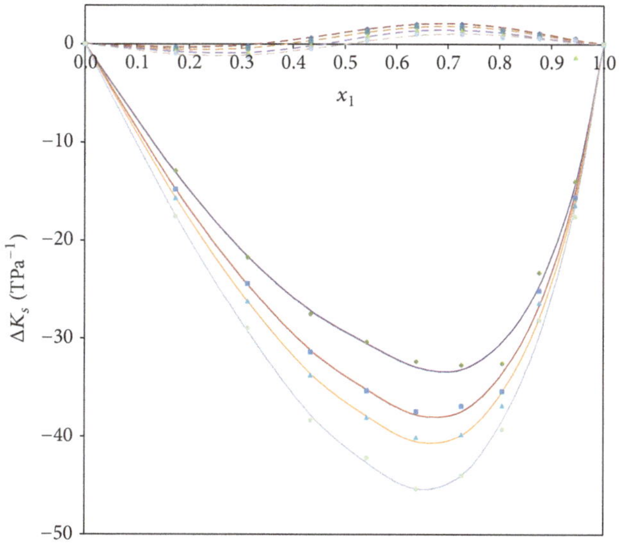

FIGURE 4: Experimental and calculated deviations in isentropic compressibility for (i) DMSO(1) + acetophenone(2) and (ii) DMSO(1) + p-xylene(2) at 313.15 K, ◆, 318.15 K, ■, 323.15 K, ▲, 328.15 K, ●; and symbols represent the experimental values; dotted lines represent DMSO-Acetophenone mixture and solid lines represent DMSO-*p*-xylene mixture, both optimised by Redlich-Kister parameters.

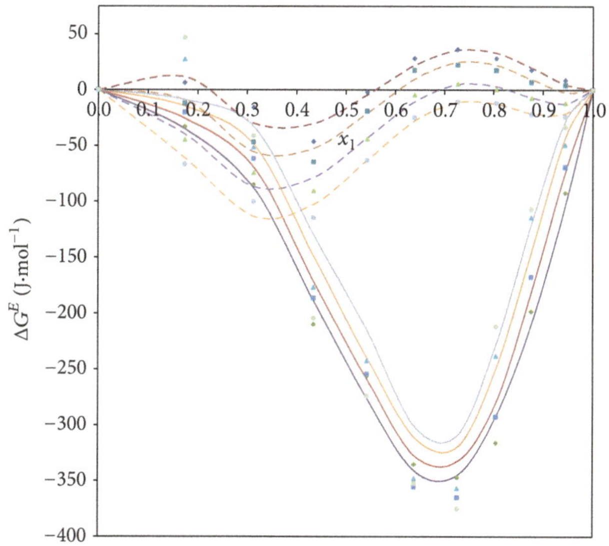

FIGURE 5: Experimental and calculated deviations in Gibbs free energy of activation for (i) DMSO(1) + acetophenone(2) and (ii) DMSO(1) + p-xylene(2) at 313.15 K, ◆, 318.15 K, ■, 323.15 K, ▲, 328.15 K, ●; and symbols represent the experimental values; solid lines represent DMSO-Acetophenone mixture and dotted lines represent DMSO-*p*-xylene mixture, both optimised by Redlich-Kister parameters.

R: Universal gas constant (8.314 $\text{Jmol}^{-1}\text{K}^{-1}$)
T: Absolute temperature (K)
d_{12}: Grunberg-nissan parameter
Φ_i: Volume fraction (dimensionless).

Conflict of Interests

The authors declare that there is no conflict of interests regarding the publication of this paper.

References

[1] M. M. Palaiologou, G. K. Arianas, and N. G. Tsierkezos, "Thermodynamic investigation of dimethyl sulfoxide binary mixtures at 293.15 and 313.15 K," *Journal of Solution Chemistry*, vol. 35, no. 11, pp. 1551–1565, 2006.

[2] K. Zhang, J. Yang, X. Yu, J. Zhang, and X. Wei, "Densities and viscosities for binary mixtures of poly(ethylene glycol) 400 + dimethyl sulfoxide and poly(ethylene glycol) 600 + water at different temperatures," *Journal of Chemical and Engineering Data*, vol. 56, no. 7, pp. 3083–3088, 2011.

[3] A. Ali, A. K. Nain, D. Chand, and R. Ahmad, "Viscosities and refractive indices of binary mixtures of dimethylsulphoxide with some aromatic hydrocarbons at different temperatures: an experimental and theoretical study," *Journal of the Chinese Chemical Society*, vol. 53, no. 3, pp. 531–543, 2006.

[4] J. A. Riddick, W. B. Bunger, and T. K. Sakano, *Organic Solvents: Physical Properties and Methods of Purifications*, vol. 2 of *Techniques of Chemistry*, John Wiley & Sons, New York, NY, USA, 1986.

[5] V. K. Rattan, S. Kapoor, and K. Tochigi, "Viscosities and densities of binary mixtures of toluene with acetic acid and propionic acid at (293.15, 303.15, 313.15, and 323.15) K," *Journal of Chemical and Engineering Data*, vol. 47, no. 5, pp. 1182–1184, 2002.

[6] R. A. McAllister, "The viscosity of liquid mixtures," *AIChE Journal*, vol. 6, pp. 427–431, 1960.

[7] L. Grunberg and A. H. Nissan, "The energies of vaporisation, viscosity and cohesion and the structure of liquids," *Transactions of the Faraday Society*, vol. 45, pp. 125–137, 1949.

[8] G. C. Benson and O. Kiyohara, "Evaluation of excess isentropic compressibilities and isochoric heat capacities," *The Journal of Chemical Thermodynamics*, vol. 11, no. 11, pp. 1061–1064, 1979.

[9] G. Douhéret, M. I. Davis, I. J. Fjellanger, and H. Høiland, "Ultrasonic speeds and volumetric properties of binary mixtures of water with poly(ethylene glycol)s at 298.15 K," *Journal of the Chemical Society - Faraday Transactions*, vol. 93, no. 10, pp. 1943–1949, 1997.

[10] A. J. Treszczanowicz, O. Kiyohara, and G. C. Benson, "Excess volumes for n-alkanols +n-alkanes IV. Binary mixtures of decan-1-ol +n-pentane, +n-hexane, +n-octane, +n-decane, and +n-hexadecane," *The Journal of Chemical Thermodynamics*, vol. 13, no. 3, pp. 253–260, 1981.

[11] A. H. Roux and J. E. Desnoyers, "Association models for alcohol-water mixtures," *Journal of Proceedings of the Indian Academy of Sciences: Chemical Sciences*, vol. 98, no. 5-6, pp. 435–451.

6

Radiation and Mass Transfer Effects on MHD Free Convection Flow of a Micropolar Fluid past a Stretching Surface Embedded in a Non-Darcian Porous Medium with Heat Generation

S. Mohammed Ibrahim,[1] T. Sankar Reddy,[2] and N. Bhaskar Reddy[3]

[1] *Department of Mathematics, Priyadarshini College of Engineering and Technology, Nellore 524004, Andhra Pradesh, India*
[2] *Deparment of Mathematics, Annamacharya College of Engineering and Technology, Cuddapa 516001, Andhra Pradesh, India*
[3] *Department of Mathematics, Sri Venkateswara University, Tirupathi 517502, Andhra Pradesh, India*

Correspondence should be addressed to S. Mohammed Ibrahim; ibrahimsvu@gmail.com

Academic Editors: G. L. Aranovich, T. M. Inerbaev, and M. Manciu

A comprehensive study of mass transfer and thermal radiation on a steady two-dimensional laminar flow of a viscous incompressible electrically conducting micropolar fluid past a stretching surface embedded in a non-Darcian porous medium in the presence of heat generation is analyzed numerically. The governing equations of momentum, angular momentum, energy, and species equations are solved numerically using Runge-Kutta fourth order method with the shooting technique. The effects of various parameters on the velocity, microrotation, temperature and concentration field as well as skin friction coefficient, Nusselt number and Sherwood number are shown graphically and tabulated. It is observed that the micropolar fluid helps in the reduction of drag forces and also acts as a cooling agent.

1. Introduction

Micropolar fluids are fluid with microstructure. They belong to a class of fluids with nonsymmetric stress tensor that we will call polar fluids. Micropolar fluids may also represent fluids consisting of rigid, randomly oriented (or spherical) particles suspended in a viscous medium, where the deformation of the particle is ignored. This constitutes a substantial generalization of the Navier-Stokes model and opens a new field of potential applications. The attractiveness and power of the model of micropolar fluids are come from the fact that it is both a significant and a simple generalization of the classical Navier-Stokes model. The theory of micropolar fluids developed by Eringen [1] and has been a field of very active research for the last few decades as this class of fluids represents, mathematically, many industrial important fluids such as paints, body fluids, polymers, colloidal fluids, suspension fluids, animal blood, liquid crystal, and so forth among the various non-Newtonian fluids models. Eringen [2] has also developed the theory of thermomicropolar fluids by the extending theory of micropolar fluids. A thorough review of the subject of the application of micropolar fluid mechanics has been given by Lukaszewicz [3] and Ariman et al. [4]. Ahmadi [5] obtained a similarity solution for micropolar boundary layer flow over a semi-infinite plate. Jena and Mathur [6] further studied the laminar free convection flow of thermomicropolar fluids past a nonisothermal vertical plate.

Boundary-layer flow and heat transfer over a continuously stretched surface have received considerable attention in recent years. This stems from various possible engineering and metallurgical applications such as a hot rolling, wire drawing, metal and plastic extrusion, continuous casting, glass fiber production, crystal growing, and paper production. The continuous surface concept was introduced by sakiadis [7, 8]. RajaGopal et al. [9] studied a boundary layer flow non-Newtonian over a stretching sheet with a

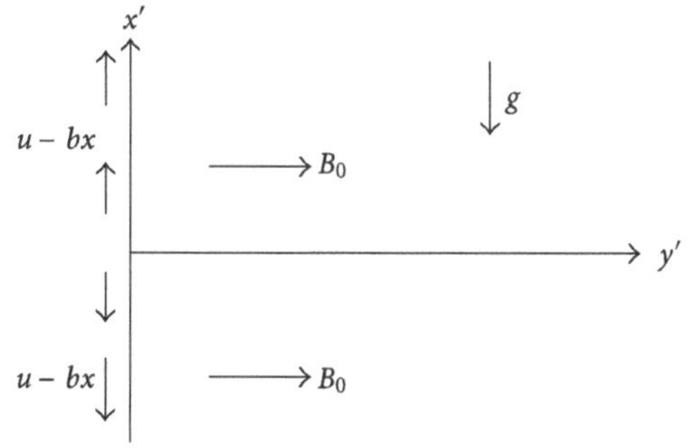

FIGURE 1: Sketch of the physical model.

uniform free stream. Hady [10] studied the solution of a heat transfer to a micropolar fluid from a nonisothermal stretching sheet with injection. Na and Pop [11] investigated the boundary layer flow of micropolar fluid due to a stretching wall. Hassanien et al. [12] studied a numerical solution for heat transfer in a micropolar fluid over a stretching sheet. Desseaux and Kelson [13] studied the flow of micropolar fluid bounded by stretching sheet. In all the above studies, the authors have taken the stretching sheet be an oriented in horizontal direction. However, of late the effects of MHD to the micropolar fluids problem are very important. Abo-Eldahab and Ghonaim [14] investigated the convective heat transfer in an electrically conducting micropolar fluid at a stretching surface with uniform free stream. Pavlov [15] studied the boundary layer flow of an electrically conducting fluid due to a stretching of a plane elastic surface in the presence of a uniform transverse magnetic field. Chakrabarti and Gupta [16] extended Pavlov's work to study the heat transfer when a uniform suction is applied at the stretching surface.

Combined heat and mass transfer in fluid-saturated porous media finds applications in a variety of engineering processes such as heat exchanger devices, petroleum reservoirs, chemical catalytic reactors and processes, geothermal and geophysical engineering, moisture migration in a fibrous insulation and nuclear waste disposal, and others. Bejan and Khair [18] investigated the free convection boundary layer flow in a porous medium owing to combined heat and mass transfer. Many problems of Darcian and non-Darcian mixed convection about a vertical plate had been reported, as in Hsu and Cheng [19]; Tien and Vafai [20] investigated the free convection boundary layer flow in a porous medium owing to combined heat and mass transfer. The open literature is rich with references dealing with natural convection flow in porous medium [21–23].

In many engineering and physical problems in which fluid undergoes exothermic or endothermic reaction, it is highly important to study the effect of heat generation and absorption. Abo-Eldahb and El Aziz [17] found the heat transfer in a micropolar fluid past a stretching surface embedded in a non-Darcian porous medium with heat generation. Ishak [24] investigated the MHD flow of a micropolar fluid past a stretched permeable surface with heat generation.

Many processes in engineering areas occur at high temperatures and knowledge of radiating heat transfer becomes very important for the design of the pertinent equipment. Nuclear power plants, gas turbines, and the various propulsion devices for aircraft, missiles, satellites, and space vehicles are examples of such engineering areas. Khedr et al. [25] analyzes the radiation effects on heat transfer of a micropolar fluids through a porous medium. Abo-Eldahad and Ghonaim [26] studied the thermal boundary layer flow over a stretching sheet in a micropolar fluid with radiation. Gnaneswara [27] studied the heat generation and thermal radiation effects over a stretching sheet in a micropolar fluid. Olanrewaju et al. [28] found radiation effects on MHD flow of micropolar fluid towards a stagnation point on a vertical plate.

However the interaction of radiation with mass transfer of an electrically conducting micropolar fluid past a stretching surface has received little attention. Hence an attempt is made to investigate the radiation effects on a steady free convection flow near an isothermal vertical stretching sheet in the presence of a magnetic field, a non-Darcian porous medium and heat generation/absorption. The governing equations are transformed by using similarity transformation and the resultant dimensionless equations are solved numerically using the Runge-Kutta fourth order method with the shooting technique. The effects of various governing parameters on the velocity, temperature, concentration, skin-friction coefficient, the Nusselt number, and Sherwood number are shown in the figures and tables and analyzed in detail.

2. Mathematical Formulation

Let us consider a steady, two-dimensional laminar, free convection boundary layer flow of an electrically conducting and heat generating/absorbing micropolar fluid through a porous medium bounded by a vertical isothermal sheet coinciding with the plane $y = 0$, where the flow confined to $y > 0$. Two equal and opposite forces are introduced along the x'-axis so that the sheet is linearly stretched keeping the origin fixed (see Figure 1). A uniformly distributed transverse magnetic field of strength B_0 is imposed along the y'-axis. The magnetic Reynolds number of the flow is taken to be small enough so that the induced distortion of the applied magnetic field can be neglected. The viscous dissipative heat is also assumed to be negligible. It is also assumed that microscopic inertia term involving J (where J is the square of the characteristic length of microstructure) can be neglected for steady two-dimensional boundary layer flow in a micropolar fluid without introducing any appreciable error in the solution. Under the above assumptions and upon treating the fluid saturated porous medium as continuum [20], including the non-Darcian inertia effects, and assuming that the Boussinesq approximation is valid, the boundary layer form of the governing equations can be written as [29, 30]

$$\frac{\partial u'}{\partial x'} + \frac{\partial v'}{\partial y'} = 0, \tag{1}$$

$$u'\frac{\partial u'}{\partial x'}+v'\frac{\partial v'}{\partial y'}=\nu\frac{\partial^2 u'}{\partial y'^2}+g\beta\left(T'-T'_\infty\right)+g\beta^*\left(C'-C'_\infty\right)$$
$$+k_1\frac{\partial\sigma'}{\partial y'}-\frac{\sigma_0 B_0^2}{\rho}u'-\frac{\nu}{K}u'-Cu'^2, \quad (2)$$

$$G_1\frac{\partial^2\sigma}{\partial y'^2}-2\sigma-\frac{\partial u'}{\partial y'}=0, \quad (3)$$

$$u'\frac{\partial T'}{\partial x'}+v'\frac{\partial T'}{\partial y'}=\frac{k_e}{\rho c_p}\frac{\partial^2 T'}{\partial y'^2}-\frac{1}{\rho c_p}\frac{\partial q_r}{\partial y'}+\frac{Q_0}{\rho c_p}\left(T'-T'_\infty\right), \quad (4)$$

$$u'\frac{\partial C'}{\partial x'}+v'\frac{\partial C'}{\partial y'}=D\frac{\partial^2 C'}{\partial y'^2}, \quad (5)$$

subject to the boundary conditions

$$u'=bx,\quad v'=0,\quad T'=T'_w,\quad C'=C'_w,$$
$$\sigma=0\quad \text{at } y=0,$$
$$u'\longrightarrow u'_\infty,\qquad T'\longrightarrow T'_\infty,\qquad C'\longrightarrow C'_\infty,$$
$$\sigma=0\quad \text{as } y\longrightarrow\infty, \quad (6)$$

where x' and y' are the coordinates along and normal to the sheet. u' and v' are the components of the velocity in the x'- and y'-directions, respectively. σ, k_1, and G_1 are the microrotation component, coupling constant, and microrotation constant, respectively. k_e, C, K, T', and C' are the effective thermal conductivity, permeability of the porous medium, transport property related to the inertia effect, fluid temperature, and fluid concentration, respectively. β, β^*, Q_0, U_∞, and g are the coefficient of thermal expansion, coefficient of concentration expansion, volumetric rate of heat generation, free stream velocity, and acceleration due to gravity, respectively. σ_0, ρ, ν, and c_p are the electrical conductivity, density, apparent kinematic viscosity, and specific heat at constant pressure of the fluid, respectively.

By using the Rosseland approximation [31], the radiative heat flux in y' direction is given by

$$q_r=-\frac{4\sigma^*}{3k^*}\frac{\partial T'^4}{\partial y'}, \quad (7)$$

where σ^* is the Stefan-Boltzmann constant and k^* is the mean absorption coefficient. By using (7), the energy equation (4) becomes

$$u'\frac{\partial T'}{\partial x'}+v'\frac{\partial T'}{\partial y'}=\frac{k_e}{\rho c_p}\frac{\partial^2 T'}{\partial y'^2}+\frac{4\sigma^*}{3k^*\rho c_p}\frac{\partial^2 T'^4}{\partial y'^2}+\frac{Q_0}{\rho c_p}\left(T'-T'_\infty\right). \quad (8)$$

It is convenient to make the governing equations and conditions dimensionless by using

$$x=\frac{bx'}{u'_\infty},\qquad y=\frac{by'}{u'_\infty}R,\qquad u=\frac{u'}{u'_\infty},\qquad v=\frac{v'}{u'_\infty}R,$$
$$R=\frac{u_\infty}{\sqrt{cv}},\qquad \theta=\frac{T'-T''_\infty}{T'_w-T'_\infty},$$
$$\phi=\frac{C'-C'_\infty}{C'_w-C'_\infty},\qquad M=\frac{\sigma_0 B_0^2}{\rho b},$$
$$\mathrm{Gr}=\frac{g\beta\left(T'_w-T'_\infty\right)}{bu'_\infty},\qquad \mathrm{Gc}=\frac{g\beta^*\left(C'_w-C'_\infty\right)}{bu'_\infty},$$
$$Q=\frac{Q_0}{b\rho c_p},\qquad \mathrm{Sc}=\frac{\nu}{D},\qquad \mathrm{Pr}=\frac{\mu c_p}{k},\qquad F=\frac{kk^*}{4\sigma^* T_\infty^3},$$
$$r=\frac{T'_w-T'_\infty}{T'_\infty},\qquad \gamma=\frac{Cu_\infty}{b},\qquad \sigma=\frac{\nu\sigma'}{u_\infty^2}R,$$
$$\mathrm{Da}^{-1}=\frac{\nu}{Kb}. \quad (9)$$

In view of (9), the equations (1), (2), (3), (5), and (8) reduce to the following nondimensional form:

$$\frac{\partial u}{\partial x}+\frac{\partial v}{\partial y}=0, \quad (10)$$

$$u\frac{\partial u}{\partial x}+v\frac{\partial u}{\partial y}=\frac{\partial^2 u}{\partial y^2}+N_1\frac{\partial\sigma}{\partial y}-Mu-\frac{1}{Da}u-\gamma u^2$$
$$+\mathrm{Gr}\,\theta+\mathrm{Gc}\,\phi \quad (11)$$

$$G\frac{\partial^2\sigma}{\partial y^2}-2\sigma-\frac{\partial u}{\partial y}=0, \quad (12)$$

$$u\frac{\partial\theta}{\partial x}+v\frac{\partial\theta}{\partial y}$$
$$=\frac{1}{\mathrm{Pr}}\frac{\partial^2\theta}{\partial y^2}+\frac{4}{3F\,\mathrm{Pr}}$$
$$\times\left((1+r\theta)^3\frac{\partial^2\theta}{\partial y^2}+3r(1+r\theta)^2\left(\frac{\partial\theta}{\partial y}\right)^2\right)+Q\theta, \quad (13)$$

$$u\frac{\partial\phi}{\partial x}+v\frac{\partial\phi}{\partial y}=\frac{1}{\mathrm{Sc}}\frac{\partial^2\phi}{\partial y^2}. \quad (14)$$

The corresponding boundary conditions are

$$u=x,\quad v=0,\quad \sigma=0,\quad \theta=1,\quad \phi=1\quad \text{at } y=0,$$
$$u=1,\quad \sigma=0,\quad \theta=0,\quad \phi=0\quad \text{as } y\longrightarrow\infty, \quad (15)$$

where R is the Reynolds number. Proceeding with the analysis, we define a stream function $\psi(x, y)$ such that

$$u = \frac{\partial \psi}{\partial \overline{y}}, \qquad v = -\frac{\partial \psi}{\partial \overline{x}}. \tag{16}$$

Now, let us consider the stream function as if

$$\begin{aligned}\psi(x, y) &= f(y) + x g(y) \\ \sigma &= x h(y).\end{aligned} \tag{17}$$

In view of (16)-(17), the continuity equation (10) is identically satisfied and the momentum equation (11), angular momentum equation (12), energy equation (13), and concentration equation (14) become

$$\begin{aligned}
&\frac{\partial \psi}{\partial y}\frac{\partial^2 \psi}{\partial x \partial y} - \frac{\partial \psi}{\partial x}\frac{\partial^2 \psi}{\partial y^2} \\
&\quad = \frac{\partial^3 \psi}{\partial y^3} + \left(M + \frac{1}{\mathrm{Da}}\right)\frac{\partial \psi}{\partial y} \\
&\quad + \gamma\left(\frac{\partial \psi}{\partial y}\right)^2 - \mathrm{Gr}\,\theta - \mathrm{Gc}\,\phi, \\
&\frac{\partial \psi}{\partial y}\frac{\partial \theta}{\partial x} - \frac{\partial \psi}{\partial x}\frac{\partial \theta}{\partial y} \\
&= \frac{1}{\Pr}\frac{\partial^2 \theta}{\partial y^2} + \frac{4}{3F\Pr}\left((1 + r\theta)^3\frac{\partial^2 \theta}{\partial y^2} + 3r(1 + r\theta)^2\left(\frac{\partial \theta}{\partial y}\right)^2\right) \\
&+ Q\theta. \\
&G\frac{\partial^2 \sigma}{\partial y^2} - 2\sigma - \frac{\partial u}{\partial y} = 0 \\
&\frac{\partial \psi}{\partial y}\frac{\partial \phi}{\partial x} - \frac{\partial \psi}{\partial x}\frac{\partial \phi}{\partial y} = \frac{1}{\mathrm{Sc}}\frac{\partial^2 \phi}{\partial y^2}
\end{aligned} \tag{18}$$

and the boundary conditions (15) become

$$\begin{aligned}
&\frac{\partial \psi}{\partial y} = x, \quad \frac{\partial \psi}{\partial x} = 0, \quad h = 0, \\
&\theta = 1, \quad \phi = 1, \quad \text{at } y = 0 \\
&\frac{\partial \psi}{\partial y} \longrightarrow 1, \quad h \longrightarrow 0, \quad \theta \longrightarrow 0, \\
&\phi \longrightarrow 0, \quad \text{as } y \longrightarrow \infty
\end{aligned} \tag{19}$$

in (18) and equating coefficient of x^0 and x^1, we obtain the coupled nonlinear ordinary differential equations

$$\begin{aligned}
&f''' + f''g - f'g' - \left(M + \frac{1}{\mathrm{Da}}\right)f' - \gamma f'^2 + \mathrm{Gr}\,\theta \\
&\quad + \mathrm{Gc}\,\phi = 0, \\
&g''' + gg'' - \left(g'\right)^2 - \left(M + \frac{1}{\mathrm{Da}}\right)g' - 2\gamma f'g' \\
&\quad + N1h' = 0, \\
&Gh'' - 2h - g'' = 0, \\
&\left(3F + 4(1 + r\theta)^3\right)\theta'' + 3\Pr Fg\theta' \\
&\quad + 12r(1 + r\theta)^2\theta'^2 + 3F\Pr Q\theta = 0, \\
&\phi'' + \mathrm{Sc}\,g\phi' = 0,
\end{aligned} \tag{20}$$

where a prime denotes differentiation with respect to y.

In view of (17), the boundary conditions (19) reduce to

$$\begin{aligned}
&f = 0, \quad f' = 0, \quad g = 0, \quad g' = 1, \quad h = 1, \\
&\theta = 1, \quad \phi = 1, \quad \text{at } y = 0 \\
&f' \longrightarrow 1 \quad g' \longrightarrow 0 \quad h \longrightarrow 0 \quad \theta \longrightarrow 0 \\
&\phi \longrightarrow 0, \quad \text{as } y \longrightarrow \infty.
\end{aligned} \tag{21}$$

Of special significance in free convection problems are the skin-friction coefficient, the Nusselt number, and the Sherwood number.

The shear stress at the stretching surface is given by

$$\begin{aligned}
\tau_w &= (\mu + k)\left(\frac{\partial u'}{\partial y'}\right)_{y'=0} + k(\sigma)_{y'=0} = \rho v b R\left(\frac{\partial u}{\partial y}\right)_{y=0} \\
&= \rho v b R\left[f''(0) + x g''(0)\right].
\end{aligned} \tag{22}$$

The skin-friction coefficient C_f is given by

$$C_f = \frac{\tau_w}{(1/2)\rho u_\infty^2} = \frac{2}{R}\left[f''(0) + x g''(0)\right]. \tag{23}$$

The wall heat flux is given by

$$\begin{aligned}
q_w &= -k\left(\frac{\partial T'}{\partial y'}\right)_{y'=0} \\
&= -k\frac{bR}{u_\infty}\left(\frac{\partial T'}{\partial y'}\right)_{y'=0} \\
&= -k\frac{bR}{u_\infty}\left(T'_w - T'_\infty\right)\theta'(0).
\end{aligned} \tag{24}$$

The wall mass flux is given by

$$M_w = -D\left(\frac{\partial C'}{\partial y'}\right)_{y'=0} = -k\frac{bR}{u_\infty}\left(\frac{\partial C'}{\partial y'}\right)_{y'=0} = -k\frac{bR}{u_\infty}\left(C'_w - C'_\infty\right)\phi'(0). \quad (25)$$

The Nussel number

$$\text{Nu} = \frac{q_w}{k\left(T'_w - T'_\infty\right)} = \frac{bR}{u_\infty}\theta'(0). \quad (26)$$

The Sherwood number

$$\text{Sh} = \frac{M_w}{D\left(C'_w - C'_\infty\right)} = \frac{bR}{u_\infty}\phi'(0). \quad (27)$$

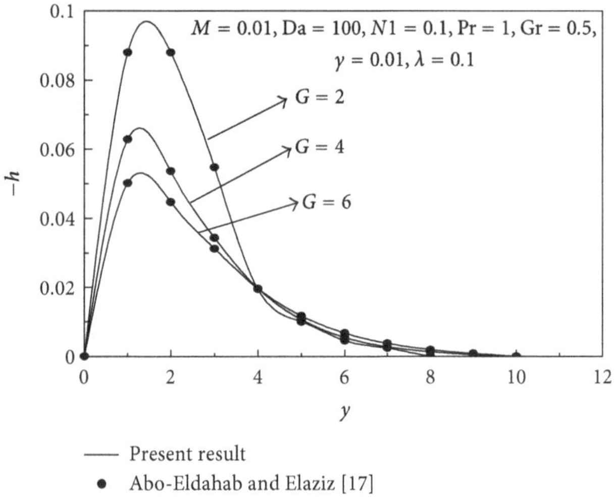

FIGURE 2: Comparison of angular velocity Profiles.

3. Numerical Procedure

The shooting method for linear equations is based on replacing the boundary value problem by two initial value problems and the solutions of the boundary value problem is a linear combination between the solutions of the two initial value problems. The shooting method for the nonlinear boundary value problem is similar to the linear case, except that the solution of the nonlinear problem cannot be simply expressed as a linear combination of the solutions of the two initial value problems. Instead, we need to use a sequence of suitable initial values for the derivatives such that the tolerance at the end point of the range is very small. This sequence of initial values is given by the secand method, and we use the fourth order Runge-Kutta method to solve the initial value problems.

Following Rosenhead [32] and Carnahan et al. [33], the value of y at infinity is fixed at 5. The full equations (20) with the boundary conditions (21) were solved numerically using the Runge-Kutta method algorithm with a systematic guessing $f''(0)$, $g''(0)$, $h'(0)$, $\theta'(0)$, and $\phi'(0)$ by the shooting technique until the boundary conditions at infinity $f'(y)$ decay exponentially to one, also $g'(y)$, $h(y)$, $\theta(y)$, and $\phi(y)$ to zero. The functions f', g', $-h$, θ, and ϕ are shown in the figures.

4. Results and Discussion

As a result of the numerical calculations, the dimensionless velocity, angular velocity, temperature, and concentration distributions for the flow under consideration are obtained and their behavior has been discussed for variations in the governing parameters, namely, the thermal Grashof number Gr, solutal Grashof number Gc, magnetic field parameter M, radiation parameter F, the parameter of relative difference between the temperature of the sheet and temperature far away from the sheet r, the Prandtl number Pr, heat generation parameter Q, Da Darcy number, porous medium inertia coefficient γ, vortes viscosity parameter $N1$, microrotation parameter G, and the Schmidt number Sc. In the present study, the following default parametric values are adopted: Gr = 0.5, Gc = 0.5, $M = 0.01$, Pr = 0.71, $F = 1.0$, $r = 0.05$, Q = 0.1, $\gamma = 0.01$, $N1 = 0.1$, G = 2.0, Da = 100, and Sc = 0.6. All graphs therefore correspond to these unless specifically indicated on the appropriate graph.

In order to assess the accuracy of our computed results, the present results have been compared with Abo-Eldahab and ELaziz [17] for different values of G as shown in Figure 2 with Gc = 0.0, $F = 0.0$, $r = 0.0$, and Sc = 0.0. It is observed that the agreements with the solution of angular velocity profiles are excellent.

Figure 3(a) shows the variation of the dimensionless velocity component f' for several sets of values of thermal Grashof number Gr. As expected, it is observed that there is a rise in the velocity due to enhancement of thermal buoyancy force. The variation of the dimensionless velocity component f' for several sets of values of solutal Grashof number Gc is depicted in Figure 3(b). As expected, the fluid velocity increases and the peak value is more distinctive due to the increase in the species buoyancy force. It should be mentioned herein that the profiles of g', $-h$, θ and ϕ were found to be insensible to change in Gr and Gc, therefore, not shown herein for brevity.

The effect of variation of the magnetic parameter M on the velocity (f' and g'), angular velocity $-h$, temperature θ, and concentration ϕ profiles is presented in Figures 4(a)–4(e) respectively. It is well known that the application of a uniform magnetic field normal to the flow direction gives rise to a force called Lorentz. This force has the tendency to slow down the velocity of the fluid and angular velocity of microrotation in the boundary layer and to increase its temperature and concentration. This is obvious from the

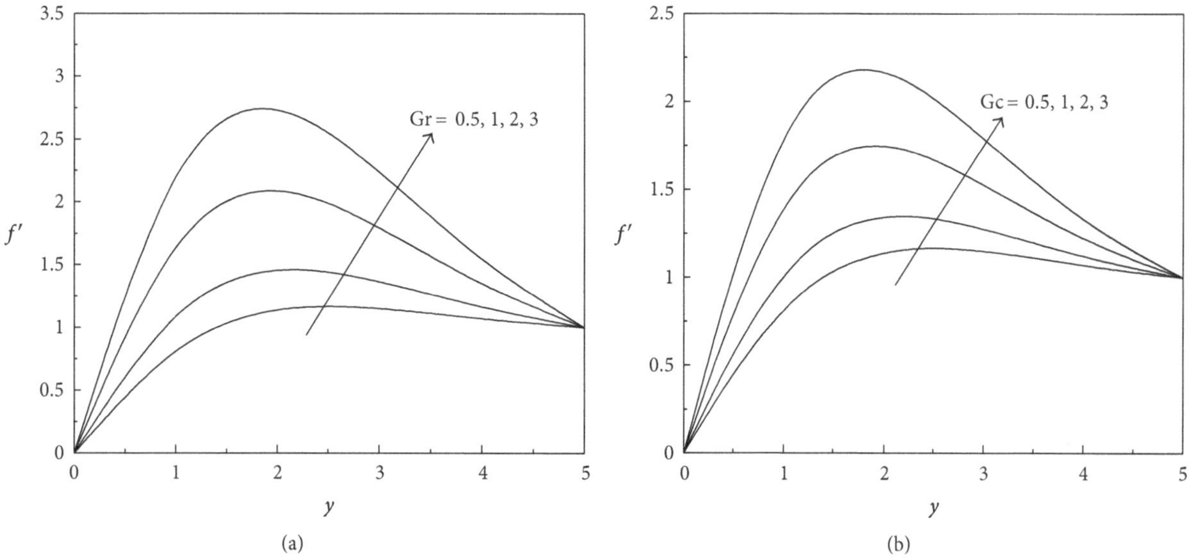

FIGURE 3: (a) Variation of the velocity component f' with Gr. (b) Variation of the velocity component f' with Gc.

TABLE 1: Comparison of the present results with the literature results given by Abo-Eldahab and El Aziz [17], Gc = 0.0, F = 0.0, r = 0.0, Sc = 0.0. Pr = 1.0, Da = 100, γ = 0.01, $N1$ = 0.1, G = 2, Gr = 0.5, and Q = 0.1 .

M	$f''(0)$		$g''(0)$	
	Present	Abo-Eldahab [17]	Present	Abo-Eldahab [17]
0.1	0.67059875	0.6708520	−1.0002147	−0.998951
0.2	0.57169871	0.5717493	−1.0002075	−0.998905

decreases in the velocity profiles, angular velocity of microrotation profiles, while temperature profiles and concentration profiles increase, presented in Figures 4(a)–4(e) respectively.

Figures 5(a)–5(e) present typical profiles for the variables of the fluid's x-component of velocity (f' and g'), angular velocity $-h$, temperature θ, and concentration ϕ for different values of Darcy number Da. It is noted that values of Da increase the fluid velocities and angular velocity increases, while temperature and concentration of the fluid decrease.

Figures 6(a)–6(e) present the typical profiles for the variables of the fluid's x-component of velocity (f' and g'), angular velocity $-h$, temperature θ and concentration ϕ for different values of the porous medium inertia coefficient γ. Obviously, the porous medium inertia effects constitute resistance to the flow. Thus as the inertia coefficient increases, the resistance to the flow increases, causing the fluid flow in the porous medium to slow down and the temperature and concentration increase and, therefore, as γ increases f', g', and $-h$ decreases while the temperature θ and concentration ϕ increase.

Figures 7(a)–7(e) present the typical profiles for the variables of the fluid's x-component of velocity (f' and g'), angular velocity $-h$, temperature θ, and concentration ϕ for different values of the vortex viscosity parameter $N1$. Increases in the values of $N1$ have a tendency to increase f', $-h, \theta$, and ϕ and to decrease g'.

Figure 8 is a plot of the dimensionless angular velocity –h profiles for different values of the presence of the microrotation parameter G. The curves illustrate that, as the values of G increases, the angular velocity $-h$, as expected, decreases with an increase in the boundary layer thickness as the maximum moves away from the sheet. Of course, when the viscosity of the fluid decreases the angular velocity of additive increases.

Figures 9(a) and 9(b) illustrate the respective changes in the profiles of f' and θ as the heat generation/absorption coefficient Q is changed. It is clear from Figures 9(a) and 9(b) that increasing in the values of Q produces increases in the velocity f' and temperature θ distributions of the fluid. This is expected since heat generation (Q > 0) causes the thermal boundary layer to become thicker and the fluid to be warmer. this enhances the effects of thermal buoyancy of the driving body force due to mass density variations which are coupled to the temperature distribution and therefore increasing the fluid velocity distribution. No figures for g', $-h$ and ϕ are presented for the same reason as mentioned before.

Figure 10(a) Illustrates the dimensionless velocity component f' for different values of the Prandtl number Pr. The numerical results show that the effect of increasing values of the Prandtl number results in a decreasing velocity. From Figure 10(b), it is observed that an increase in the Prandtl number results in a decrease of the thermal boundary layer thickness and in general lower average temperature within

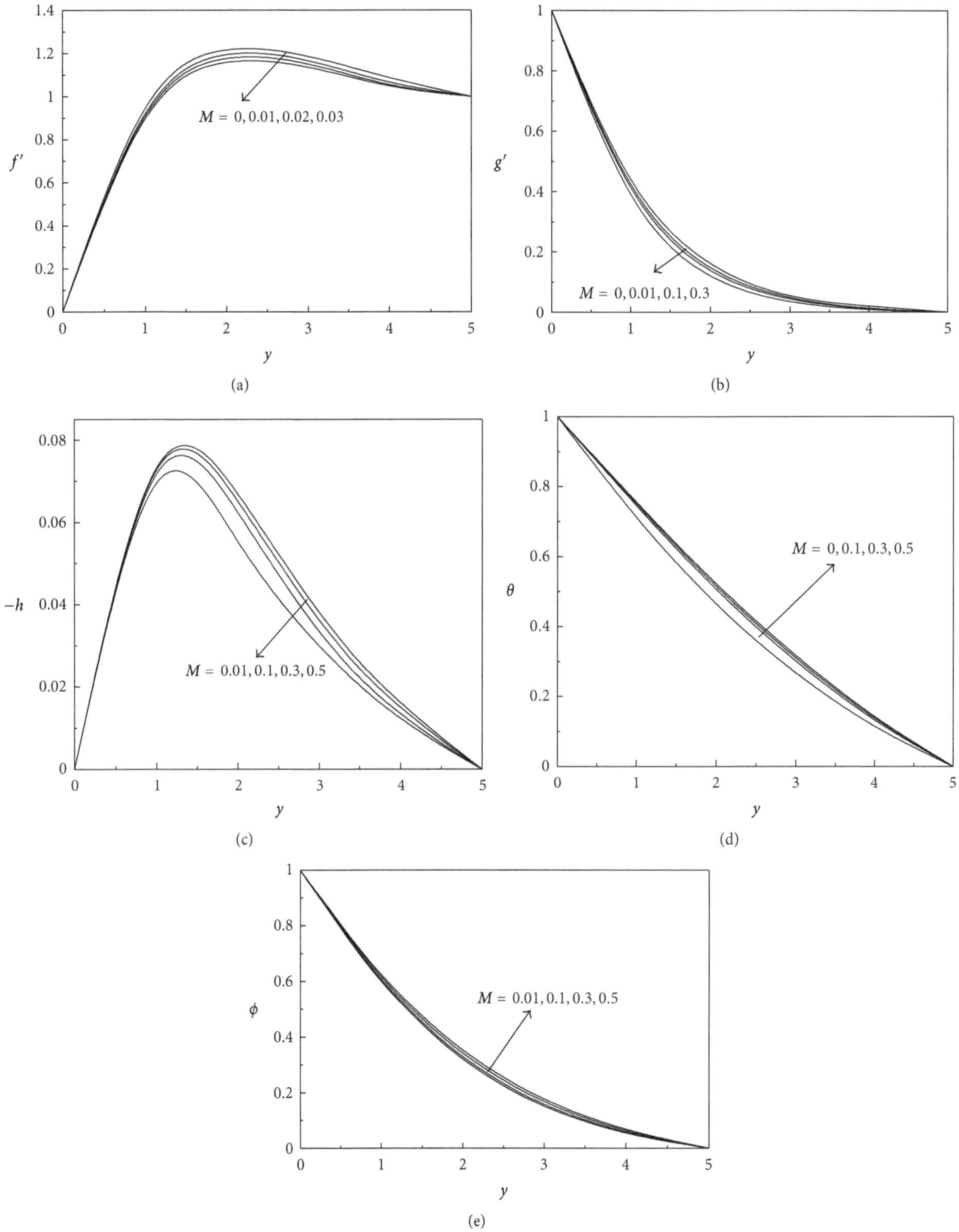

FIGURE 4: (a) Variation of the velocity component f' with M. (b) Variation of the velocity component g' with M. (c) Variation of the velocity component $-h$ with M. (d) Variation of the velocity component θ with M. (e) Variation of the velocity component ϕ with M.

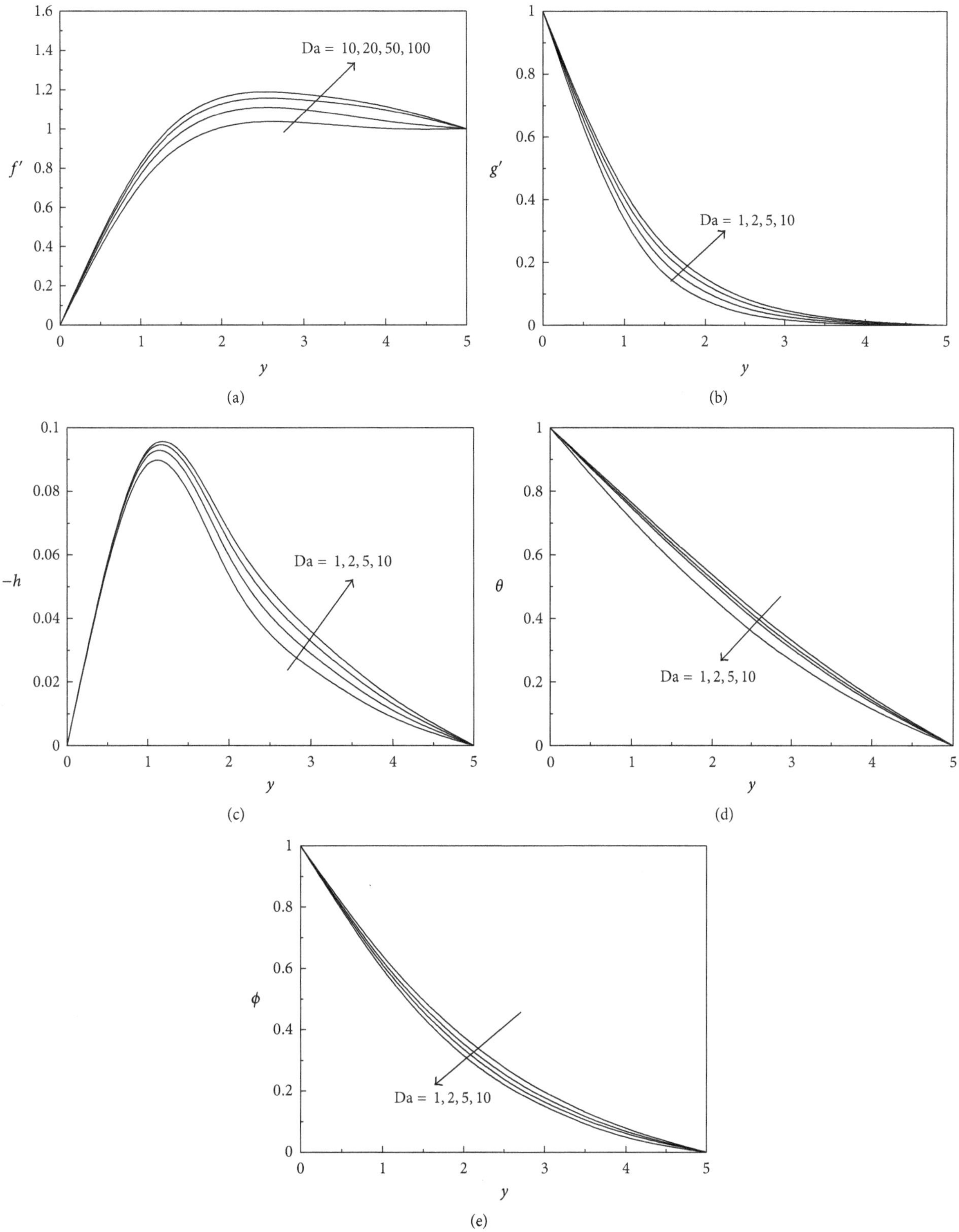

FIGURE 5: (a) Variation of the velocity component f' with Da. (b) Variation of the velocity component g' with Da. (c) Variation of the velocity component $-h$ with Da. (d) Variation of the velocity component θ with Da. (e) Variation of the velocity component ϕ with Da.

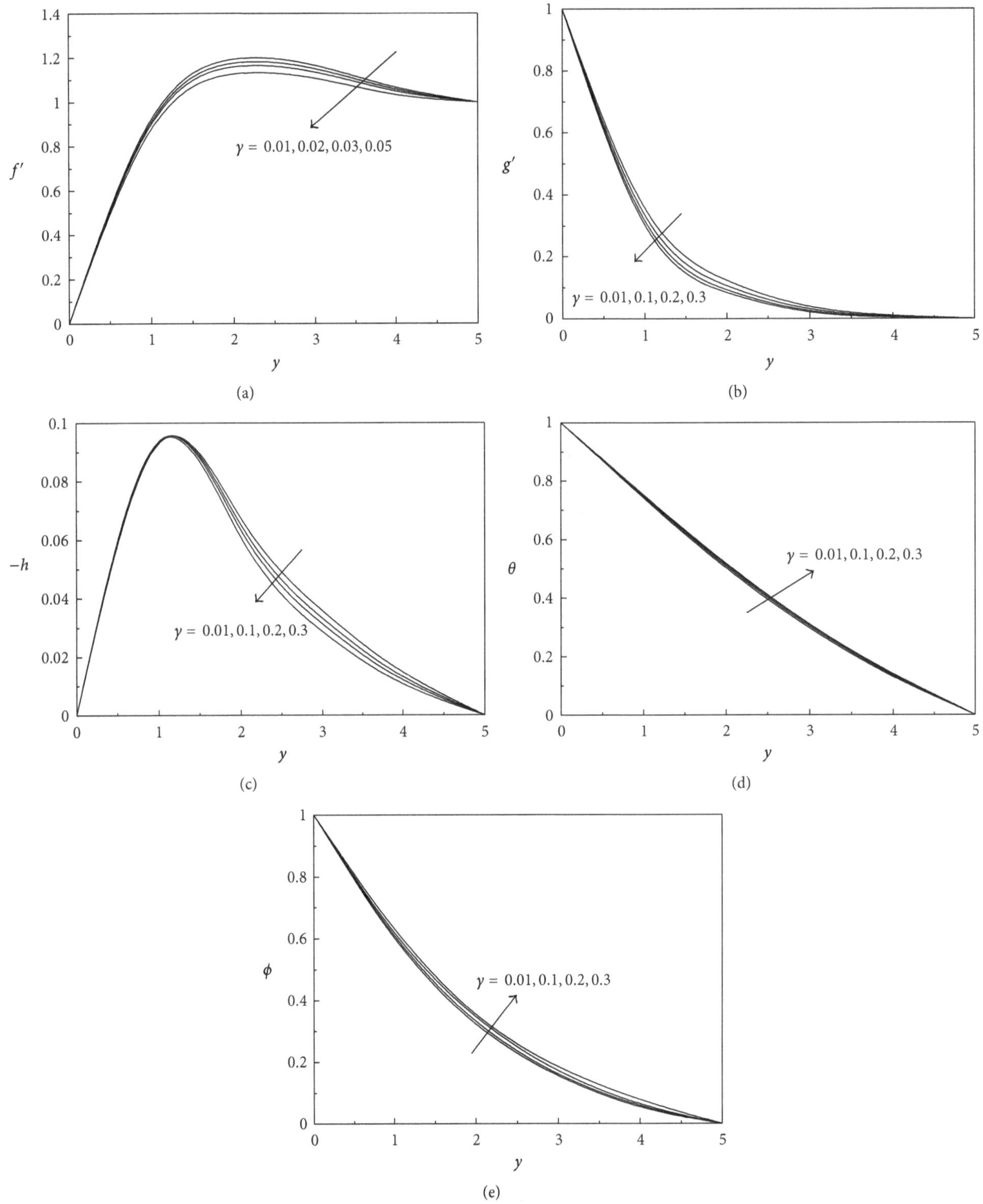

FIGURE 6: (a) Variation of the velocity component f' with γ. (b) Variation of the velocity component g' with γ. (c) Variation of the velocity component $-h$ with γ. (d) Variation of the velocity component θ with γ. (e) Variation of the velocity component ϕ with γ.

the boundary layer. The reason is that smaller values of Pr are equivalent to increasing the thermal conductivities, and therefore heat is able to diffuse away from the heated plate more rapidly than for higher values of Pr. Hence in the case of smaller Prandtl numbers as the boundary layer is thicker, the rate of heat transfer is reduced.

The effect of the radiation parameter F on the dimensionless velocity component f' and dimensionless temperature is

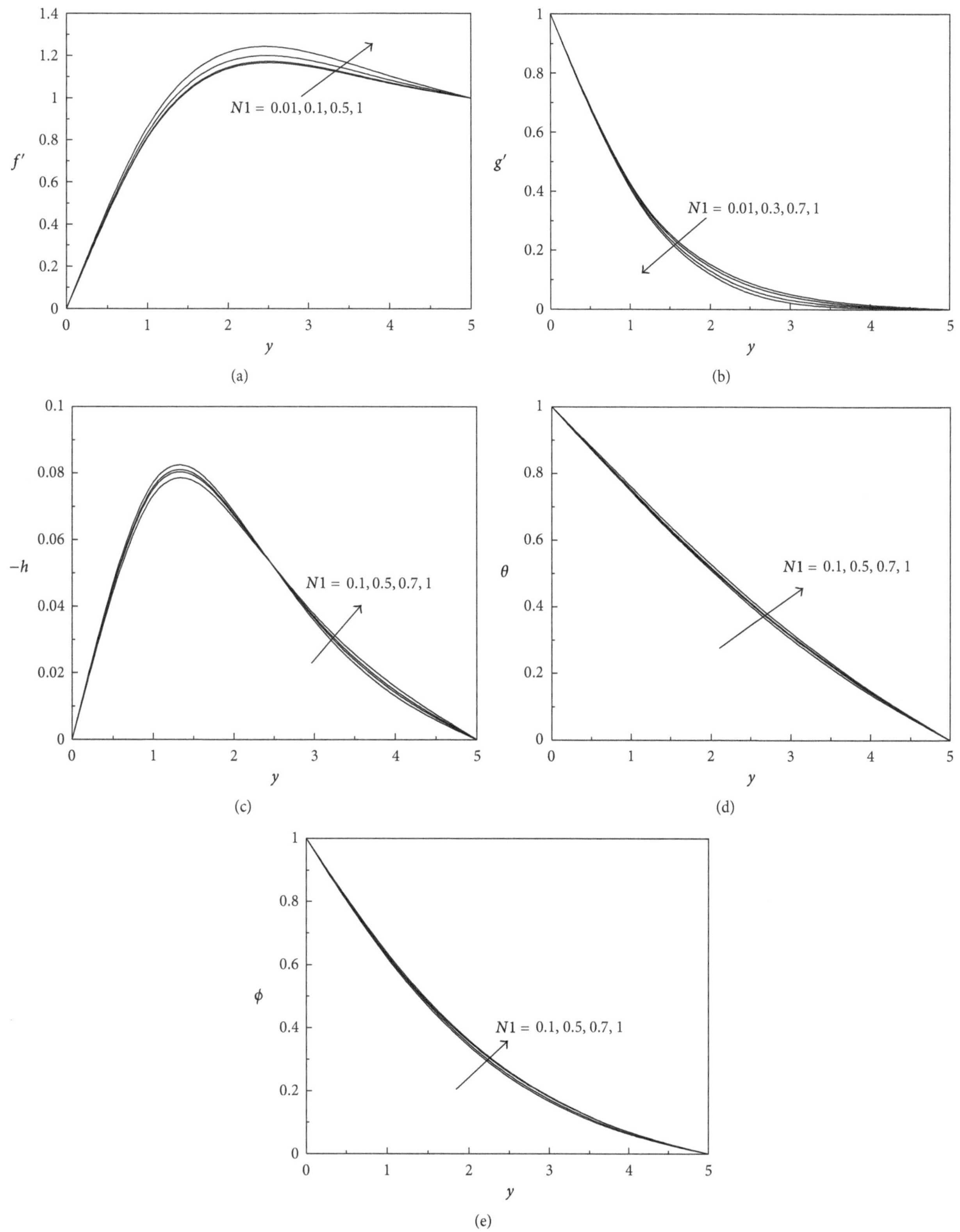

FIGURE 7: (a) Variation of the velocity component f' with $N1$. (b) Variation of the velocity component g' with $N1$. (c) Variation of the velocity component $-h$ with $N1$. (d) Variation of the velocity component θ with $N1$. (e) Variation of the velocity component ϕ with $N1$.

Table 2: Variation of f'', g'', $-h'$, θ', and ϕ' at the plate with Gr, Gc, M, Da, and γ for Pr = 0.71, F = 1.0, r = 0.05, Q = 0.1, and Sc = 0.6.

Gr	Gc	M	Da	γ	$f''(0)$	$g''(0)$	$h'(0)$	$-\theta'(0)$	$-\phi'(0)$
0.5	0.5	0.1	100	0.01	1.21865	−1.05339	0.25808	0.27721	0.42336
1.0	0.5	0.1	100	0.01	1.72256	−1.05339	0.25833	0.276989	0.42336
0.5	1.0	0.1	100	0.01	1.64847	−1.05339	0.25827	0.277042	0.42336
0.5	0.5	0.2	100	0.01	1.11425	−1.05275	0.26313	0.272843	0.41783
0.5	0.5	0.1	10	0.01	1.12368	−1.09429	0.26264	0.273265	0.41836
0.5	0.5	0.1	20	0.01	1.17407	−1.09973	0.26014	0.275425	0.42110
0.5	0.5	0.1	100	0.1	1.14886	−1.08451	0.26392	0.271914	0.41678

Table 3: Variation of f'', g'', $-h'$, θ', and ϕ' at the plate with G, Pr, $N1$, F, Q, and Sc for Gr = 0.5, Gc = 0.5, M = 0.1, and Da = 100.

G	Pr	$N1$	F	λ	Sc	$f''(0)$	$g''(0)$	$-h'(0)$	$\theta'(0)$	$\phi'(0)$
2	0.71	0.1	1.0	0.1	0.6	1.12865	−1.05339	0.25808	0.27721	0.42336
4	0.71	0.1	1.0	0.1	0.6	1.21747	−1.05339	0.15006	0.27721	0.42336
2	1.0	0.1	1.0	0.1	0.6	1.1752	−1.05339	0.25805	0.33548	0.42336
2	0.71	0.4	1.0	0.1	0.6	1.22994	−1.03823	0.25937	0.27557	0.42164
2	0.71	0.1	2.0	0.1	0.6	1.18077	−1.05339	0.25805	0.32959	0.42336
2	0.71	0.1	1.0	0.3	0.6	1.28848	−1.05366	0.25814	0.13497	0.42336
2	0.71	0.1	1.0	0.1	0.78	1.18289	−1.05339	0.25806	0.27723	0.49329

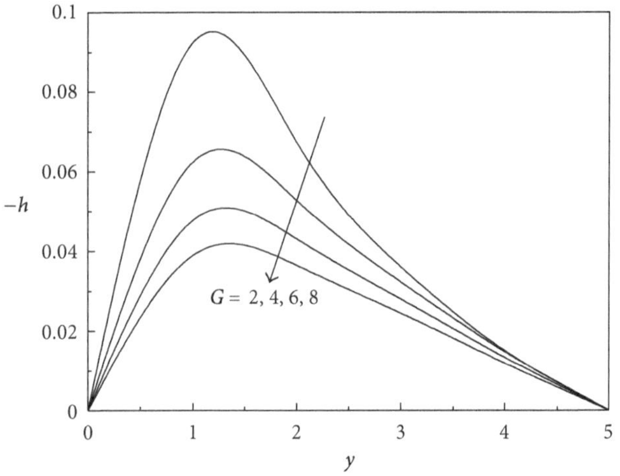

Figure 8: Variation of the velocity component $-h$ with G.

shown in Figures 11(a) and 11(b) respectively. Figure 11(a) shows that velocity component f' decreases with an increase in the radiation parameter F. From Figure 11(b) it is seen that the temperature decreases as the radiation parameter F increases. This result qualitatively agrees with expectations, since the effect of radiation is to decrease the rate of energy transport to the fluid, thereby decreasing the temperature of the fluid.

The influence of the parameter of relative difference between the temperature of the sheet and the temperature far away from the sheet r on dimensionless velocity f' and temperature profiles are plotted in Figures 12(a) and 12(b), respectively. Figure 12(a) shows that dimensionless velocity f' increases with an increase in r. It is observed that the temperature increases with an increase in r (Figure 12(b)).

The influence of the Schmidt number Sc on the dimensionless velocity f' and concentration profiles are plotted in Figures 13(a) and 13(b), respectively. As the Schmidt number increases the concentration decreases. This causes the concentration buoyancy effects to decrease yielding a reduction in the fluid velocity. The reductions in the velocity and concentration profiles are accompanied by simultaneous reductions in the velocity and concentration boundary layers. These behaviors are clear from Figures 13(a) and 13(b).

The accuracy of the aforementioned numerical method was also validated by direct comparisons with the numerical results reported by Abo-Eldahab and El Aziz [17]. Table 1 presents comparisons for the velocities of $f''(0)$ and $g''(0)$ values for various M values. These comparisons show excellent agreement between the results.

Table 2 illustrates the missing wall functions for velocity, angular velocity, temperature, and concentration functions. These quantities are useful in the evaluation of wall shear stresses, gradient of angular velocity, surface heat transfer rate, and mass transfer rate. The results are obtained for r = 0.05 and different values of the thermal Grashof number Gr, solutal Grashof number Gc, magnetic field parameter M, radiation parameter F, the parameter of relative difference between the temperature of the sheet and temperature far away from the sheet r, Prandtl number Pr, heat generation parameter λ, Darcy number Da, porous medium inertia coefficient γ, vortes viscosity parameter $N1$, microrotation parameter G, and Schmidt number Sc. Table 1 indicates that increasing the values of the Grashof number Gr and solutal Grashof number Gc results in an increase in the values of $f''(0)$. This is because as Gr and Gc increase, the

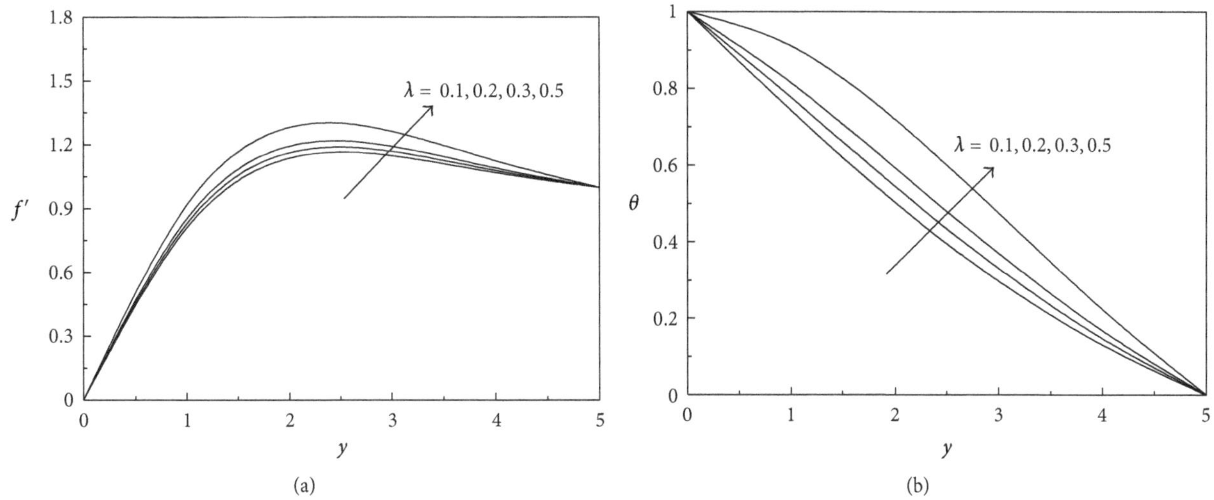

FIGURE 9: (a) Variation of the velocity component f' with Q. (b) Variation of the velocity component θ with Q.

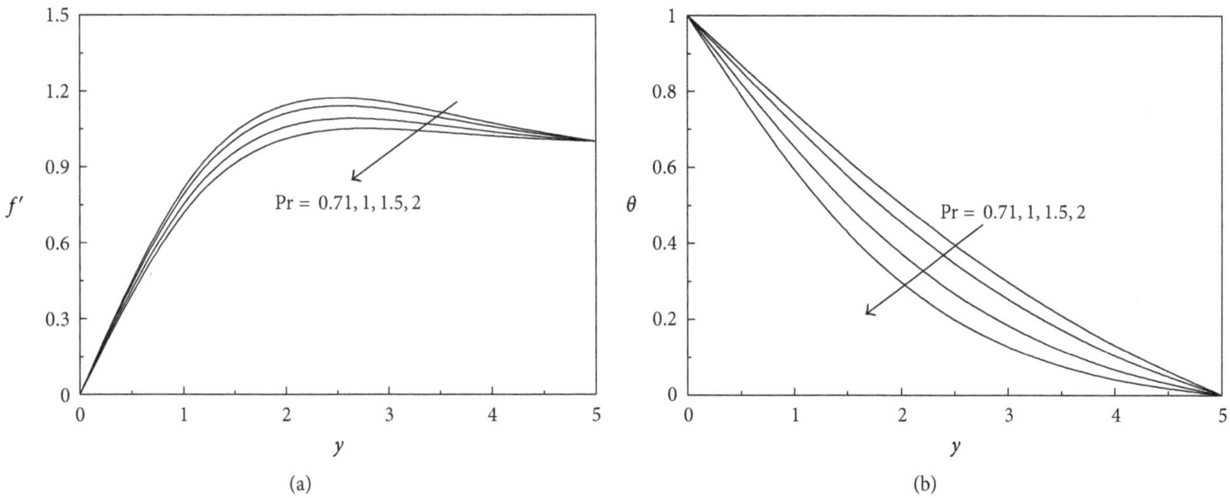

FIGURE 10: (a) Variation of the velocity component f' with Pr. (b) Variation of the velocity component θ with Pr.

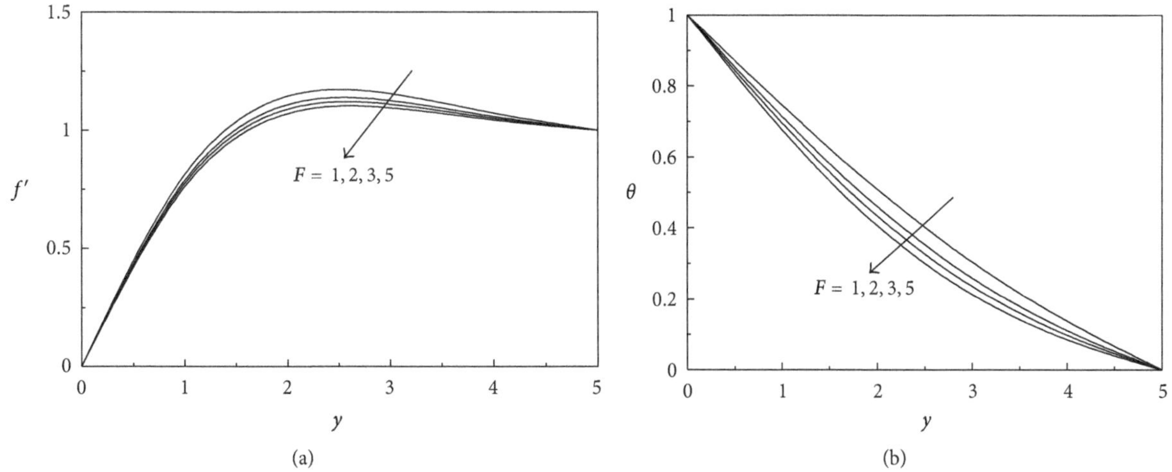

FIGURE 11: (a) Variation of the velocity component f' with F. (b) Variation of the velocity component θ with F.

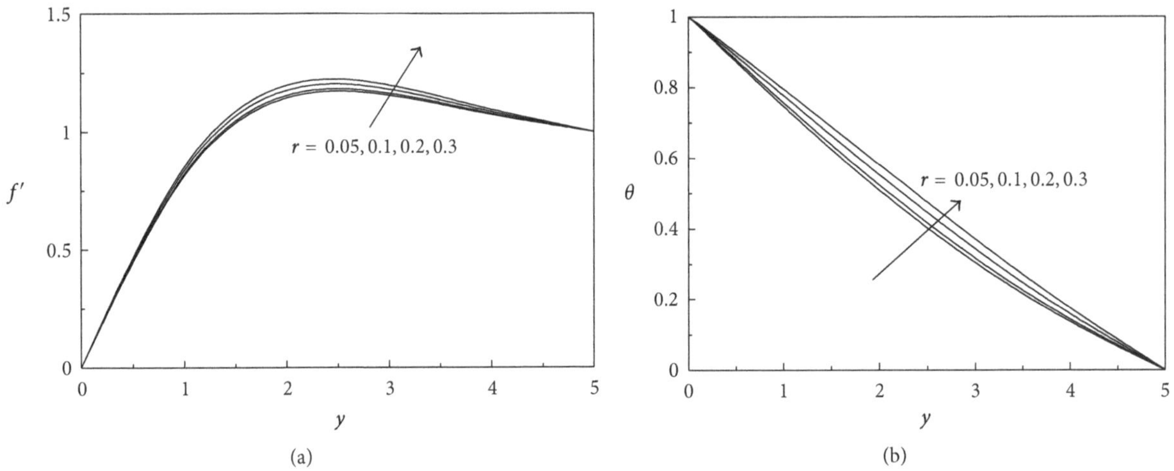

FIGURE 12: (a) Variation of the velocity component f' with r. (b) Variation of the velocity component θ with r.

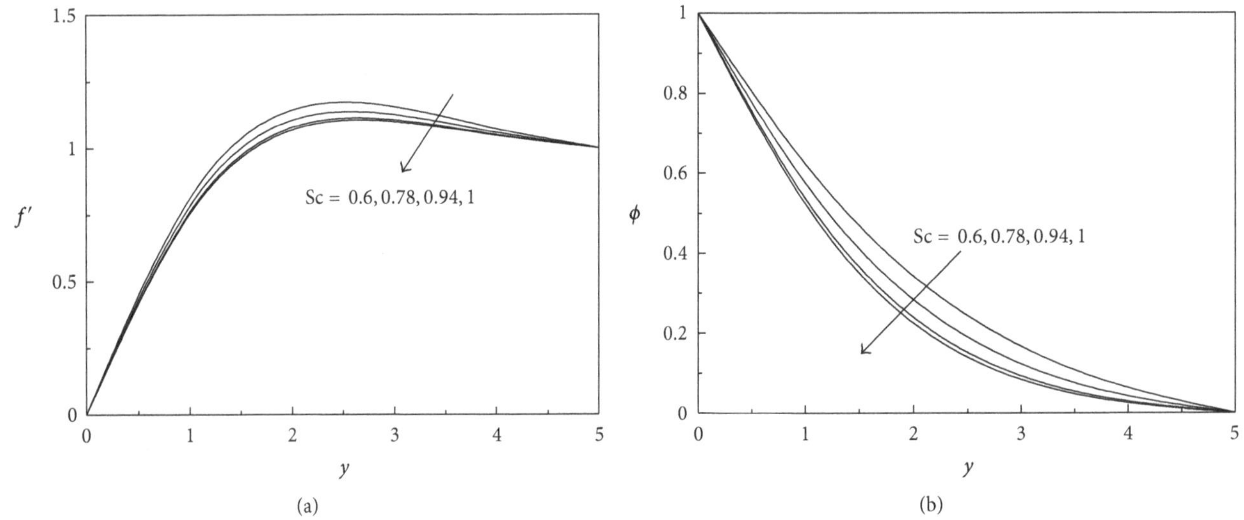

FIGURE 13: (a) Variation of the velocity component f' with Sc. (b) Variation of the velocity component ϕ with Sc.

momentum boundary layer thickness decreases and, therefore, an increase in the values of $f''(0)$ occurs. The results indicate that a distinct fall in the skin-friction coefficient in the x-direction ($f''(0)$ and $g''(0)$), the surface heat transfer rate $-\theta'(0)$, and mass transfer rate $-\phi'(0)$, while gradient of angular velocity $h'(0)$ increases accompanies a rise in the magnetic field parameter M. Increases in the values of Da have the effect of increasing the skin-friction function $f''(0)$, heat transfer rate $-\theta'(0)$ and mass transfer rate $-\phi'(0)$ while the gradient of angular velocity $h'(0)$ and the skin-friction function $g''(0)$ slightly decreases as Da increases. Further, the influence of the porous medium inertia coefficient γ on the wall shear stresses, gradient of angular velocity, surface heat transfer, and surface mass transfer rate is the same as that of the inverse Darcy number Da^{-1} since it also represents resistance to the flow. Namely, as γ increases, $f''(0)$, $\theta'(0)$, $\phi'(0)$ decrease while $g''(0)$, $-h'(0)$ slightly increases, respectively.

From Table 3 for given values of Gr, Gc, M, Da, γ, an increase in the values of microrotation parameter $N1$ leads to reduction in the skin-friction function $g''(0), \theta'(0)$ and $\phi'(0)$ while the skin-friction function $f''(0)$ and gradient of angular velocity $h'(0)$, increase as $N1$ increases. The skin friction $f''(0)$ increases and the gradient of angular velocity $h'(0)$ is decreased as the microrotation parameter G increases, while the skin-friction coefficient in the x-directions $g''(0)$ heat transfer rate $-\theta'(0)$, and mass transfer rate $-\phi'(0)$ are insensible to change in G. Increasing the values of heat generation parameter Q results in an increase in the values of $f''(0)$ and the heat transfer rate $-\theta'(0)$ decreases. It is observed that the magnitude of the wall temperature gradient increases as Prandtl number Pr or radiation parameter F increases. The magnitude of the wall concentration increases with an increase in the Schmidt number Sc. Furthermore, the negative values of the wall temperature and concentration gradients, for all values of the dimensionless parameters, are indicative of the physical fact that the heat flows from the sheet surface to the ambient fluid.

5. Conclusions

The problem of steady, laminar, free convection boundary layer flow of micropolar fluid from a vertical stretching surface embedded in a non-Darcian porous medium in the presence of thermal radiation, mass transfer, uniform magnetic field, heat generation, and free stream velocity was investigated. A similarity transformation was employed to change the governing partial differential equations into ordinary one. These equations were solved numerically by the fourth order Rung-Kutta along with the shooting technique. A wide selection of numerical results have been presented giving the evolution of the velocity, microrotation, temperature, and concentration profiles as well as the skin-friction coefficient, heat transfer rate, and mass transfer rate. It was found that the skin-friction coefficient, heat transfer rate and mass transfer rate are decreased and the gradient of angular velocity increases as the inverse Darcy number, porous medium inertia coefficient, or magnetic field parameter is increased. Increases in the heat generation/absorption coefficient caused increases in the skin-friction coefficient and decrease the heat transfer rate. It was noticed that the increase in radiation parameter or Prandtl number caused decrease in the skin-friction coefficient and increase in heat transfer rate. In addition it was found that the increase in Schmidt number caused decrease in skin-friction coefficient and increase in mass transfer rate.

Nomenclature

C : Transport property related to the inertia effect
C': Fluid concentration
c_p: Specific heat at constant pressure
Da: Darcy number
f: Dimensionless stream function
F: Radiation parameter
G: Microrotation constant
Gc: Solutal Grashof number
Gr: Grashof number
h: Dimensionless microrotation
J: Length of microstructure
k_1: Coupling constant
k_e: Thermalconductivity
K: Permeability of porous medium
m: Boundary parameter
M: Magnetic field parameter
$N1$: Vortes viscosity parameter
Pr: Prandtl number
q_r: Radiative heat flux
Q_0: Heat generation constant
Q: Heat generation parameter
r : Parameter of relative difference between the temperature of the sheet and temperature far away from the sheet
T: Fluidtemperature
T_w: Surface temperature
T_∞: Ambient temperature
Sc: Schmidt number
u, v: Velocity components in the x- and y-directions, respectively
Uw: Velocity of the stretching sheet
x, y : Cartesian coordinates along the sheet andnormal to it, respectively.

Greek Letters

α: Thermaldiffusivity
β: Thermal expansion coefficient
γ: Spin gradient viscosity
η: Similarity variable
θ: Dimensionless temperature
ϕ: Dimensioless Concentration
κ: Vortexviscosity
ν: Kinematic viscosity
μ: Dynamic viscosity
ρ: Fluid density
σ^*: Stefan-Boltzmann constant
ψ: Stream function.

Subscripts

w Condition at the solid surface
∞: Ambient condition.

Superscript

( )_: Differentiatiation with respect to η.

Acknowledgments

The authors are very much thankful to the editor and the reviewers for their encouraging and constructive suggestions to improve the presentation of this paper.

References

[1] A. C. Eringen, "Theory of anisotropic micropolar fluids," *International Journal of Engineering Science*, vol. 18, no. 1, pp. 5–17, 1980.

[2] A. C. Eringen, "Theory of thermomicrofluids," *Journal of Mathematical Analysis and Applications*, vol. 38, no. 2, pp. 480–496, 1972.

[3] G. Lukaszewicz, *Micropolar Fluids-Theory and Applications*, Birkhäauser, Boston, Mass, USA, 1999.

[4] T. Ariman, M. A. Turk, and N. D. Sylvester, "Microcontinuum fluid mechanics—a review," *International Journal of Engineering Science*, vol. 11, no. 8, pp. 905–930, 1973.

[5] G. Ahmadi, "Self-similar solution of imcompressible micropolar boundary layer flow over a semi-infinite plate," *International Journal of Engineering Science*, vol. 14, no. 7, pp. 639–646, 1976.

[6] S. K. Jena and M. N. Mathur, "Similarity solutions for laminar free convection flow of a thermomicropolar fluid past a non-isothermal vertical flat plate," *International Journal of Engineering Science*, vol. 19, no. 11, pp. 1431–1439, 1981.

[7] B. C. Sakiadis, "Boundary-layer on continuous solid surface," *AIChE Journal*, vol. 7, no. 1, pp. 26–28, 1961.

[8] B. C. Sakiadis, "Boundary-layer behavior on continuous moving solid surface," *AIChE Journal*, vol. 7, no. 2, pp. 221–225, 1961.

[9] K. R. RajaGopal, T. Y. Na, and A. S. Gupta, "A non-similar boundary layer on a stretching sheet in a non-Newtonian fluid with uniform free stream," *Journal of Mathematical and Physical Sciences*, vol. 21, no. 2, pp. 189–200, 1987.

[10] F. M. Hady, "On the solution of heat transfer to micropolar fluid from a non-isothermal stretching sheet with injection," *International Journal of Numerical Methods for Heat and Fluid Flow*, vol. 6, no. 6, pp. 99–104, 1996.

[11] T. Y. Na and I. Pop, "Boundary-layer flow of a micropolar fluid due to a stretching wall," *Archive of Applied Mechanics*, vol. 67, no. 4, pp. 229–236, 1997.

[12] I. Hassanien, R. S. R. Gorla, and A. A. Abdulah, "Numerical solution for heat transfer in micropolar fluids over a stretching sheet," *Applied Mechanics For Engineers*, vol. 3, pp. 3–12, 1998.

[13] A. Desseaux and N. A. Kelson, "Flow of micropolar fluid bounded by stretching sheet," *ANZIAM Journal*, vol. 42, pp. 532–560, 2000.

[14] E. M. Abo-Eldahab and A. F. Ghonaim, "Convective heat transfer in an electrically conducting micropolar fluid at a stretching surface with uniform free stream," *Applied Mathematics and Computation*, vol. 137, no. 2-3, pp. 323–336, 2003.

[15] K. B. Pavlov, "Magnetohydrodynamic flow of an incompressible viscous fluid caused by deformation of a surface," *Magnitnaya Gidrodinamika*, vol. 4, pp. 146–147, 1974.

[16] A. Chakrabarti and A. S. Gupta, "Hydromagnetic flow and heat transfer over a stretching sheet," *Quarterly of Applied Mathematics*, vol. 37, no. 1, pp. 73–78, 1979.

[17] E. M. Abo-Eldahab and M. A. El Aziz, "Flow and heat transfer in a micropolar fluid past a stretching surface embedded in a non-Darcian porous medium with uniform free stream," *Applied Mathematics and Computation*, vol. 162, no. 2, pp. 881–899, 2005.

[18] A. Bejan and K. R. Khair, "Heat and mass transfer by natural convection in a porous medium," *International Journal of Heat and Mass Transfer*, vol. 28, no. 5, pp. 909–918, 1985.

[19] C. T. Hsu and P. Cheng, "The Brinkman model for natural convection about a semi-infinite vertical flat plate in a porous medium," *International Journal of Heat and Mass Transfer*, vol. 28, no. 3, pp. 683–697, 1985.

[20] C. L. Tien and K. Vafai, "Convective and radiative heat transfer in porous media," *Advances in Applied Mechanics*, vol. 27, pp. 225–281, 1989.

[21] D. Pal and S. Chatterjee, "Heat and mass transfer in MHD non-Darcian flow of a micropolar fluid over a stretching sheet embedded in a porous media with non-uniform heat source and thermal radiation," *Communications in Nonlinear Science and Numerical Simulation*, vol. 15, no. 7, pp. 1843–1857, 2010.

[22] M. A. A. Mahmoud, "Heat generation/absorption and viscous dissipation effects on MHD flow of a micropolar fluid over a moving permeable surface embedded in a Non-darcian porous medium," *Journal of the Korean Physical Society*, vol. 54, no. 4, pp. 1526–1531, 2009.

[23] A. Nakayama and H. Koyama, "Similarity solutions for buoyancy-induced flows over a non-isothermal curved surface in a thermally stratified porous medium," *Applied Scientific Research*, vol. 46, no. 4, pp. 309–322, 1989.

[24] A. Ishak, "Thermal boundary layer flow over a stretching sheet in a micropolar fluid with radiation effect," *Meccanica*, vol. 45, no. 3, pp. 367–373, 2010.

[25] M. E. M. Khedr, A. J. Chamkha, and M. Bayomi, "MHD flow of a micropolar fluid past a stretched permeable surface with heat generation or absorption," *Nonlinear Analysis: Modelling and Control*, vol. 14, no. 1, pp. 27–40, 2009.

[26] E. M. Abo-Eldahab and A. F. Ghonaim, "Radiation effect on heat transfer of a micropolar fluid through a porous medium," *Applied Mathematics and Computation*, vol. 169, no. 1, pp. 500–510, 2005.

[27] M. Gnaneswara Reddy, "Heat generation and thermal radiation effects over a stretching sheet in a micropolar fluid," *ISRN Thermodynamics*, vol. 2012, Article ID 795814, 6 pages, 2012.

[28] P. O. Olanrewaju, G. T. Okedayo, and J. A. Gbadeyan, "Effects of thermal radiation on magnetohydrodynamic flow of a micropolar fluid towards a stagnation point on a vertical plate," *International Journal of Applied Science and Technology*, vol. 1, no. 6, pp. 1–6, 2011.

[29] A. J. Willson, "Boundary layer in micropolar liquids," *Mathematical Proceeding of the Cambridge Philosophical Society*, vol. 67, no. 2, pp. 469–476, 1970.

[30] D. A. Nield and A. Bejan, *Convection in Porous Media*, Springer, New York, NY, USA, 1999.

[31] M. Q. Brewster, *Thermal Radiative Transfer and Properties*, John Wiley & Sons, New York, NY, USA, 1992.

[32] L. Rosenhead, *Laminar Boundary*, Oxford University Press, Oxford, UK, 1963.

[33] B. Carnahan, H. A. Luther, and O. W. James, *Applied Numerical Methods*, John Wiley & Sons, New York, NY, USA, 1969.

7

Studies on Excess Volume, Viscosity, and Speed of Sound of Binary Mixtures of Methyl Benzoate in Ethers at $T = (303.15, 308.15, \text{and } 313.15)$ K

M. V. Rathnam,[1] Devappa R. Ambavadekar,[1] and M. Nandini[2]

[1] *Physical Chemistry Research Laboratory, B.N. Bandodkar College of Science, Thane 400601, India*
[2] *Department of Chemistry, Dr. P.R.Ghogrey Science College, Deopur, Dhule 424005, India*

Correspondence should be addressed to M. V. Rathnam; mvrathnam58@rediffmail.com

Academic Editor: Felix Sharipov

Densities, viscosities, and speed of sound have been determined at $T = (303.15, 308.15, \text{and } 313.15)$ K for the binary mixtures of methyl benzoate with tetrahydrofuran, 1,4-dioxane, anisole, and butyl vinyl ether over the entire range of composition. Using these measured values, excess volume V^E, deviation in viscosities $\Delta\eta$, excess Gibb's free energy of activation for viscous flow ΔG^{*E}, and deviation in isentropic compressibility Δk_s have been calculated. These calculated binary data have been fitted to Redlich-Kister equation to determine the appropriate coefficients. The values of excess volume V^E and deviation in viscosities $\Delta\eta$ are negative over the entire range of composition for all the binary systems at the studied temperatures. The behavior of these parameters with composition of the mixture has been discussed in terms of molecular interactions between the components of liquids.

1. Introduction

The excess or deviation properties of liquid-liquid mixtures are useful in several industrial applications owing to their influence upon the effectiveness of the operations. Nowadays, the determination of ultrasonic velocity and the related acoustical parameters derived from it has attracted the attention of several researchers. Much work has been done in solutions of polymers [1–3], pharmamaterials [4], electrolytes [5–7], and nonelectrolytes [8–10].

Tetrahydrofuran is a polar ($\mu = 1.75D$), aprotic solvent. It's main use is a precursor to polymers. It is also used as an industrial solvent for PVC and in varnishes. Dye carrier formulations based on methyl benzoate are useful in textile processing. Further methyl benzoate finds its primary uses as a perfume or flavouring agent and also used as a source of benzoyl radical. While ethers are excellent solvents, because they are relatively unreactive, yet they solvate a wide variety of compounds. Tetrahydrofuran, a cyclic ether which is one of the most polar simple ethers, is used as a solvent for polymers and also commercially used for the production of several compounds. Anisole, an aryl ether, is a major constituent of essential oil of anise seed.

In view of the abovementioned significances, these solvents are chosen with an aim to investigate the effect of the presence of two lone pairs of electrons on the oxygen atom of ether, when it is mixed with methyl benzoate. Moreover, it would be interesting to know the behavior of ether in an environment of aromatic ester molecules. The main objective of the present study is to determine the density ρ, viscosity η, and speed of sound u for the binary mixtures formed by the methyl benzoate with ether compounds.

In our earlier papers [11–13], we have reported some data on thermodynamic, transport, acoustical, and optical properties on mixtures of methyl benzoate with hydrocarbons and ketones and analysed the data in terms of molecular interactions. Several authors have also published density, viscosity, and speed of sound data [14–20] of binary mixtures of methyl benzoate with different types of organic solvents. In the present study, we report the results of density ρ, viscosity η, and speed of sound u for the binary mixtures of methyl benzoate with ethers, namely, tetrahydrofuran, 1,4-dioxane,

Table 1: Comparison of experimental density ρ and viscosity η of pure liquids with the literature values at (303.15, 308.15, and 313.15) K.

Liquid	T/K	ρ/g·cm^{-3} Exptl.	ρ/g·cm^{-3} Lit.	η/mPa·s Exptl.	η/mPa·s Lit.
Methyl benzoate	303.15	1.0785	1.0788 [18]	1.678	1.673 [28]
			1.0790 [22]		1.656 [18]
	308.15	1.0743	1.0740 [18]	1.517	1.510 [18]
			1.0741 [19]		
					1.504 [20]
			1.07399 [20]		1.510 [21]
	313.15	1.0696	1.0690 [27]	1.373	1.365 [27]
Tetrahydrofuran	303.15	0.8787	0.8771 [22]	0.439	
	308.15	0.8730	0.87214 [22]	0.429	
	313.15	0.8669	0.86719 [22]	0.390	
1,4-Dioxane	303.15	1.0227	1.02271 [23]	1.090	1.102 [23]
					1.095 [24]
	308.15	1.0178	1.0172 [24]	0.999	1.008 [24]
	313.15	1.0116	1.01132 [23]	0.946	0.946 [23]
Anisole	303.15	0.9853	0.984374 [25]	0.923	0.931 [25]
	308.15	0.9792	0.9788 [26]	0.849	0.849 [26]
	313.15	0.9728		0.764	
Butyl vinyl ether	303.15	0.7741		0.387	
	308.15	0.7682		0.365	
	313.15	0.7633		0.354	

anisole, and butyl vinyl ether measured at (303.15, 308.15, and 313.15) K over the entire mixture composition. To the best of our knowledge, such data on the abovementioned mixtures are not available in the earlier literature. Using the experimental data of ρ, η, and u, various parameters such as excess volume V^E, deviation in viscosity $\Delta\eta$, excess Gibb's free energy of activation for viscous flow ΔG^{*E}, and deviation in isentropic compressibility Δk_s were determined. These computed data are discussed to study the nature of behaviors between the components of the mixtures.

2. Experimental

2.1. Materials. Methyl benzoate (Fluka AG >0.996), tetrahydrofuran, 1,4-dioxane, anisole, and butyl vinyl ether (all Sigma-Aldrich, AR grade) of mass purity >0.997 were used without further purification. These chemicals were kept over molecular sieves for several days before use. The mass fraction purities as determined by gas chromatography (HP 8610) using FID were as follows: methyl benzoate (>0.996), tetrahydrofuran, (>0.996), 1,4-dioxane, (>0.998), anisole, (>0.998), and butyl vinyl ether (>0.997). The purity of the samples was further checked by comparing the measured density ρ and viscosity η with the literature values [18–28] as shown in Table 1.

Table 2: Values of density ρ, excess volume V^E, viscosity η, speeds of sound u, and isentropic compressibility k_s for the binary liquid mixtures.

x_1	ρ/g·cm^{-3}	V^E/cm^3·mol^{-1}	η/mPa·s	u/m·s^{-1}	k_s/Tpa^{-1}
Methyl benzoate (1) + tetrahydrofuran (2)					
T = 303.15 K					
0.0689	0.9020	−0.261	0.496	1260	698
0.1417	0.9246	−0.496	0.562	1272	669
0.2202	0.9468	−0.699	0.638	1284	641
0.3030	0.9680	−0.858	0.727	1296	615
0.3932	0.9885	−0.943	0.829	1308	591
0.4930	1.0084	−0.938	0.951	1320	569
0.6018	1.0272	−0.822	1.096	1332	549
0.7225	1.0452	−0.590	1.262	1344	530
0.8535	1.0623	−0.277	1.446	1356	512
T = 308.15 K					
0.0689	0.8964	−0.269	0.484	1232	735
0.1417	0.9192	−0.510	0.546	1236	712
0.2202	0.9413	−0.719	0.616	1244	686
0.3030	0.9627	−0.882	0.699	1256	659
0.3932	0.9833	−0.969	0.790	1268	633
0.4930	1.0033	−0.964	0.904	1284	605
0.6018	1.0222	−0.847	1.032	1300	679
0.7225	1.0403	−0.612	1.181	1316	555
0.8535	1.0575	−0.294	1.343	1336	530
T = 313.15 K					
0.0689	0.8905	−0.283	0.443	1216	759
0.1417	0.9134	−0.538	0.505	1220	736
0.2202	0.9358	−0.749	0.567	1228	709
0.3030	0.9572	−0.913	0.641	1240	679
0.3932	0.9779	−0.999	0.729	1252	652
0.4930	0.9980	−0.990	0.830	1268	623
0.6018	1.0171	−0.877	0.943	1284	596
0.7225	1.0354	−0.645	1.078	1300	572
0.8535	1.0527	−0.317	1.220	1320	545
Methyl benzoate (1) + 1,4-dioxane (2)					
T = 303.15 K					
0.0898	1.0307	−0.059	1.130	1328	550
0.1468	1.0355	−0.105	1.157	1336	541
0.2280	1.0421	−0.185	1.198	1344	531
0.3145	1.0485	−0.254	1.244	1352	521
0.4054	1.0543	−0.281	1.293	1360	513
0.5056	1.0599	−0.278	1.352	1368	504
0.6161	1.0653	−0.242	1.418	1372	499
0.7316	1.0701	−0.161	1.491	1372	496
0.8597	1.0749	−0.062	1.573	1372	494
T = 308.15 K					
0.0898	1.0261	−0.093	1.038	1324	556
0.1468	1.0311	−0.157	1.066	1332	547
0.2280	1.0377	−0.237	1.106	1340	537
0.3145	1.0440	−0.297	1.147	1348	530
0.4054	1.0499	−0.333	1.192	1352	521
0.5056	1.0555	−0.329	1.246	1356	515
0.6161	1.0608	−0.282	1.308	1360	510
0.7316	1.0655	−0.189	1.376	1360	507
0.8597	1.0702	−0.077	1.449	1360	505

Table 2: Continued.

x_1	ρ/g·cm^{-3}	V^E/cm^3·mol^{-1}	η/mPa·s	u/m·s^{-1}	k_s/Tpa^{-1}
		T = 313.15 K			
0.0898	1.0205	−0.132	0.984	1320	562
0.1468	1.0257	−0.207	1.006	1328	553
0.2280	1.0326	−0.304	1.040	1336	543
0.3145	1.0390	−0.363	1.075	1340	536
0.4054	1.0450	−0.398	1.110	1348	527
0.5056	1.0507	−0.392	1.157	1352	521
0.6161	1.0560	−0.331	1.208	1356	515
0.7316	1.0608	−0.234	1.263	1356	513
0.8597	1.0655	−0.106	1.324	1356	510
		Methyl benzoate (1) + Anisole (2)			
		T = 303.15 K			
0.0860	0.9950	−0.052	0.974	1388	522
0.1821	1.0056	−0.113	1.038	1380	522
0.2756	1.0157	−0.179	1.101	1372	523
0.3978	1.0282	−0.229	1.187	1368	520
0.4637	1.0347	−0.248	1.234	1368	516
0.5659	1.0442	−0.238	1.307	1368	512
0.6701	1.0534	−0.202	1.391	1368	507
0.7776	1.0624	−0.140	1.480	1368	503
0.8850	1.0710	−0.063	1.572	1368	499
		T = 308.15 K			
0.0860	0.9891	−0.062	0.894	1368	540
0.1821	1.0000	−0.145	0.949	1356	544
0.2756	1.0102	−0.209	1.007	1352	542
0.3978	1.0229	−0.264	1.086	1348	538
0.4637	1.0295	−0.285	1.129	1348	535
0.5659	1.0392	−0.282	1.195	1348	530
0.6701	1.0485	−0.242	1.272	1348	525
0.7776	1.0576	−0.175	1.355	1348	520
0.8850	1.0663	−0.093	1.439	1348	516
		T = 313.15 K			
0.0860	0.9830	−0.082	0.800	1356	553
0.1821	0.9941	−0.170	0.850	1344	557
0.2756	1.0045	−0.241	0.902	1340	554
0.3978	1.0174	−0.297	0.972	1336	551
0.4637	1.0241	−0.317	1.012	1336	547
0.5659	1.0340	−0.318	1.074	1336	541
0.6701	1.0435	−0.282	1.144	1336	537
0.7776	1.0528	−0.218	1.220	1336	532
0.8850	1.0616	−0.127	1.301	1336	528
		Methyl benzoate (1) + Butyl vinyl ether (2)			
		T = 303.15 K			
0.1026	0.8056	−0.108	0.437	1104	1019
0.2055	0.8375	−0.260	0.497	1128	938
0.3060	0.8690	−0.423	0.571	1152	867
0.4047	0.8999	−0.544	0.657	1176	804
0.5057	0.9313	−0.605	0.761	1204	741
0.6065	0.9623	−0.596	0.891	1232	685
0.7061	0.9926	−0.524	1.038	1260	635
0.8034	1.0218	−0.388	1.205	1292	586
0.9021	1.0511	−0.201	1.422	1328	540

Table 2: Continued.

x_1	ρ/g·cm^{-3}	V^E/cm^3·mol^{-1}	η/mPa·s	u/m·s^{-1}	k_s/Tpa^{-1}
		T = 308.15 K			
0.1026	0.8001	−0.141	0.421	1092	1048
0.2055	0.8322	−0.320	0.478	1116	965
0.3060	0.8637	−0.476	0.546	1136	891
0.4047	0.8947	−0.602	0.657	1160	825
0.5057	0.9261	−0.651	0.626	1188	760
0.6065	0.9572	−0.639	0.720	1216	702
0.7061	0.9875	−0.550	0.974	1248	650
0.8034	1.0168	−0.408	1.124	1280	600
0.9021	1.0462	−0.212	1.319	1316	552
		T = 313.15 K			
0.1026	0.7954	−0.182	0.400	1084	1070
0.2055	0.8275	−0.367	0.452	1108	984
0.3060	0.8591	−0.541	0.515	1132	908
0.4047	0.8901	−0.667	0.585	1156	841
0.5057	0.9215	−0.713	0.671	1184	774
0.6065	0.9526	−0.697	0.778	1212	715
0.7061	0.9829	−0.601	0.897	1244	657
0.8034	1.0121	−0.439	1.034	1276	607
0.9021	1.0414	−0.221	1.205	1308	561

2.2. Methods. Binary mixtures were prepared by mass in airtight ground stoppered bottles. Mass measurements accurate to ±0.01 mg were made on a digital electronic balance (Mettler AE 240, Switzerland). The resulting uncertainty in mole fraction was estimated to be less than ±0.0001. Each mixture was immediately used after it was well mixed by shaking. The densities of the pure and their binary mixtures were determined by using an Anton Paar density meter (DMA 4100). The uncertainty in the density measurements was found to be less than ±0.0004 g·cm^{-3}.

Viscosities were determined using an Ubbelohde viscometer with an uncertainty of ±0.008 mPa.s. The detailed method of measurement of viscosity has been reported earlier [11]. The speeds of sound were measured with a single crystal variable path interferometer (Mittal Enterprises, New Delhi, India) at an operating frequency of 2 MHz that had been calibrated with double distilled water and benzene. The uncertainty in speed of sound was estimated to be ±1 m·s^{-1}.

3. Results and Discussion

The experimental values of densities ρ, excess volumes V^E, viscosities η, speeds of sound u, and isentropic compressibility k_s of the binary mixtures of methyl benzoate with tetrahydrofuran, 1,4-dioxane, anisole, and butyl vinyl ether at temperatures 303.15, 308.15, and 313.15 K are listed in Table 2.

The excess volume V^E of the mixtures was deduced from the measured densities using the following relation:

$$V^E = \frac{(x_1M_1 + x_2M_2)}{\rho_{12}} - \left(\frac{x_1M_1}{\rho_1} + \frac{x_2M_2}{\rho_2}\right), \quad (1)$$

where x, M, and ρ are the mole fraction, molar mass, and density, respectively, of pure components 1 and 2. ρ_{12} is

TABLE 3: Derived parameters of excess functions by Redlich-Kister equation and standard deviations σ of binary liquid mixtures.

Function	T/K	a_0	a_1	a_2	a_3	σ
Methyl benzoate (1) + tetrahydrofuran (2)						
	303.15	−3.7342	1.2905	1.1561	—	0.003
V^E	308.15	−3.8390	1.2879	1.1173	—	0.003
	313.15	−3.9542	1.2921	1.9336	—	0.004
	303.15	−0.3703	0.0783	−0.0054	—	0.002
$\Delta\eta$	308.15	−0.2820	0.0762	−0.0017	—	0.002
	313.15	−0.2206	0.0052	−0.0013	—	0.001
	303.15	1303.56	9.2800	−47.1207	−173.82	2.5
ΔG^{*E}	308.15	1332.27	−4.3204	−21.9964	−216.76	3.3
	313.15	1442.68	−177.56	50.2305	−79.613	3.8
	303.15	−17.457	7.2762	−0.2584	1.4106	0.008
Δk_s	308.15	−11.729	−0.0066	3.5145	−0.134	0.093
	313.15	−11.726	0.5719	4.7747	1.1539	0.090
Methyl benzoate (1) + 1,4-dioxane (2)						
	303.15	−1.144	0.290	0.846	—	0.002
V^E	308.15	−1.326	0.425	0.753	—	0.002
	313.15	−1.568	0.552	0.575	—	0.003
	303.15	−0.122	0.018	−0.011	—	0.001
$\Delta\eta$	308.15	−0.096	0.007	0.009	—	0.001
	313.15	−0.062	0.019	0.045	—	0.001
	303.15	152.594	48.243	−4.7463	−34.315	0.7
ΔG^{*E}	308.15	191.571	23.232	37.394	−27.738	2.8
	313.15	201.888	29.749	100.81	−116.35	2.4
	303.15	−8.7386	1.8216	3.7965	−1.7238	0.094
Δk_s	308.15	−9.2275	2.7930	−0.9078	−0.8437	0.044
	313.15	−11.303	2.1793	−5.0991	−0.8657	0.048
Methyl benzoate (1) + anisole (2)						
	303.15	−0.9866	0.2637	0.5566	—	0.003
V^E	308.15	−1.1427	−0.0217	0.4368	—	0.003
	313.15	−1.2778	−0.0899	0.2005	—	0.002
	303.15	−0.1440	0.0055	0.0234	—	0.002
$\Delta\eta$	308.15	−0.1774	−0.0067	−0.0351	—	0.001
	313.15	−0.1602	−0.0069	−0.0270	—	0.002
	303.15	177.724	−46.913	99.649	70.211	2.2
ΔG^{*E}	308.15	116.959	−70.397	12.737	80.252	2.5
	313.15	56.636	−50.389	−50.388	63.520	1.2
	303.15	3.5076	−3.8057	0.9701	5.8130	0.036
Δk_s	308.15	4.7056	−4.2092	5.0184	4.5773	0.053
	313.15	5.3764	−4.7277	6.5667	2.5075	0.050
Methyl benzoate (1) + butyl vinyl ether (2)						
	303.15	−2.4161	−0.7060	1.0922	—	0.004
V^E	308.15	−2.6110	−0.5335	0.9973	—	0.004
	313.15	−2.8690	−0.4280	0.9846	—	0.005
	303.15	−1.0856	−0.3096	−0.1033	—	0.002
$\Delta\eta$	308.15	−0.9534	−0.2594	−0.0654	—	0.003
	313.15	−0.8261	−0.2221	−0.0264	—	0.003
	303.15	−663.33	174.93	−102.07	−27.064	5.8
ΔG^{*E}	308.15	−627.07	153.92	−43.740	−3.6493	7.2
	313.15	−596.84	106.49	131.85	−19.948	5.7
	303.15	−20.411	5.1677	5.1277	−1.1243	0.080
Δk_s	308.15	−22.018	5.1564	3.8155	−3.0159	0.085
	313.15	−25.224	3.8500	−3.2901	2.8527	0.072

the density of the liquid mixture. The Gibb's free energy of activation for viscous flow ΔG^{*E} was calculated from the viscosity data using the following relation:

$$\Delta G^{*E} = RT\left[\ln(\eta v) - \{x_1 \ln(\eta_1 v_1) + x_2 \ln(\eta_2 v_2)\}\right], \quad (2)$$

where $v = (x_1 M_1/\rho_1 + x_2 M_2)/\rho_{12}$ is the molar volume of the mixture and v_1 and v_2 are the molar volumes of the pure components.

The isentropic compressibilities k_s were calculated from the densities ρ and speeds of sound u using the Newton-Laplace equation:

$$k_s = \frac{1}{u^2 \rho}. \quad (3)$$

The deviation in viscosity $\Delta\eta$ and deviation in isentropic compressibility Δk_s were calculated using the general equation:

$$\Delta Y = Y_m - x_1 Y_1 - x_2 Y_2, \quad (4)$$

where ΔY is the deviation property in question, Y_m refers to the property of the mixture, and $x_1 Y_1$ and $x_2 Y_2$ refer to the mole fraction and specific property of the pure components 1 and 2, respectively. The results of these excess or deviation properties were fitted by the method of least squares to the Redlich-Kister [29] polynomial type equation:

$$Y = x_1 x_2 \sum_{j=1}^{p} a_{j-1} (x_1 - x_2)^{j-1}, \quad (5)$$

where Y is V^E, $\Delta\eta$, ΔG^{*E}, or Δk_s and x_1 and x_2 are the mole fractions of pure components 1 and 2, respectively. a_{j-1} is the polynomial coefficient, and p is the polynomial degree. The degree of (5) was optimised by applying the F-test. The correlated results are shown in Table 3 in which the tabulated standard deviation σ was calculated using the following relation:

$$\sigma = \left[\frac{\sum \left(y^E_{\text{exp}} - y^E_{\text{cal}}\right)^2}{(n - J)}\right]^{0.5}, \quad (6)$$

where n is the number of data points and j is the number of coefficients. The subscripts exp and cal denote experimental and calculated values, respectively.

The variations of excess volume V^E with mole fraction x_1 of methyl benzoate at the studied temperatures for the binary mixtures are displayed in Figure 1. It is observed that all the studied systems exhibit negative deviations. The V^E data for different components of the mixtures vary in the sequence:

tetrahydrofuran > butyl vinyl ether > 1,4-dioxane > anisole

The effect of temperature on V^E as displayed in Figure 1 is quite significant, as the negative V^E values increase with increase in temperature. The behavior of V^E can be explained

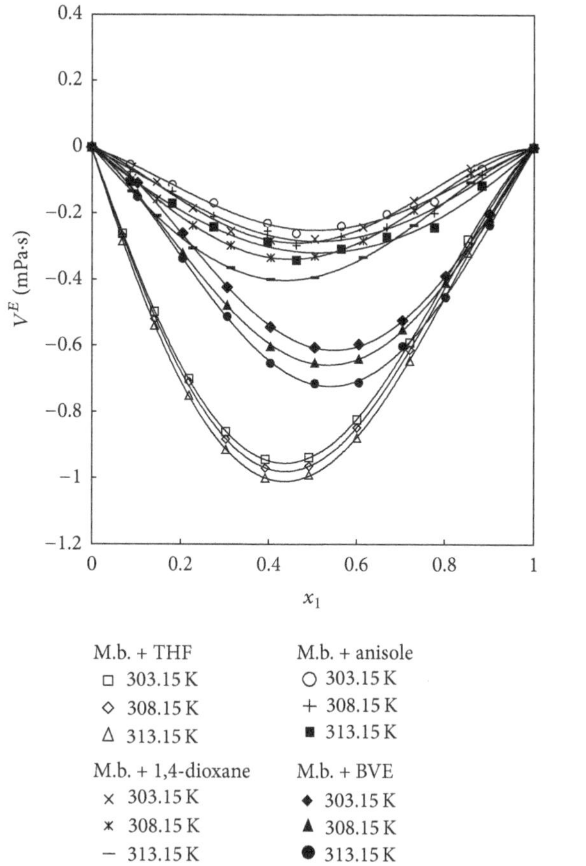

FIGURE 1: Curves of excess volumes (V^E) versus mole fraction for the binary mixtures of methyl benzoate + tetrahydrofuran at (□, 303.15; ◊, 308.15; Δ, 313.15) K, methyl benzoate + 1,4-dioxane at (×, 303.15; ∗, 308.15; —, 313.15) K, methyl benzoate + anisole at (O, 303.15; +, 308.15; ■, 313.15) K, and methyl benzoate + butyl vinyl ether at (⧫, 303.15; ▲, 308.15; •, 313.15) K.

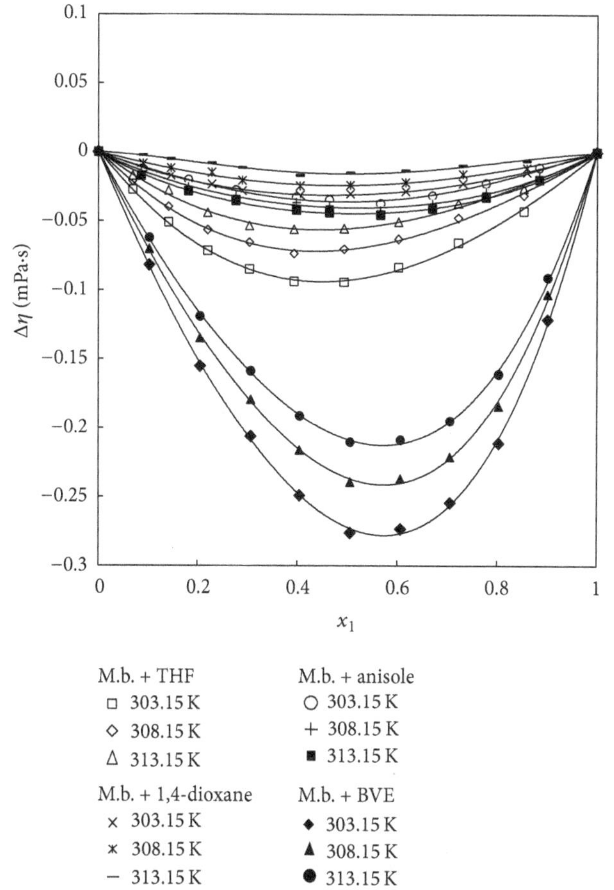

FIGURE 2: Curves of deviation in viscosity ($\Delta\eta$) versus mole fraction for the binary mixtures of methyl benzoate + tetrahydrofuran at (□, 303.15; ◊, 308.15; Δ, 313.15) K, methyl benzoate + 1,4-dioxane at (×, 303.15; ∗, 308.15; —, 313.15) K, and methyl benzoate + anisole at (O, 303.15; +, 308.15; ■, 313.15) K.

by considering the type of component molecules in pure state and in mixture. In the present case, the negative V^E values may be attributed to the differences in the dielectric constants of the components. The dielectric constants of methyl benzoate, tetrahydrofuran, 1,4-dioxane, anisole, and butyl vinyl ether are $\epsilon = 6.02, 7.261, 2.209, 4.33, 3.9$, respectively. Further, there is a possibility of electron donor-acceptor type or charge transfer interactions [30] between the highly electronegative oxygen of ether and the π-electron of ring of aromatic ester molecule resulting in negative V^E values. The less negative V^E values in case of methyl benzoate + anisole may be due to the presence of methyl ($-CH_3$) group in anisole. This observation suggests that with an increase in methyl group in mixture, the donor-acceptor interactions between the unlike molecules tend to decrease.

The results of $\Delta\eta$ as displayed in the Figure 2 are all negative over the entire range of composition and the magnitude values of $\Delta\eta$ vary according to the following sequence:

butyl vinyl ether > tetrahydrofuran > anisole > 1,4-dioxane

It is observed that negative values of $\Delta\eta$ decrease with increase in the temperature indicating the effect of temperature on $\Delta\eta$. Fort and Moore [31] observed that the negative $\Delta\eta$ values indicate the dispersion forces without involving the formation of any hetero-molecular complexes. A perusal of Figures 1 and 2 reveals that both V^E and $\Delta\eta$ exhibit the negative deviations over the entire range of composition; however, the magnitude values of the studied mixtures follow a different sequence. This type of behavior supports the observation of Rastogi et al. [32] and Kaulgud [33], suggesting that the strength of specific or dispersion forces is not the only factor influencing the $\Delta\eta$, but the molecular size and shape of the components are also equally important.

The dependence of ΔG^{*E} on the mole fraction is shown in Figure 3, where it was observed that ΔG^{*E} is positive for mixtures of methyl benzoate with tetrahydrofuran, 1,4-dioxane, and anisole, whereas for mixtures of methyl benzoate + butyl vinyl ether, the values of ΔG^{*E} are negative over the entire range of composition. The positive ΔG^{*E} values may be an indicative of specific interactions [31, 34, 35] and the negative ΔG^{*E} values indicate the presence of dispersive forces in these mixtures.

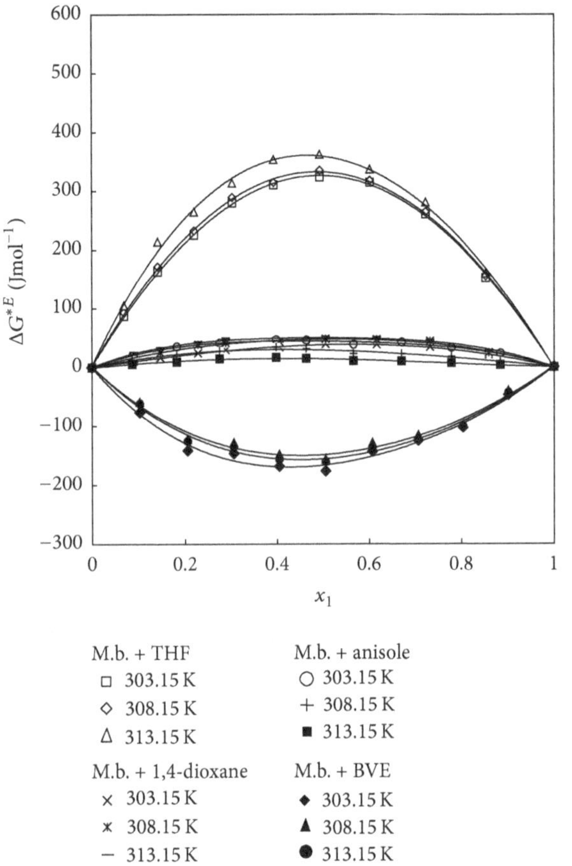

FIGURE 3: Curves of excess Gibb's free energy of activation of viscous flow (ΔG^{*E}) versus mole fraction for the binary mixtures of methyl benzoate + tetrahydrofuran at (□, 303.15; ◊, 308.15; Δ, 313.15) K, methyl benzoate + 1,4-dioxane at (×, 303.15; ∗, 308.15; —, 313.15) K, methyl benzoate + anisole at (O, 303.15; +, 308.15; ■, 313.15) K, and methyl benzoate + Butyl vinyl ether at (⧫, 303.15; ▲, 308.15; •, 313.15) K.

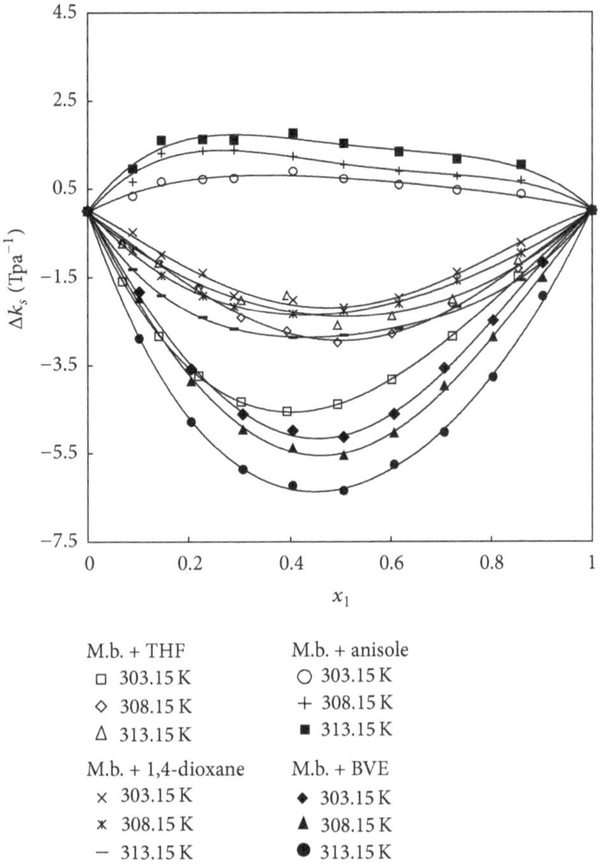

FIGURE 4: Curves of deviation in isentropic compressibility (Δk_s) versus mole fraction for the binary mixture of methyl benzoate + tetrahydrofuran at (□, 303.15; ◊, 308.15; Δ, 313.15) K, methyl benzoate + 1,4-dioxane at (×, 303.15; ∗, 308.15; —, 313.15) K, methyl benzoate + anisole at (O, 303.15; +, 308.15; ■, 313.15) K, and methyl benzoate + Butyl vinyl ether at (⧫, 303.15; ▲, 308.15; •, 313.15) K.

The plots of Δk_s, as displayed in Figure 4, show that for methyl benzoate with tetrahydrofuran, 1,4-dioxane, and butyl vinyl ether, the values of Δk_s exhibit negative deviation over the entire range of mixture composition; while for methyl benzoate + anisole, the values of Δk_s exhibit positive deviations. The positive Δk_s values are a sign of weak interaction between component molecules, which may be attributed to the mutual disruption in molecules associated with pure liquids. These positive Δk_s values are accompanied by a decrease in sound velocity over the entire range of composition of ester. The negative values of Δk_s may be attributed to the formation of weak bonds [36–38] by dipole-induced dipole interaction between unlike molecules and the geometrical fitting of component molecules in to each other's structure.

4. Conclusions

The densities ρ, excess volumes V^E, viscosities η, speeds of sound u, and data of binary mixtures methyl benzoate with tetrahydrofuran, 1,4-dioxane, anisole, and butyl vinyl ether have been reported at (303.15, 308.15 and 313.15) K. The excess molar volumes V^E, deviation in viscosity $\Delta\eta$, excess Gibb's free energies of activation for viscous flow ΔG^{*E}, and deviation in isentropic compressibility Δk_s were evaluated using the experimental data of ρ, η, and u and discussed in terms of the interactions between the components of the mixtures. Both V^E and $\Delta\eta$ exhibit negative deviations, while ΔG^{*E} and Δk_s exhibit both positive and negative deviations. It is concluded that the negative V^E may be attributed to the differences in the dielectric constants of the components and the possibility of the electron donor-acceptor type or charge transfer interactions between the electronegative oxygen of ether and the π-electrons of aromatic ester resulting into contraction in volumes.

Acknowledgments

The financial support from University Grants Commission, New Delhi, India, through Major Research Project (no. 38-24/2009 SR) to the corresponding author (M. V. Rathnam) is gratefully acknowledged. The authors also sincerely express their gratitude to the editor-in-chief and the reviewers for their valuable comments.

References

[1] J. N. Prassianakis, "Correlation of mechanical and acoustical properties of plasticized epoxy polymers," *Journal of Applied Polymer Science*, vol. 39, no. 10, pp. 2031–2041, 1990.

[2] V. Kannappan, S. Mahendran, P. Sathyamoorthy, and D. Roopsingh, "Ultrasonic method of determination of glass transition temperatures of certain thermotropic liquid crystalline copolyesters," *Journal of Polymer Materials*, vol. 18, no. 4, pp. 409–416, 2001.

[3] D. R. Godhani and P. H. Parsania, "Studies on acoustical properties of poly(4,4′-diphenylphthalidediphenyl-4,4′-disulfonate) at 30°C," *Journal of the Indian Chemical Society*, vol. 79, no. 7, pp. 620–622, 2002.

[4] M. Levina and M. H. Rubinstein, "The effect of ultrasonic vibration on the compaction characteristics of paracetamol," *Journal of Pharmaceutical Sciences*, vol. 89, no. 6, pp. 705–723, 2000.

[5] V. K. Syal, V. Bhalla, and S. Chauhan, "Ultrasonic studies of some tetraalkylammonium salts in acetonitrile+dioxane mixtures at 35°C," *Acustica*, vol. 81, no. 3, pp. 276–278, 1995.

[6] D. Geetha and P. S. Ramesh, "Ultrasonic studies on polymer blend (natural/synthetic) in strong electrolyte solutions," *Journal of Molecular Liquids*, vol. 136, no. 1-2, pp. 50–53, 2007.

[7] E. S. Balankina and A. K. Lyaschenko, "The acoustical and structural changes of aqueous solutions due to transition to molten supercooled salts," *Journal of Molecular Liquids*, vol. 101, no. 1–3, pp. 273–283, 2002.

[8] M. Gowrishankar, S. Sivarambabu, P. Venkateswarlu, and K. S. Kumar, "Excess volumes, speeds of sound, isentropic compressibilities and viscosities of binary mixtures of N-ethyl aniline with some aromatic ketones at 303.15 K," *Bulletin of the Korean Chemical Society*, vol. 33, no. 5, pp. 1686–1692, 2012.

[9] A. C. H. Chandrasekhar, S. K. RamanJaneyulu, and A. Krishnaiah, "Volumetric and ultrasonic behaviour of acetophenone and alcohol mixtures," *Physics and Chemistry of Liquids*, vol. 19, no. 3, pp. 171–179, 1989.

[10] M. Yasmin, K. P. Singh, S. Parveen, M. Gupta, and J. P. Shukla, "Thermoacoustical excess properties of binary liquid mixtures: a comparative experimental and theoretical study," *Acta Physica Polonica A*, vol. 115, no. 5, pp. 890–900, 2009.

[11] M. V. Rathnam, R. T. Sayed, K. R. Bhanushali, and M. S. S. Kumar, "Molecular interaction study of binary mixtures of methyl benzoate: viscometric and ultrasonic study," *Journal of Molecular Liquids*, vol. 166, pp. 9–16, 2012.

[12] M. V. Rathnam, S. Mankumare, and M. S. S. Kumar, "Density, viscosity, and speed of sound of (Methyl Benzoate + Cyclohexane), (Methyl Benzoate + n-Hexane), (Methyl Benzoate + Heptane), and (Methyl Benzoate + Octane) at temperatures of (303.15, 308.15, and 313.15) K," *Journal of Chemical and Engineering Data*, vol. 55, no. 3, pp. 1354–1358, 2010.

[13] M. V. Rathnam, S. Mohite, and M. S. Kumar, "Thermophysical properties of isoamyl acetate or methyl benzoate + hydrocarbon binary mixtures, at (303.15 and 313.15) K," *Journal of Chemical and Engineering Data*, vol. 54, no. 2, pp. 305–309, 2009.

[14] P. J. Raju, K. Rambabu, D. Ramchandran, M. Nagshwar Rao, and C. Rambabu, "Densities, adiabatic compressibility, freelength, viscosities and excess volumes of dimethyl sulfoxide with some aromatic esters at 303.15-318.15 K," *Physics and Chemistry of Liquids*, vol. 34, pp. 51–60, 1997.

[15] B. Garcia, R. Alcade, S. Aparicio, and J. M. Leal, "Volumetric properties, viscosities and refractive indices of binary mixed solvents containing methyl benzoate," *Physical Chemistry Chemical Physics*, vol. 4, pp. 5833–5840, 2002.

[16] Y. W. Sheu and C. H. Tu, "Densities, viscosities, refractive indices, and surface tensions for 12 flavor esters from T = 288.15 K to T = 358.15 K," *Journal of Chemical and Engineering Data*, vol. 50, no. 5, pp. 1706–1710, 2005.

[17] Y. W. Sheu and C. H. Tu, "Densities and viscosities of binary mixtures of ethyl acetoacetate, ethyl isovalerate, methyl benzoate, benzyl acetate, ethyl salicylate, and benzyl propionate with ethanol at T = (288.15, 298.15, 308.15, and 318.15) K," *Journal of Chemical and Engineering Data*, vol. 51, no. 2, pp. 545–553.

[18] T. M. Aminabhavi, S. K. Raikar, and R. H. Balundgi, "Densities, viscosities, refractive indices, and speeds of sound in methyl acetoacetate + methyl acetate, + ethyl acetate, + n-butyl acetate, + methyl benzoate, and + ethyl benzoate at 298.15, 303.15, and 308.15 K," *Journal Of Chemical And Engineering Data*, vol. 38, no. 3, pp. 441–445, 1993.

[19] A. M. Blanco, J. Ortega, B. Garcia, and J. M. Leal, "Studies on densities and viscosities of binary mixtures of alkyl benzoates in n-heptane," *Thermochimica Acta*, vol. 222, no. 1, pp. 127–136, 1993.

[20] B. Garcia, R. Alcade, S. Aparicio, and J. M. Leal, "Thermophysical behavior of methylbenzoate + n-alkanes mixed aolvents: application of cubic equations of state and viscosity models," *Industrial and Engineering Chemistry Research*, vol. 41, no. 17, pp. 4399–4408, 2002.

[21] T. M. Aminabhavi and S. K. Raikar, "Thermodynamic interactions in binary mixtures of 2-methoxyethanol with alkyl and aryl esters at 298. 15, 303. 15 and 308. 15 K," *Collection of Czechoslovak Chemical Communications*, vol. 58, pp. 1761–1776, 1993.

[22] A. K. Nain, "Densities and volumetric properties of binary mixtures of tetrahydrofuran with some aromatic hydrocarbons at temperatures from 278.15 to 318.15 K," *Journal of Solution Chemistry*, vol. 35, no. 10, pp. 1417–1439, 2006.

[23] A. G. Oskoei, N. Sataei, and J. Ghasemi, "Densities and viscosities for binary and ternary mixtures of 1, 4-dioxane + 1-hexanol + N,N-dimethylaniline from T = (283.15 to 343.15) K," *Journal of Chemical and Engineering Data*, vol. 53, no. 2, pp. 343–349, 2008.

[24] J. G. Baragi, M. I. Aralaguppi, T. M. Aminabhavi, M. Y. Kariduraganavar, and S. S. Kulkarni, "Density, viscosity, refractive index, and speed of sound for binary mixtures of 1,4-dioxane with different organic liquids at (298.15, 303.15, and 308.15) K," *Journal of Chemical and Engineering Data*, vol. 50, no. 3, pp. 917–923, 2005.

[25] J. A. Al-Kandary, A. S. Al-Jimaz, and A.-H. M. Abdul-Latif, "Viscosities, densities, and speeds of sound of binary mixtures of benzene, toluene, o-xylene, m-xylene, p-xylene, and mesitylene with anisole at (288.15, 293.15, 298.15, and 303.15) K," *Journal of Chemical and Engineering Data*, vol. 51, no. 6, pp. 2074–2082, 2006.

[26] J. N. Nayak, M. I. Aralaguppi, and T. M. Aminabhavi, "Density, viscosity, refractive index, and speed of sound in the binary mixtures of ethyl chloroacetate with aromatic liquids at 298.15, 303.15, and 308.15 K," *Journal of Chemical and Engineering Data*, vol. 47, no. 4, pp. 964–969, 2002.

[27] T. M. Aminabhavi, H. T. S. Phayde, R. S. Khinnavar, B. Gopalakrishna, and K. C. Hansen, "Densities, refractive indices, speeds of sound, and shear viscosities of diethylene glycol dimethyl ether with ethyl acetate, methyl benzoate, ethyl benzoate, and diethyl succinate in the temperature range from 298.15 to 318.15 K," *Journal of Chemical and Engineering Data*, vol. 39, no. 2, pp. 251–260, 1994.

[28] J. A. Reddick, W. B. Bunger, and T. K. Sakano, *Organic Solvents, Physical Properties and Method of Purification, Techniques of Chemistry, II*, Wiley-Interscience, New York, NY, USA, 1986.

[29] O. Redlich and A. T. Kister, "Algebraic representation of thermodynamic properties and the classification of solutions," *Industrial and Engineering Chemistry Research*, vol. 40, no. 2, pp. 345–348.

[30] C. Yang, P. Ma, and Q. Zhouy, "Excess molar volumes and viscosities of binary mixtures of sulfolane with benzene, toluene, ethylbenzene, *p*-Xylene, *o*-Xylene, and *m*-Xylene at 303.15 and 323.15 K and atmospheric pressure," *Journal of Chemical and Engineering Data*, vol. 49, pp. 881–885, 2004.

[31] R. J. Fort and W. R. Moore, "Viscosities of binary liquid mixtures," *Transactions of the Faraday Society*, vol. 62, pp. 1112–1119, 1966.

[32] R. P. Rastogi, J. Nath, and J. Misra, "Thermodynamics of weak interactions in liquid mixtures. I. Mixtures of carbon tetrachloride, benzene, toluene, and p-xylene," *Journal of Physical Chemistry*, vol. 71, no. 5, pp. 1277–1286, 1967.

[33] M. V. Kaulgud, "Ultrasonic velocity and compressibility in Binary liquid mixtures," *Zeitschrift Für Physikalische Chemie*, vol. 36, pp. 365–370, 1963.

[34] R. PalePu, J. Oliver, and B. Mackinnon, "Viscosities and densities of binary liquid mixtures of *m*-cresol with substituted anilines," *Canadian Journal of Chemistry*, vol. 63, no. 5, pp. 1024–1030, 1985.

[35] T. M. Reed III and T. E. Taylor, "Viscosities of liquid mixtures," *Journal of Physical Chemistry*, vol. 63, no. 1, pp. 58–67, 1959.

[36] R. Mehra and M. Pancholi, "Effect of temperature variation on acoustic and thermodynamic properties in the multicomponent system of heptane + 1-dodecanol + cyclohexane at 298, 308 and 318 K and suitability of different theories of sound speed," *Journal of the Indian Chemical Society*, vol. 83, no. 11, pp. 1149–1152, 2006.

[37] A. Ali, A. K. Nain, D. Chand, and B. Lal, "Molecular interactions in binary mixtures of anisole with benzyl chloride, chlorobenzene and nitrobenzene at 303.15 K: an ultrasonic, volumetric, viscometric and refractive index study," *Indian Journal of Chemistry A*, vol. 44, no. 3, pp. 511–515, 2005.

[38] R. Palani and K. Meenakshi, "Investigation of molecular interactions in ternary liquid mixtures using ultrasonic velocity," *Indian Journal of Chemistry A*, vol. 46, no. 2, pp. 252–257, 2007.

8

Batch Sorption Experiments: Langmuir and Freundlich Isotherm Studies for the Adsorption of Textile Metal Ions onto Teff Straw (*Eragrostis tef*) Agricultural Waste

Mulu Berhe Desta

Department of Chemistry, College of Natural and Computational Sciences, Mekelle University, P.O. Box 231, Mekelle, Ethiopia

Correspondence should be addressed to Mulu Berhe Desta; muller.desta@yahoo.com

Academic Editor: Angelo Lucia

Adsorption of heavy metals (Cr, Cd, Pb, Ni, and Cu) onto Activated Teff Straw (ATS) has been studied using batch-adsorption techniques. This study was carried out to examine the adsorption capacity of the low-cost adsorbent ATS for the removal of heavy metals from textile effluents. The influence of contact time, pH, Temperature, and adsorbent dose on the adsorption process was also studied. Results revealed that adsorption rate initially increased rapidly, and the optimal removal efficiency was reached within about 1 hour. Further increase in contact time did not show significant change in equilibrium concentration; that is, the adsorption phase reached equilibrium. The adsorption isotherms could be fitted well by the Langmuir model. The R_L value in the present investigation was less than one, indicating that the adsorption of the metal ion onto ATS is favorable. After treatment with ATS the levels of heavy metals were observed to decrease by 88% (Ni), 82.9% (Cd), 81.5% (Cu), 74.5% (Cr), and 68.9% (Pb). Results indicate that the freely abundant, locally available, low-cost adsorbent, Teff straw can be treated as economically viable for the removal of metal ions from textile effluents.

1. Introduction

Demands of clothing and apparel increase with the improving sense of fashion and lifestyle; thus textiles are manufactured to meet the growing demands. In developing countries such as Ethiopia, textile production becomes their source of income that contributes to their gross domestic product (GDP). However, this has brought both consequences to such countries either in a positive way which is an improvement of economy or in a negative way which led to an increased anthropogenic impact on the biosphere.

Heavy metals, particularly, Pb, Cr, Cd, and Cu are widely used for the production of color pigments of textile dyes. These heavy metals which have transferred to the environment are highly toxic and can bioaccumulate in the human body aquatic life and natural water bodies and also possibly get trapped in the soil [1]. These toxic heavy metals entering in the aquatic environment are adsorbed onto particulate matter, although they can form free metal ions and soluble complexes that are available for uptake by biological organisms [2].

Various methods of treating effluents containing heavy metals have been developed over years, such as chemical precipitation, chemical oxidation or reduction, electrochemical treatment, ion exchange, reverse osmosis, filtration, evaporation recovery, and electrocoagulation [3–11]. These methods have significant disadvantages, including high energy requirements, inefficient metal removal, generation of toxic sludge, and expensive equipment. Therefore, there is a need to develop an efficient, rapid, cost-effectively and environment friendly method for the removal of heavy metals from effluents. Adsorption of heavy metals is a new technology for treatment of textile effluents containing different types of heavy metals. Adsorption is the adhesion of a chemical substance (adsorbate) onto the surface of a solid (adsorbent). The most widely used adsorbent is activated carbon [12]. The adsorption process is being widely used by various researchers for the removal of heavy metals [13–18] from

(a) (b)

FIGURE 1: Teff straw before treatment (a) and after treatment (b).

waste streams, and activated carbon has been frequently used as an adsorbent.

Adsorption isotherm is an empirical relationship used to predict how much solute can be adsorbed by activated carbon [19]. Adsorption isotherm is defined as a graphical representation showing the relationship between the amount adsorbed by a unit weight of adsorbent (e.g., activated carbon) and the amount of adsorbate remaining in a test medium at equilibrium, and it shows the distribution of adsorbable solute between the liquid and solid phases at various equilibrium concentrations [20]. The three well-known isotherms are (a) Freundlich, (b) Langmuir, and (c) BET adsorption isotherm [19].

In this study a new, environment friendly, cost-effective, and locally available agrowaste *Eragrostis tef* adsorbent was investigated for the adsorption of heavy metals (Cr, Cd, Pb, Ni, and Cu) from textile effluents. The effects of different parameters such as the pH, the initial metal concentration, the dose of the adsorption, the contact time, and temperature were studied.

2. Materials and Methods

2.1. Description of the Study Area. The study was carried at the MAA Garment and Textile Factory built near Alula Abanega airport, due southeast of Mekelle city, Tigray, Northern Ethiopia. The study area lies between latitudes $13^\circ 28' 56''$N longitudes $39^\circ 31' 59''$E. The factory is owned by Kebire Enterprises Plc. and is fully equipped with state-of-art production facilities acquired from the most reputable manufacturers of the world with over 825 sewing machines. The industrial area is spread over 210,000 m^2 of land and it utilizes about 24,000 m^3/day fresh water. The effluent discharge amounts to about 13,050 m^3/day, that is, 54.4% of the total industrial effluents.

2.2. Adsorbent. Teff straw was used as biomass sample in this study for adsorption of heavy metals from textile effluents. Teff (*Eragrostis tef*) is obviously one of Ethiopia's most important cereal crops. It is indigenous to the country, and it is a fundamental part of the culture, tradition, and food. Recently, the taxonomy of Teff has been clarified by numerical taxonomy techniques, cytology, and biochemistry, including leaf flavonoids and seed protein electrophoretic patterns [21–23]. The Teff straw was provided by Mekelle University College of Agriculture and treated in the Physical Chemistry laboratory. Figure 1 shows some samples of Teff straw before treatment and after treatment.

2.3. Sorbent Preparation. Teff straw was washed several times with double-distilled water to remove all dirt in its original particle size followed by filtration and was dried at 100°C. The cleaned and dried Teff straw was oven-dried at 400°C for three hours to form carbonized Teff straw (CTS). The CTS was treated with 1 M sulfuric acid at room temperature. Acidified Teff Straw was dried overnight at 70°C and the dried Teff straw was heated to 120°C. The reacted product was washed repeatedly with distilled water. Finally the cleaned Teff straw was oven-dried overnight at 100°C and packed for utilization.

2.4. Sorption Studies

2.4.1. Batch Experiments and Adsorption Isotherms. Sorption studies were conducted by the batch technique and column test using wastewater from MAA Garment. Batch experiments were carried out to determine the adsorption isotherms of metal ions onto the adsorbents in 250 mL glass flask. The flasks were shaken at a constant rate, allowing sufficient time for adsorption equilibrium. It was assumed that the applied shaking speed allows all the surface area to come in contact with heavy metal ions over the course of the experiments. The study was performed at room temperature to be representative of environmentally relevant condition. All experiments were carried out in duplicate and the average value was used for further calculation. The pH of the solution was measured with a HACH-pH meter. The flasks were plugged and kept closed to avoid the fluctuation of pH due to the exchange of gases during the experiment. The effects of various parameters on the rate of adsorption process were observed by varying contact time, adsorbent concentration, temperature, and pH of the solution. The solution volume (V) was kept constant. The amount of metal adsorbed per unit mass is calculated as

$$Q_e = \frac{(C_i - C_e)V}{m}, \quad (1)$$

TABLE 1: Mean concentration (mg/L) with St. dev. of HMs in MAA Garment and Textile Factory effluents.

Sample	Parameters	Ni	Cu	Cd	Cr	Pb
Untreated	Mean ± St. dev.	3.53 ± 0.33	2.54 ± 0.74	1.23 ± 0.22	0.98 ± 0.07	0.12 ± 0.02
Treated	Mean ± St. dev.	0.43 ± 0.19	0.47 ± 0.12	0.02 ± 0.01	0.25 ± 0.05	0.04 ± 0.01
Percentage removed		88%	81.5%	82.9%	74.5%	68.9%

where C_i and C_e are the initial and equilibrium concentration (mg/L), m is the mass of the adsorbent (g) and V is the volume of the solution (mL). Percent metal ion removal (%MR) was calculated using the equation

$$\%\text{MR} = \frac{(C_i - C_e)}{C_i} * 100. \quad (2)$$

2.4.2. Column Test. The adsorption experiments were carried out in columns that were equipped with a stopper for controlling the column flow rate. This experiment is useful in understanding and predicting the behavior of the process. The sample solution was passed through the adsorption column with a known amount of ATS at a flow rate of 6 mL/min by gravity. The flow rate was kept constant by controlling the stopper valve.

The concentration of residual individual heavy metal in the sorption medium was determined using fully automated PC-controlled true double-beam AAS with fast sequential operation (Varian AA50 FS, Australia) for fast multielement flame AA determinations with features 4 lamp positions and automatic lamp selection. Statistical analysis was performed using the statistical software packages. Detection limits for each metal on the instrument are: Cr: 0.006 mg/Kg, Cd: 0.002 mg/Kg, Pb: 0.010 mg/Kg, Ni: 0.010 mg/Kg, and Cu: 0.003 mg/Kg.

3. Results and Discussion

3.1. Adsorption of the Tested Heavy Metals onto ATS. Concentrations of heavy metals of the factory's effluent are presented in Table 1. Nickel concentration in the wastewater sample constituted the major portion of the total metal ions determined (3.53) followed by Cu (2.54) and Cd (1.23), while Cr (0.98) and Pb (0.12) concentrations were the lowest. The concentration of nickel in untreated effluent was found to be 3.53 ± 0.33 and in treated sample was 0.43 ± 0.19; it was 88% less as compared to untreated effluent (Table 1). The level of copper was 2.54 ± 0.74 in untreated samples and 0.47 ± 0.12 in treated samples which means 81.5% less in treated sample. And the concentrations of cadmium before and after treatment were found to be 1.23 ± 0.22 and 0.02 ± 0.01 respectively, while the levels of the concentration of Pb, 0.12 ± 0.02 and 0.04 ± 0.01. The obtained results revealed that ATS has good adsorption capacity and is effective for the removal of heavy metals from textile effluent.

3.2. Effect of Adsorbent Dose on Adsorption. The effect of adsorbent dose on the adsorption of the heavy metals onto ATS was investigated. The concentration of heavy metals in

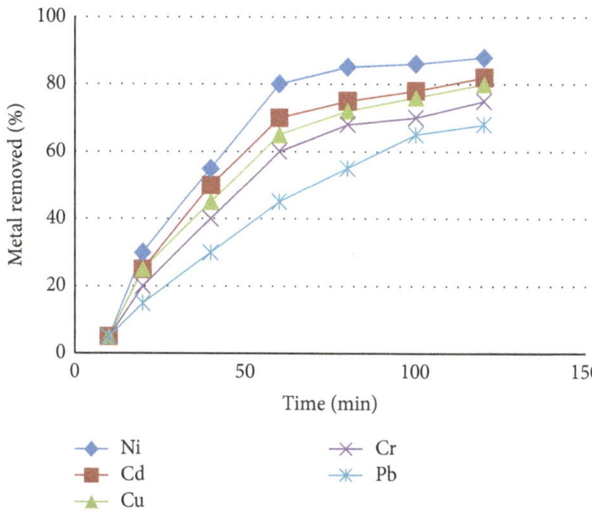

FIGURE 2: Effect of contact time on adsorption efficiency.

the textile effluent was reduced by 88% (Ni), 82.9% (Cd), 81.5% (Cu), 74.5% (Cr), and 68.9% (Pb) after treatment. Metals adsorption efficiency was increased with increase in adsorbent dose. This revealed that the adsorption sites remain unsaturated during the adsorption reaction whereas the number of sites available for adsorption site increases by increasing the adsorbent dose.

3.3. Effect of Contact Time on Adsorption. The effect of contact time on the adsorption of metal was investigated. Figure 2 shows the percent metal ion removal with respect to time. Contact time is one of the most effective factors in batch adsorption process. Adsorption rate initially increased rapidly, and the optimal removal efficiency was reached within about 60 min. Further increase in contact time did not show significant change in equilibrium concentration; that is, the adsorption phase reached equilibrium.

3.4. Effect of Temperature on Adsorption. Effect of temperature on adsorption of the heavy metals onto ATS was studied by conducting different sets of experiments at different temperatures, that is, 298, 308, 318, 328, and 343°K. It was observed that adsorption of metal ions increases with the increase in the temperature.

3.5. Effect of pH on Adsorption. The pH of a solution is an important parameter in the adsorption process. During this study, results revealed that the removal of metal ions was strongly dependent on the pH of the solution. The effect of pH on adsorption of metal ions onto ATS was studied at

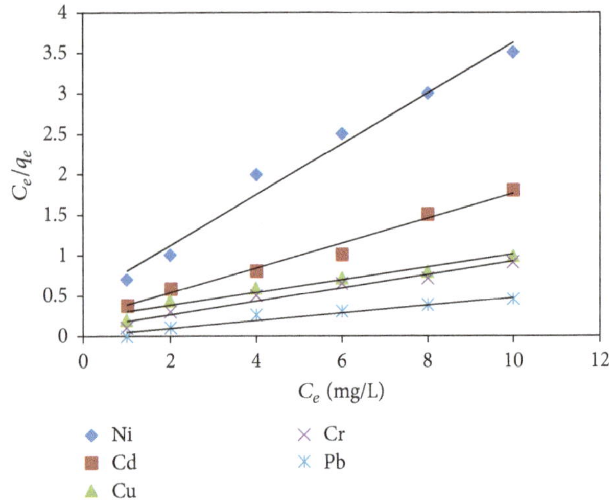

FIGURE 3: Langmuir isotherm showing the variation of adsorption (C_e/q_e) against the equilibrium concentration (C_e) for adsorption of metal ions onto ATS.

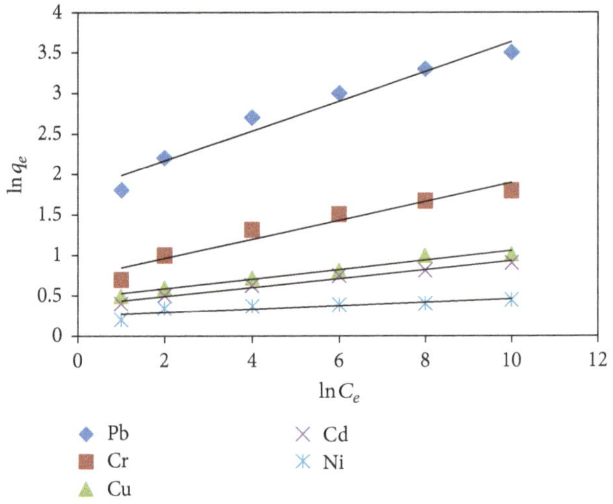

FIGURE 4: Freundlich isotherm representing variation of $\ln(q_e)$ with respect to $\ln(C_e)$ for adsorption of metal ions onto ATS.

TABLE 2: Isotherm parameters for adsorption of metal ions onto ATS.

Isotherm model	Metal ion	Estimated isotherm parameters		
		q_{max} (mg/g)	b (L/mg)	R^2
Langmuir	Ni	41.2	0.189	0.998
	Cu	34	0.173	0.982
	Cd	27.2	1.908	0.940
	Cr	21.3	0.143	0.962
	Pb	17.5	0.114	0.945
	Metal ion	n	K_f	R^2
Freundlich	Ni	1.10	1.24	0.748
	Cd	2.11	3.52	0.977
	Cu	1.53	2.01	0.967
	Cr	1.50	1.77	0.939
	Pb	1.51	1.82	0.957

pH 1.0–11.0, and the maximum removal capacity of ATS was found to be at pH 6.5. Percent metal ions removal increased rapidly with the increase in pH of the solution initially, and the optimal pH was observed at pH 6.5. Further increase in pH causes a drastic decrease in the adsorption percentage. This might be due to the weakening of electrostatic force of attraction between the oppositely charged adsorbate and adsorbent which ultimately leads to the reduction in sorption capacity [24].

3.6. Adsorption Isotherm. Adsorption is usually described through isotherms, that is, functions which connect the amount of adsorbate on the adsorbent. Distribution of metal ions between the liquid phase and the solid phase can be described by several isotherm models such as Langmuir and Freundlich. The Langmuir isotherm assumes monolayer adsorption onto a surface containing a finite number of adsorption sites of uniform strategies with no transmigration of adsorbate in the plane surface [25]. Once a site is filled, no further sorption can take place at that site. This indicates that the surface reaches a saturation point where the maximum adsorption of the surface will be achieved. The isotherm is represented by

$$\frac{C_e}{q_e} = \frac{1}{bq_{\max}} + \frac{C_e}{q_{\max}}. \quad (3)$$

The linear plot of specific adsorption (C_e/q_e) against the equilibrium concentration (C_e) (Figure 3) shows that the adsorption obeys the Langmuir model. The constants b and $q_{\max}$ relate to the energy of adsorption and maximum adsorption capacity, and their values are obtained from the slope and interception of the plot and are presented in Table 2.

The Freundlich isotherm is introduced as an empirical model, where q_e represents the amount adsorbed per amount of adsorbent at the equilibrium (mg/g), C_e represents the equilibrium concentration (mg/L), and K_f and n are parameters that depend on the adsorbate and adsorbent.

Consider

$$q_e = K_f C_e^{1/n}. \quad (4)$$

The equation can be linearized and the temperature dependent constants K_f and $1/n$ found by linear regression:

$$\ln q_e = \ln K_f - \frac{1}{n}\ln C_e, \quad (5)$$

where K_f and n are Freundlich constants which correspond to adsorption capacity and adsorption intensity, respectively. Freundlich equilibrium constants were determined from the plot of $\ln q_e$ versus $\ln C_e$ from Figure 4, on the basis of the linear form of Freundlich equation. The n value indicates the degree of nonlinearity between solution concentration and adsorption as follows: if $n = 1$, then adsorption is linear; if $n < 1$, then adsorption is a chemical process; if $n > 1$, then adsorption is a physical process. The n value in Freundlich equation was found to be 1.10–2.11, (Table 2). The situation $n > 1$ is most common and may be due to a distribution of surface sites or any factor that causes a decrease in adsorbent-adsorbate interaction with increasing surface density [26] and the values of n within the range of 1–10 represent good adsorption [27, 28]. In the present study, since n lies between 1 and 10 it indicates the physical adsorption of metal ions onto ATS.

In both cases, linear plots were obtained, which reveal the applicability of these isotherms on the ongoing adsorption process. Figures 3 and 4 exhibit Langmuir and Freundlich plots, respectively, for the adsorption of metal ions onto ATS, and different Langmuir and Freundlich constants derived from these plots are presented in Table 2. Nearly similar results have been reported in the literature [29–31] for Langmuir and Freundlich models.

Langmuir and Freundlich adsorption constants and correlation coefficients (R^2) are presented in Table 2. To find the most appropriate model for the metal ions adsorption; data were fitted to Langmuir and Freundlich isotherm models. Results revealed that Langmuir adsorption isotherm was the best model for the metal ions adsorption onto ATS with R^2 of 0.998. The essential features of Langmuir adsorption isotherm parameter can be used to predict the affinity between the sorbate and sorbent using a dimensionless constant called separation factor or equilibrium parameter (R_L), which is expressed by the following relationship [32, 33]:

$$R_L = \frac{1}{(1 + bC_i)}, \quad (6)$$

where b is the Langmuir constant and C_i is the initial concentration. The value of R_L indicated the type of Langmuir isotherm to be irreversible ($R_L = 0$), linear ($R_L = 1$), unfavorable ($R_L > 1$), or favorable ($0 < R_L < 1$) [34]. The R_L values between 0 and 1 indicate favorable adsorption. The R_L value in the present investigation was found to be 0.298–0.986, indicating that the adsorption of the metal ion onto ATS is favorable.

4. Conclusions

Teff (*Eragrostis tef*) straw is environment friendly, cost-effective, and locally available agrowaste adsorbent for the adsorption of metal ions from textile effluents. Results from this study showed that adsorption of heavy metals onto ATS increases with the amount of adsorbent. This may be explained because of the fact that adsorption is a surface phenomenon where adsorbate molecules occupy specific sites on the adsorbent. These sites are commonly known as active centers. The concentration of these active centers on the surface is further related to the pore size and pore volume available after impregnation. This explains why increasing the quantity of adsorbent results in increased adsorption. Results revealed that initially the % removal increase, rapidly; however, after some times the rate becomes almost constant. This is because all the available active centers on the adsorbent have been occupied and there are no further sites and hence no further adsorption is possible. The time when this phenomenon occurs, therefore, may be termed optimum time.

The adsorption isotherms could well be fitted by the Langmuir model. The R_L value in the present investigation was less than one, indicating that the adsorption of the metal ion onto ATS is favorable. In treatment with ATS, the removal of metal ions from textile wastewater sample is high whereby the percentage of removal is more than 80 percent of the initial concentration. Since Teff Straw is freely abundant, locally available, low-cost adsorbent and has a considerable high adsorption capacity, it may be treated as economically viable for removal of metal ions from textile effluents.

Acknowledgments

This study was done in the Department of Chemistry, Mekelle University, Ethiopia. The author is grateful to Dr. Abraha G/Kidan for providing column adsorption apparatus. The author would like to thank Mr. Mehari (the factory manager), Mr. Amanuel (textile engineer), and Mr. Mesfin (treatment plant supervisor) for providing the necessary facilities to carry out this work.

References

[1] N. Mathur, P. Bhatnagar, and P. Bakre, "Assessing mutagenicity of textile dyes from pali (Rajasthan) using ames bioassay," *Applied Ecology and Environmental Research*, vol. 4, no. 1, pp. 111–118, 2006.

[2] P. U. Singare, R. S. Lokhande, and K. U. Naik, "A case study of some lakes located at and around thane city of Maharashtra, India, with special reference to physico-chemical properties and heavy metal content of lake water," *Interdisciplinary Environmental Review*, vol. 11, no. 1, pp. 90–107, 2010.

[3] J. O. Esalah, M. E. Weber, and J. H. Vera, "Removal of lead, cadmium and zinc from aqueous solutions by precipitation with sodium di-(n-octyl) phosphinate," *Canadian Journal of Chemical Engineering*, vol. 78, no. 5, pp. 948–954, 2000.

[4] A. I. Zouboulis, K. A. Matis, B. G. Lanara, and C. L. Neskovic, "Removal of cadmium from dilute solutions by hydroxy apatite. II. floatation studies," *Separation Science and Technology*, vol. 32, no. 10, pp. 1755–1767, 1997.

[5] A. I. Zouboulis, K. A. Matis, B. G. Lanara, and C. L. Neskovic, "Removal of cadmium from dilute solutions by hydroxy apatite. II: floatation studies," in *The Protocols*, K. R. Fall and W. R. Stevens, Eds., vol. 1 of *TCP/IP Illustrated*, Addison-Wesley, Reading, Mass, USA, 2nd edition, 2011.

[6] L. Canet, M. Ilpide, and P. Seta, "Efficient facilitated transport of lead, cadmium, zinc, and silver across a flat-sheet-supported liquid membrane mediated by lasalocid A," *Separation Science and Technology*, vol. 37, no. 8, pp. 1851–1860, 2002.

[7] K. Dermentiz, A. Christoforidis, E. Valsamidou, A. Loucas, and K. Greece, "Removal of nickel, copper, zinc and chromium from synthetic and industrial wastewater by electrocoagulation," *International Journal of Environmental Sciences*, vol. 1, no. 5, pp. 697–710, 2011.

[8] V. J. Inglezakis, M. D. Loizidou, and H. P. Grigoropoulou, "Ion exchange of Pb^{2+}, Cu^{2+}, Fe^{3+}, and Cr^{3+} on natural clinoptilolite: selectivity determination and influence of acidity on metal uptake," *Journal of Colloid and Interface Science*, vol. 261, no. 1, pp. 49–54, 2003.

[9] P. T. Bolger and D. C. Szlag, "Electrochemical treatment and reuse of nickel plating rinse waters," *Environmental Progress*, vol. 21, no. 3, pp. 203–208, 2002.

[10] K. Dermentzis, "Removal of nickel from electroplating rinse waters using electrostatic shielding electrodialysis/electrodeionization," *Journal of Hazardous Materials*, vol. 173, no. 1–3, pp. 647–652, 2010.

[11] M. A. Amer, F. I. Khaili, and A. M. Awwad, "Adsorption of lead, zinc and cadmium ions on polyphosphate-modified kaolinite clay," *Journal of Environmental Chemistry and Ecotoxicology*, vol. 2, no. 1, pp. 1–8, 2010.

[12] N. O. Reuben and J. A. Miebaka, "Chromium (VI) adsorption rate in the treatment of liquid phase oil based drill cuttings," *African Journal of Environmental Science and Technology*, vol. 2, no. 4, pp. 68–674, 2008.

[13] R. Ahmed, T. Yamin, M. S. Ansari, and S. M. Hasany, "Sorption behaviour of lead(II) ions from aqueous solution onto Haro river sand," *Adsorption Science and Technology*, vol. 24, no. 6, pp. 475–486, 2006.

[14] D. F. Aloko and E. A. Afolabi, "Titanium dioxide as a cathode material in a dry cell," *Leonardo Electronics Journal of Practices and Technologies*, vol. 11, pp. 97–108, 2007.

[15] D. F. Aloko and E. A. Afolabi, "Model development of the adsorption of cations on manganese dioxide (MnO_2) used in a Leclanche dry cell," *Leonardo Journal of Sciences*, vol. 8, pp. 13–20, 2006.

[16] C. C. Chen and Y. C. Chung, "Arsenic removal using a biopolymer chitosan sorbent," *Journal of Environmental Science and Health - Part A Toxic/Hazardous Substances and Environmental Engineering*, vol. 41, no. 4, pp. 645–658, 2006.

[17] N. Amin, S. Kaneco, T. Kitagawa et al., "Removal of arsenic in aqueous solutions by adsorption onto waste rice husk," *Industrial and Engineering Chemistry Research*, vol. 45, no. 24, pp. 8105–8110, 2006.

[18] A. M. M. Khaled, *A Comparative study for distribution of some heavy metals in aquatic organisms fished from Alexandria region [Ph.D. thesis]*, Chemistry Department, Faculty of Science, Alexandria University, Alexandria, Egypt, 1998.

[19] K. Steve, T. Erika, T. Reynold, and M. Paul, "Activated carbon: a unit operations and processes of activated carbon," in *Environmental Engineering*, vol. 25, pp. 350–749, PWS Publishing, 2nd edition, 1998.

[20] C. Ng, J. N. Losso, W. E. Marshall, and R. M. Rao, "Freundlich adsorption isotherms of agricultural by-product-based powdered activated carbons in a geosmin-water system," *Bioresource Technology*, vol. 85, no. 2, pp. 131–135, 2002.

[21] B. M. G. Jones, J. Ponti, A. Tavassoli, and P. A. Dixon, "Relationships of the ethiopian cereal t'ef (*Eragrostis tef* (Zucc.) Trotter): evidence from morphology and chromosome number," *Annals of Botany*, vol. 42, no. 6, pp. 1369–1373, 1978.

[22] S. H. Costanza, J. M. J. Dewet, and J. R. Harlan, "Literature review and numerical taxonomy of *Eragrostis tef* (T'ef)," *Economic Botany*, vol. 33, no. 4, pp. 413–424, 1979.

[23] E. Bekele and R. N. Lester, "Biochemical assessment of the relationships of *Eragrostis tef* (Zucc.) trotter with some wild Eragrostis species (Gramineae)," *Annals of Botany*, vol. 48, no. 5, pp. 717–725, 1981.

[24] S. S. Baral, S. N. Das, and P. Rath, "Hexavalent chromium removal from aqueous solution by adsorption on treated sawdust," *Biochemical Engineering Journal*, vol. 31, no. 3, pp. 216–222, 2006.

[25] B. H. Hameed, A. T. M. Din, and A. L. Ahmad, "Adsorption of methylene blue onto bamboo-based activated carbon: kinetics and equilibrium studies," *Journal of Hazardous Materials*, vol. 141, no. 3, pp. 819–825, 2007.

[26] B. E. Reed and M. R. Matsumoto, "Modeling cadmium adsorption by activated carbon using the Langmuir and Freundlich isotherm expressions," *Separation Science and Technology*, vol. 28, no. 13-14, pp. 2179–2195, 1993.

[27] G. McKay, M. S. Otterburn, and A. G. Sweeney, "The removal of colour from effluent using various adsorbents. III. Silica: rate processes," *Water Research*, vol. 14, no. 1, pp. 15–20, 1980.

[28] A. Özer and H. B. Pirinççi, "The adsorption of Cd(II) ions on sulphuric acid-treated wheat bran," *Journal of Hazardous Materials*, vol. 137, no. 2, pp. 849–855, 2006.

[29] A. Mittal, L. Kurup, and J. Mittal, "Freundlich and Langmuir adsorption isotherms and kinetics for the removal of Tartrazine from aqueous solutions using hen feathers," *Journal of Hazardous Materials*, vol. 146, no. 1-2, pp. 243–248, 2007.

[30] M. Malakootian, J. Nouri, and H. Hossaini, "Removal of heavy metals from paint industry's wastewater using Leca as an available adsorbent," *International Journal of Environmental Science and Technology*, vol. 6, no. 2, pp. 183–190, 2009.

[31] H. Liu, Y. Dong, H. Wang, and Y. Liu, "Adsorption behavior of ammonium by a bioadsorbent—Boston ivy leaf powder," *Journal of Environmental Sciences*, vol. 22, no. 10, pp. 1513–1518, 2010.

[32] K. R. Hall, L. C. Eagleton, A. Acrivos, and T. Vermeulen, "Pore- and solid-diffusion kinetics in fixed-bed adsorption under constant-pattern conditions," *Industrial and Engineering Chemistry Fundamentals*, vol. 5, no. 2, pp. 212–223, 1966.

[33] P. K. Malik, "Dye removal from wastewater using activated carbon developed from sawdust: adsorption equilibrium and kinetics," *Journal of Hazardous Materials*, vol. 113, no. 1–3, pp. 81–88, 2004.

[34] G. McKay, H. S. Blair, and J. R. Gardner, "Adsorption of dyes on chitin," *Journal of Applied Polymer Science*, vol. 27, no. 8, pp. 3043–3057, 1982.

9

Computation of Isobaric Vapor-Liquid Equilibrium Data for Binary and Ternary Mixtures of Methanol, Water, and Ethanoic Acid from T, p, x, and H_m^E Measurements

Daming Gao,[1,2] Hui Zhang,[1,2] Peter Lücking,[2,3] Hong Sun,[1,2] Jingyu Si,[1] Dechun Zhu,[1,2] Hong Chen,[1] and Jianjun Shi[1,2]

[1] *Department of Chemistry and Materials Engineering, Hefei University, Anhui, Hefei 230022, China*
[2] *Sino-German Research Center for Process Engineering and Energy Technology, Anhui, Hefei 230022, China*
[3] *Department of Engineering, Jade University of Applied Science, 26389 Wilhelmshaven, Germany*

Correspondence should be addressed to Daming Gao, dmgao@hfuu.edu.cn

Academic Editor: Ahmet Z. Sahin

Vapor-liquid equilibrium (VLE) data for the strongly associated ternary system methanol + water + ethanoic acid and the three constituent binary systems have been determined by the total pressure-temperature-liquid-phase composition-molar excess enthalpy of mixing of the liquid phase (p, T, x, H_m^E) for the binary systems using a novel pump ebulliometer at 101.325 kPa. The vapor-phase compositions of these binary systems had been calculated from Tpx and H_m^E based on the Q function of molar excess Gibbs energy through an indirect method. Moreover, the experimental T, x data are used to estimate nonrandom two-liquid (NRTL), Wilson, Margules, and van Laar model parameters, and these parameters in turn are used to calculate vapor-phase compositions. The activity coefficients of the solution were correlated with NRTL, Wilson, Margules, and van Laar models through fitting by least-squares method. The VLE data of the ternary system were well predicted from these binary interaction parameters of NRTL, Wilson, Margules, and van Laar model parameters without any additional adjustment to build the thermodynamic model of VLE for the ternary system and obtain the vapor-phase compositions and the calculated bubble points.

1. Introduction

New strategies for the correlation and accurate prediction of the vapor-liquid equilibrium (VLE) data play a vital role in the distillation and separation process in chemical industry. The common technique for obtaining VLE data is by direct measurement on the system. That is to say, when the VLE is established, and phases are sampled and analyzed. Consequently, the experimental technique must be rather highly accurate to ensure meaningful results in the operation of equilibrium stills. Actually, when the vapor-phase components are sampled and analyzed, the whole compositions of components in solution and vapor have been changed. Accordingly, the behavior of the systems has been changed with the amount of compositions. Moreover, it has been long realized that the analysis of vapor-liquid composition for the infinite dilute solution is very difficult. In addition, for VLE measurements of mixtures containing a highly volatile compound, the accurate measurement of the vapor-phase composition can be difficult. This fact, coupled with the necessity for much analytical work, tends to enhance interest in exploring new methods for the determination of equilibrium data that do not involve sampling and analysis of the vapor-phase components.

Several methods have been explored for the calculation of component behavior from gross solution behavior. Van Ness and coworkers have suggested the classification of these methods into two categories, direct and indirect methods [1]. The direct methods involve calculation of vapor compositions by integration of the coexistence equation, a first-order differential equation derived from the Gibbs-Duhem equation relating phase compositions at equilibrium. Hala and coworkers have given a detailed discussion of the basic direct method [2], and Van Ness and coworkers have discussed techniques for handling nonconstant temperature or pressure conditions as well as nonideal vapor phase behavior [1, 3]. Moreover, in the total pressure method one can calculate y from T, p, x measurements using an indirect method discussed by Mixon et al. [4].

Of all the methods, the indirect methods involve first the measurement, by some appropriate means, the temperature and total pressure of the system, the liquid-phase compositions, and subsequent calculation of vapor-phase compositions there from. These methods usually involve ascertaining which of selected solution equations to the Gibbs-Duhem equation lead to the best fit to the experimental data, and of the determination of the parametric values producing the best fit. For example, Barker has developed a procedure based on the assumption that the excess free energy can be represented as a polynomial function of composition [5].

There is a basic difference in the degree of rigor associated with the direct method and the indirect method of Barker. In the former, one makes no assumptions about the solution behavior for the nature of molecular interactions. Solution behavior is determined directly from the experimental data. The Barker method necessitates the assumption of a particular model and the estimation of its parameters, this deficiency in the method of Barker has been recognized by Tao, who has presented another indirect method in which the necessity for the a priori assumption of a particular functional form for the excess free energy has been removed. Tao's procedure involves calculation of the activity coefficient essentially by integration of an equation resembling the coexistence equation. His procedure, though indirect, retains the rigor usually associated with the direct method [6]. The method of Tao appears specific to binary systems and does not seem to be easily generalized. However, an indirect method such as that of Barker is readily and easily generalized to ternary and higher-ordered systems, but this method retains the disadvantage of lacking rigor as compared with the direct method.

This paper presents the vapor-liquid compositions calculated based on the measurements of VLE data for temperature-total pressure-liquid-phase composition-molar excess enthalpy energy of mixing of the liquid phase (T, p, x, H_m^E) at 101.325 kPa according to the Q function of molar excess Gibbs energy through an indirect method. We know that the reaction of methanol carboxylated with carbon monoxide is the most common and important technology for synthesis of ethanoic acid in the chemical industrial process. Modeling the thermodynamic properties and correlating and predicting the phase equilibria of a mixture involving associating components forming hydrogen bonding such as carboxylic acid remain a challenging problem, since such systems show extremely nonideal behavior and the formation of monomer, dimer, and even trimer in vapor and liquid phase. In addition, for the VLE measurements of vapor-phase components containing a highly volatile compound such as carboxylic acid, the accurate measurement of the vapor-phase composition can be difficult. Many attempts have been made to describe the vapor-liquid equilibria of carboxylic acid containing mixtures using the concept of multiscale association [7]. Arlt reported the isothermal VLE data of a new apparatus for phase equilibria in reaction mixtures containing water with ethanoic acid and propanoic acid at (333 to 363) K [8]. Xu and Chuang have developed a new correlation for the prediction of the vapor-liquid equilibrium of methyl acetate-methanol-water-ethanoic acid mixtures [9]. Sawistowski and Pilavakis explored the vapor-liquid equilibrium behavior of the quaternary system methyl acetate-methanol-water-ethanoic acid modeled by using the Margules equation in combination with Marek's method for the association of ethanoic acid [10]. Moreover, Guan et al. investigated that the isobaric vapor-liquid equilibria for water + ethanoic acid + n-pentyl acetate, isopropyl acetate, N-methyl pyrrolidone, or N-methyl acetamide were correlated and predicted by both the nonrandom two-liquid (NRTL) and universal quasichemical activity coefficient (UNIQUAC) models used in combination with the Hayden-O'Connell (HOC) method [11–14]. In our recent work, we have concluded that the VLE data for the associating systems containing the carboxylic acid system can be correlated and predicted [15, 16]. Although the VLE data of the mixture containing the associating systems were previously reported by the different research groups [17–21], respectively, the VLE data for the associating binary system containing the carboxylic acid have been still extensively studied because of the extensive association effects occurring in them and the difficulty of properly calculating the activity coefficients [22–25]. Nominally, the system is binary but in practice, it is multicomponent. Because the carboxylic acid monomer undergoes partial dimerization and even higher polymerization. This association is attributed to the formation of hydrogen bonds and occurs in both the vapor and liquid phase. Therefore, the challenge for the VLE data of the associating systems has evoked more and more researchers to focus on new strategies for exploring them. The VLE data for methanol + water + ethanoic acid ternary system and the constituent binary systems are indispensable in distillation separation process to the product of methanol carboxylation through the correlation and prediction by the new method, while some of the isobaric VLE data for these binary and ternary systems are correlated and predicted earlier [9–25]. To provide the new correlation and prediction for some necessary basic thermodynamic data on the separation process of methanol carboxylation, therefore, it is very indispensable for these systems studied on the VLE data of the constituent binary and ternary systems using the new method. This paper reports a novel correlation and prediction for the VLE data for these systems containing the associating component that has been developed. The VLE data for methanol + water + ethanoic acid system and constituent binary systems were measured by the total pressure-temperature-liquid-phase composition-molar excess enthalpy of mixing of the liquid phase (p, T, x, H_m^E) for the binary systems using the novel pump-ebulliometer at 101.325 kPa. Owing to the association of ethanoic acid molecules, Marek's method in combination with Hayden-O'Connell (HOC) model was used to deal with the associating properties of the liquid and vapor phase and the nonideality of vapor phase, respectively [26–28]. However, the nonideality of liquid phase was corrected by the calculation of its activity coefficient, which was obtained based on NRTL, Wilson, Margules, and van Laar models as the function of T, x through the nonlinear fit of the least-squares method. NRTL, Wilson, Margules, and van Laar models were applied to correlate the VLE data for the three constituent binary systems, and the model parameters together with the deviations of temperature and vapor-phase molar fractions calculated from T, p, x, H_m^E according to the Q function of molar excess Gibbs energy by the indirect method were obtained by the least-squares method. The VLE data of the ternary system were well predicted from these binary interaction parameters of NRTL, Wilson, Margules, and van Laar models without any additional adjustment to build the thermodynamic model of VLE for the ternary system and obtain the vapor-phase compositions and the calculated bubble points. The excess Gibbs free energies for these binary systems as the function of liquid-phase composition and activity coefficient were calculated through the activity coefficient correlation to NRTL model parameters with the experimental data.

2. Modeling Section

There is an added complexity when working with carboxylic acids because they associate in the vapor and liquid phases. This association can be represented by assuming that the organic acid exists as monomer and dimer according to the Marek's method [26, 27]. This fact, coupled with the necessity for much analytical work, tends to enhance interest in exploring the new strategies for correlation and prediction of the VLE data for the systems containing the associating carboxyl acid.

According to the Marek's chemical theory, there are monomer and dimer carboxylic acid molecules in both liquid and vapor phases. The equilibrium constant of vapor-phase dimerization, C_E^V, is calculated by the expression

$$C_E^V = \frac{y_D^*}{\left(y_M^*\right)^2 p} \tag{1}$$

and the equilibrium constant of liquid-phase dimerization, C_E^L, is calculated by the expression

$$C_E^L = \frac{x_D^*}{\left(x_M^*\right)^2}. \tag{2}$$

In equations mentioned above, y_D^* and x_D^* are the mole fractions of dimers of ethanoic acid molecules in both vapor and liquid phases, respectively, and C_E^V can be defined as the function of temperature by the expression obtained from the literature [27]

$$C_E^V = \exp\left(\frac{7290}{T} - 21.980\right), \tag{3}$$

where C_E^V was presented in kPa^{-1} and T in K.

When the dimers of ethanoic acid molecules mainly exist, the binary system for the methanol or water (1) + ethanoic acid (2) is nominally binary; however, it is actually ternary for the methanol or water (1) + monomer ethanoic acid (M) + dimer ethanoic acid (D) system. In this system, the mole fractions of vapor-liquid equilibrium phases are y_1^*, y_M^*, y_D^*, x_1^*, x_M^*, and x_D^*, respectively. Therefore, the VLE relations of the nonassociating and associating components are individually calculated by the expressions

$$\begin{aligned} p y_1^* \Phi_1 &= p_1^s x_1^* \gamma_1, \\ p y_M^* \Phi_M &= p_M^s x_M^* \gamma_M, \\ p y_D^* \Phi_D &= p_D^s x_D^* \gamma_D. \end{aligned} \tag{4}$$

In (4), the actual mole fractions can be denoted by the apparent mole fractions of easily determined components (y_1, y_2, x_1, and x_2 multiplied by a modified coefficient. Likewise, the measured apparent vapor pressure saturated of ethanoic acid (p_2^s) multiplied by a modified coefficient can also denote the actual vapor pressures saturated of monomer and dimer ethanoic acid (p_M^s, p_D^s). Consequently, the apparent compositions and vapor pressures saturated substitute for the actual ones, and the VLE relation of the nonassociating component, methanol or water (1), is expressed by

$$p y_1 \alpha_1 \Phi_1 = p_1^s x_1 \beta_1 \gamma_1, \tag{5}$$

where

$$\alpha_1 = \frac{2}{(2-y_2)} \frac{1 - y_2 - \sqrt{1 + 4C_E^V p y_2 (2-y_2)}}{1 + \sqrt{1 + 4C_E^V p y_2 (2-y_2)}}, \tag{6}$$

$$\beta_1 = \frac{2}{(2-x_2)} \frac{1 - x_2 - \sqrt{1 + 4C_E^L p x_2 (2-x_2)}}{1 + \sqrt{1 + 4C_E^L p x_2 (2-x_2)}}, \tag{7}$$

and for ethanoic acid (2), which is the associating component, its relation is expressed by

$$p y_2 \alpha_2 \Phi_2 = p_2^s \theta_2^s x_2 \beta_2 \gamma_2, \tag{8}$$

where

$$\alpha_2 = \frac{2}{1 + \sqrt{1 + 4C_E^V p y_2 (2-y_2)}}, \tag{9}$$

$$\beta_2 = \frac{2}{1 + \sqrt{1 + 4C_E^L x_2 (2-x_2)}}, \tag{10}$$

$$\theta_2^s = \frac{-1 + \sqrt{1 + 4C_E^V p_2^s}}{2C_E^V p_2^s}. \tag{11}$$

In (5) to (11), α_1 and α_2 can be regarded as modified coefficients for the deviation from ideality in vapor phase on account of association. The fugacity of coefficients Φ_1 and Φ_2 is not negligible, and their values were obtained through the HOC model [28]. From another point of view, θ_2^s can be viewed as a modified coefficient for the vapor pressure of the associating component. Moreover, β_1 and β_2 denote modified coefficients of the deviation from ideality in the liquid phase by reason of the existence of association, and γ_1 and γ_2 express the deviation from ideality in liquid

Table 1: Physical properties of the pure compounds: densities ρ, refractive indexes n_D at 298.15 K and normal boiling points T_b.

Compound	ρ/kg·m^{-3}		n_D		T_b/K	
	expt	lit[a]	expt	lit[a]	expt	lit[a]
Methanol	786.44	786.37	1.3264	1.3265	337.71	337.69
Water	997.01	997.05	1.3324	1.3325	373.16	373.15
Ethanoic acid	1043.90	1043.92	1.3716	1.3718	391.53	391.15

[a] Riddick et al. [29].

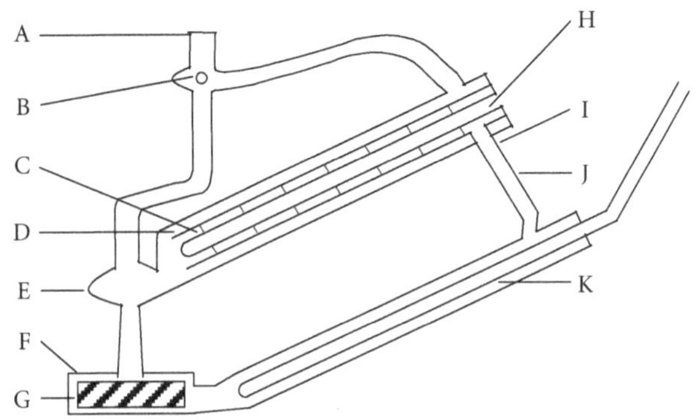

Figure 1: Schematic illustration for the structure of pump ebulliometer. (A) Normal distilling adaptor; (B) feed inlet; (C) coiled glass thread; (D) constant temperature introducing tube; (E) mixed vial section; (F) magnetic recirculating bump; (G) magnetic stirrer; (H) measured thermowell (filled with conducting oil); (I) tube window to observe in VLE status; (J) internal heating tube; (K) part-heating tube (resistance heating wire inserted).

phase because of other factors. Herein, in liquid phase, the modified coefficients of deviation from ideality, β_1 and β_2, can be incorporated to the activity coefficients, γ_1 and γ_2, respectively. So (5) and (8) are obtained by the expressions

$$py_1\alpha_1\Phi_1 = p_1^s x_1 \gamma_1, \quad (12)$$

$$py_2\alpha_2\Phi_2 = p_2^s \theta_2^s x_2 \gamma_2. \quad (13)$$

3. Experimental Section

3.1. Materials. Methanol (99.8 mass %) and ethanoic acid (99.8 mass %) were purchased from Sigma. The purities of the chemicals are provided by the manufacture's specifications. Ultra sound was used to dispel the solvent air in the materials, which were dried on molecular sieve (pore diameter 30 nm from Fluka) to remove all possible traces of moisture before use, but no other treatments were applied. The densities and refractive indices at 298.15 K and normal boiling points at 101.325 kPa of the pure component were compared with the literature values of Riddick et al. [29]. The results show that the measured values are approximately in agreement with those of the literature, as presented in Table 1. The measurement method of the composition dependence of densities and refractive indices has been previously reported [30].

3.2. Apparatus and Procedure. A new type of magnetic pump ebulliometer described in detail by Qiu et al. [31] was used to measure the boiling points with different liquid phase compositions at the 101.325 kPa. The experimental main apparatus for pump-ebulliometer is schematically shown in Figure 1. The recirculation still is entirely constructed from borosilicate glass. The main parts are a magnetic recirculating bump (F), a magnetic stirrer (G), a thermo-well (filled with a conducting oil) provided to enable good measurement of equilibrium temperature (H), a part-heating tube (resistance heating wire inserted) (K), and a tube window to observe in VLE status (I). The apparatus is an all-glass dynamic recirculation still with total volume of about 1.00×10^{-4} m^3. During the run, to avoid the over heating, the still was submerged in a constant temperature bath at about 3°C below the equilibrium boiling point, which was obtained by the Nichrome wire in the part-heating tube (K) to partially heat the known mass of the material. The equilibrium pressure was measured using a Fischer digital manometer with a precision of ±0.01 kPa. The pressure measurement for the manometer had an uncertainty of ±0.07 kPa, as provided by the manufacturer's specifications. The total uncertainty of calibration and pressure measurement is estimated to be ±0.15 kPa because of the uncertainty of the calibration and measurement errors. Since the barometric pressure changed slightly, the experimental temperatures of the systems were automatically calibrated to that at 101.325 kPa with self-adjusted pressure system. The temperature was measured using a Heraeus QuaT100 quartz thermometer with a thermosensor, with an accuracy of ±0.01 K. The calibration of the thermometers was carried out at the accredited calibration laboratory (Quality and Technique Bureau, Anhui). The total uncertainty of calibration and temperature measurement is evaluated to be ±0.085 K because of the uncertainty of the calibration, the probe's position, and the pressure fluctuations. The equilibrium temperature, T, was measured by means of the thermosensor inserted into the thermowell (filled with conducting oil) (H). In each experiment, a known mass of the material was introduced from the injector into the still from feed inlet (B) and heated to equilibrium boiling point of the system at a fixed pressure of 101.325 kPa by an automatic pressure regulation system. The liquid mixtures of required composition were prepared gravimetrically, with the use of a Sartorius electronic analytic balance (model ER-182A) with an accuracy of ±0.0001 g. The values of mole fraction were reproducible to ±0.0001 and have uncertainty of 0.01%. The ebulliometer was charged with the mixture of desired composition, and the boiler was then heated by nichrome wire wound around the boiler. After the liquid mixture started boiling, the bubbles along with the drops of liquid slowly spurted on the thermowell one by one through the tube window observed in VLE status (I). When the VLE state was attained by adjusting the pressure to 101.325 kPa, it remained constant for 20 min to ensure the stationary state, and then the boiling temperature was measured.

A flow microcalorimeter (model 2107, LKB produkter, Bromma, Sweden) was applied to measure the molar excess enthalpies, H_m^E, of the mixtures. The electrical-calibration apparatus and its operating procedure have been described elsewhere by Francesconi and Comelli [34]. Two automatic burets (ABU, Radiometer, Copenhagen, Denmark) were used to pump liquids through the mixing cell of the calorimeter. The performance of the calorimeter was checked by measuring the H_m^E of the test mixture hexane + cyclohexane system reported by Gmehling [35]. Agreement with literature is better than 0.5% over the central range of mole fraction of hexane.

The uncertainties in mole fraction and H_m^E are estimated to be 0.0005 and 0.5%, respectively. The liquid-phase mole fraction of component i, x_i, could be calculated from the known mass of the material added to the still. The vapor-phase mole fraction of component i, y_i, was calculated from the experimental T, p, x, and H_m^E data based on the Q function (the function of molar excess Gibbs energy) by the indirect method.

4. Results and Discussion

4.1. Calculation of Vapor-Phase Mole Fraction y for the Binary Systems from Tpx and H_m^E. The vapor-phase mole fraction y_i of the components was calculated from

$$y_i \Phi_i p = x_i \gamma_i p_i^s, \tag{14}$$

and the fugacity of coefficients Φ_i of the components was obtained by the expression

$$\Phi_i = \frac{\hat{\varphi}_i^V}{\varphi_i^s} \exp\left[-\frac{V_i^L (p - p_i^s)}{RT}\right], \tag{15}$$

where $\hat{\varphi}_i^V$ is the fugacity coefficient of component i in the vapor mixture, φ_i^s is the fugacity coefficient of component i at saturation, V_i^L is the molar volume of component i in the liquid phase, R is the universal gas constant, T is the experimental temperature, p is the total pressure (101.325 kPa), and p_i^s is the vapor pressure of pure component i. The Antoine equations were applied to calculate the values of these vapor pressures. The Antoine constants A_i, B_i, C_i, and the values of T_c, p_c, V_c, Z_c and ω were obtained from Shi et al. [36], as shown in Table 2. The Poynting correction factor was also included in the calculation of fugacity at the reference state. The liquid molar volumes were evaluated from the modified Rackett equation [37].

According to the Gibbs-Duhem equation, any extensive molar thermodynamic property of a given phase, such as the Gibbs and Helmholtz energies, the enthalpy, and molar volume, must satisfy the following differential relation:

$$\left(\frac{\partial M}{\partial T}\right)_{p,x_{j\neq i}} dT + \left(\frac{\partial M}{\partial p}\right)_{T,x_{j\neq i}} dp - \sum_{i=1}^{C} x_i d\overline{M}_i = 0. \tag{16}$$

In (16) M is a general molar property, x_i the molar fraction of component i in the phase under consideration, $x_{j\neq i}$ the pertinent set of compositions, C the number of components, and $\overline{M}_i$ the partial contribution of component i to M.

When considering VLE, the molar excess Gibbs energy, G^E, can be evaluated from measurable (T, p, x) data using activity coefficient relations. Replacing M by G^E/RT in the following equation yields the well-known relation

$$-\frac{H_m^E}{RT^2} dT + \frac{V_m^E}{RT} dp - \sum_{i=1}^{C} x_i d\ln\gamma_i = 0, \tag{17}$$

where H_m^E and V_m^E are the molar excess enthalpy and volume of mixing of the liquid phase. According to the thermodynamic principles, the activity coefficients γ_i of the components were calculated from the expression as follows:

$$Q = \frac{G^E}{RT} = \sum_{i=1}^{C} x_i \ln\gamma_i, \tag{18}$$

where Q is the function of molar excess Gibbs energy and combining (17) and (18) yields

$$d\left(\frac{G^E}{RT}\right) - \sum_{i=1}^{C} \ln\gamma_i dx_i = \sum_{i=1}^{C} x_i d\ln\gamma_i = -\frac{H_m^E}{RT^2} dT + \frac{V_m^E}{RT} dp. \tag{19}$$

Application to a binary system gives

$$d\left(\frac{G^E}{RT}\right) - \ln\frac{\gamma_1}{\gamma_2} dx_1 = -\frac{H_m^E}{RT^2} dT + \frac{V_m^E}{RT} dp. \tag{20}$$

Simultaneous solution of (18) and (20) yields

$$\begin{aligned} \ln\gamma_1 &= \frac{G^E}{RT} + x_2\left[\frac{d}{dx_1}\left(\frac{G^E}{RT}\right) + \frac{H_m^E}{RT^2}\frac{dT}{dx_1} - \frac{V_m^E}{RT}\frac{dp}{dx_1}\right], \\ \ln\gamma_2 &= \frac{G^E}{RT} - x_1\left[\frac{d}{dx_1}\left(\frac{G^E}{RT}\right) + \frac{H_m^E}{RT^2}\frac{dT}{dx_1} - \frac{V_m^E}{RT}\frac{dp}{dx_1}\right]. \end{aligned} \tag{21}$$

Equation (18) substituted into (21) reduced it to

$$\begin{aligned} \gamma_1 &= \exp\left[Q + (1 - x_1)\left(\frac{dQ}{dx_1} + \frac{H_m^E}{RT^2}\frac{dT}{dx_1} - \frac{V_m^E}{RT}\frac{dp}{dx_1}\right)\right], \\ \gamma_2 &= \exp\left[Q - x_1\left(\frac{dQ}{dx_1} + \frac{H_m^E}{RT^2}\frac{dT}{dx_1} - \frac{V_m^E}{RT}\frac{dp}{dx_1}\right)\right]. \end{aligned} \tag{22}$$

For a binary system comprised of species 1 and 2 at VLE state, at constant pressure, $(dp/dx_1) = 0$, substitution into (22) reduced it to

$$\gamma_i = \exp\left[Q + (1 - x_i)\left[\left(\frac{dQ}{dx_i}\right) + \frac{H_m^E}{RT^2}\left(\frac{dT}{dx_i}\right)\right]\right], \quad (i = 1, 2). \tag{23}$$

The right-hand side, $(H_m^E/RT^2)(dT/dx_i)$, of (23) cannot be neglected. Proper use of (23) requires the availability

TABLE 2: Antoine coefficients of the compounds and published parameters [36] used to calculate fugacity coefficients: critical temperature T_c, critical pressure p_c, critical volume V_c, critical compression Z_c, and acentric factor ω of pure compounds.

Compound	A_i	B_i	C_i	T_c/K	p_c/MPa	V_c/m^3·kmol^{-1}	Z_c	ω
Methanol	7.19736	1574.99	−34.29	512.6	8.096	0.118	0.224	0.559
Water	7.07404	1657.46	−46.13	647.3	22.048	0.056	0.229	0.344
Ethanoic acid	6.42452	1479.02	−56.34	594.4	5.786	0.171	0.200	0.454

of heats of mixing as a function of composition and temperature. The activity coefficients γ_i of the components as functions of the excess Gibbs energy are as follows:

$$\gamma_1 = \exp\left\{Q + (1 - x_1)\left[\left(\frac{dQ}{dx_1}\right) + \frac{H_m^E}{RT^2}\left(\frac{dT}{dx_1}\right)\right]\right\},$$
$$\gamma_2 = \exp\left\{Q - x_1\left[\left(\frac{dQ}{dx_1}\right) + \frac{H_m^E}{RT^2}\left(\frac{dT}{dx_1}\right)\right]\right\}. \tag{24}$$

From (14), this equation is rearranged to obtain

$$y_i = \frac{x_i \gamma_i p_i^s}{\Phi_i p}. \tag{25}$$

Because $\sum y_i = 1$, for the binary system, the equation may be summed to give

$$\frac{x_1\gamma_1\varphi_1^s p_1^s \exp\left[V_1^L(p - p_1^s)/RT\right]}{\hat{\varphi}_1^V p} + \frac{x_2\gamma_2\varphi_2^s p_2^s \exp\left[V_2^L(p - p_2^s)/RT\right]}{\hat{\varphi}_2^V p} = 1. \tag{26}$$

In (26), solved for y_i by difference method. Suppose that $[0, 1]$ is subdivided into n subintervals $[x_k, x_{k+1}]$ of equal step size $h = 1/n$ by using $x_k = kh$ for $k = 0, 1, \ldots, n$. In k difference point, we obtain

$$F_k = 1 - \left\{\frac{x_1\gamma_1\varphi_1^s p_1^s \exp\left[V_1^L(p - p_1^s)/RT\right]}{\hat{\varphi}_1^V p}\right\}_k - \left\{\frac{x_2\gamma_2\varphi_2^s p_2^s \exp\left[V_2^L(p - p_2^s)/RT\right]}{\hat{\varphi}_2^V p}\right\}_k = 0. \tag{27}$$

Meanwhile, (23) may be shown as follows:

$$\gamma_i|_k = \exp\left\{Q_k + (1 - x_i|_k)\left[\frac{Q|_{k+1} - Q|_{k-1}}{2/h} + \frac{H_m^E}{RT^2}\left(\frac{dT}{dx_1}\right)\bigg|_k\right]\right\}. \tag{28}$$

And (27) is linearized to obtain:

$$-F_k = \Delta Q|_{k-1}\left(\frac{\delta F_k}{\delta Q|_{k-1}}\right) + \Delta Q|_k\left(\frac{\delta F_k}{\delta Q|_k}\right) + \Delta Q|_{k+1}\left(\frac{\delta F_k}{\delta Q|_{k+1}}\right). \tag{29}$$

The number of n linear equation from (28) is solved for $\Delta Q|_k$ by chasing method

$$Q_k^{j+1} = Q_k^j + t\Delta Q_k^j, \tag{30}$$

where t is relaxation factor, finally, y_{ik} $(k = 1 \sim n)$ is obtained by difference method. The block diagram for calculation procedure was detailedly shown in Figure 2.

4.2. Calculation of Vapor-Phase Mole Fraction y for the Binary Systems from Model. There are many methods concerning the correlation and prediction of VLE data. The model-free approach data treatment of vapor-liquid equilibrium is also one of the best strategies for the correlation and prediction of VLE data. Wisniak's group has developed that the novel model-free computation techniques and limiting conditions have been applied to VLE data for azeotropic systems [42]. Moreover, Segura and coworkers reported that a model-free approach dealt with VLE data in application of ternary systems [38, 43]. For the three binary systems of this study, the activity coefficients were correlated with the NRTL [39], Wilson [40], Margules [41], and van Laar [44] equations, respectively. The interaction parameters optimized were achieved by the objective function (OF) minimized using the least-squares fitting as follows:

$$\mathrm{OF} = \sum_{i=1}^{N}\left(x_{i,\mathrm{cal}} - x_{i,\mathrm{exp}}\right)^2, \tag{31}$$

where N is the number of experimental points $x_{i,\mathrm{cal}}$ and $x_{i,\mathrm{exp}}$ are the liquid-phase mole fraction of component i calculated and experimental values from the (12) or (13) and from measured data, respectively.

Because carboxylic acids are always present in an associated form, like a dimer or trimer, in both the vapor and liquid phases even at low pressures, a significant deviation in fugacity coefficient may exist using the ideal gas assumption. To illustrate the deviation from ideal behavior, Marek's chemical theory and HOC model were applied to deal with the associating component and modify the deviation from idealities of vapor phase [26–28], respectively. The Poynting correction factor was also included in the calculation of fugacity at the reference state. The liquid molar volumes were evaluated from the modified Rackett equation [37]. However, under isobaric conditions, the most volatile component cannot exist in the liquid state, only as superheated vapor. Hence, there is no way to calculate or measure this property for the molar volumes of the pure liquids. Therefore, the correct procedure for isobaric measurements is to calculate the overall values of A_{ij} and A_{ij}, as adjustable parameters,

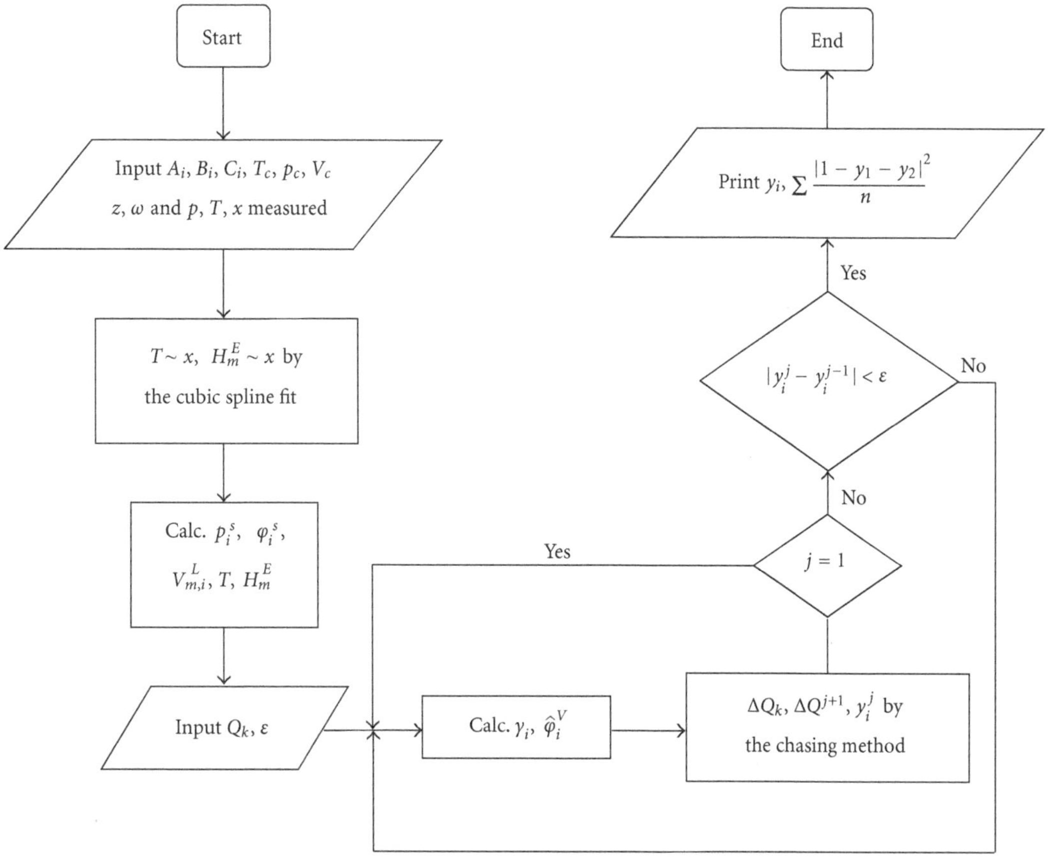

FIGURE 2: Block diagram for the calculation y by T, p, x, and H_m^E.

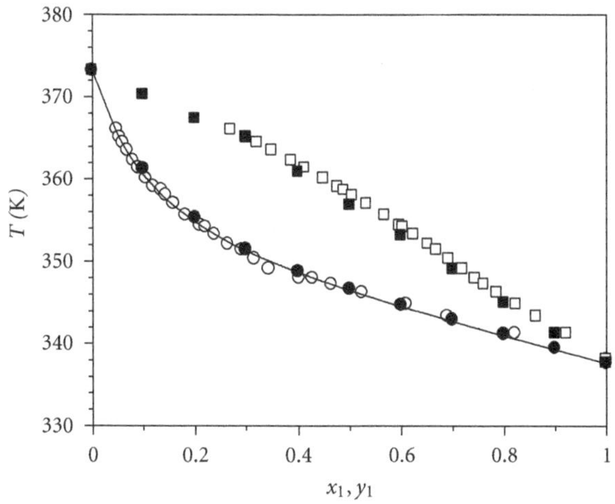

FIGURE 3: T-x_1-y_1 diagram for methanol (1) + water (2) at 101.325 kPa: □, vapor-phase mole fraction y_1 from Tpx and H_m^E; ■, vapor-phase mole fraction y_1 from the literature [32, 33]; ∘, liquid-phase experimental temperature; •, liquid-phase mole fraction x_1 from the literature [32, 33]; —, NRTL correlation temperature.

and not the values of the interaction excess energy. Hence, for Wilson model, A_{ij} and A_{ij} were reported as adjustable parameters.

The activity coefficients computed on the basis of NRTL model were used to evaluated dimensionless excess Gibbs function at 101.325 kPa for three binary systems over the overall range of composition. Liquid-phase mole fraction x_1, experimental boiling point temperature T_{exp}, calculated bubble point temperature T_{cal}, vapor-phase mole fraction y_1 from the model, activity coefficients γ_1 and γ_2 using NRTL equation correlation, fugacity coefficients $\hat{\varphi}_1^V$ and $\hat{\varphi}_2^V$, and the molar excess enthalpies of mixing of the liquid phase H_m^E are included in Table 3. The T-x_1-y_1 diagrams for the methanol (1) + water (2), methanol (1) + ethanoic acid (2), and water (1) + ethanoic acid (2) three binary systems at 101.325 kPa are shown in Figures 3, 4, and 5, respectively. The plot of excess Gibbs energy function G^E/RT versus liquid-phase mole fraction x_1 for the three binary systems is shown in Figure 6. All the mixtures exhibit deviations from ideality with a range that may be attributed to interactions leading to the formation of various associated aggregates. Observed nonideal behavior is indicative of the magnitude of experimental activity coefficients γ_i, as well as of the variation of excess Gibbs energy function, G^E/RT, with composition, as depicted in Figure 6. The obtained absolute maximum values of G^E/RT for the methanol (1) + water (2), methanol (1) + ethanoic acid (2), and water (1) + ethanoic acid (2) three binary systems are 0.1703, 0.0180, and 0.0892, respectively. The values of excess Gibbs energy

Table 3: VLE data for the methanol (1) + water (2), methanol (1) + ethanoic acid (2), and water (1) + ethanoic acid (2). Three binary systems at 101.325 kPa: liquid-phase mole fraction x_1, experimental boiling point temperature T_{exp}, calculated bubble point temperature T_{cal}, vapor-phase mole fraction y_1, activity coefficients γ_1 and γ_2 using NRTL equation correlation, fugacity coefficients $\hat{\varphi}_1^V$ and $\hat{\varphi}_2^V$, and molar excess enthalpies of mixing of the liquid phase H_m^E.

x_1	T_{exp}/K	T_{cal}/K	y_1	γ_1	$\hat{\varphi}_1^V$	γ_2	$\hat{\varphi}_2^V$	H_m^E/J·mol^{-1}
				Methanol (1) + water (2)				
0.0000	373.15	373.15	0.0000			1.0000	0.9882	0.00
0.0484	366.00	365.54	0.2704	2.1436	0.9852	1.0034	0.9875	225.74
0.0539	365.05	364.88	0.3009	2.1147	0.9849	1.0042	0.9875	248.74
0.0607	364.40	364.12	0.3213	2.0799	0.9846	1.0053	0.9874	276.88
0.0696	363.50	363.19	0.3492	2.0358	0.9843	1.0069	0.9874	312.53
0.0806	362.25	362.13	0.3866	1.9835	0.9839	1.0092	0.9873	354.36
0.0911	361.35	361.21	0.4126	1.9360	0.9836	1.0117	0.9872	392.94
0.1056	360.05	360.05	0.4492	1.8739	0.9832	1.0155	0.9871	443.59
0.1197	359.05	359.04	0.4762	1.8172	0.9829	1.0197	0.9870	489.29
0.1354	358.60	358.02	0.4885	1.7584	0.9826	1.0248	0.9869	538.32
0.1436	358.00	357.53	0.5047	1.7293	0.9824	1.0277	0.9869	562.20
0.1586	356.95	356.70	0.5323	1.6788	0.9822	1.0334	0.9868	602.83
0.1815	355.55	355.56	0.5677	1.6083	0.9819	1.0428	0.9867	659.52
0.2092	354.35	354.35	0.5969	1.5325	0.9815	1.0555	0.9865	719.43
0.2202	354.10	353.91	0.6031	1.5049	0.9814	1.0609	0.9865	741.25
0.2380	353.25	353.24	0.6237	1.4633	0.9812	1.0701	0.9864	772.62
0.2648	352.05	352.32	0.6519	1.4066	0.9809	1.0847	0.9862	812.77
0.2918	351.35	351.48	0.6681	1.3561	0.9807	1.1005	0.9861	846.82
0.3166	350.30	350.76	0.6923	1.3148	0.9805	1.1157	0.9860	870.15
0.3442	349.05	350.02	0.7197	1.2740	0.9803	1.1336	0.9859	889.06
0.4015	347.95	348.61	0.7431	1.2038	0.9800	1.1730	0.9856	909.80
0.4282	347.90	348.00	0.7443	1.1767	0.9798	1.1924	0.9854	911.28
0.4644	347.20	347.21	0.7613	1.1449	0.9796	1.2194	0.9853	903.04
0.5240	346.20	345.98	0.7860	1.1026	0.9793	1.2656	0.9850	869.55
0.6091	344.80	344.33	0.8231	1.0595	0.9789	1.3338	0.9845	783.83
0.6899	343.31	342.85	0.8633	1.0325	0.9785	1.3995	0.9840	666.92
0.8213	341.20	340.56	0.9221	1.0086	0.9779	1.5038	0.9832	419.31
1.0000	338.10	337.66	1.0000	1.0000	0.9771			0.00
				Methanol (1) + ethanol acid (2)				
0.0000	390.15	391.04	0.0000			1.0000	0.9703	0.00
0.0359	387.32	386.23	0.1574	0.9388	0.9765	0.9999	0.9685	77.28
0.0410	386.21	385.56	0.1664	0.9393	0.9768	0.9999	0.9682	86.70
0.0648	383.89	382.68	0.2719	0.9419	0.9779	0.9997	0.9666	134.05
0.0741	382.73	381.90	0.3133	0.9428	0.9782	0.9996	0.9660	149.55
0.0859	381.20	380.60	0.3333	0.9441	0.9787	0.9995	0.9652	171.14
0.1040	380.03	379.06	0.3830	0.9460	0.9791	0.9992	0.9639	205.37
0.1280	378.46	377.35	0.4602	0.9485	0.9797	0.9988	0.9622	245.43
0.1629	375.62	374.70	0.5213	0.9521	0.9802	0.9981	0.9595	299.80
0.1919	373.21	372.43	0.5873	0.9551	0.9805	0.9973	0.9572	341.32
0.2510	369.31	368.44	0.6495	0.9609	0.9808	0.9954	0.9528	411.44
0.3391	364.35	363.56	0.7538	0.9692	0.9809	0.9916	0.9461	490.73
0.4401	358.54	357.71	0.8232	0.9776	0.9805	0.9857	0.9388	536.56
0.5829	351.77	350.75	0.8999	0.9874	0.9797	0.9748	0.9292	526.43
0.7050	346.47	346.39	0.9389	0.9936	0.9790	0.9633	0.9217	449.37
0.7509	344.90	344.06	0.9533	0.9955	0.9787	0.9583	0.9189	401.45
0.8109	342.81	342.35	0.9693	0.9974	0.9783	0.9515	0.9155	330.61
0.8711	341.60	340.75	0.9766	0.9988	0.9779	0.9442	0.9122	241.41
0.9161	339.65	340.00	0.9859	0.9995	0.9776	0.9384	0.9097	163.78
1.0000	338.15	337.66	1.0000	1.0000	0.9771			0.00

TABLE 3: Continued.

x_1	T_{exp}/K	T_{cal}/K	y_1	γ_1	$\hat{\varphi}_1^V$	γ_2	$\hat{\varphi}_2^V$	H_m^E/J·mol^{-1}
Water (1) + ethanoic acid (2)								
0.0000	390.15	391.04	0.0000			1.0000	0.9703	0.00
0.0665	387.22	387.67	0.1140	1.4161	0.9899	1.0025	0.9693	245.31
0.0749	387.00	387.31	0.1288	1.4046	0.9898	1.0031	0.9692	273.17
0.0855	386.48	386.88	0.1603	1.3903	0.9897	1.0040	0.9691	307.50
0.0995	385.86	386.34	0.1938	1.3722	0.9896	1.0054	0.9689	350.32
0.1507	384.16	384.60	0.2802	1.3117	0.9892	1.0121	0.9683	489.63
0.1722	383.44	383.96	0.3155	1.2888	0.9891	1.0156	0.9681	540.35
0.1926	382.99	383.39	0.3375	1.2684	0.9890	1.0193	0.9679	585.57
0.2168	382.39	382.77	0.3676	1.2458	0.9890	1.0242	0.9677	632.34
0.2502	381.53	381.98	0.4101	1.2171	0.9889	1.0316	0.9674	689.92
0.2936	380.94	381.06	0.4407	1.1838	0.9887	1.0425	0.9670	750.61
0.3303	380.20	380.36	0.4815	1.1589	0.9887	1.0528	0.9667	788.98
0.3756	379.40	379.57	0.5253	1.1318	0.9886	1.0666	0.9663	821.39
0.4465	378.27	378.48	0.5890	1.0965	0.9885	1.0908	0.9658	841.59
0.4946	377.61	377.82	0.6274	1.0768	0.9885	1.1087	0.9654	836.35
0.5615	376.80	376.99	0.6777	1.0543	0.9884	1.1355	0.9649	802.60
0.6036	376.24	376.51	0.7137	1.0427	0.9884	1.1533	0.9645	767.63
0.6517	375.59	376.00	0.7546	1.0315	0.9884	1.1746	0.9642	714.78
0.7067	375.11	375.46	0.7849	1.0213	0.9884	1.1998	0.9637	639.01
0.7045	375.11	375.48	0.7848	1.0217	0.9884	1.1988	0.9638	642.76
0.7507	374.71	375.05	0.8173	1.0148	0.9883	1.2207	0.9634	568.24
0.8008	374.31	374.62	0.8501	1.0091	0.9883	1.2451	0.9630	475.20
0.8723	373.80	374.05	0.9011	1.0035	0.9883	1.2809	0.9624	323.52
0.9116	373.54	373.76	0.9315	1.0016	0.9883	1.3009	0.9620	230.13
0.9546	373.24	373.45	0.9630	1.0004	0.9883	1.3231	0.9616	121.95
1.0000	373.15	373.15	1.0000	1.0000	0.9882			0.00

TABLE 4: Correlation parameters for activity coefficients, average deviation for studied systems and vapor-phase composition mean absolute deviation from the literatures.

Equation	Parameters or deviations	Methanol (1) + water (2)	Methanol (1) + ethanoic acid (2)	Water (1) + ethanoic acid (2)
NRTL[a]	$(g_{12} - g_{11})$/J·mol^{-1}	−436.50	−85.50	−352.42
	$(g_{21} - g_{22})$/J·mol^{-1}	1159.55	0.98	715.43
	α_{12}	0.241	0.30	0.23
	dT/K	0.28	0.51	0.33
	dy	0.0072	0.0054	0.0096
Wilson[a]	Λ_{12}/J·mol^{-1}	389.37	50.46	644.71
	Λ_{21}/J·mol^{-1}	368.45	95.58	−87.59
	dT/K	0.43	0.47	0.64
	dy	0.0036	0.0059	0.0053
Margules[b]	A_{12}	0.95	0.12	0.29
	A_{21}	0.65	−0.46	0.57
	dT/K	0.37	0.32	0.56
	dy	0.0049 (0.0057)[c]	0.0075 (0.0234)[c]	0.0086 (0.0123)[c]
van Laar[b]	A_{12}	0.86	0.24	0.32
	A_{21}	0.56	0.03	0.55
	dT/K	0.27	0.58	0.47
	dy	0.0069	0.0078	0.0098

[a]Wilson's interaction parameters (J·mol^{-1}), NRTL's interaction parameters (J·mol^{-1}). [b]Margules and van Laar interaction parameters (dimensionless). $dT = \sum |T_{exp} - T_{cal}|/N$; N: number of data points; T_{cal}: calculated bubble point from model, K; T_{exp}: experimental boiling point temperature, K. $dy = \sum |y_{cal} - y_{mod}|/N$; N: number of data points; y_{cal}: calculated vapor-phase mole fraction from Tpx; y_{mol}: calculated vapor-phase mole fraction from model. [c]The values in parentheses from the literatures [38–41].

function G^E/RT are positive for methanol (1) + water (2) and water (1) + ethanoic acid (2) binary systems. However, for methanol (1) + ethanoic acid (2) system, the values of those are negative in the overall range of mole fraction. G^E/RT values follow the order methanol (1) + water (2) > water (1) + ethanoic acid (2) > methanol (1) + ethanoic acid (2). The absolute maximum value of G^E/RT is approximate at an equimolar fraction in three binary systems. Figures 3–5 show that the comparison of the predicted vapor-phase and experimental liquid-phase compositions with those of the literature [10, 11, 32, 33]. Comparing with the values of vapor-phase and liquid-phase component from the literatures, the values of those from the paper are in good agreement with the literature, as shown in Figures 3–5. The results have demonstrated that the methods for Tpx and H_m^E are appropriate for representing the experimental data of the three binary systems. The optimum model interaction parameter of liquid activity coefficient and the absolute average deviations for the different models, and from Tpx and H_m^E are listed in Table 4. Herein, we obtained the results by the four different types of correlations for the prediction of activity coefficients in these systems, which reveal that the deviations of NRTL, Wilson, Margules, and van Laar equations are reasonably small in Table 4. For comparison, the mean deviations obtained by Gmehling and Onken [32, 33, 45, 46] are also shown in the Table 4. It can seem that two sets of deviation values are comparable. Since the superiority of one method over the others is not always obvious, practice must rely on experience and analogy. The comprehensive comparisons of four of the methods (NRTL, Wilson, van Laar, and Margules) were made in Table 4. From the data analysis, the temperature deviations between the experimental and calculated values of four different types of model are very similar in the three binary systems, and the vapor-phase mole fraction deviations between calculated values from Tpx and H_m^E, and from the model are very similar. Therefore, the activity coefficient models are appropriate for representing the experimental data of the three binary systems. In Table 4, the absolute average deviations dT of the difference between boiling point temperature from experiment and bubbling point temperature from calculation by NRLT model parameters for the three binary systems are 0.28°C, 0.51°C, and 0.33°C, respectively. And the absolute average deviations dy of the difference between vapor-phase mole fraction from Tpx and H_m^E calculation and from NRTL model calculation are 0.0072, 0.0054, and 0.0096, respectively.

4.3. Correlation and Prediction of VLE of Ternary System. The binary interaction parameters of the NRTL, Wilson, Margules, and van Laar model given in Table 4 were used to correlate and predict the VLE data of the ternary system. VLE data for methanol (1) + water (2) + ethanoic acid (3) at 101.325 kPa included liquid-phase mole fraction x_1, x_2, and x_3, experimental boiling point temperature T_{exp}, calculated bubble point temperature T_{cal}, calculated vapor-phase mole fraction y_1, y_2, and y_3, activity coefficients γ_1, γ_2, and γ_3, the average deviation in the bubble temperatures of the ternary system using NRTL equation correlation listed in Table 5. The absolute average and maximum deviation between the boiling point from experimental data and the bubble point from NRTL model calculation are 0.77°C and 1.94°C, respectively. The average and maximum deviations using Wilson, Margules, and van Laar equation individually are 0.79°C, 1.90°C, 0.85°C, 1.89°C, and 0.96°C, 1.91°C. Diagram of VLE for the ternary system methanol (1) + water (2) + ethanoic acid (3) at 101.325 kPa is shown in Figure 7.

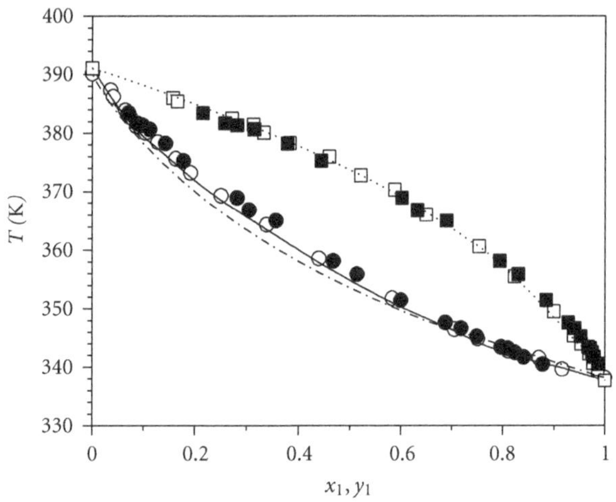

Figure 4: T-x_1-y_1 diagram for methanol (1) + ethanoic acid (2) at 101.325 kPa: □, vapor-phase mole fraction y_1 from Tpx and H_m^E; ■, vapor-phase mole fraction y_1 from literature [10]; —, dot line, vapor-phase mole fraction y_1 with Block diagram for the calculation y by T, p, x, and H_m^E neglected from our previous work [15]; ∘, liquid-phase experimental temperature; •, liquid-phase experimental temperature from literature [10]; -·-·-, dash dot, liquid-phase experimental temperature from our previous work [15]; —, NRTL correlation temperature.

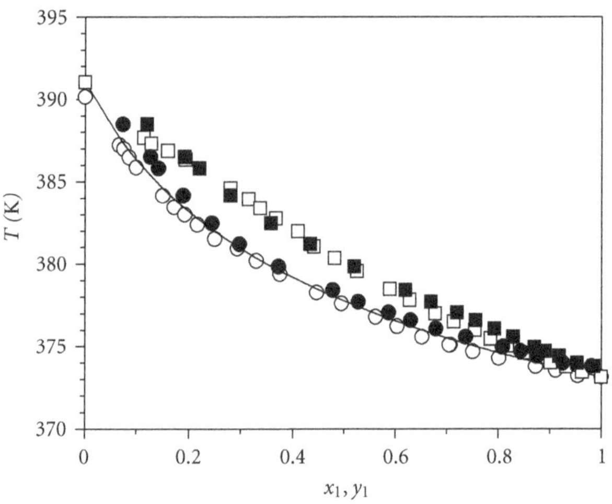

Figure 5: T-x_1-y_1 diagram for water (1) + ethanoic acid (2) at 101.325 kPa: □, vapor-phase mole fraction y_1 from Tpx and H_m^E; ■, vapor-phase mole fraction y_1 from literature [11]; •, liquid-phase experimental temperature from the literature; ∘, liquid-phase experimental temperature [11]; —, NRTL correlation temperature.

TABLE 5: VLE data for the methanol (1) + water (2) + ethanoic acid (3) Ternary system at 101.325 kPa: liquid-phase mole fraction x_1, x_2, and x_3, experimental boiling point temperature T_{exp}, calculated bubble point temperature T_{cal}, vapor-phase mole fraction y_1, y_2, and y_3, activity coefficients γ_1, γ_2, and γ_3 using NRTL equation correlation.

x_1	x_2	x_3	T_{exp}/K	T_{cal}/K	y_1	y_2	y_3	γ_1	γ_2	γ_3
0.7919	0.2080	0.0001	341.08	340.94	0.9108	0.0870	0.0022	1.0152	1.5155	0.9163
0.7502	0.1967	0.0531	342.79	342.12	0.9024	0.0882	0.0094	1.0156	1.5068	0.9294
0.6923	0.1816	0.1261	344.60	343.90	0.8886	0.0872	0.0242	1.0147	1.4953	0.9444
0.6623	0.1841	0.1536	345.45	344.77	0.8783	0.0910	0.0307	1.0152	1.4838	0.9493
0.5917	0.1645	0.2438	348.36	347.27	0.8561	0.0895	0.0544	1.0114	1.4719	0.9632
0.4619	0.4554	0.0827	349.12	348.30	0.7566	0.2242	0.0191	1.1037	1.2749	0.9612
0.4122	0.3196	0.2682	350.38	352.17	0.7339	0.1929	0.0732	1.0461	1.3352	0.9808
0.5333	0.1483	0.3184	350.67	349.55	0.8338	0.0881	0.0781	1.0072	1.4623	0.9722
0.4263	0.4203	0.1534	351.53	350.19	0.7359	0.2254	0.0387	1.0874	1.2853	0.9763
0.4946	0.1375	0.3679	351.93	351.19	0.8167	0.0869	0.0964	1.0040	1.4558	0.9772
0.3166	0.4774	0.2060	352.84	354.29	0.6481	0.2893	0.0627	1.1178	1.2310	1.0116
0.3959	0.3903	0.2138	353.19	351.89	0.7162	0.2258	0.0580	1.0733	1.2939	0.9847
0.2537	0.5812	0.1651	354.90	355.93	0.5880	0.3565	0.0555	1.1972	1.1675	1.0536
0.2148	0.6454	0.1398	356.16	357.08	0.5479	0.4013	0.0508	1.2669	1.1308	1.0932
0.1750	0.4020	0.4230	363.36	363.66	0.4673	0.3513	0.1814	1.0664	1.2364	1.0246
0.0585	0.9055	0.0360	364.18	364.83	0.2965	0.6797	0.0238	1.9547	1.0150	1.5148
0.1563	0.3589	0.4848	365.42	365.67	0.4367	0.3420	0.2213	1.0459	1.2516	1.0186
0.0478	0.9228	0.0294	366.09	365.86	0.2627	0.7164	0.0209	2.0514	1.0104	1.5759
0.1413	0.3246	0.5341	366.99	367.35	0.4109	0.3321	0.2570	1.0318	1.2630	1.0145
0.0404	0.9348	0.0248	367.10	366.66	0.2359	0.7454	0.0187	2.1259	1.0077	1.6231
0.0350	0.9435	0.0215	367.41	367.30	0.2142	0.7689	0.0169	2.1843	1.0059	1.6602
0.1290	0.2963	0.5747	368.56	368.80	0.3888	0.3218	0.2895	1.0214	1.2720	1.0116
0.0241	0.9592	0.0167	368.82	368.78	0.1626	0.8230	0.0144	2.2986	1.0032	1.7357
0.0203	0.9657	0.0140	369.63	369.35	0.1425	0.8449	0.0126	2.3510	1.0023	1.7687
0.0175	0.9704	0.0121	370.14	369.79	0.1266	0.8622	0.0112	2.3906	1.0017	1.7938
0.0154	0.9740	0.0106	370.54	370.13	0.1141	0.8759	0.0100	2.4220	1.0013	1.8136
0.0884	0.2313	0.6803	371.87	373.49	0.3017	0.3004	0.3979	1.0004	1.2864	1.0068
0.0779	0.2038	0.7183	373.44	375.08	0.2771	0.2814	0.4415	0.9933	1.2933	1.0050
0.0690	0.1805	0.7505	375.20	376.50	0.2548	0.2629	0.4822	0.9878	1.2987	1.0037
0.0625	0.1635	0.7740	375.65	377.59	0.2376	0.2480	0.5144	0.9841	1.3022	1.0030
0.0413	0.1802	0.7785	379.37	379.33	0.1656	0.2871	0.5473	0.9859	1.2885	1.0044
0.0359	0.1567	0.8074	379.87	380.59	0.1486	0.2617	0.5897	0.9812	1.2942	1.0032
0.0314	0.1369	0.8317	381.03	381.69	0.1338	0.2380	0.6282	0.9775	1.2986	1.0023
0.0280	0.1222	0.8498	381.64	382.55	0.1220	0.2192	0.6588	0.9749	1.3016	1.0018
Deviations				$dT = 0.77$						

5. Conclusions

VLE data for the ternary system methanol + water + ethanoic acid and three constituent binary systems at 101.325 kPa: methanol + water, methanol + ethanoic acid, and water + ethanoic acid were determined at different liquid-phase compositions using a novel pump ebulliometer. The equilibrium composition of the vapor phase was calculated from T, p, x, and H_m^E based on the Q function of excess Gibbs free energy by the indirect method. The experimental data were correlated using the NRTL, Wilson, Margules, and van Laar equations. It was shown that the deviations of NRTL, Wilson, Margules, and van Laar equations are reasonably small. The VLE data of ternary system were predicted by NRTL, Wilson, Margules, and van Laar equation. The calculated bubble points accorded well with the experimental data. The results show that the calculated bubble point is fitted by the models, which satisfy the need for the design and operation of separation process in chemical industry. Moreover, the method will provide theoretical guidance for the research of VLE data of strongly associating system of vapor and liquid phase in nonideal behavior.

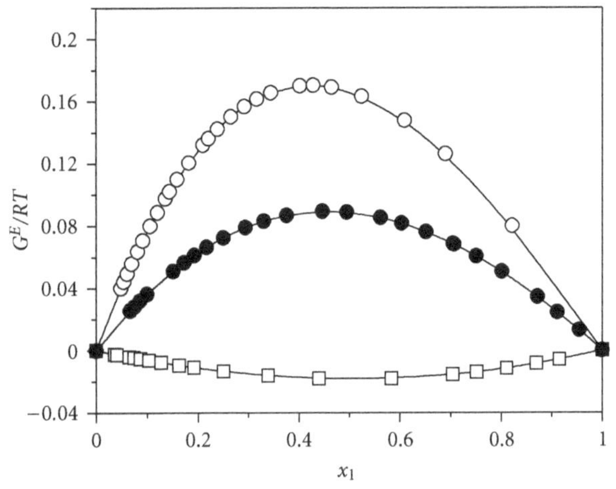

FIGURE 6: Excess Gibbs energy functions (G^E/RT) versus liquid-phase mole fraction of component 1 (x_1) diagram: ∘, methanol (1) + water (2); •, water (1) + ethanoic acid (2); □, methanol (1) + ethanoic acid (2).

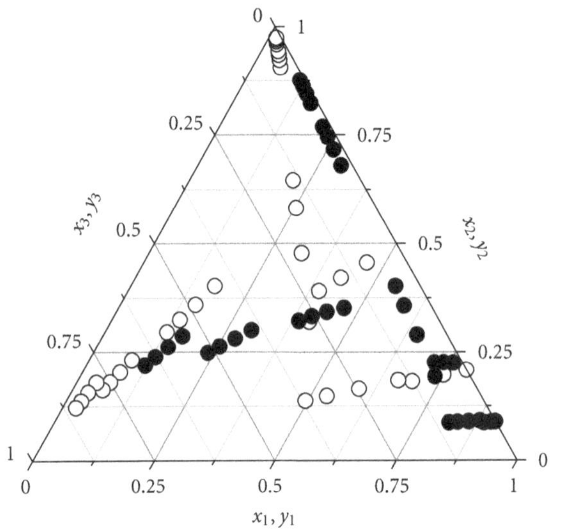

FIGURE 7: Diagram of VLE for the ternary system methanol (1) + water (2) + ethanoic acid (3) at 101.325 kPa: ∘, liquid-phase mole fraction; •, vapor-phase mole fraction.

Acknowledgments

This work was supported by the Deutscher Akademischer Austausch Dienst (DAAD) (Ref. Code: A/11/06441), the National Natural Science Foundation of China (Grant no. 21075026), and the Natural Science Foundation of the Higher Education Institutions of Anhui Province (Grant no. ZD200902).

References

[1] J. J. Ljunglin and H. C. van Ness, "Calculation of vapour-liquid equilibria from vapour pressure data," *Chemical Engineering Science*, vol. 17, no. 7, pp. 531–539, 1962.

[2] E. Hala, J. Pick, V. Fried, and O. Vilim, *Vapor-Liquid Equilibrium*, Pergamon, New York, NY, USA, 1958.

[3] H. C. Van Ness, "On the integration of the coexistence equation for binary vapor-liquid equilibrium," *AIChE Journal*, vol. 16, no. 1, pp. 18–22, 1970.

[4] F. O. Mixon, B. Gumowski, and B. H. Carpenter, "Computation of vapor-liquid equilibrium data from solution vapor pressure measurements," *Industrial and Engineering Chemistry Fundamentals*, vol. 4, no. 4, pp. 455–459, 1965.

[5] J. A. Barker, "Determination of activity coefficients from total pressure measurements," *Australian Journal of Chemistry*, vol. 6, no. 3, pp. 207–210, 1953.

[6] L. C. Tao, "How to compute binary vapor-liquid equilibrium compositions from experimental *P-x* or *t-x* data," *Industrial and Engineering Chemistry*, vol. 53, no. 4, pp. 307–309, 1961.

[7] J. M. Prausnitz, T. Anderson, E. Grens, C. Eckert, R. Hsieh, and J. P. O. 'Connell, *Computer Calculations For Multicomponent Vapor-Liquid and Liquid-Liquid Equilibria*, Prentice-Hall, Englewood Cliffs, NJ, USA, 1980.

[8] W. Arlt, "A new apparatus for phase equilibria in reacting mixtures," *Fluid Phase Equilibria*, vol. 158–160, pp. 973–977, 1999.

[9] Z. P. Xu and K. T. Chuang, "Correlation of vapor-liquid equilibrium data for methyl acetate methanol-water-acetic acid mixtures," *Industrial and Engineering Chemistry Research*, vol. 36, no. 7, pp. 2866–2870, 1997.

[10] H. Sawistowski and P. A. Pilavakis, "Vapor-liquid equilibrium with association in both phases. Multicomponent systems containing acetic acid," *Journal of Chemical and Engineering Data*, vol. 27, no. 1, pp. 64–71, 1982.

[11] W. Chang, G. Guan, X. Li, and H. Yao, "Isobaric vapor-liquid equilibria for water + acetic acid + (n-pentyl acetate or isopropyl acetate)," *Journal of Chemical and Engineering Data*, vol. 50, no. 4, pp. 1129–1133, 2005.

[12] C. Zhang, H. Wan, L. Xue, and G. Guan, "Investigation on isobaric vapor liquid equilibrium for acetic acid+water+(n-propyl acetate or iso-butyl acetate)," *Fluid Phase Equilibria*, vol. 305, no. 1, pp. 68–75, 2011.

[13] Q. Xie, H. Wan, M. Han, and G. Guan, "Investigation on isobaric vapor-liquid equilibrium for acetic acid + water + methyl ethyl ketone + isopropyl acetate," *Fluid Phase Equilibria*, vol. 280, no. 1-2, pp. 120–128, 2009.

[14] W. Chang, H. Wan, G. Guan, and H. Yao, "Isobaric vapor-liquid equilibria for water + acetic acid + (N-methyl pyrrolidone or N-methyl acetamide)," *Fluid Phase Equilibria*, vol. 242, no. 2, pp. 204–209, 2006.

[15] D. Gao, D. Zhu, H. Sun, L. Zhang, H. Chen, and J. Si, "Isobaric vapor-liquid equilibria for binary and ternary mixtures of methanol, ethanoic acid, and propanoic acid," *Journal of Chemical and Engineering Data*, vol. 55, no. 9, pp. 4002–4009, 2010.

[16] D. Gao, D. Zhu, L. Zhang et al., "Isobaric vapor—liquid equilibria for binary and ternary mixtures of propanal, propanol, and propanoic acid," *Journal of Chemical and Engineering Data*, vol. 55, no. 12, pp. 5887–5895, 2010.

[17] A. Tamir and J. Wisniak, "Vapour-liquid equilibria in associating solutions," *Chemical Engineering Science*, vol. 30, no. 3, pp. 335–342, 1975.

[18] A. Apelblat and F. Kohler, "Excess Gibbs energy of methanol + propionic acid and of methanol + butyric acid," *The Journal of Chemical Thermodynamics*, vol. 8, no. 8, pp. 749–756, 1976.

[19] A. Rius, J. L. Otero, and A. Macarron, "Equilibres liquide-vapeur de mélanges binaires donnantune réaction chimique: systèmes méthanol-acide acétique; éthanol-acide acétique; n-propanol-acide acétique; n-butanol-acide acétique," *Chemical Engineering Science*, vol. 10, no. 1-2, pp. 105–111, 1959.

[20] Z. H. Wang, B. Wu, J. W. Zhu, K. Chen, and W. Q. Wang, "Isobaric vapor-liquid-liquid equilibrium for isopropenyl acetate plus water plus acetic acid at 101 kPa," *Fluid Phase Equilibria*, vol. 314, no. 2, pp. 152–155, 2012.

[21] S. Bernatová, K. Aim, and I. Wichterle, "Isothermal vapour-liquid equilibrium with chemical reaction in the quaternary water + methanol + acetic acid + methyl acetate system, and in five binary subsystems," *Fluid Phase Equilibria*, vol. 247, no. 1-2, pp. 96–101, 2006.

[22] A. Tamir and J. Wisniak, "Association effects in ternary vapour-liquid equilibria," *Chemical Engineering Science*, vol. 31, no. 8, pp. 625–630, 1976.

[23] A. Tamir and J. Wisniak, "Vapor equilibrium in associating systems (water-formic acid-propionic acid)," *Industrial and Engineering Chemistry Fundamentals*, vol. 15, no. 4, pp. 274–280, 1976.

[24] A. Tamir and J. Wisniak, "Activity coefficient calculations in multicomponent associating systems," *Chemical Engineering Science*, vol. 33, no. 6, pp. 651–656, 1978.

[25] J. Wisniak and A. Tamir, "Vapor-liquid equilibria in the ternary systems water-formic acid-acetic acid and water-acetic acid-propionic acid," *Journal of Chemical and Engineering Data*, vol. 22, no. 3, pp. 253–260, 1977.

[26] J. Marek and G. Standart, "Vapor-liquid equilibria in mixtures containing an associating substance. I. Equilibrium relationships for systems with an associating component," *Collection of Czechoslovak Chemical Communications*, vol. 19, pp. 1074–1084, 1954.

[27] J. Marek, "Vapor-liquid equilibria in mixtures containing an associating substance. II. Binary mixtures of acetic acid at atmospheric pressure," *Collection of Czechoslovak Chemical Communications*, vol. 20, pp. 1490–1502, 1955.

[28] J. G. Hayden and J. P. O'Conell, "A generalized method for predicting second virial coefficient," *Industrial & Engineering Chemistry Process Design and Development*, vol. 14, no. 3, pp. 209–215, 1975.

[29] J. A. Riddick, W. B. Bunger, and T. K. Sakano, *Organic Solvents: Physical Properties and Methods of Purification*, Wiley-Interscience, New York, NY, USA, 4th edition, 1986.

[30] M. V. P. Rao and P. R. Naidu, "Excess volumes of binary mixtures of alcohols in methylcohexane," *Canadian Journal of Chemistry*, vol. 52, no. 5, pp. 788–790, 1974.

[31] Z. M. Qiu, Z. C. Luo, and Y. Hu, "The pump ebullionmeter," *Journal of Chemical Engineering of Chinese Universities*, vol. 11, no. 1, pp. 74–77, 1997.

[32] J. Gmehling and U. Onken, *Vapor-Liquid Equilibrium Data Collection-Aqueous Organic Systems*, vol. 1, part 1 of *DECHEMA Chemistry Data Series*, DECHEMA, Frankfurt, Germany, 1977.

[33] J. Gmehling and U. Onken, *Vapor-Liquid Equilibrium Data Collection-Organic Hydroxy Compounds: Alcohols*, vol. 1, part 2 of *DECHEMA Chemistry Data Series*, DECHEMA, Frankfurt, Germany, 1977.

[34] R. Francesconi and F. Comelli, "Liquid-phase enthalpy of mixing for the system 1,3-dioxolane-chlorobenzene in the temperature range 288.15–313.15 K," *Journal of Chemical and Engineering Data*, vol. 31, no. 2, pp. 250–252, 1986.

[35] J. Gmehling, "Excess enthalpies for 1,1,1-trichloroethane with alkanes, ketones, and esters," *Journal of Chemical and Engineering Data*, vol. 38, no. 1, pp. 143–146, 1993.

[36] J. Shi, J. D. Wang, and G. Z. Yu, *Handbook of Chemical Engineering*, Chemical Industry Press, Beijing, China, 2nd edition, 1996.

[37] C. Tsonopoulos, "An empirical correlation of second virial coefficients," *AIChE Journal*, vol. 20, no. 2, pp. 263–272, 1974.

[38] E. Lam, A. Mejía, H. Segura, J. Wisniak, and S. Loras, "A model-free approach data treatment of vapor-liquid equilibrium data in ternary systems. 1. Theory and numerical procedures," *Industrial and Engineering Chemistry Research*, vol. 40, no. 9, pp. 2134–2148, 2001.

[39] H. Renon and J. M. Prausnitz, "Estimation of parameters for the NRTL equation for excess gibbs energies of strongly nonideal liquid mixtures," *Industrial & Engineering Chemistry Process Design and Development*, vol. 8, no. 3, pp. 413–419, 1969.

[40] G. M. Wilson, "Vapor-liquid equilibrium. XI. A new expression for the excess free energy of mixing," *Journal of the American Chemical Society*, vol. 86, no. 2, pp. 127–130, 1964.

[41] M. S. Margules, "On the composition of saturated vapors of mixtures," *Akademie der Wissenschaften in Wien, Mathematisch-Naturwissenschaftliche Klasse Abteilung II*, vol. 104, pp. 1234–1239, 1895.

[42] J. Wisniak, A. Apelblat, and H. Segura, "Application of model-free computation techniques and limiting conditions for azeotropy. An additional assessment of experimental data," *Chemical Engineering Science*, vol. 52, no. 23, pp. 4393–4402, 1997.

[43] E. Lam, A. Mejía, H. Segura, J. Wisniak, and S. Loras, "A model-free approach data treatment of vapor-liquid equilibrium data in ternary systems. 2. Applications," *Industrial and Engineering Chemistry Research*, vol. 40, no. 9, pp. 2149–2159, 2001.

[44] J. J. van Laar, "The vapor pressure of binary mixtures," *Zeitschrift für Physikalische Chemie*, vol. 72, pp. 723–751, 1910.

[45] J. Gmehling and U. Onken, *Vapor-Liquid Equilibrium Data Collection-Carboxylic Acids, Anhydrides, Esters*, vol. 1, part 5 of *DECHEMA Chemistry Data Series*, DECHEMA, Frankfurt, Germany, 1977.

[46] J. Gmehling and U. Onken, *Vapor-Liquid Equilibrium Data Collection-Organic Hydroxy Compounds: Alcohols (Supplement 3)*, vol. 1, part 2 of *DECHEMA Chemistry Data Series*, DECHEMA, Frankfurt, Germany, 1977.

10

Excess Molar Volumes and Viscosities for the Binary Mixtures of n-Octane, n-Decane, n-Dodecane, and n-Tetradecane with Octan-2-ol at 298.15 K

Arvind R. Mahajan and Sunil R. Mirgane

P. G. Department of Chemistry, Jalna Education Society's R. G. Bagdia Arts, R. Benzonji Science College, S. B. Lakhotia Commerce, Jalna, Maharashtra 431 203, India

Correspondence should be addressed to Arvind R. Mahajan; marvind22@yahoo.co.in

Academic Editor: M. A. Rosen

Experimental values of densities (ρ) and viscosities (η) in the binary mixtures of n-octane, n-decane, n-dodecane, and n-tetradecane with octan-2-ol are presented over the whole range of mixture composition at T = 298.15 K. From these data, excess molar volume (V_m^E), deviations in viscosity ($\Delta\eta$), and excess Gibbs free energy of activation ΔG^{*E} have been calculated. These results were fitted to Redlich-Kister polynomial equations to estimate the binary coefficients and standard errors. Jouyban-Acree model is used to correlate the experimental values of density and viscosity at T = 298.15 K. The values of V_m^E have been analyzed using Prigogine-Flory-Patterson (PFP) theory. The results of the viscosity composition are discussed in the light of various viscosity equations suggested by Grunberg-Nissan, Tamara and Kurata, Hind et al., Katti and Chaudhri, Heric, Heric and Brewer, and McAllister multibody model. The values of $\Delta \ln \eta$ have also been analyzed using Bloomfield and Dewan model. The experiments on the constituted binaries are analyzed to discuss the nature and strength of intermolecular interactions in these mixtures.

1. Introduction

Density and viscosity data for liquid mixtures are important from practical and theoretical points of view. Experimental measurements of these properties for binary mixtures have gained much importance in many chemical industries and engineering disciplines [1]. Experimental liquid viscosities of pure hydrocarbons and their mixtures are needed for the design of chemical processes where heat and mass transfer and fluid mechanics are important. Prediction of the liquid behavior of hydrocarbon mixture viscosities is not yet possible within the experimental uncertainty. Therefore, experimental measurements are needed to understand the fundamental behavior of this property and then to develop new models [2]. Alkanes are important series of homologous, nonpolar, and organic solvents. They have often been used in the study of solute dynamics because their physicochemical properties as a function of chain length are well-known [3]. They are also employed in a large range of chemical processes [4]. The physicochemical properties play an important role in the understanding of several industrial processes. Properties such as viscosity or surface tension are required in many empirical equations for different operations such as mass and heat transfer processes. For example, it is necessary to know the mass transfer coefficient to design gas-liquid contactors. To determine the equations that modelize the mass transfer process requires knowledge of the density, viscosity, and surface tension of the liquid phase [5].

Viscosities and excess molar volumes of binary mixtures of methylbenzene [6] and methylcyclohexane [7] with octan-2-ol at T = 298.15 K are only reported. To the best of our knowledge, the properties of the binary mixtures of n-octane, n-decane, n-dodecane, and n-tetradecane with octan-2-ol have not been reported earlier.

In the present paper, we report density and viscosity data for the binary mixtures of n-octane, n-decane, n-dodecane, and n-tetradecane with octan-2-ol T = 298.15 K. This work will also provide a test of various semiempirical equations to correlate viscosity of binary mixtures. The types of used relations are Tamara and Kurata, Heric, Heric and Brewer, Hind et al., Katti and Chaudhri, and McAllister multibody model.

2. Experimental

Chemicals used in the present study were of analytical grade and supplied by S.D. Fine Chemicals Pvt Ltd. Mumbai with quoted mass fraction purities: n-octane (>0.99), n-decane (>99.6), n-dodecane (>99.8), and n-tetradecane (>99.7). Octan-2-ol (purity >99.3) was supplied by E-Merck. Prior to use, all liquids were stored over 0.4 nm molecular sieves to reduce the water content and were degassed. The binary mixtures of varying composition were prepared by mass in special air-tight bottles. The masses were recorded on a Mettler balance to an accuracy of $\pm 1 \times 10^{-5}$ g. Care was taken to avoid evaporation and contamination during mixing. The estimated uncertainty in mole fraction was $< 1 \times 10^{-4}$.

The densities of the solutions were measured using a single capillary pycnometer made up of borosil glass with a bulb of 8 cm^3, and capillary with internal diameter of 0.1 cm was chosen for the present work. The detailed pertaining to calibration, experimental set up, and operational procedure has been previously described [6–10]. An average of triplicate measurement was taken in to account. The reproducibility of density measurement was $\pm 3 \times 10^{-5}$ g/cm^3.

The dynamic viscosities were measured using an Ubbelohde suspended level viscometer [6–10] calibrated with conductivity water. An electronic digital stop watch with readability of ±0.01 s was used for the flow time measurements. At least three repetitions of each data reproducible to ±0.05 s were obtained, and the results were averaged. Since all flow times were greater than 300 s and capillary radius (0.1 mm) was far less than its length (50 to 60) mm, the kinetic energy and end corrections, respectively, were found to be negligible. The uncertainties in dynamic viscosities are of the order of ±0.003 mPa·s.

The purity of the samples and accuracy of data were checked by comparing the measured densities and viscosities of the pure compounds with the literature values, which are given in Table 1. Thus, our results are in good agreement with those listed in the literature.

Table 1: Comparison of experimental densities (ρ), viscosities (η), and speeds of sound (u) of pure liquids with the literature values at $T = 298.15$ K.

Pure liquids	$\rho \times 10^{-3}$ (Kg m^{-3})		η (mPa·s)	
	Experimental	Literature	Experimental	Literature
n-Octane	0.69867	0.69866[a]	0.512	0.515[b]
n-Decane	0.72635	0.7267[c]	0.845	0.861[d]
n-Dodecane	0.74518	0.74172[e]	1.336	1.332[f]
n-Tetradecane	0.75913	0.75911[g]	2.081	2.077[g]
Octan-2-ol	0.81705	0.81708[h]	6.429	6.435[h]

[a][39], [b][37], [c][48], [d][42], [e][41], [f][38], [g][38], [h][6].

3. Results and Discussion

Experimental values of densities (ρ) and viscosities (η) of mixtures at 298.15 K are listed as a function of mole fraction in Table 2. The density values have been used to calculate excess molar volumes (V^E) using the following equation:

$$V_m^E/\text{m}^3 \cdot \text{mol}^{-1} = \frac{(x_1 M_1 + x_2 M_2)}{\rho_{12}} - \left(\frac{x_1 M_1}{\rho_1}\right) - \left(\frac{x_2 M_2}{\rho_2}\right), \quad (1)$$

where ρ_{12} is the densities of the mixture and x_1, M_1, ρ_1, and x_2, M_2, ρ_2 are the mole fractions, the molecular weights, and the densities of pure components 1 and 2, respectively.

The variation in excess molar volumes, V_m^E with mole fraction of the binary mixtures of n-octane, n-decane, n-dodecane, and n-tetradecane with octan-2-ol at $T = 298.15$ K, is displayed in Figure 1. The V_m^E curve for the mixture of n-octane and octan-2-ol is sigmoidal and tends to change to positive values at higher mole fractions ($x_1 \geq 0.65$) of n-octane while V_m^E values for n-decane, n-dodecane, and n-tetradecane with octan-2-ol mixtures show positive deviation over the entire composition range.

Generally, V_m^E can be considered as arising from three types of interactions between component molecules of liquid mixtures [11–13]: (1) physical interactions consisting of mainly of dispersion forces or weak dipole-dipole interaction making a +ve contribution, hereby the contraction volume and compressibility of the mixtures, (2) chemical or specific interactions, which include charge transfer, forming of H-bonds and other complex forming interactions, resulting in a −ve contribution, and (3) structural contribution due to differences in size and shape of the component molecules of the mixtures, due to fitting of component molecules into each other's structure, hereby reducing the volume and compressibility of the mixtures, resulting in a −ve contribution.

The large positive V_m^E values (Figure 1) for n-decane, n-dodecane, and n-tetradecane with octan-2-ol mixture are attributed to the breaking up of three-dimensional H-bonded network of octan-2-ol due to the addition of solute, which is not compensated by the weak interactions between unlike molecules. The V_m^E exhibits an inversion in sign in the mixtures f (n-octane + 2-octanol). The values of V_m^E are negative in the lower region of x_1 due to interstitial accommodation is more as compared to the de-clustering of 1-octanol molecules and beyond ($x_1 \geq 0.65$) as the amount of n-octane increases in the mixture, due to dispersion forces, thereby making a positive contribution to V_m^E.

The viscosity deviations ($\Delta\eta$) were calculated using

$$\Delta\eta/\text{mPa} \cdot \text{s} = \eta_{12} - x_1\eta_1 - x_2\eta_2, \quad (2)$$

where η_{12} is the viscosities of the mixture and x_1, x_2 and η_1, η_2 are the mole fraction and the viscosities of pure components 1 and 2, respectively.

Figure 2 depicts the variation of $\Delta\eta$ with mole fraction of the binary mixtures of n-octane, n-decane, n-dodecane, and n-tetradecane with octan-2-ol at $T = 298.15$ K.

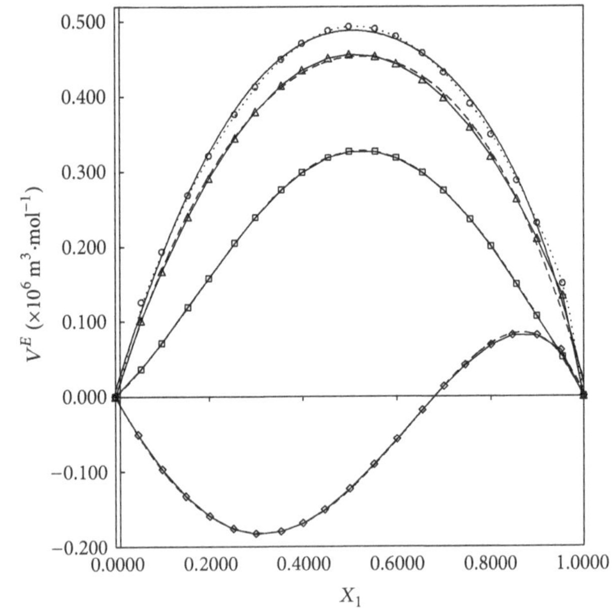

FIGURE 1: Plot of excess molar volumes (V_m^E) against mole fraction of octan-2-ol with (◊) n-octane; (□) n-decane; (△) n-dodecane, and n-tetradecane (○) at T = 298.15 K. The corresponding dotted (- - -) curves have been derived from PFP theory.

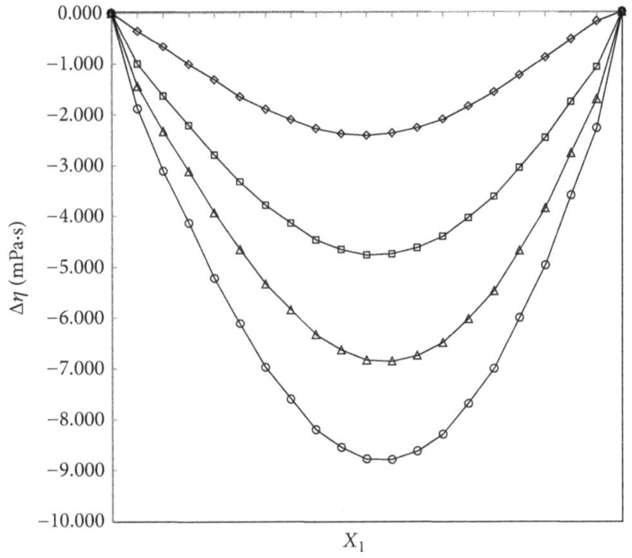

FIGURE 2: Plot of viscosity deviations ($\Delta\eta$) against mole fraction of octan-2-ol with (◊) n-octane; (□) n-decane; (△) n-dodecane and n-tetradecane (○) at T = 298.15 K.

The deviations viscosity may be generally explained by considering the following factors [14]. (1) The difference in size and shape of the component molecules and the loss of dipolar association to a decrease in viscosity; (2) specific interactions between unlike molecules such as H-bond formation and charge transfer complexes may cause increase in viscosity in mixtures rather than in pure component. The former effect produces negative in excess viscosity, and latter effect produces positive in excess viscosity. Positive values of $\Delta\eta$ are indicative of strong interactions whereas negative values indicate weaker interactions [15].

The negative deviations in viscosity support the main factor of breaking of the self-associated alcohols and weak interactions between unlike molecules. The negative values viscosity deviation decreases in the following sequence: n-octane > n-decane > n-dodecane > n-tetradecane.

Excess Gibbs free energies of activation of viscous flow ΔG^{*E} for binary mixtures can be calculated as

$$\Delta G^{*E} = RT\left[\ln\left(\frac{\eta v}{\eta_2 v_2}\right) - x_1 \ln\left(\frac{\eta_1 v_1}{\eta_2 v_2}\right)\right], \quad (3)$$

where v is the molar volume of the mixture, v_i is the molar volume of the pure component, R is the gas constant, T is the absolute temperature, and η is the dynamic viscosity of the mixture, respectively. η_i is the dynamic viscosity of the pure component i and x_1 the mole fraction in component. The ΔG^{*E} values of all binary systems are shown in Table 2. The values of ΔG^{*E} for all binary mixtures are negative over entire mole fraction. According to Meyer et al. [16], negative values of ΔG^{*E} correspond to the existence of solute-solute association.

The excess molar volumes and deviations in viscosity were fitted to Redlich and Kister [17] equation of the type

$$Y = x_1 x_2 \sum_i^n a_i (x_1 - x_2)^i, \quad (4)$$

where Y is either V^E or $\Delta\eta$ and n is the degree of polynomial. Coefficient a_i was obtained by fitting (5) to experimental results using a least-squares regression method. In each case, the optimum number of coefficients is ascertained from an examination of the variation in standard deviation (σ).

σ was calculated using the relation

$$\sigma(Y) = \left[\frac{\sum\left(Y_{\text{expt}} - Y_{\text{calc}}\right)^2}{N - n}\right]^{1/2}, \quad (5)$$

where N is the number of data points and n is the number of coefficients. The calculated values of the coefficients a_i along with the standard deviations (σ) are given in Table 3.

4. Theoretical Analysis

4.1. Semiempirical Models for Analyzing Viscosity of Liquid Mixtures. Several empirical and semiempirical relations have been used to represent the dependence of viscosity on concentration of components in binary liquid mixtures, and these are classified according to the number of adjustable parameters used to account for the deviation from some average [18, 19]. We will consider here some of the most commonly used semiempirical models for analyzing viscosity of liquid mixtures based on one, two, and three parameters. An attempt has been made to check the suitability of equations for experimental data fits by taking into account the number of empirical adjustment coefficients.

Table 2: Densities (ρ), viscosities (η), excess molar volumes V^E, viscosities deviations $\Delta\eta$, and excess Gibbs free energy ΔG^{*E} of binary mixtures at T = 298.15 K.

x_1	$\rho \times 10^{-3}$ kg·m^{-3}	$V_m^E \times 10^6$ m^3·mol^{-1}	η mPa·s	$\Delta\eta$ mPa·s	ΔG^{*E} KJmol^{-1}
		n-Octane(1) + octan-2-ol(2)			
0	0.81705	0	6.429	0	0
0.0487	0.8114	−0.051	5.956	−0.185	115
0.1002	0.8054	−0.097	5.301	−0.535	149
0.1498	0.79959	−0.133	4.654	−0.888	137
0.2006	0.79361	−0.16	3.998	−1.244	78
0.2516	0.78756	−0.177	3.364	−1.576	−30
0.3003	0.78176	−0.183	2.796	−1.856	−183
0.3538	0.77534	−0.18	2.228	−2.108	−410
0.4003	0.76975	−0.169	1.79	−2.27	−661
0.4466	0.76415	−0.151	1.411	−2.375	−960
0.5001	0.75767	−0.123	1.049	−2.421	−1359
0.5517	0.75142	−0.091	0.778	−2.387	−1776
0.6002	0.74554	−0.057	0.59	−2.287	−2157
0.6539	0.73905	−0.018	0.454	−2.106	−2469
0.6994	0.73357	0.013	0.39	−1.901	−2560
0.7449	0.72811	0.042	0.366	−1.656	−2431
0.8	0.72157	0.068	0.376	−1.32	−2019
0.8465	0.7161	0.081	0.402	−1.018	−1571
0.9003	0.70985	0.081	0.434	−0.668	−1034
0.9521	0.70394	0.061	0.441	−0.355	−670
1	0.69867	0	0.512	0	0
		n-Decane(1) + octan-2-ol(2)			
0	0.81705	0	6.429	0	0
0.0554	0.81077	0.036	5.222	−0.898	−233
0.0998	0.80583	0.071	4.638	−1.234	−301
0.1554	0.79974	0.119	3.972	−1.589	−403
0.1999	0.79499	0.158	3.491	−1.822	−497
0.2554	0.7892	0.205	2.951	−2.052	−632
0.2999	0.78467	0.239	2.564	−2.19	−755
0.3555	0.77916	0.275	2.136	−2.308	−927
0.3998	0.7749	0.298	1.836	−2.361	−1078
0.4554	0.76969	0.318	1.509	−2.377	−1285
0.4999	0.76564	0.326	1.284	−2.354	−1461
0.5555	0.76074	0.326	1.046	−2.281	−1690
0.5998	0.75694	0.318	0.889	−2.191	−1871
0.6555	0.7523	0.298	0.73	−2.039	−2080
0.7	0.7487	0.274	0.63	−1.89	−2223
0.7554	0.74432	0.236	0.538	−1.673	−2337
0.7998	0.74091	0.2	0.487	−1.476	−2362
0.8554	0.73674	0.149	0.45	−1.203	−2281
0.8999	0.73347	0.106	0.438	−0.966	−2126
0.9554	0.72948	0.052	0.445	−0.649	−1811
1	0.72635	0	0.845	0	0
		n-Dodecane(1) + octan-2-ol(2)			
0	0.81705	0	6.429	0	0
0.0554	0.81098	0.101	5.513	−0.634	−159
0.0999	0.80637	0.167	4.916	−1.004	−296
0.1554	0.80091	0.24	4.254	−1.383	−404

Table 2: Continued.

x_1	$\rho \times 10^{-3}$ kg·m^{-3}	$V_m^E \times 10^6$ m^3·mol^{-1}	η mPa·s	$\Delta\eta$ mPa·s	ΔG^{*E} KJmol^{-1}
0.1998	0.79676	0.291	3.786	−1.625	−517
0.2555	0.7918	0.345	3.272	−1.856	−659
0.2998	0.78805	0.38	2.916	−1.986	−770
0.3554	0.78355	0.415	2.532	−2.087	−902
0.3998	0.78012	0.435	2.27	−2.123	−999
0.4554	0.77603	0.451	1.995	−2.115	−1102
0.4999	0.7729	0.456	1.813	−2.071	−1167
0.5555	0.76916	0.453	1.627	−1.973	−1219
0.5998	0.76631	0.444	1.508	−1.866	−1236
0.6555	0.76288	0.422	1.391	−1.699	−1221
0.6998	0.76027	0.398	1.32	−1.545	−1180
0.7554	0.75713	0.359	1.253	−1.328	−1095
0.7998	0.75473	0.32	1.213	−1.142	−1005
0.8555	0.75185	0.263	1.176	−0.896	−869
0.8999	0.74965	0.21	1.152	−0.694	−750
0.9555	0.747	0.134	1.124	−0.439	−599
0.9997	0.74519	0	1.337	0	0
		n-Tetradecane(1) + octan-2-ol(2)			
0	0.81705	0	6.429	0	0
0.0555	0.81134	0.126	5.619	−0.568	−144
0.0999	0.80721	0.193	5.163	−0.831	−230
0.1555	0.80239	0.268	4.632	−1.121	−330
0.1998	0.7988	0.321	4.238	−1.321	−417
0.2554	0.79459	0.376	3.786	−1.533	−531
0.2998	0.79144	0.413	3.455	−1.67	−628
0.3555	0.78774	0.449	3.081	−1.802	−751
0.3999	0.78496	0.47	2.813	−1.876	−850
0.4554	0.78171	0.487	2.519	−1.93	−968
0.4999	0.77925	0.492	2.314	−1.941	−1054
0.5555	0.77637	0.49	2.098	−1.915	−1144
0.5998	0.7742	0.48	1.957	−1.863	−1196
0.6554	0.77163	0.457	1.82	−1.759	−1226
0.7	0.76969	0.432	1.741	−1.644	−1217
0.7555	0.76741	0.39	1.683	−1.46	−1154
0.8	0.76569	0.348	1.668	−1.282	−1060
0.8555	0.76365	0.287	1.688	−1.02	−887
0.9	0.76211	0.23	1.736	−0.779	−703
0.9553	0.76029	0.15	1.835	−0.439	−424
0.9997	0.75914	0	2.081	0	0

The equation of Grunberg-Nissan, Tamara and Kurata Hind et al., and Katti and Chaudhri has one adjustable parameter.

Gruenberg-Nissan provided the following empirical equation containing one adjustable parameter [20]. The equation is

$$\ln \eta_{12} = x_1 \ln \eta_1 + x_2 \ln \eta_2 + x_1 x_2 G_{12}, \quad (6)$$

where G_{12} may be regarded as a parameter proportional to the interchange energy also an approximate measure of the strength of the interaction between the components.

The one-parameter equation due to Tamura and Kurata [21] gave the equation of the form

$$\eta_m = x_1 \eta_1 \Phi + x_2 \eta_2 \Phi_2 + 2\left(x_1 x_2 \Phi \Phi_2\right)^{1/2} T_{12}, \quad (7)$$

where Φ and Φ_2 are the volume fractions of components 1 and 2, respectively; T_{12} is Tamura and Kurata constant.

TABLE 3: Coefficients a_i of (4) and corresponding standard deviation (σ) of (5) at $T = 298.15$ K.

System	a_0	a_1	a_2	a_3	σ
n-Octane + octan-2-ol					
$V_m^E \times 10^6/(\mathrm{m^3 \cdot mol^{-1}})$	0.125	0.1577	−0.41	0.1523	0.943
$\Delta\eta/(\mathrm{mPa \cdot s})$	−9.6807	0.4921	4.661	−2.7989	0.024
n-Decane + octan-2-ol					
$V_m^E \times 10^6/(\mathrm{m^3 \cdot mol^{-1}})$	0.129	0.167	−0.462	0.1437	0.0023
$\Delta\eta/(\mathrm{mPa \cdot s})$	−8.4704	2.213	−6.8274	−1.0134	0.1273
n-Dodecane + octan-2-ol					
$V_m^E \times 10^6/(\mathrm{m^3 \cdot mol^{-1}})$	0.363	−1.276	0.289	0.812	0.529
$\Delta\eta/(\mathrm{mPa \cdot s})$	−7.989	3.365	−2.905	8.812	0.060
n-Tetradecane + octan-2-ol					
$V_m^E \times 10^6/(\mathrm{m^3 \cdot mol^{-1}})$	0.880	−0.116	0.009	0.708	0.020
$\Delta\eta/(\mathrm{mPa \cdot s})$	−6.322	−4.730	−10.25	15.66	0.377

Hind et al. [22] proposed the following equation:

$$\eta_m = x_1^2\eta_1 + x_2^2\eta_2 + 2x_1x_2H_{12}, \quad (8)$$

where H_{12} is attributed to unlike pair interactions.

Katti and Chaudhri [23] derived the following equation

$$\ln \eta V = x_1 \ln \eta V_1^0 + x_2 \ln \eta V_2^0 + \frac{x_1x_2W_{\mathrm{vis}}}{RT}, \quad (9)$$

where W_{vis} is an interaction term and v_i is the molar volume of pure component i.

Heric [24] expression is

$$\begin{aligned}\ln \eta_m = {} & x_1 \ln \eta_1 + x_2 \ln \eta_2 + x_1 \ln \eta_1 \\ & + x_2 \ln \eta_2 + \ln (x_1\eta_1 + x_2\eta_2) + \delta_{12},\end{aligned} \quad (10)$$

where $\delta_{12} = \alpha_{12}x_1x_2$ is a term representing departure from a noninteracting system and $\alpha_{12} = \alpha_{21}$ is the interaction parameter. Either α_{12} or α_{21} can be expressed as a linear function of composition:

$$\alpha_{12} = \beta'_{12} + \beta''_{12}(x_1 - x_2). \quad (11)$$

From an initial guess of the values of coefficients β'_{12} and β''_{12}, the values of α_{12} are computed.

Heric and Brewer [25] equation is

$$\begin{aligned}\ln \nu = {} & x_1 \ln \nu_1 + x_2 \ln \nu_2 + x_1 \ln M_1 + x_2 \ln M_2 \\ & - \ln [x_1M_1 + x_2M_2] + x_1x_2 [\alpha_{12} + \alpha_{21}(x_1 - x_2)].\end{aligned} \quad (12)$$

M_1 and M_2 are molecular weights of components 1 and 2, and α_{12} and α_{21} are interaction parameters, and other terms involved have their usual meaning. α_{12} and α_{21} are parameters, which can be calculated from the least-squares method.

McAllister's multibody interaction model [26] was widely used to correlate kinematic viscosity (ν) data. The two-parameter McAllister equation based on Eyring's [27] theory of absolute reaction rates has taken into account interaction of both like and unlike molecules by two-dimensional three-body model. The three-body interaction model is

$$\begin{aligned}\ln \nu_m = {} & x_1^3 \ln \nu_1 + 3x_1^2x_2 \ln Z_{12} + 3x_1x_2^2 \ln Z_{21} \\ & + x_2^3 \ln \nu_2 - \ln \left[x_1 + \frac{(x_2M_2)}{M_1}\right] \\ & + 3x_1^2x_2 \ln \left[\frac{2}{3} + \frac{M_2}{(3M_1)}\right] \\ & + 3x_1x_2^2 \ln \left[\frac{1}{3} + \frac{2M_2}{(3M_1)}\right] x_2^3 \ln \left(\frac{M_2}{M_1}\right).\end{aligned} \quad (13)$$

And four-body model was given by

$$\begin{aligned}\ln \nu_m = {} & x_1^4 \ln \nu_1 + 4x_1^3x_2 \ln Z_{1112} \\ & + 6x_1^2x_2^2 \ln Z_{1122} + 4x_1x_2^3 \ln Z_{2221} + x_2^4 \ln \nu_2 \\ & - \ln \left[x_1 + x_2\left(\frac{M_2}{M_1}\right)\right] \\ & + 4x_1^3x_2 \ln \left[3 + \frac{(M_2/M_1)}{4}\right] \\ & + 6x_1^2x_2^2 \ln \left[1 + \frac{(M_2/M_1)}{2}\right] \\ & + 4x_1x_2^3 \ln \left[1 + \frac{(3M_2/M_1)}{4}\right] + x_2^4 \ln \left(\frac{M_2}{M_1}\right),\end{aligned} \quad (14)$$

where Z_{12}, Z_{21}, Z_{1112}, Z_{1122}, and Z_{2221} are interaction parameters and M_i and ν_i are the molecular mass and kinematic viscosity of pure component i, respectively.

The correlating ability of each of (6)–(14) was tested as well as their adjustable parameters and standard deviations (σ):

$$\sigma\,(\%) = \left[\left(\frac{1}{(n-k)}\sum \frac{100\left(\eta_{\mathrm{exptl}} - \eta_{\mathrm{calcd}}\right)}{\eta_{\mathrm{exptl}}}\right)^2\right]^{1/2}, \quad (15)$$

where n represents the number of data points and k is the number of numerical coefficients given in Table 4. The interaction parameter G_{12} is negative for binary systems. Nigam and Mahl [28] concluded from the study of binary mixtures that (1) if $\Delta\eta > 0$, $G_{12} > 0$ and magnitude of both are large then strong specific interaction would be present; (2) if $\Delta\eta < 0$, $G_{12} > 0$ then weak specific interaction would be present; (3) if $\Delta\eta < 0$, $G_{12} < 0$ magnitude of both are large then the dispersion force would be dominant. According to Fort and Moore [29] and Ramamoorty and Varadachari [30], system exhibits strong interaction if the G_{12} is positive; if it is negative they show weak interaction. On this basis, we can say that there is a weak interaction in the system studied.

Interaction parameter W_{vis} shows almost the same trend as that of G_{12}. In fact, one could say that the parameters

TABLE 4: Adjustable parameters of (6)–(13) and standard deviations of binary mixture viscosities for x_1 n-alkanes + $(1 - x_1)$ octan-2-ol at $T = 298.15$ K.

Equation	System including n-alkanes + octan-2-ol			
	n-Octane	n-Decane	n-Dodecane	n-Tetradecane
	Grunberg-Nissan			
G_{12}	−2.398	−3.226	−1.997	−2.002
σ	3.759	3.904	0.858	1.075
	Tamura and Kurata			
T_{12}	0.872	−1.117	0.184	0.675
σ	1.737	1.884	0.812	0.782
	Hind et al.			
H_{12}	−0.91	−1.409	−0.401	0.079
σ	1.758	2.145	1.172	0.711
	Katti and Chaudhri			
W_{vis}	−2.402	−3.198	−1.96	−1.871
σ	3.755	3.905	0.857	1.081
	Heric and Brewer			
α_{12}	−2.381	−3.936	−2.109	−2.104
α_{21}	−4.123	−5.529	−1.506	−0.827
σ	−0.926	2.449	−0.559	0.866
	McAllister's three-body			
Z_{12}	0.185	0.094	0.879	1.439
Z_{21}	6.363	7.068	3.929	3.544
σ	0.926	2.449	0.559	0.866
	McAllister's four-body			
Z_{1112}	0.269	0.51	0.817	1.140
Z_{1122}	0.936	19.584	3.639	7.334
Z_{2221}	6.926	1.998	3.598	2.871
σ	5.149	5.865	0.622	1.622

G_{12} and W_{vis} exhibit almost similar behaviour, which is not unlikely in view of logarithmic nature of both equations.

Tamara and Kurata and Hind et al. represent the binary mixture satisfactory as compared to Gruenberg-Nissan and Katti and Chaudhri. Use of three parameters equation reduces the σ values significantly below that of two parameters equation. From this study, it can be concluded that the correlating ability significantly improves for these nonideal systems as number of adjustable parameters is increased. From Table 4, it is clear that McAllister's three-body interaction model is suitable for correlating the kinematic viscosities of the binary mixtures studied.

4.2. Prigogine-Flory-Patterson (PFP) Theory. The Prigogine-Flory-Patterson (PFP) theory [31–34] has been commonly employed to estimate and analyze excess thermodynamic functions theoretically. This theory has been described in details by Patterson and coworkers [35, 36]. According to PFP theory, V_m^E can be separated into three factors: (1) an interactional contribution, V_m^E (int.) (2) a free volume contribution, $V_m^E(\text{fv})$, and (3) an internal pressure contribution, $V_m^E(P^*)$. The expression for these three contributions are given as V_m^E:

$$V_m^E(\text{int}) = \left[\frac{\left(v^{1/3}-1\right)v^{2/3}\psi_1\theta_2}{\left(4/3v^{-1/3}-1\right)P_1^*\chi_{12}}\right],$$

$$V_m^E(\text{fv}) = \frac{\left[\left(v_1-v_2\right)^2\left(14/9v^{-1/3}-1\right)\psi_1\psi_2\right]}{\left[\left(14/9v^{-1/3}-1\right)v\right]}, \quad (16)$$

$$V_m^E(P^*) = \frac{\left[\left(v_1-v_2\right)\left(P_1^*+P_2^*\right)\psi_1\psi_2\right]}{\left(P_1^*\psi_1+P_2^*\psi_2\right)}.$$

Thus, the excess molar volume V_m^E is given as

$$\frac{V_m^E}{\left(x_1V_1+x_2V_2\right)} = V_m^E(\text{int}) - V_m^E(\text{fv}) + V_m^E(P^*), \quad (17)$$

where ψ, θ, and P^* represent the contact energy fraction, surface site fraction, and characteristic pressure, respectively, and are calculated as

$$\psi_1 = \left(1-\psi_2\right) = \frac{\Phi_1P_1^*}{\left(\Phi_1P_1^*+\Phi_2P_2^*\right)}, \quad (18)$$

$$\theta_2 = \left(1-\theta_1\right) = \frac{\Phi_2}{\left[\Phi_1\left(V_2^*/V_1^*\right)\right]}, \quad (19)$$

$$P^* = \frac{Tv^2\alpha}{\kappa_T}. \quad (20)$$

The details of the notations and terms used in (16)–(19) may be obtained from the literature [31–34, 37, 38]. The other parameters pertaining to pure liquids and the mixtures are obtained from the Flory theory [7, 31, 38] while α and κ_T values are taken from the literature [39–44]. The interaction parameter χ_{12} required for the calculation of V_m^E using PFP theory has been derived by fitting the V_m^E expression to the experimental equimolar value of V_m^E for each system under study.

The values of χ_{12}, θ_2, three PFP contributions interactional, free volume, P^* effect, and experimental and calculated (using PFP theory) V_m^E values at near equimolar composition are presented in Table 5. Study of the data presented in Table 6 reveals that the interactional and free volume contributions are positive, whereas P^* contributions are negative for all the three systems under investigation. For these binary mixtures, it is only the interactional contribution which dominates over the remaining two contributions. The P^* contribution, which depends both on the differences of internal pressures and differences of reduced volumes of the components, has little significance for the studied binary mixtures as compared to the other two.

Furthermore, in order to check whether χ_{12}, derived from nearly equimolar V_m^E values, can predict the correct composition dependence, V_m^E has been calculated theoretically using χ_{12} over the entire composition range. The theoretically calculated values are plotted in Figure 1 for comparison with the experimental results. Figure 1 show that the PFP theory is

Table 5: Flory parameters of the pure compounds.

Components	$10^6 V^*/(\mathrm{m^3 \cdot mol^{-1}})$	$10^6 P^*$ $(\mathrm{J \cdot m^{-3}})$	T^*/K
Octan-2-ol	129.9790	535	5563
n-Octane	127.4844	444	4826
n-Decane	155.6091	453	5091
n-Dodecane	183.7700	455	5290
n-Tetradecane	212.1200	460	5479

quite successful in predicting the trend of the dependence of V_m^E on composition for the present systems.

In order to perform a numerical comparison of the estimation capability of the PFP theory, we calculated the standard percentage deviations ($\sigma\%$) using the relation

$$\sigma\% = \left[\sum\left\{\frac{100\,(\mathrm{expt}-\mathrm{theor.})}{\mathrm{expt}}\right\}^2 (n-1)\right]^{1/2}, \quad (21)$$

where n represents the number of experimental data points.

4.3. Bloomfield and Dewan Model. There are different expressions available in the literature to calculate η. Here, Bloomfield and Dewan [45] model have been applied to compare calculated $\Delta \ln \eta$ values using experimental data for each binary mixture by the following relation:

$$\Delta \ln \eta = \ln \eta - (x_1 \ln \eta_1 + x_2 \ln \eta_2). \quad (22)$$

Bloomfield and Dewan [45] developed the expression from the combination of the theories of free volumes and absolute reaction rate

$$\Delta \ln \eta = f(v) - \frac{\Delta G^R}{RT}, \quad (23)$$

where $f(v)$ is the characteristic function of the free volume defined by

$$f(v) = \frac{1/(v-1)\,x_1}{(v_1-1)} - \frac{x_2}{v_2-1} \quad (24)$$

and ΔG^R is the residual energy of mixing, calculated with the following expression [46]:

$$\Delta G^R = \Delta G^E + RT\left\{x_1 \ln\left(\frac{x_1}{\Phi_1}\right) + x_2 \ln\left(\frac{x_2}{\Phi_2}\right)\right\}, \quad (25)$$

where Φ_1 and Φ_2 are segment fractions defined by

$$\Phi_2 = 1 - \Phi_1 = \frac{x_2}{[x_2 + x_1(r_1/r_2)]}, \quad (26)$$

where r_1 and r_2 are in the ratio of respective molar core volumes V_1^* and V_2^*.

The excess energy can be obtained from the statistical theory of liquid mixtures proposed by Flory and coworkers [31, 32] and is given by

$$\begin{aligned}\Delta G^E = {} & x_1 P_1^* V_1^* \left[\frac{1}{(v_1)} - \frac{1}{(v)}\right] \\ & + 3\tilde{T}_1 \ln\left\{\frac{(v_1^{1/3}-1)}{(v^{1/3}-1)}\right\} \\ & + x_2 P_2^* V_2^* \left[\frac{1}{(v_2)} - \frac{1}{(v)}\right] \\ & + \left[3\tilde{T}_2 \ln\left\{\frac{(v_2^{1/3}-1)}{(v^{1/3}-1)}\right\}\right] \\ & + \frac{(x_1\theta_2 V_1^* \chi_{12})}{v},\end{aligned} \quad (27)$$

where the various symbols used have their usual meanings.

Using the χ_{12} values from fitting values of V_m^E and the values of the parameters for the pure liquid components, we have calculated ΔG^R and $f(v)$ and finally $\Delta \ln \eta$, according to the Bloomfield and Dewan relationship, which is compared with the experimental data. Figure 3 shows that the good agreement between the estimated and experimental curves occurs for given binary systems.

4.4. Jouyban and Acree Model. Hasan et al. [6, 47, 48] has used this model for various binary system proposed a model for correlating the density and viscosity of liquid mixtures at various temperatures. The proposed equation is

$$\ln y_{m,T} = f_1 \ln y_{1,T} + f_2 \ln y_{2,T} + f_1 f_2 \sum\left[\frac{A_j (f_1 - f_2)^j}{T}\right], \quad (28)$$

where $y_{m,T}$, $y_{1,T}$, and $y_{2,T}$ is density or viscosity of the mixture and solvents 1 and 2 at temperature T, respectively, f_1 and f_2 are the volume fractions of solvents in case of density and mole fraction in case of viscosity, and A_j are the model constants.

The correlating ability of the Jouyban-Acree model was tested by calculating the average percentage deviation (APD) between the experimental and calculated density and viscosity as

$$\mathrm{APD} = \left(\frac{100}{N}\right)\sum\left[\frac{(|y_{\mathrm{exptl}} - y_{\mathrm{calcd}}|)}{y_{\mathrm{exptl}}}\right], \quad (29)$$

where N is the number of data points in each set. The optimum numbers of constants A_j, in each case, were determined from the examination of the average percentage deviation value.

TABLE 6: Calculated values of the three contributions to the excess molar volume from the PFP theory with interaction parameter $T = 298.15$ K.

Component	$\chi_{12} \times 10^6$ (J·m^{-3})	$V^E \times 10^6$ (m^3·mol^{-1}) at $x = 0.5$		Calculated contribution		
		Experimental	PEP	$V^E \times 10^6$ (int)	$V^E \times 10^6$ (fv)	$V^E \times 10^6$ (P^*)
n-Octane + octan-2-ol	11.59	0.123	0.124	0.0331	0.1087	−0.4194
n-Decane + octan-2-ol	12.57	0.326	0.328	0.5999	0.0644	−0.2283
n-Dodecane + octan-2-ol	16.57	0.456	0.458	0.5524	0.0175	−0.0666
n-Ttetradecane + octan-2-ol	15.08	0.492	0.494	0.5157	0.0116	−0.0187

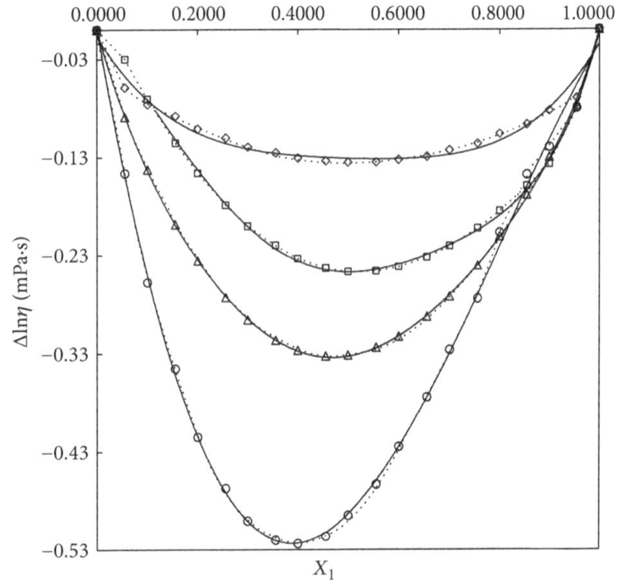

FIGURE 3: Plot of $\Delta \ln \eta$ against mole fraction of octan-2-ol with (◊) n-octane; (□) n-decane; (△) n-dodecane, and n-tetradecane (○) at $T = 298.15$ K. The corresponding dotted (- - -) curves have been derived from Bloomfield and Dewan model.

The constants A_j calculated from the least square analysis are presented in Table 7 along with the average percentage deviation (APD). The proposed model provides reasonably accurate calculations for the density, viscosity, and ultrasonic velocity of binary liquid mixtures at 298.15 K, and the model could be used in data modeling.

5. Conclusions

The present paper is a continuing effort towards the understanding of the mixing behavior of binary liquid mixtures comprising of n-alkanes + octan-2-ol. Excess molar volumes and deviations in viscosity were calculated and fitted to the Redlich-Kister equation to test the quality of the experimental values. The negative deviations in viscosity support the main factor of breaking of the self-associated alcohols and weak interactions between unlike molecules. Several empirical and semiempirical relations have been used to represent the dependence of viscosity on concentration of components in binary liquid mixtures. An attempt has been made to check the suitability of empirical and semiempirical relations for experimental viscosities data of n-alkanes + octan-2-ol fits by taking into account the number of empirical

TABLE 7: Parameters of Jouyban-Acree model and average percentage deviation for densities, viscosities, and ultrasonic velocities at of binary mixtures at $T = 298.15$ K.

System	A_0	A_1	A_2	APD
Density				
n-Octane + octan-2-ol	3.7446	−1.8027	0.616	0.0293
n-Decane + octan-2-ol	−9.0926	1.0266	0.6448	0.1183
n-Dodecane + octan-2-ol	−11.8504	0.0401	0.4771	0.0816
n-Tetradecane + octan-2-ol	−12.6833	3.5123	−4.3389	0.0559
Viscosity				
n-Octane + octan-2-ol	−7.1847	−6.9259	3.3215	0.1423
n-Decane + octan-2-ol	−4.2926	11.2857	−2.7614	0.7668
n-Dodecane + octan-2-ol	−7.5692	−10.9008	4.4397	0.1031
n-Tetradecane + octan-2-ol	−6.4032	−26.9478	−7.7829	0.4296

adjustment coefficients. Bloomfield and Dewan model and McAllister's three-body interaction model are suitable for the above binary system. PFP theory is also quite successful in predicting the trend of the dependence of V_m^E on composition for the present systems.

Acknowledgments

Authors are thankful to Prof B. R. Arbad, Dr. B. A. M. university for their valuable suggestions and discussion. Authors are also thankful to Principal Dr. R. S. Agarwal, J. E. S. College, Jalna for the facilities provided.

References

[1] K. Lal, N. Tripathi, and G. P. Dubey, "Densities, viscosities, and refractive indices of binary liquid mixtures of hexane, decane, hexadecane, and squalane with benzene at 298.15 K," *Journal of Chemical and Engineering Data*, vol. 45, no. 5, pp. 961–964, 2000.

[2] J. A. Estrada-Baltazar, F. J. Juan, and A. Gustavo, "Experimental liquid viscosities of decane and octane + decane from 298.15 K to 373.15 K and up to 25 MPa," *Journal of Chemical and Engineering Data*, vol. 43, no. 3, pp. 441–446, 1998.

[3] Y. Zhang, R. M. Venable, and R. W. Pastor, "Molecular dynamics simulations of neat alkanes: the viscosity dependence of rotational relaxation," *Journal of Physical Chemistry*, vol. 100, no. 7, pp. 2652–2660, 1996.

[4] T. M. Aminabhavi, M. I. Aralaguppi, B. Gopalakrishna, and R. S. Khinnavar, "Densities, shear viscosities, refractive indices, and speeds of sound of bis(2-methoxyethyl) ether with hexane, heptane, octane, and 2,2,4-trimethylpentane in the temperature

interval 298.15-318.15 K," *Journal of Chemical and Engineering Data*, vol. 39, no. 3, pp. 522–528, 1994.

[5] D. Gómez-Díz, J. C. Mejuto, J. M. Navaza, and A. Rodríguez-Álvarez, "Viscosities, densities, surface tensions, and refractive indexes of 2,2,4-trimethylpentane + cyclohexane + decane ternary liquid systems at 298.15 K," *Journal of Chemical and Engineering Data*, vol. 47, no. 4, pp. 872–875, 2002.

[6] M. Hasan, D. F. Shirude, A. P. Hiray, A. B. Sawant, and U. B. Kadam, "Densities, viscosities and ultrasonic velocities of binary mixtures of methylbenzene with hexan-2-ol, heptan-2-ol and octan-2-ol at T = 298.15 and 308.15 K," *Fluid Phase Equilibria*, vol. 252, no. 1-2, pp. 88–95, 2007.

[7] H. Iloukhani, B. Samiey, and M. A. Moghaddasi, "Speeds of sound, isentropic compressibilities, viscosities and excess molar volumes of binary mixtures of methylcyclohexane + 2-alkanols or ethanol at T = 298.15 K," *Journal of Chemical Thermodynamics*, vol. 38, no. 2, pp. 190–200, 2006.

[8] A. Pal and R. K. Bhardwaj, "Excess molar volumes and viscosities of binary mixtures of diethylene glycol dibutyl ether with chloroalkanes at 298.15 K," *Indian Journal of Chemistry A*, vol. 41, no. 4, pp. 706–711, 2002.

[9] A. R. Mahajan, S. R. Mirgane, and S. B. Deshmukh, "Volumetric, Viscometric and Ultrasonic studies of some amino acids in aqueous 0. 2M. $LiClO_4$. $3H_20$ solutions at 298.15 K," *Material Science Research India*, vol. 4, no. 2, pp. 345–352, 2007.

[10] A. R. Mahajan, S. R. Mirgane, and S. B. Deshmukh, "Ultrasonic studies of some amino acids in aqueous 0.02M. $LiCl0_4$. $3H_20$ solutions at 298.15 K," *Material Science Research India*, vol. 4, no. 2, pp. 373–378, 2007.

[11] A. J. Treszczanowicz and G. C. Benson, "Excess volumes for *n*-alkanols + *n*-alkanes II. Binary mixtures of n-pentanol, *n*-hexanol, *n*-octanol, and *n*-decanol + *n*-heptane," *The Journal of Chemical Thermodynamics*, vol. 10, no. 10, pp. 967–974, 1978.

[12] A. Ali, A. K. Nain, V. K. Sharma, and S. Ahmad, "Molecular interactions in binary mixtures of tetrahydrofuran with alkanols (C6,C8,c10): an ultrasonic and volumetric study," *Indian Journal of Pure and Applied Physics*, vol. 42, no. 9, pp. 666–673, 2004.

[13] H. Iloukhani, M. Rezaei-Sameti, and J. Basiri-Parsa, "Excess molar volumes and dynamic viscosities for binary mixtures of toluene + n-alkanes (C_5 − C_{10}) at T = 298.15 K—Comparison with Prigogine-Flory-Patterson theory," *Journal of Chemical Thermodynamics*, vol. 38, no. 8, pp. 975–982, 2006.

[14] R. Mehra and M. Pancholi, "Temperature-dependent studies of thermo-acoustic parameters in hexane + 1-dodecanol and application of various theories of sound speed," *Indian Journal of Physics*, vol. 80, no. 3, pp. 253–263, 2006.

[15] C. Yang, W. Xu, and P. Ma, "Thermodynamic properties of binary mixtures of *p*-xylene with cyclohexane, heptane, octane, and *N*-methyl-2-pyrrolidone at several temperatures," *Journal of Chemical and Engineering Data*, vol. 49, no. 6, pp. 1794–1801, 2004.

[16] R. Meyer, M. Meyer, J. Metzger, and A. Peneloux, "Thermodynamic and physicochemical properties of binary solvent," *Journal de Chimie Physique et de Physico-Chimie Biologique*, vol. 68, pp. 406–412, 1971.

[17] O. Redlich and A. T. Kister, "Thermodynamics of nonelectrolyte solutions. Algebraic representation of thermodynamic properties and the classification of solutions," *Industrial & Engineering Chemistry*, vol. 40, pp. 345–348, 1948.

[18] J. B. Irving, "Viscosity of binary liquid mixtures, a survey of mixture equations," NEL Report 630, National Eng Lab, East Kilbride, UK, 1977.

[19] J. B. Irving, "The effectiveness of mixture equations," NEL Report 631, National Eng Lab, East Kilbride, UK, 1977.

[20] L. Grunberg and A. H. Nissan, "Vaporisation, viscosity, cohesion and structure of the liquids," *Nature*, vol. 164, pp. 799–800, 1949.

[21] M. Tamura and M. Kurata, "Viscosity of binary mixture of liquids," *Bulletin of the Chemical Society of Japan*, vol. 25, pp. 32–37, 1952.

[22] R. K. Hind, E. McLaughlin, and A. R. Ubbelohde, "Structure and viscosity of liquids camphor+pyrene mixtures," *Transactions of the Faraday Society*, vol. 56, pp. 328–330, 1960.

[23] P. K. Katti and M. M. Chaudhri, "Viscosities of binary mixtures of benzyl acetate with dioxane, aniline, and m-cresol," *Journal of Chemical and Engineering Data*, vol. 9, no. 3, pp. 442–443, 1964.

[24] E. L. Heric, "On the viscosity of ternary mixtures," *Journal of Chemical and Engineering Data*, vol. 11, no. 1, pp. 66–68, 1966.

[25] E. L. Heric and J. G. Brewer, "Viscosity of some binary liquid nonelectrolyte mixtures," *Journal of Chemical and Engineering*, vol. 12, no. 4, pp. 574–583, 1967.

[26] R. A. McAllister, "The viscosities of lquid mixtures," *AIChE Journal*, vol. 6, pp. 427–431, 1960.

[27] S. Glasstone, K. J. Laidler, and H. Eyring, *The Theory of Rate Process*, McGraw-Hill, New York, NY, USA, 1941.

[28] R. K. Nigam and B. S. Mahl, "Molecular interaction in binary liquid mixtures of dimethylslphoxide with chlroehanes & chlroehenes," *Indian Journal of Chemistry*, vol. 9, p. 1255, 1971.

[29] R. J. Fort and W. R. Moore, "Viscosities of binary liquid mixtures," *Transactions of the Faraday Society*, vol. 62, pp. 1112–1119, 1966.

[30] K. Ramamoorty and P. S. Varadachari, "Study of some binary liquid mixtures," *Indian Journal of Pure and Applied Physics*, vol. 11, p. 238, 1973.

[31] P. J. Flory, "Statistical thermodynamics of liquid mixtures," *Journal of the American Chemical Society*, vol. 87, no. 9, pp. 1833–1838, 1965.

[32] A. Abe and P. J. Flory, "The thermodynamic properties of mixtures of small, nonpolar molecules," *Journal of the American Chemical Society*, vol. 87, no. 9, pp. 1838–1846, 1965.

[33] I. Prigogine, *Molecular Theories of Solutions*, North-Holland, Amsterdam, The Netherlands, 1957.

[34] D. Patterson and G. Delmas, "Corresponding states theories and liquid models," *Discussions of the Faraday Society*, vol. 49, pp. 98–105, 1970.

[35] P. Tancrède, P. Bothorel, P. De St. Romain, and D. Patterson, "Interactions in alkane systems by depolarized Rayleigh scattering and calorimetry. Part 1.—Orientational order and condensation effects in n-hexadecane + hexane and nonane isomers," *Journal of the Chemical Society, Faraday Transactions 2*, vol. 73, no. 1, pp. 15–28, 1977.

[36] P. De St. Romain, H. T. Van, and D. Patterson, "Effects of molecular flexibility and shape on the excess enthalpies and heat capacities of alkane systems," *Journal of the Chemical Society, Faraday Transactions 1*, vol. 75, pp. 1700–1707, 1979.

[37] A. T. Rodriguez and D. Patterson, "Excess thermodynamic functions of n-alkane mixtures. Prediction and interpretation through the corresponding states principle," *Journal of the Chemical Society, Faraday Transactions 2*, vol. 78, no. 3, pp. 501–523, 1982.

[38] T. M. Aminabhavi, K. Banerjee, and R. H. Balundgi, "Thermodynamic interactions in binary mixtures of 1-chloronaphthalene

and monocyclic aromatics," *Indian Journal of Chemistry A*, vol. 38, no. 8, pp. 768–777, 1999.

[39] J. A. Riddick, W. B. Bunger, and T. K. Sakano, *Techniques of Chemistry, Organic Solvents. Physical Properties and Methods of Purifications*, vol. 2, John Wiley & Sons, New York, NY, USA, 1986.

[40] A. Krishnaiah and P. R. Naidu, "Excess thermodynamic properties of binary liquid mixtures of 1,2-dichloroethane with normal alkanes," *Journal of Chemical and Engineering Data*, vol. 25, no. 2, pp. 135–137, 1980.

[41] E. Aicart, G. Tardajos, and M. Díaz Peña, "Isothermal compressibility of cyclohexane + n-hexane, cyclohexane + n-heptane, cyclohexane + n-octane, and cyclohexane + n-nonane," *Journal of Chemical and Engineering Data*, vol. 25, no. 2, pp. 140–145, 1980.

[42] J. D. Pandey and N. Pant, "Surface tension of ternary polymeric solution," *Journal of the American Chemical Society*, vol. 104, no. 12, pp. 3299–3302, 1982.

[43] A. S. Al-Jimaz, J. A. Al-Kandary, and A.-H. M. Abdul-Latif, "Acoustical and excess properties of {chlorobenzene + 1-hexanol, or 1-heptanol, or 1-octanol, or 1-nonanol, or 1-decanol} at (298.15, 303.15, 308.15, and 313.15) K," *Journal of Chemical and Engineering Data*, vol. 52, no. 1, pp. 206–214, 2007.

[44] E. Aicart, G. Tardajos, and M. Diaz Peña, "Isothermal compressibility of cyclohexane-n-decane, cyclohexane-n-dodecane, and cyclohexane-n-tetradecane," *Journal of Chemical and Engineering Data*, vol. 26, no. 1, pp. 22–26, 1981.

[45] V. A. Bloomfield and R. K. Dewan, "Viscosity of liquid mixtures," *Journal of Physical Chemistry*, vol. 75, no. 20, pp. 3113–3119, 1971.

[46] A. Jouyban, M. Khoubnasabjafari, Z. Vaez-Gharamaleki, Z. Fekari, and W. E. Acree Jr., "Calculation of the viscosity of binary liquids at various temperatures using Jouyban-Acree model," *Chemical and Pharmaceutical Bulletin*, vol. 53, no. 5, pp. 519–523, 2005.

[47] A. Jouyban, A. Fathi-Azarbayjani, M. Khoubnasabjafari, and W. E. Acree Jr., "Mathematical representation of the density of liquid mixtures at various temperatures using Jouyban-Acree model," *Indian Journal of Chemistry A*, vol. 44, no. 8, pp. 1553–1560, 2005.

[48] A. Krishnaiah and P. R. Naidu, "Excess thermodynamic properties of binary liquid mixtures of 1,2-dichloroethane with normal alkanes," *Journal of Chemical and Engineering Data*, vol. 25, no. 2, pp. 135–137, 1980.

11

Effects of Binary Chemical Reaction and Activation Energy on MHD Boundary Layer Heat and Mass Transfer Flow with Viscous Dissipation and Heat Generation/Absorption

Kh. Abdul Maleque

American International University-Bangladesh, House 23, 17, Kamal Ataturk Avenue, Banani, Dhaka-1213, Bangladesh

Correspondence should be addressed to Kh. Abdul Maleque; maleque@aiub.edu

Academic Editors: G. L. Aranovich, C. D. Daub, P. Espeau, and Z. Xu

We study an unsteady MHD free convection heat and mass transfer boundary layer incompressible fluid flow past a vertical porous plate in the presence of viscous dissipation, heat generation/absorption, chemical reaction, and Arrhenius activation energy. The plate is moving with uniform velocity. The chemical reaction rate in the function of temperature is also considered. The governing partial differential equations are reduced to ordinary differential equations by introducing local similarity transformation (Maleque (2010)) and then are solved numerically by shooting method using the Nachtsheim-Swigert iteration technique. The results of the numerical solution are then presented graphically as well as the tabular form for difference values of the various parameters.

1. Introduction

In free convection boundary layer flows with simultaneous heat mass transfer, one important criterion that is generally not encountered is the species chemical reactions with finite Arrhenius activation energy. The Arrhenius law is usually of the following form [1]:

$$K = B(T - T_\infty)^w \exp\left[\frac{-E_a}{k(T - T_\infty)}\right], \quad (1)$$

where K is the rate constant of chemical reaction and B is the preexponential factor, simply prefactor (constant), is based on the fact that increasing the temperature frequently causes a marked increase in the rate of reactions. E_a is the activation energy and $k = 8.61 \times 10^{-5}$ eV/K is the Boltzmann constant which is the physical constant relating energy at the individual particle level with temperature observed at the collective or bulk level.

In areas such as geothermal or oil reservoir engineering, the above phenomenon is usually applicable. Apart from experimental works in these areas, it is also important to make some theoretical efforts to predict the effects of the activation energy in flows mentioned above. But in this regard the very few theoretical works are available in the literature. The reason is that the chemical reaction processes involved in the system are quite complex and generally the mass transfer equation that is required for all the reactions involved also becomes complex. Theoretically, such an equation is rather impossible to tackle. From chemical kinetic viewpoint, this is a very difficult problem, but if the reaction is restricted to binary type a lot of progress can be made. The thermomechanical balance equations for a mixture of general materials were first formulated by Truesdell [2, 3]. Thereafter, Mills [4] and Beevers and Craine [5] have obtained some exact solutions for the boundary layer flow of a binary mixture of incompressible Newtonian fluids. Several problems relating to the mechanics of oil and water emulsions, particularly with regard to applications in lubrication practice, have been considered within the context of a binary mixture theory by Al-Sharif et al. [6] and Wang et al. [7].

A simple model involving binary reaction was studied by Bestman [8]. He considered the motion through the plate to be large which enabled him to obtain analytical solutions (subject to same restrictions) for various values of activation energy by employing the perturbation technique proposed by Singh and Dikshit [9]. Bestman [10] and Alabraba et al. [11] took into account the effect of the Arrhenius activation energy under the different physical conditions.

Recently Kandasamy et al. [12] studied the combined effects of chemical reaction, heat and mass transfer along a wedge with heat source, and concentration in the presence of suction or injection. Their result shows that the flow field is influenced appreciably by chemical reaction, heat source, and suction or injection at the wall of the wedge. Recently, Makinde et al. [13] studied the problem of unsteady convection with chemical reaction and radiative heat transfer past a flat porous plate moving through a binary mixture in an optically thin environment. More recently Maleque [14] studied the effects of exothermic chemical reaction and activation energy on free convective heat and mass transfer flow past a vertical plate. In the present paper, we investigate a numerical solution of unsteady natural convection heat and mass transfer boundary layer flow past a vertical porous plate taking into account the effects of viscous dissipation, heat generation/absorption, and chemical reaction with Arrhenius activation energy in the presence of uniform magnetic field. The plate is moving with uniform velocity. The chemical reaction rate in the function of temperature is also considered. The governing partial differential equations are reduced to ordinary differential equations by introducing local similarity transformation (Maleque [15]). Numerical solutions to the reduced nonlinear similarity equations are then obtained by adopting Runge-Kutta and shooting methods using the Nachtsheim-Swigert iteration technique. The results of the numerical solution are then presented graphically as well as the tabular form for difference values of the various parameters.

2. Governing Equations

We consider the boundary wall to be of infinite extension so that all quantities are homogeneous in x and hence all derivatives with respect to x are neglected. The x-axis is taken along the plate and y-axis is perpendicular to the plate. Assuming a uniform magnetic field, B_0 is applied perpendicular to the plate and the plate is moving with uniform velocity U_0 in its own plane. The system of governing equations in the presence of viscous dissipation, heat generation/absorption, and a binary chemical reaction with Arrhenius activation energy is of the form

$$\frac{\partial v}{\partial y} = 0, \tag{2}$$

$$\frac{\partial u}{\partial t} + v\frac{\partial u}{\partial y} = \upsilon\frac{\partial^2 u}{\partial y^2} + g\beta(T - T_\infty) + g\beta^*(C - C_\infty) - \frac{\sigma B_0^2}{\rho}u, \tag{3}$$

$$\rho c_p\left(\frac{\partial T}{\partial t} + v\frac{\partial T}{\partial y}\right) = \kappa\frac{\partial^2 T}{\partial y^2} + \mu\left(\frac{\partial u}{\partial y}\right)^2 + Q_0(T - T_\infty), \tag{4}$$

$$\frac{\partial C}{\partial t} + v\frac{\partial C}{\partial y} = D_m\frac{\partial^2 C}{\partial y^2} - k_r^2(T - T_\infty)^w \exp\left[\frac{-E_a}{k(T - T_\infty)}\right](C - C_\infty). \tag{5}$$

The boundary conditions of the above system are

$$\begin{gathered} u = U_0, \quad v = v(t), \quad T = T_w, \quad C = C_w \quad \text{at } y = 0, \\ u = 0, \quad T \longrightarrow T_\infty, \quad C \longrightarrow C_\infty, \quad \text{at } y = \infty, \end{gathered} \tag{6}$$

where (u, v) is the velocity vector, T is the temperature, C is the concentration of the fluid, υ is the kinematic coefficient of viscosity, κ is the heat diffusivity coefficient, β and β^* are the coefficients of volume expansions for temperature and concentration, respectively, D_m is the coefficient of mass diffusivity, k_r^2 is the chemical reaction rate constant, and $(T - T_\infty)^w \exp[-E_a/k(T - T_\infty)]$ is the Arrhenius function where w is a unit less constant exponent fitted rate constants typically lie in the range $-1 < w < 1$.

3. Mathematical Formulations

In order to solve the governing equations (2) to (5) under the boundary conditions (6), we adopt the well-defined similarity technique to obtain the similarity solutions. For this purpose the following nondimensional variables are now introduced:

$$\begin{gathered} \eta = \frac{y}{\delta(t)}, \qquad \frac{u}{U_0} = f(\eta), \\ \frac{T - T_\infty}{T_w - T_\infty} = \theta(\eta), \qquad \frac{C - C_\infty}{C_w - C_\infty} = \phi(\eta). \end{gathered} \tag{7}$$

From the equation of continuity (2), we have

$$v(t) = -\frac{v_0\upsilon}{\delta(t)}, \tag{8}$$

where v_0 is the dimensionless suction/injection velocity at the plate, $v_0 > 0$ corresponds to suction, and $v_0 < 0$ corresponds to injection.

Introducing the dimensionless quantities from (7) and v from (8) in (3), (4), and (5), we finally obtain the nonlinear ordinary differential equations as

$$f'' + \left(\eta\frac{\delta\delta'}{\upsilon} + v_0\right)f' - Mf + G_r\theta + G_m\phi = 0, \tag{9}$$

$$P_r^{-1}\theta'' + \left(\eta\frac{\delta\delta'}{\upsilon} + v_0\right)\theta' + \gamma\theta + E_c f'^2 = 0, \tag{10}$$

$$S_c^{-1}\phi'' + \left(\eta\frac{\delta\delta'}{\upsilon} + v_0\right)\phi' - \lambda^2 e^{-E/\theta}\theta^w\phi = 0. \tag{11}$$

Here, Grashof number $G_r = g\beta(T_w - T_\infty)\delta^2/\upsilon U_0$, modified (Solutal) Grashof number $G_m = g\beta^*(C_w - C_\infty)\delta^2/\upsilon U_0$, Prandtl number $P_r = \rho\upsilon c_p/\kappa$, magnetic parameter $M = \sigma'\delta^2 B_0^2/\mu$, Eckert number $E_c = U_0^2/\Delta T c_p$, the chemical reaction rate constant $\lambda^2 = k_r^2\delta^2(T_w - T_\infty)^w/\upsilon$, Schmidt number $S_c = \upsilon/D_m$, the nondimensional activation energy $E = E_a/k(T_w - T_\infty)$, the dimensionless heat generation, or absorption coefficient $\gamma = \delta^2 Q_0/\upsilon\rho c_p$.

Equations (9) to (11) are similar except for the term $\delta\delta'/\upsilon$, where time t appears explicitly. Thus, the similarity condition requires that $\delta\delta'/\upsilon$ must be a constant quantity. Hence, following Maleque [15], one can try a class of solutions of (9) to (11) by assuming the following:

$$\frac{\delta\delta'}{\upsilon} = A \ (\text{Constant}). \tag{12}$$

From (12) we have

$$\delta(t) = \sqrt{2A\upsilon t + L^2}, \tag{13}$$

where the constant of integration L is determined through the condition that $\delta = L$ when $t = 0$. Here $A = 0$ implies that $\delta = L$ represents the length scale for steady flow and $A \neq 0$, that is, δ represents the length scale for unsteady flow. Since δ is a scaling factor as well as similarity parameter, any other values of A in (13) would not change the nature of the solution except that the scale would be different. Finally introducing (12) in (9) to (11), respectively, we have the following dimensionless nonlinear ordinary differential equations:

$$f'' + (\eta A + v_0) f' - Mf + G_r\theta + G_m\phi = 0, \tag{14}$$

$$P_r^{-1}\theta'' + (\eta A + v_0)\theta' + \gamma\theta + E_c f'^2 = 0, \tag{15}$$

$$S_c^{-1}\phi'' + (A\eta + v_0)\phi' - \lambda^2 \exp\left(-\frac{E}{\theta}\right)\theta^w\phi = 0. \tag{16}$$

The boundary condition equation (6) then becomes

$$\begin{aligned} &f(0) = 1, \quad &\theta(0) = 1, \quad &\phi(0) = 1, \\ &f(\infty) = 0, \quad &\theta(\infty) = 0, \quad &\phi(\infty) = 0. \end{aligned} \tag{17}$$

In all equations primes denote the differentiation with respect to η. Equations (14) to (16) are solved numerically under the boundary conditions (17) using Nachtsheim-Swigert iteration technique.

4. Numerical Solutions

Equations (14)–(16) are solved numerically under the boundary conditions (17) using Nachtsheim Swigert [16] iteration technique. In (17), there are three asymptotic boundary conditions and hence follows three unknown surface conditions $f'(0)$, $\theta'(0)$, and $\phi'(0)$. Within the context of the initial-value method and the Nachtsheim-Swigert iteration technique, the outer boundary conditions may be functionally represented by the first-order Taylor's series as follows:

$$\begin{aligned} f(\eta_{\max}) &= f(X, Y, Z) \\ &= f_c(\eta_{\max}) + \Delta X f_X + \Delta Y f_Y + \Delta Z f_Z = \delta_1, \end{aligned} \tag{18}$$

$$\begin{aligned} \theta(\eta_{\max}) &= \theta(X, Y, Z) \\ &= \theta_c(\eta_{\max}) + \Delta X \theta_X + \Delta Y \theta_Y + \Delta Z \theta_Z = \delta_2, \end{aligned} \tag{19}$$

$$\begin{aligned} \phi(\eta_{\max}) &= \phi(X, Y, Z) \\ &= \phi_c(\eta_{\max}) + \Delta X \phi_X + \Delta Y \phi_Y + \Delta Z \phi_Z = \delta_3, \end{aligned} \tag{20}$$

with the asymptotic convergence criteria given by

$$\begin{aligned} f'(\eta_{\max}) &= f'(X, Y, Z) \\ &= f_c'(\eta_{\max}) + \Delta X f_X' + \Delta Y f_Y' + \Delta Z f_Z' = \delta_4, \\ \theta'(\eta_{\max}) &= \theta'(X, Y, Z) \\ &= \theta_c'(\eta_{\max}) + \Delta X \theta_X' + \Delta Y \theta_Y' + \Delta Z \theta_Z' = \delta_5, \\ \phi'(\eta_{\max}) &= \phi'(X, Y, Z) \\ &= \phi_c'(\eta_{\max}) + \Delta X \phi_X' + \Delta Y \phi_Y' + \Delta Z \phi_Z' = \delta_6, \end{aligned} \tag{21}$$

where $X = f'(0)$, $Y = \theta'(0)$, $Z = \phi'(0)$, and X, Y, Z subscripts indicate partial differentiation, for example, $f_Y' = \partial f'(\eta_{\max})/\partial\theta'(0)$. The subscript c indicates the value of the function at $\eta_{\max}$ to be determined from the trial integration.

Solutions of these equations in a least-square sense requires determining the minimum value of $E = \delta_1^2 + \delta_2^2 + \delta_3^2 + \delta_4^2 + \delta_5^2 + \delta_6^2$ with respect to X, Y, and Z. To solve ΔX, ΔY and ΔZ we require to differentiate E with respect to X, Y, and Z respectively. Thus, adopting this numerical technique, a computer program was set up for the solutions of the basic nonlinear differential equations of our problem, where the integration technique was adopted as the six ordered Range-Kutta method of integration. The results of this integration are then displayed graphically in the form of velocity, temperature, and concentration profiles in Figures 1–11. In the process of integration, the local skin-friction coefficient, the local rates of heat, and mass transfer to the surface, which are of chief physical interest, are also calculated out. The equation defining the wall skin-friction is

$$\tau = -\mu\left(\frac{\partial u}{\partial y}\right)_{y=0}. \tag{22}$$

Hence, the skin-friction coefficient is given by

$$C_f = \frac{\tau}{\rho U_0^2}\alpha - f'(0). \tag{23}$$

The heat flux (q_w) and the mass flux (M_w) at the wall are given by

$$\begin{aligned} q_w &= -\kappa\left(\frac{\partial T}{\partial y}\right)_{y=0}, \\ M_w &= -D_M\left(\frac{\partial C}{\partial y}\right)_{y=0}. \end{aligned} \tag{24}$$

Hence the Nusselt number (N_u) and the Sherwood number (S_h) are obtained as

$$N_u\alpha - \theta'(0), \qquad S_h\alpha - \phi'(0). \tag{25}$$

These above coefficients are then obtained from the procedure of the numerical computations and are sorted in Table 1.

TABLE 1: Effects of G_r, λ, w, γ, E, and v_0 on the skin friction coefficient, the Nusselt number, and the Sherwood number for $G_r = 10$, $G_m = 1$, $P_r = 0.71$, $v_0 = 3$, $S_c = 0.6$, $\lambda = 5$, $E = 1$, $w = 1$, $M = E_c = 0$, and $\gamma = 0$.

	$-f'(0)$	$-\theta'(0)$	$-\phi'(0)$
$G_r =$			
−10.00000	6.28618	2.58611	2.83618
−5.00000	4.71973	2.59209	2.83672
0.00000	3.19774	2.59299	2.83998
5.00000	1.70224	2.59852	2.83923
10.00000	0.24525	2.60349	2.83849
$\lambda =$			
1.00000	0.24033	2.60608	2.26552
5.00000	0.24525	2.60349	2.83849
10.00000	0.30840	2.60610	4.37298
$w =$			
−0.750000	0.06827	2.58234	3.21792
0.00000	0.16991	2.59250	3.01768
0.75000	0.20504	2.59811	2.87688
$E =$			
0.00000	0.34340	2.60624	4.35339
0.50000	0.16287	2.58891	3.37658
1.00000	0.24525	2.60349	2.83849
5.00000	0.10981	2.58889	2.26497
$v_0 =$			
−3.00000	−3.88025	0.10081	1.62314
0.00000	−3.66969	0.95110	1.95706
3.00000	0.24525	2.60349	2.83849
$\gamma =$			
−4.00000	0.68965	3.31240	2.75431
−2.00000	0.41067	2.98095	2.78286
0.00000	0.24525	2.60349	2.83849
2.00000	−0.48425	2.05361	2.92446
4.00000	−0.77199	1.46498	3.02587

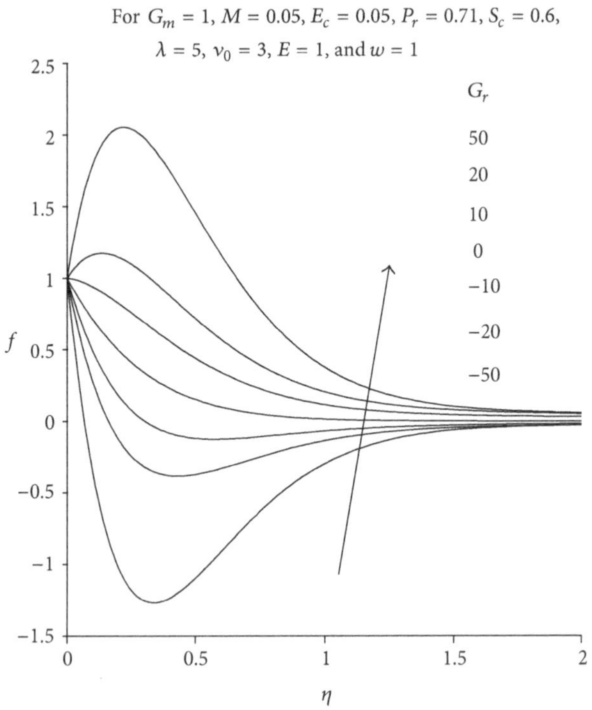

FIGURE 1: Effects of G_r on the velocity profiles.

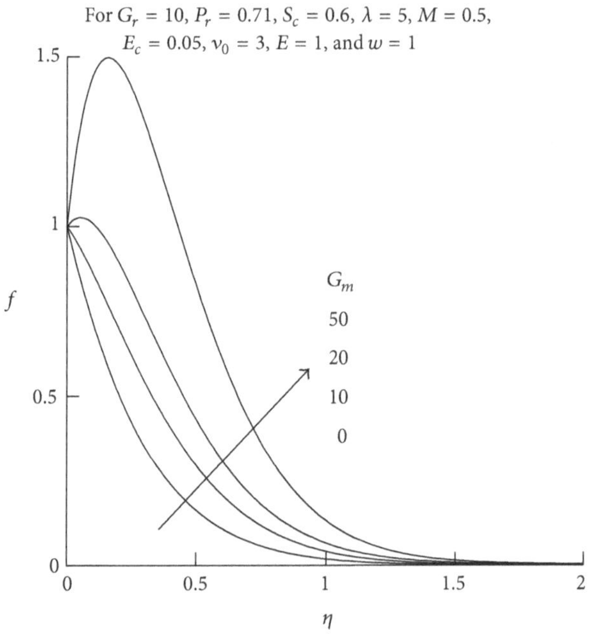

FIGURE 2: Effects of G_m on the velocity profiles.

5. Results and Discussions

The parameters entering into the fluid flow are Grashof number G_r, Solutal (modified) Grashof number G_m, magnetic interaction parameter M, suction parameter v_0, Prandtl number P_r, Eckert number E_c, the dimensionless heat generation/absorption coefficient γ, the nondimensional chemical reaction rate constant λ^2, Schmidt number S_c and the nondimensional activation energy E.

It is, therefore, pertinent to inquire the effects of variation of each of them when the others are kept constant. The numerical results are thus presented in the form of velocity profiles, temperature profiles, and concentration profiles in Figures 1–11 for the different values of G_r, G_m, M, E_c, γ, λ, E, and v_0. The values of G_r are taken to be large ($G_r = 10$) since the value corresponds to a cooling problem that is generally encountered in nuclear engineering in connection with the cooling of reactors. In air ($P_r = 0.71$) the diluting chemical species of most common interests have Schmidt number in the range from 0.6 to 0.75. Therefore, the Schmidt number $S_c = 0.6$ is considered. In particular, 0.6 corresponds to water vapor that represents a diffusing chemical species of the most common interest in air. The values of the suction parameter v_0 are taken to be large. Apart from the above figures and tables, the representative velocity, temperature and concentration

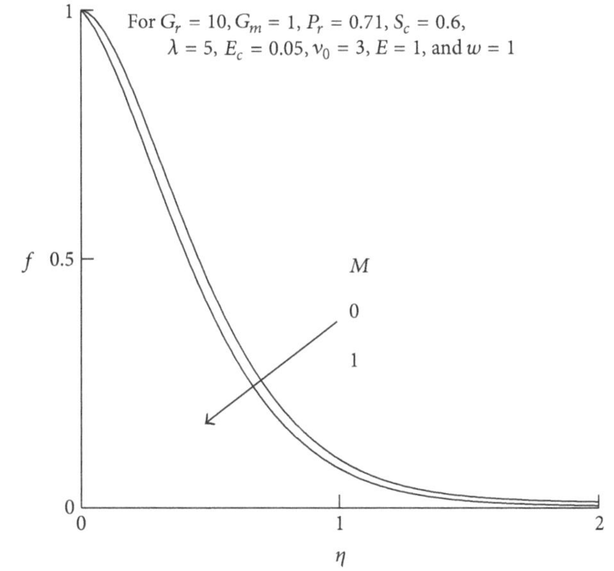

FIGURE 3: Effects of M on the velocity profiles.

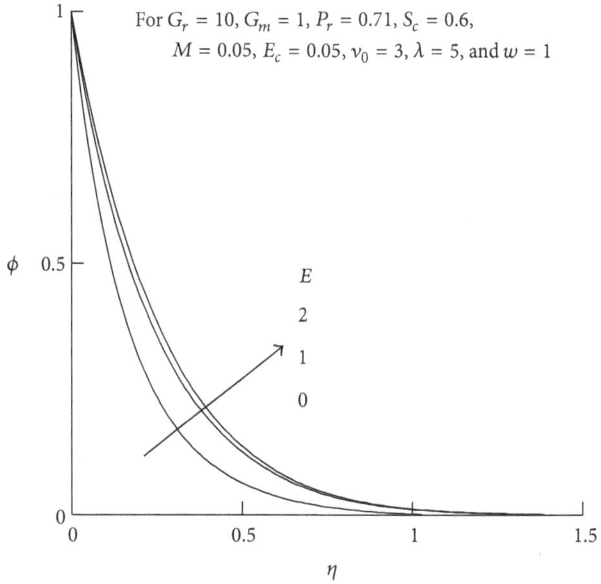

FIGURE 5: Effects of E on the concentration profiles.

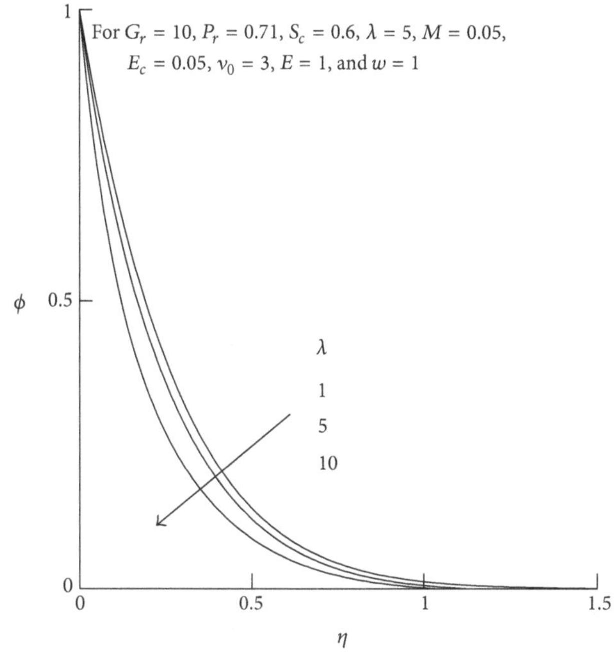

FIGURE 4: Effects of λ on the concentration profiles.

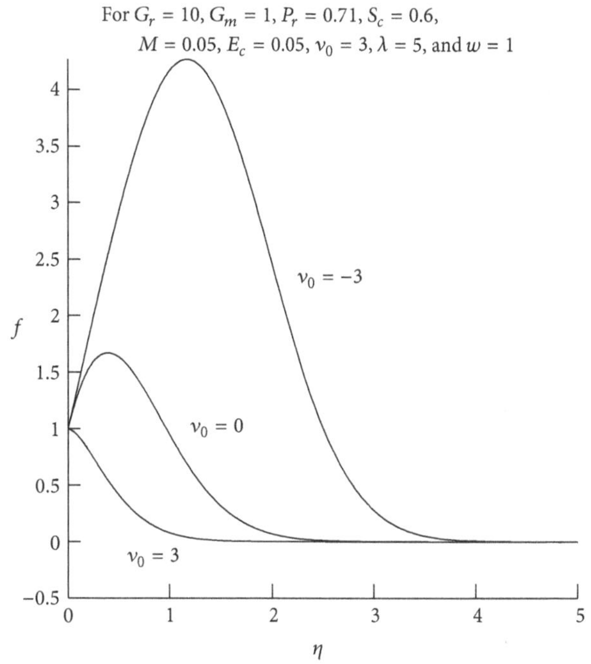

FIGURE 6: Effects of v_0 on the velocity profiles.

profiles, and the values of the physically important parameters, that is, the local shear stress, the local rates of heat, and mass transfers, are illustrated for uniform wall temperature and species concentration in Figures 1–11 and in Table 1.

5.1. Effects of G_r, G_m, and M on the Velocity Profiles. The value of G_r is taken to be both positive and negative, since these values represent, respectively, cooling and heating of the plate.

The velocity profiles generated due to impulsive motion of the plate are plotted in Figure 1 for both cooling ($G_r > 0$) and heating ($G_r < 0$) of the plate keeping other parameters fixed ($G_m = 1.0$, $M = E_c = 0.5$, $P_r = 0.71$, $S_c = 0.6$, $\lambda = 5$, $v_0 = 3.0$, $E = 1.0$, and $w = 1$). In Figure 1, velocity profiles are shown for different values of G_r. We observe that velocity increases with increasing values of G_r for the cooling of the plate. From this figure, it is also observed that the negative increase in the Grashof number leads to the decrease in the velocity field. That is, for heating of the plate (Figure 1), the effects of the Grashof number G_r on the velocity field have also opposite effects, as compared to the cooling of the plate.

Solutal Grashof number $G_m > 0$ corresponds to the fact that the chemical species concentration in the free stream region is less than the concentration at the boundary surface.

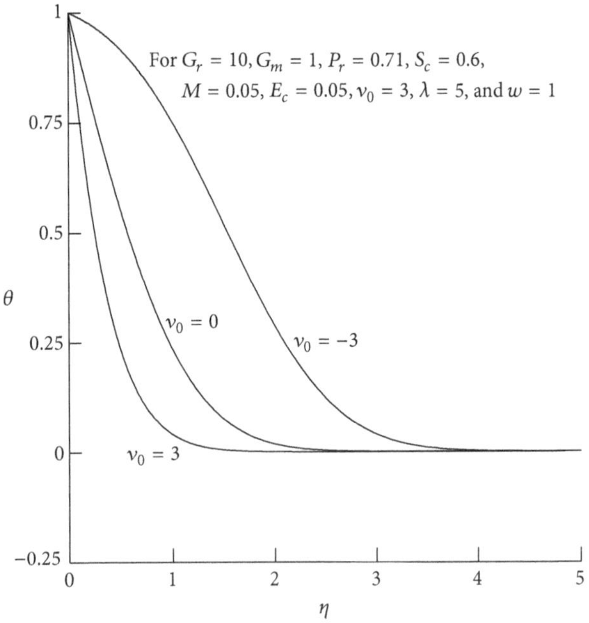

FIGURE 7: Effects of v_0 on the temperature profiles.

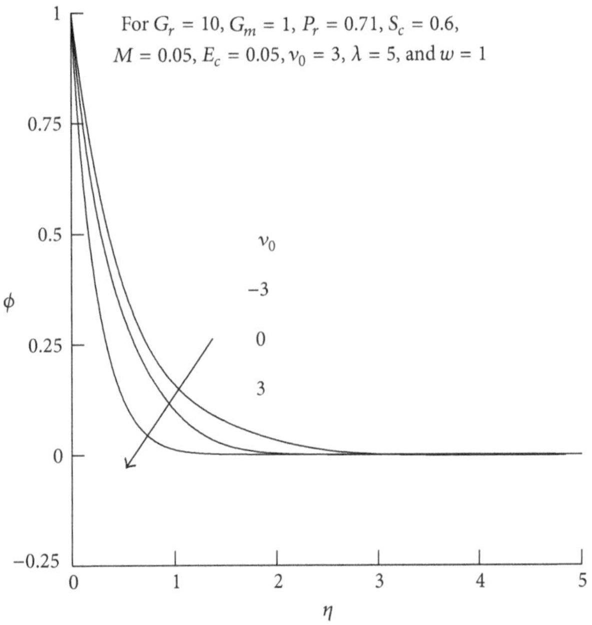

FIGURE 8: Effects of v_0 on the concentration profiles.

Figure 2 presents the effects of Solutal Grashof number G_m on the velocity profiles. It is observed that the velocity profile increases with the increasing values of Solutal Grashof number G_m.

Imposition of a magnetic field to an electrically conducting fluid creates a drag-like force called Lorentz force. The force has the tendency to slow down the flow around the plate at the expense of increasing its temperature. This is depicted by decreases in velocity profiles as magnetic parameter M increases as shown in Figure 3.

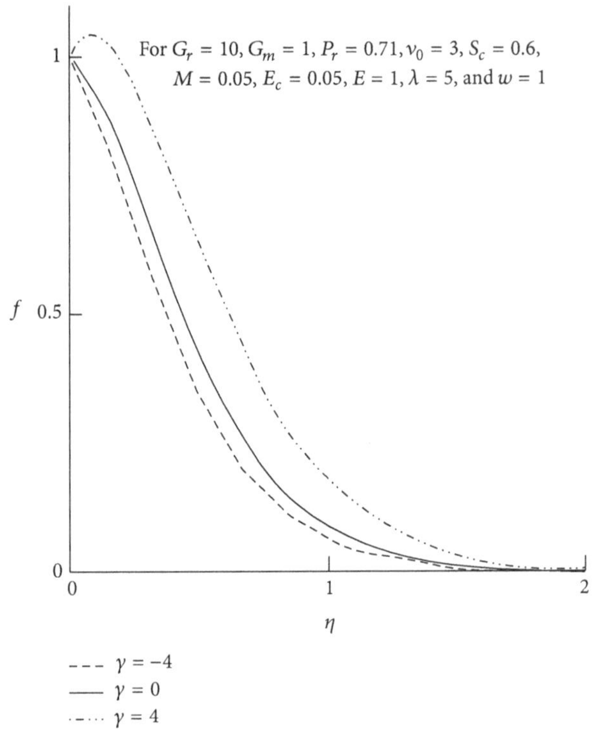

FIGURE 9: Effects of γ on the velocity profiles.

5.2. Effects of λ and Activation Energy (E) on the Concentration Profiles. Considering chemical reaction rate constant λ^2 is always positive. Figure 4 represents the effect of chemical rate constant λ on the concentration. We observe from the last part of (18) that $\lambda^2 \exp(-E/\theta)$ increases with the increasing values of λ. We also observe from this equation that increase in $\lambda^2 \exp(-E/\theta)$ means increase in λ, leads to the decrease in the concentration profiles. This is in great agreement with Figure 3. The parameter λ does not enter directly into the momentum equation but its influence comes through the mass equation. λ has no mark effect on the velocity and temperature profiles.

In chemistry, *activation energy* is a term introduced in 1889 by the Swedish scientist Svante Arrhenius which is defined as the energy that must be overcome for a chemical reaction to occur. Activation energy may also be defined as the minimum energy required starting a chemical reaction. The activation energy of a reaction is usually denoted by E_a and given in units of kilojoules per mole.

Activation energy can be thought of as the height of the potential barrier (sometimes called the energy barrier) separating two minima of potential energy (of the reactants and products of a reaction). For a chemical reaction to proceed at a reasonable rate, there should exist an appreciable number of molecules with energy equal to or greater than the activation energy. Effects of activation energy (E) on the concentration are shown in Figure 5. From (1), we observe that chemical reaction rate (K) decreases with the increasing values of activation energy (E_a). We also observe from (18) that increase in activation energy (E) leads to decrease in

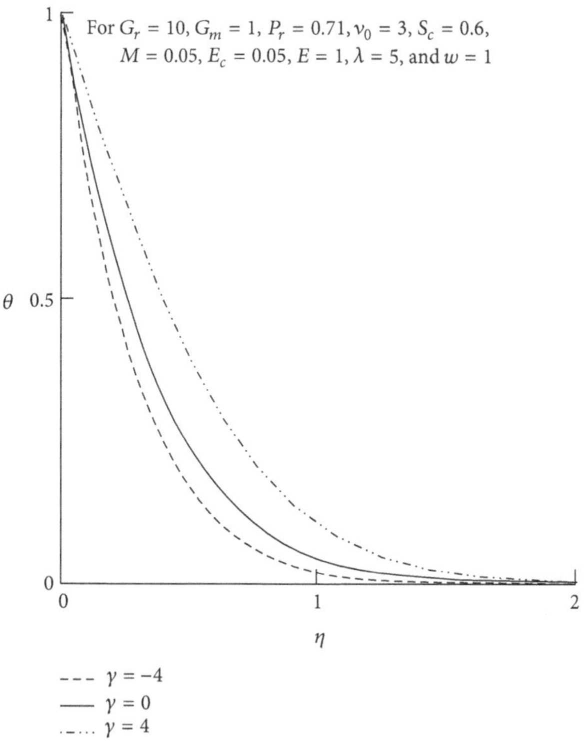

FIGURE 10: Effects of γ on the temperature profiles.

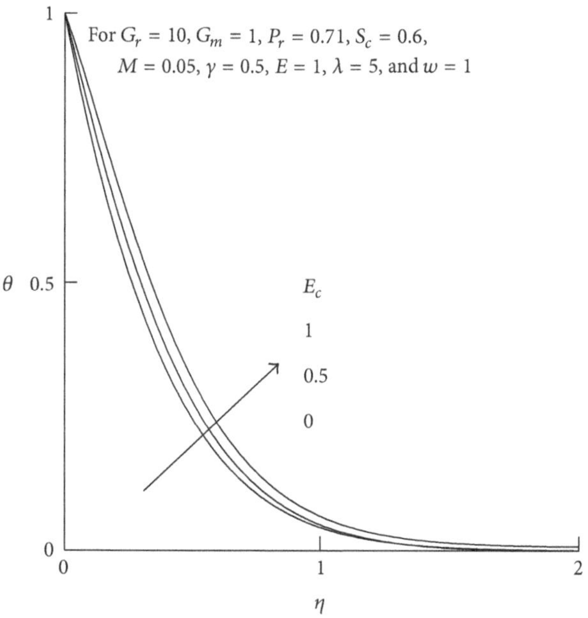

FIGURE 11: Effects of E_c on the temperature profiles.

$\lambda^2 \exp(-E/\theta)$ as well as increase in the concentration profiles shown in Figure 5.

5.3. Effects of Suction/Injection (ν_0) on the Velocity, the Temperature, and the Concentration Profiles. The effects of suction and injection (ν_0) for $\lambda = 5$, $E = 1$, $G_r = 5$, $G_m = 1$, $S_c = 0.6$, and $P_r = 0.7$ on the velocity profiles, temperature profiles and

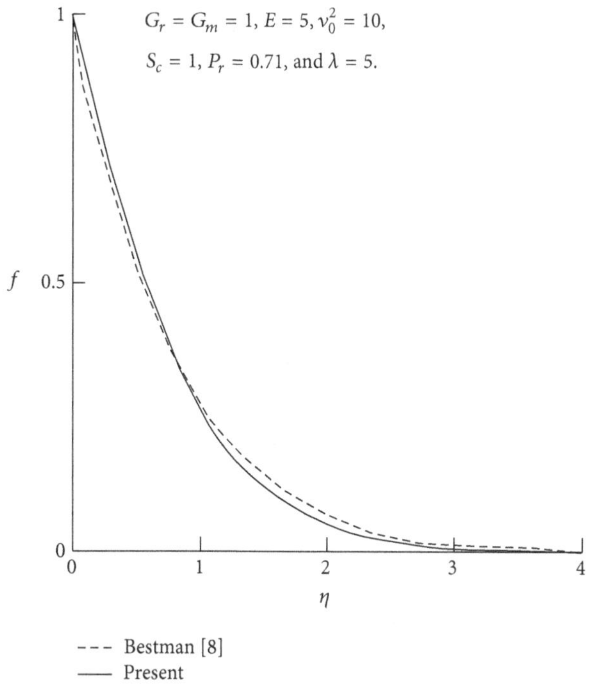

FIGURE 12: Comparison of our calculated velocity profile and the velocity profile of Bestman [8].

concentration profiles are shown, respectively, in Figures 6 to 8. For strong suction ($\nu_0 > 0$), the velocity, the temperature, and the concentration profiles decay rapidly away from the surface. The fact that suction stabilizes the boundary layer is also apparent from these figures. As for the injection ($\nu_0 < 0$), from Figures 6 to 8 it is observed that the profiles overshoot and are close to the boundary. From Figure 6 it is observed that for strong suction, the velocity is small while for injection the boundary layer is increasingly blown away from the surface of the plate to form an interlayer between the injection and the outer flow regions. Figure 6 also depicts that for high values of injection parameter ($\nu_0 < 0$), the velocity near the plate (for small values of η) is lower than that for smaller values of injection parameter ($\nu_0 < 0$). Figure 7 shows the usual decay of temperature for strong suction while the thermal boundary layer increases rapidly for strong injection. It is seen from Figure 8 that the species concentration within the boundary layer is higher for injection and lower for suction.

5.4. Effects of the Heat Generation and Absorption (γ) on the Velocity and the Temperature Profiles. The positive value and the negative value of γ represent the heat generation and heat absorption, respectively. The dimensionless heat generation/absorption coefficient γ does not enter directly into the equation of continuity, Navier Stokes equations and concentration equation its influence come through the energy equation. In the presence of the buoyancy force, it appears that the heat generation/absorption coefficient γ has a marked effect on the boundary layer and the velocity profiles of the fluid is shown in Figure 9. From this figure it has been observed that the effect of heat generation coefficient is to expand the boundary layer thickness and opposite effect is found for the effect of heat absorption. Thus, the dimensionless heat generation/absorption coefficient has the same effects on skin-friction coefficients shown in Table 1.

The effects of heat absorption coefficient ($\gamma < 0$) on temperature profiles and the Nusselt number are shown in Figure 10 and Table 1. In general, the effect of heat absorption coefficient ($\gamma < 0$) is to reduce the thickness of the thermal boundary layer. The temperature profiles decrease as the heat absorption coefficient γ (<0) increases as shown in Figure 10. On the other hand, the heat generation coefficient γ (<0) leads to increasing the temperature profiles. Thermal boundary layer thickness expands for increasing values of the heat generation coefficient γ (>0) as shown in Figure 10.

Finally, the effects of heat generation/absorption coefficient (γ) on the dimensionless Nusselt number N_u are shown in Table 1. The Nusselt number increases with the increasing values of the heat absorption coefficient ($\gamma < 0$). Because the heat absorption increases, the disk tends to supply more energy through its wall to maintain the constant temperature of the wall, and hence the Nusselt number increases. Opposite effects are found for the increasing values of the heat generation coefficient ($\gamma > 0$).

5.5. Effects of the Eckert Number E_c on the Temperature Profiles. Figure 11 presents the effects of Eckert number (E_c) on the temperature profiles for $G_r = 10$, $G_m = 1.0$, $E = 1$, $w = 1$, $\lambda = 5$, $M = 0.05$, $\gamma = 0.5$, $v_o = 3$, and $P_r = 0.71$. From this figure it has been observed that the parameter E_c has marked effect on the temperature profiles. The temperature profiles increase for increasing values of E_c. The dimensionless Eckert number (E_c) does not enter directly into the equation of continuity and momentum equation and concentration equation, its influence come through the energy equation. The Eckert number (E_c) has no marked effect on velocity profiles and concentration profiles.

5.6. The Skin-Friction, the Heat Transfer, and the Mass Transfer Coefficients. The skin-friction, the heat transfer, and the mass transfer coefficients are tabulated in Table 1 for different values of G_r, λ, E, w, v_0, and γ. We observe from Table 1 that the skin-friction coefficient ($-f'(0)$) decreases with increase in the Grash of number G_r. It is also observed from Table 1 that the rate of mass transfer coefficient increases with the increasing values of λ, and the mass transfer coefficient decreases with the increasing values of E and w. It is observed from this table that this is an increasing effect of suction parameter v_0 on the skin friction coefficient and heat and mass transfer coefficients. Finally, the effects of heat generation/absorption coefficient (γ) on the dimensionless Nusselt number N_u are shown in Table 1. The Nusselt number increases with the increasing values of the heat absorption coefficient ($\gamma < 0$). Because the heat absorption increases, the disk tends to supply more energy through its wall to maintain the constant temperature of the wall, and hence the Nusselt number increases. Opposite effects are found for the increasing values of the heat generation coefficient ($\gamma > 0$).

5.7. Comparison between Our Numerical Result and the Analytical Result of Bestman [8]. Bestman studied steady natural convective boundary layer flow with large suction. He solved his problem analytically by employing the perturbation technique proposed by Singh and Dikshit [9]. In our present work, we took $A = 0$ in (14) for considering steady flow. The values of the suction parameter $v_0^2 = 10$ are taken to see the effects of large suction. $G_r = G_m = 1.0$, $\lambda = 5$, $M = E_c = \gamma = 0$, $E = 5$, and $w = 1$ are also chosen with a view to compare our numerical results with the analytical results of Bestman [8]. The comparison of velocity profiles as seen in Figure 12 highlights the validity of the numerical computations adapted in the present investigation.

6. Conclusions

In this paper, we investigate the effects of chemical reaction rate and Arrhenius activation energy on an unsteady natural convection heat and mass transfer boundary layer flow past a flat porous plate in the presence of uniform magnetic field viscous dissipation and heat generation/absorption.

The Nachtsheim-Swigert [16] iteration technique based on the sixth-order Range-Kutta and shooting methods has been employed to complete the integration of the resulting solutions.

The following conclusions can be drawn as a result of the computations.

(1) Velocity increases with increasing values of G_r for the cooling of the plate, and the negative increase in the Grashof number leads to the decrease in the velocity field. That is, for heating the plate, the effects of the Grashof number G_r on the velocity field have also opposite effects, as compared to the cooling of the plate.

(2) Solutal Grashof number $G_m > 0$ that corresponds to the chemical species concentration in the free stream region is less than the concentration at the boundary surface. It is observed that the velocity profile increases with the increasing values of Solutal Grashof number G_m.

(3) We observe that the velocity field decreases due to magnetic interaction parameter M. This is because the application of a transverse magnetic field normal to the flow direction will result in a resistive force (Lorentz force) similar to a drag force which tends to resist the fluid flow and thus reduces its velocity.

(4) Eckert number (E_c) has marked effect on velocity and temperature fields. The velocity and the temperature profiles increase with the increasing values of E_c.

(5) Increase in λ leads to the decrease in the concentration profiles. It is observed from (12) that increasing temperature frequently causes a marked increase in the rate of reactions λ. The velocity profile slightly increases with the increasing values of λ.

(6) Increase in activation energy (E) leads to increasing the concentration, temperature, and velocity profiles.

For small changes of activation energy that must occur to convert the reactants into the product of the reaction, energy must be invested to begin breaking the reactant bond so that product bond be can formed.

(7) For strong suction ($v_0 > 0$), the velocity, the temperature, and the concentration profiles decay rapidly away from the surface. As for the injection ($v_0 > 0$), it is observed that the boundary layer is increasingly blown away from the plate to form an interlayer between the injection and the outer flow regions.

(8) The Nusselt number increases with the increasing values of the heat absorption coefficient ($\gamma < 0$). Because the heat absorption increases, the disk tends to supply more energy through its wall to maintain the constant temperature of the wall, and hence the Nusselt number increases.

Nomenclature

(u, v): components of velocity field
c_p: specific heat at constant pressure
N_u: Nusselt number
P_r: Prandtl number
T: temperature of the flow field
T_∞: temperature of the fluid at infinity
T_w: temperature at the plate
f: dimensionless similarity functions
E_a: the activation energy
k: the Boltzmann constant
S_c: Schmidt number
E: the nondimensional activation energy
G_r: Grashof number
G_m: modified (Solutal) Grashof number.

Greek Symbols

θ: Dimensionless temperature
ϕ: dimensionless concentration
η: dimensionless similarity variable
δ: scale factor
ν: kinematic viscosity
ρ: density of the fluid
κ: thermal conductivity
τ: shear stress
μ: fluid viscosity
λ^2: the nondimensional chemical reaction rate constant
β and β^*: the coefficients of volume expansions for temperature, and concentration respectively.

References

[1] M. Tencer, J. S. Moss, and T. Zapach, "Arrhenius average temperature: the effective temperature for non-fatigue wearout and long term reliability in variable thermal conditions and climates," *IEEE Transactions on Components and Packaging Technologies*, vol. 27, no. 3, pp. 602–607, 2004.

[2] C. Truesdell, "Sulle basi della thermomeccanica," *Accademia Nazionale del Lincei, Rendiconti della Classe di Scienze Fisiche*, vol. 22, no. 8, pp. 33–38, 1957.

[3] C. Truesdell, "Sulle basi della thermomeccanica," *Accademia Nazionale del Lincei, Rendiconti della Classe di Scienze Fisiche*, vol. 22, no. 8, pp. 158–166, 1957.

[4] N. Mills, "Incompressible mixtures of newtonian fluids," *International Journal of Engineering Science*, vol. 4, no. 2, pp. 97–112, 1966.

[5] C. E. Beevers and R. E. Craine, "On the determination of response functions for a binary mixture of incompressible newtonian fluids," *International Journal of Engineering Science*, vol. 20, no. 6, pp. 737–745, 1982.

[6] A. Al-Sharif, K. Chamniprasart, K. R. Rajagopal, and A. Z. Szeri, "Lubrication with binary mixtures: liquid-liquid emulsion," *Journal of Tribology*, vol. 115, no. 1, pp. 46–55, 1993.

[7] S. H. Wang, A. Al-Sharif, K. R. Rajagopal, and A. Z. Szeri, "Lubrication with binary mixtures: liquid-liquid emulsion in an EHL conjunction," *Journal of Tribology*, vol. 115, no. 3, pp. 515–522, 1993.

[8] A. R. Bestman, "Natural convection boundary layer with suction and mass transfer in a porous medium," *International Journal of Energy Research*, vol. 14, no. 4, pp. 389–396, 1990.

[9] A. K. Singh and C. K. Dikshit, "Hydromagnetic flow past a continuously moving semi-infinite plate for large suction," *Astrophysics and Space Science*, vol. 148, no. 2, pp. 249–256, 1988.

[10] A. R. Bestman, "Radiative heat transfer to flow of a combustible mixture in a vertical pipe," *International Journal of Energy Research*, vol. 15, no. 3, pp. 179–184, 1991.

[11] M. A. Alabraba, A. R. Bestman, and A. Ogulu, "Laminar convection in binary mixture of hydromagnetic flow with radiative heat transfer, I," *Astrophysics and Space Science*, vol. 195, no. 2, pp. 431–439, 1992.

[12] R. Kandasamy, K. Periasamy, and K. K. S. Prabhu, "Effects of chemical reaction, heat and mass transfer along a wedge with heat source and concentration in the presence of suction or injection," *International Journal of Heat and Mass Transfer*, vol. 48, no. 7, pp. 1388–1394, 2005.

[13] O. D. Makinde, P. O. Olanrewaju, and W. M. Charles, "Unsteady convection with chemical reaction and radiative heat transfer past a flat porous plate moving through a binary mixture," *Afrika Matematika*, vol. 21, no. 1, pp. 1–17, 2011.

[14] Kh. A. Maleque, "Unsteady natural convection boundary layer heat and mass transfer flow with exothermic chemical reactions," *Journal of Pure and Applied Mathematics*, vol. 9, no. 1, pp. 17–41, 2013.

[15] Kh. A. Maleque, "Effects of combined temperature- and depth-dependent viscosity and hall current on an unsteady MHD laminar convective flow due to a rotating disk," *Chemical Engineering Communications*, vol. 197, no. 4, pp. 506–521, 2010.

[16] P. R. Nachtsheim and P. Swigert, "Satisfaction of asymptotic boundary conditions in numerical solution of system of non-linear of boundary layer type," NASA TN-D3004, 1965.

12

Nonlinear Closure Relations for Electron Transport in Hydrodynamical Models

A. Salhoumi[1,2]

[1] *Faculté des Sciences Juridiques Economiques et Sociales Ain Sebaa, Université Hassan II-Mohammedia, Beausite, BP 2634, Ain sebaa, Casablanca, Morocco*
[2] *Laboratoire de Physique de la Matière Condensée (LPMC), Faculté des Sciences Ben M'sik, Université Hassan II-Mohammedia, BP 7955, Casablanca, Morocco*

Correspondence should be addressed to A. Salhoumi; ahmedsalhoumi@gmail.com

Academic Editor: Brian J. Edwards

Closure relations problem of hydrodynamical models in semiconductors is considered by expressing third- and fourth-order closure relations for the moments of the distribution function in terms of second-order Lagrange multipliers using a generalized Maxwell-Boltzmann distribution function within information theory. Calculation results are commented and compared with others to justify the accuracy of the approach developed in this paper. The comparison involves, in the first part with good agreements, the closure relations results obtained within extended thermodynamics which were checked by means of Monte Carlo simulations, in the second part, the results obtained by Grad's method which expands the distribution function up to fourth-order in Hermite polynomials. It is seen that the latter method cannot give any restriction on closure relations for higher-order moments, within the same conditions proposed in our approach. The important role of Lagrange multipliers for the determination of all closure relations is asserted.

1. Introduction

The analysis of transport in small semiconductor devices is essential for the optimization of their functioning. Such transport could in principle be described by means of Boltzmann transport equation (BTE) for charge carriers. However, in small devices the electric fields are extremely large, and therefore nonlinear effects are unavoidable [1, 2] which leads to insurmountable difficulties to obtain solutions. Notwithstanding this, BTE contains more information than needed in practical applications. It is common in practice to consider only the lowest-order moments of the distribution function, which are directly related to density, charge flux, kinetic energy, heat flux, and so on. These variables are measured and controlled. This kind of approach is called a hydrodynamical approach [3–5].

The basic model, in which the various steps and approximations are derived and discussed in detail, is due to Blotekjaer [4]. So as to close the set of balance equations considered by Blotekjaer, one assumes that higher-order moments have the value appropriate for a displaced Maxwellian. A slightly different model has been suggested by Hänsch and Miura-Mattausch [6]. In their model, the distribution function is expanded in Legendre polynomials and only the first two terms in the expansion are retained. Only the five balance equations for particle number, momentum, and energy are considered; then the closure is accomplished by means of the Wiedemann-Franz law for heat flux. Both models [4, 6] are then further simplified in order to provide a manageable set of equations appropriate for the device simulation. However, for more accurate results, the full models must be registered. In approaches similar to those cited in references [4, 6], Woolard et al. [7] and Thoma et al. [8] have proposed models taking into account the nonparabolicity of the band structure of the crystal. All these approaches have in common the assumption that some higher-order moments can be calculated by means of a displaced Maxwellian, such an approximation is rather rough and imprecise and its range of validity needs to be assessed [9].

Another method suitable for deriving hydrodynamical-like equations is Grad's method of moments [10]. This method yields, with an appropriate truncation, a set of evolution equations for the thirteen fields comprising, beside the five balance laws corresponding to particle number, momentum, and energy, rate-type equations for heat flux and anisotropic stress. These equations are known to describe dilute gases only near thermal equilibrium and fail in nonequilibrium situations. Note that Banach and Larecki [11] proposed a generalized Grad's method, which begins by expanding the phase density about an anisotropic Planck function, in order to close systems of moment equations in nonequilibrium situations.

Extended irreversible thermodynamics [1, 2] and extended thermodynamics [12, 13] are relatively recent approaches to nonequilibrium thermodynamics phenomena [14, 15], which at variance with classical irreversible thermodynamics [16] incorporate higher-order moments (to be interpreted as flux of fluxes) in the thermodynamic state description of system. Though both approaches coincide at least up to second-order in the fluxes, we adopt in this paper extended thermodynamics in order to derive hydrodynamical equations for carrier transport in semiconductors and try to close the system by means of a generalized distribution function within information theory (IT) [17]. Furthermore, our approach deals with the higher-order moments based models which still require a lot of fine tuning and a detailed understanding of the underlying physical phenomena as mentioned by Grasser et al. [18]. Furthermore, two recent contributions based on the maximum entropy principle (MEP) and validated through numerical results are to underline at this stage; the first one concerns an exact closure obtained of the 8-moment model for silicon semiconductors [19], and, the second one is a technique for the construction of realizable 5-moment closures whose extension to a fully three-dimensional gas was not necessarily simple. In fact, the cost of the accurate numerical integration of multidimensional distribution functions required for the resynchronization step was expected to be overwhelming [20].

Some rudimentary results of our approach were published previously [21] which contain some difficulties in comparing with other works based on Monte Carlo simulations, specially [22]. Here, we revisit it completely to overcome difficulties and to provide accurate results. It will be done by assuming nonvanishing higher-order Lagrange multipliers (L-Multipliers). Such hypothesis is not the first one to this field but there is some relevant contributions in quantum regime due to Trovato and Reggiani [23–26], in which they used quantum maximum entropy principle (QMEP) and dealt with closure problems by assuming that L-Multipliers could be expanded in powers of $\hbar^2$; $\hbar$ being the reduced Planck constant, in order to determine the reduced Wigner function [27] for equilibrium and nonequilibrium conditions. Thus, we compare and comment briefly, the aims and the main results of the recent unified physical approach [25] with those of the present paper. Finally, an attempt to solve the closure problem by means of Grad's method will show within the same conditions proposed to deal with closure relations (CR) in our approach.

The plan of this paper is as follows: in Section 2, we recall the basic formalism of the hydrodynamic description in order to write the CR, that is, third- and fourth-order moment tensors in terms of lower-order quantities. In Section 3, we summarize some foregoing works dealing with the CR problem. Section 4 presents the treatment of the closure problem, in the case of an ideal gas under heat flux or subject to both heat flux and viscous pressure by expanding a generalized distribution function up to second-order. Furthermore, we compare our results with other works. In Section 5, we summarize for comparison some results of the unified physical approach underlined above, particularly those concerning nonlinear CR. Section 6 is devoted to concluding remarks. The detail of some expressions used in this paper and some useful integrals are included as Appendices A, B, and C.

2. Hydrodynamic Description

The evolution equations for the moments of the distribution function are directly obtained from the semiclassical BTE for charge carriers [3]. In the case of electrons in the conduction band of a semiconductor it is written as

$$\frac{\partial f}{\partial t} + \mathbf{v}(\mathbf{k}) \cdot \nabla f - e\mathbf{E} \cdot \nabla_{\mathbf{k}} f = P, \qquad (1)$$

with $f(\mathbf{x}, \mathbf{k}, t)$ the distribution function, $\mathbf{k}$ the electron momentum, $\mathbf{v}(\mathbf{k})$ the electron group velocity given by $\mathbf{v}(\mathbf{k}) = \nabla_{\mathbf{k}}\varepsilon$, where ε is the energy, e the absolute value of electron charge, $\mathbf{E}$ the electric field, and P the collision term.

In the effective mass approximation [28], the energy is given by

$$\varepsilon(\mathbf{k}) = \frac{\mathbf{k}^2}{2m^*} \quad \text{with } \mathbf{v}(\mathbf{k}) = \frac{\mathbf{k}}{m^*}, \qquad (2)$$

where m^* is the effective electron mass which in silicon is $m^* = 0.26m_e$ with m_e the electron mass and *the reduced* Planck's constant is taken as $\hbar = 1$ for convenience.

By multiplying (1) by several products of the components of the momentum vector $\mathbf{k}$ and integrating, we obtain a hierarchy of equations for the different moments of the distribution function. Thus, we obtain different hydrodynamical models which differ in the choice of selected variables and on the level at which the truncation is made in the hierarchy [1, 2]. We will summarize these models in Section 3. For instance, let us write this hierarchy of equations [9, 14, 15, 29].

For consistency, the boundary of the first Brillouin zone is moved toward infinity, the particle density $n(\mathbf{x}, t)$ and the mean velocity $\mathbf{u}(\mathbf{x}, t)$ are defined, respectively, as $n(\mathbf{x}, t) = \int d\mathbf{k} f(\mathbf{x}, t, \mathbf{k})$ and $\mathbf{u}(\mathbf{x}, t) = \int d\mathbf{k}\mathbf{v}(\mathbf{k}) f(\mathbf{x}, t, \mathbf{k})$, where $\mathbf{J} = n\mathbf{u}$ defines the particle flux. Assuming as usual that $f(\mathbf{x}, t, \mathbf{k})$ vanishes sufficiently fast at infinity and integrating (1) in $\mathbf{k}$-space, we obtain the particle continuity equation

$$\frac{\partial n}{\partial t} + \nabla \cdot (n\mathbf{u}) = 0. \qquad (3)$$

Multiplying (1) by k^i and integrating yield the momentum balance equation

$$\frac{\partial (nu^i)}{\partial t} + \frac{\partial \theta^{ij}}{\partial x^j} + \frac{neE^i}{m^*} = P^i. \quad (4)$$

Multiplying (1) by k^ik^j and integrating, we find the expression of stress equation for the traceless part of $\widehat{\theta}_{\langle ij\rangle}$

$$\frac{\partial \widehat{\theta}_{\langle ij\rangle}}{\partial t} + \frac{\partial \widehat{\theta}_{\langle ij\rangle r}}{\partial x^r} + \frac{2neE_{\langle iuj\rangle}}{m^*} = P_{\langle ij\rangle}. \quad (5)$$

We denote the completely symmetric and traceless part of a tensor A_{ij} as

$$A_{\langle ij\rangle} = \frac{1}{2}\left[A_{ij} + A_{ji} - \frac{2}{3}\delta_{ij}A_k^k\right]. \quad (6)$$

From the trace of $\widehat{\theta}_{ij}$, we obtain the balance equation for the energy W, with P_W the energy production,

$$\frac{\partial W}{\partial t} + \nabla \cdot \mathbf{S} + ne\mathbf{E} \cdot \mathbf{u} = \mathbf{P}_W. \quad (7)$$

The energy flux equation for S_i is governed by the following equation, with P_i' as the energy-flux production:

$$\frac{\partial S_i}{\partial t} + \frac{\partial S_{ij}}{\partial x^j} + e\left[E_i\widehat{\theta}_{ij} + \frac{1}{m^*}WE_i\right] = P_i'. \quad (8)$$

Here S_{ij} denotes the flux of energy flux which is written as

$$\begin{aligned} S_{ij} &= Wu_iu_j + 2nk_BTu_iu_j + \frac{1}{2}u^2nk_BT\delta_{ij} \\ &+ \frac{m^*}{2}\left(2u_ru_j\widehat{\theta}_{\langle ir\rangle} + 2u_iu_r\widehat{\theta}_{\langle jr\rangle} + u^2\widehat{\theta}_{\langle ij\rangle}\right) \\ &+ u_jQ_i + u_iQ_j + u_r\widehat{\theta}_{ijr} + \widehat{\theta}_{ijrr}, \end{aligned} \quad (9)$$

with Q_i the heat flux. $\widehat{\theta}_{ijr}$ and $\widehat{\theta}_{ijrr}$ are, respectively, the random parts of the third- and fourth-order moment tensors, (10).

3. Closure Problem and Extended Thermodynamics

As we have mentioned above, different hydrodynamical models have tried to close the hierarchy of equations for the different moments of the distribution function. The variables appearing in the thirteen-moment equations (3)–(8) are the particle number density n, the momentum nu^i, the temperature T, the stress tensor $\widehat{\theta}_{\langle ij\rangle}$, the heat flux Q_i, and plus the random parts of the third- and fourth-order moment tensors $\widehat{\theta}_{ijr}$ and $\widehat{\theta}_{ijrr}$ given by

$$\begin{aligned} \widehat{\theta}_{ijr} &= \frac{m^*}{2}\int fc_ic_jc_r d\mathbf{k}, \\ \widehat{\theta}_{ijrs} &= \frac{m^*}{2}\int fc_ic_jc_rc_s d\mathbf{k}, \end{aligned} \quad (10)$$

where the random component $\mathbf{c}$ of the moment vector $\mathbf{k}$ is used for convenience and which is defined by $\mathbf{k} = m^*(\mathbf{u} + \mathbf{c})$ with $\mathbf{u}$ the mean velocity.

So as to close the system at this level of approximation, we need explicit expressions for $\widehat{\theta}_{ijr}$ and $\widehat{\theta}_{ijrr}$. One way is based on the assumption that the distribution function could be approximated by a drifted-Maxwellian [3, 4, 8, 30]. This approach implies the following closure relations for the random parts of the third-order and fourth-order moment tensors, (10),

$$\widehat{\theta}_{ijr} = \frac{1}{5}\left(Q_i\delta_{jr} + Q_j\delta_{ir} + Q_k\delta_{ij}\right), \quad (11)$$

$$\widehat{\theta}_{ijrr} = \frac{5}{2}\frac{n(k_BT)^2}{m^*}\delta_{ij}. \quad (12)$$

However, this model presents some limitations which makes a sounder analysis of the possible truncation of equations to optimize the description of the system desirable [1, 2]. Therefore, CR related to the hierarchy of moment equations of the BTE take an essential role in building hydrodynamical models of carrier transport in semiconductors, and it deserves a thorough investigation starting from first principles [12, 13]. This is done in framework of extended thermodynamics [1, 9, 12–15, 29]. The critical assumption is that $\widehat{\theta}_{ijr}$ and $\widehat{\theta}_{ijrr}$ can be considered as functions of the lower-order moments of particle density, momentum, energy, components of the pressure tensor, and heat flux. Within the definition and the justification of the partial thermal equilibrium state, Anile and Muscato [29] derived the results (11) and (12) as $\widehat{\theta}_{\langle ijr\rangle} = 0$ and $\widehat{\theta}_{ijrr} = (5/2)((n(k_BT)^2)/m^*)\delta_{ij}$ which were obtained with the drift-Maxwellian assumption. The last closure led to a flux-limited expression for the heat flow by using the Schwartz's inequality as

$$|Q| \le Q_{\max} = 3\left(\frac{5}{2}\right)^{1/2}\frac{n}{\sqrt{m^*}}(k_BT)^{3/2}. \quad (13)$$

Let us mention that other attempts to investigate the flux-limited phenomena are well known, for instance, in radiations hydrodynamics [9, 31, 32], in plasma physics [33, 34] and in electronics [1, 2, 29, 35–38]. Furthermore, Anile et al. have applied [9, 14, 15, 29, 39, 40] a method based on extended thermodynamics [1, 2, 13] in order to establish expressions corresponding to third- and fourth-order moment tensors, that is, $\widehat{\theta}_{ijr}$ and $\widehat{\theta}_{ijrr}$, in the second-order of Q_i and $\widehat{\theta}_{ij}$. They obtained the following constitutive equations up to second-order about partial thermal equilibrium:

$$\begin{aligned} \widehat{\theta}_{\langle ijk\rangle} = \frac{2}{5}\frac{m^*}{nk_BT}\Big(&Q_i\widehat{\theta}_{jk} + Q_j\widehat{\theta}_{ki} + Q_k\widehat{\theta}_{ij} \\ &- \frac{2}{15}Q_l\left(\widehat{\theta}_{\langle li\rangle}\delta_{jk} + \widehat{\theta}_{\langle lj\rangle}\delta_{ki} + \widehat{\theta}_{\langle lk\rangle}\delta_{ij}\right)\Big), \end{aligned} \quad (14)$$

$$\begin{aligned} \widehat{\theta}_{illj} &= \left(\frac{5n(k_BT)^2}{2m^*} + \left(\frac{\sigma}{Mk_BT_L} - \frac{2}{5nk_BT}\right)Q^2\right)\delta_{ij} \\ &+ \frac{2\sigma}{nk_BT_L}Q_iQ_j + \frac{7k_BT}{2}\widehat{\theta}_{\langle ij\rangle} + \frac{m^*}{n}\widehat{\theta}_{\langle il\rangle}\widehat{\theta}_{\langle jl\rangle}, \end{aligned} \quad (15)$$

where T_L is the lattice temperature.

In the constitutive equation (15), only the free parameter σ can be determined by comparison with Monte Carlo results [22]. In order to check CR, (14) and (15), the authors evaluate the quantities n, u^i, T, Q_i, and $\widehat{\theta}_{\langle ij\rangle}$ by Monte Carlo simulations and by using a suitable fitting. Accordingly, they determine the free parameter σ appearing in (15). In fact, this free parameter σ is in the range -0.15–$+0.15$ and it gives a maximal error less than 7% with respect to the Monte Carlo calculations as the authors described. Furthermore, one finds that in (14), the error increases with electrical field $\mathbf{E}$. Indeed, for $E \simeq 10^4$ V/cm the error is 0.01% and for $E \approx 10^5$ V/cm, the error is 1%. In (15), the error is 7% for $E \approx 10^2$ V/cm and reduces to 2% for E between 10^3 V/cm and 10^4 V/cm.

Note that Romano [41] has dealt with another method based on a finite difference scheme of Scharfetter-Gummel type in order to simulate a consistent energy-transport model for electron transport in semiconductors devices. This last method is free of any fitting parameters. Otherwise, it is known that the maximum entropy moment systems of the gas-dynamical Boltzmann equation suffer from severe disadvantages which are related to the nonsolvability of an underlying maximum entropy moment problem unless restrictions on the choice of the macroscopic variables are made. Thus, Junk and Romano [42] showed that no such difficulties appeared in the semiconductor case if Kane's dispersion relation is used for the energy band of electrons. In addition, a hydrodynamic subband model for semiconductors had been formulated by closing the moment system derived from the Schrödinger-Poisson-Boltzmann equations on the basis of the MEP where explicit closure relations for the fluxes and the production terms were obtained taking into account scattering of electrons with acoustic and nonpolar optical phonons, as well as surface scattering. For this model, a suitable numerical scheme was presented together with simulations of a nanoscale silicon diode [43]. Evenly, Camiola et al. [44] proposed an expression of the entropy combines quantum effects and semiclassical transport by weighting the contribution of each subband with the square modulus of the envelope functions arising from the Schrödinger-Poisson subsystem. The simulations shown that their model was able to capture the relevant confining and transport features, and, assessed the robustness of the numerical scheme. By taking into account nonparabolic energy bands of Kane's type, Mascali and Romano [45] obtained explicit closure relations for fluxes and production terms, including scattering of electrons with acoustic and nonpolar optical phonons and surface scattering. Numerical simulations of a quantum diode showed the feasibility of their model and the importance of the nonparabolicity was assessed. Likewise, a recent unified physical approach developed by Trovato and Reggiani [23–26] deal with the same CR problem in quantum scheme, we will report in Section 5 some results of this latter for a comparison asserting that the knowledge of the L-Multipliers implies the determination of all CR for the system.

4. Information Theory and Closure Relations

In the microscopic description, several methods (kinetic theory and information theory) coincide in the first-order in the fluxes. Here, we want to check disagreements in higher-orders (second-order for instance). We think that production terms and closure relations are good fields to do that. In this work, we consider only the latter ones. We apply IT [17] to deal with closure relations of a gas in nonequilibrium steady state subject to heat flux and viscous pressure. The treatment proposed here differs from that we have presented in the precedent section. Indeed, there are two modifications: (i) from a microscopic point of view, it is based on a second-order expansion of a generalized distribution function obtained from IT and (ii) from a macroscopic point of view, it is based on extended thermodynamics but with a slightly different interpretation of the inequalities arising from the second law of thermodynamics.

Thus, we consider a gas in a nonequilibrium steady state characterized by numerical density n, internal energy, or local equilibrium temperature T and by a heat flux $\mathbf{Q}$ and viscous pressure $\widehat{\theta}_{ij}$. The conditions to be imposed on the nonequilibrium distribution function f_{neq} are

$$\int f_{\text{neq}} d\mathbf{k} = \int f_{\text{eq}} d\mathbf{k} = n, \tag{16}$$

$$\int f_{\text{neq}} \mathbf{c} d\mathbf{k} = \int f_{\text{eq}} \mathbf{c} d\mathbf{k} = 0, \tag{17}$$

$$\int f_{\text{neq}} \left(\frac{1}{2} m^* \mathbf{c}^2\right) d\mathbf{k} = \int f_{\text{eq}} \left(\frac{1}{2} m^* \mathbf{c}^2\right) d\mathbf{k} = \frac{3}{2} n k_B T, \tag{18}$$

$$\int f_{\text{neq}} \left(\frac{1}{2} m^* \mathbf{c}^2\right) c_i d\mathbf{k} = Q_i, \tag{19}$$

$$\int f_{\text{neq}} c_{\langle icj\rangle} d\mathbf{k} = \widehat{\theta}_{\langle ij\rangle}, \tag{20}$$

where f_{eq} is the local equilibrium distribution function corresponding to the number density n and the internal energy $(3/2)nk_BT$. Since f_{neq} differs from f_{eq}, the higher-order moments of f_{neq}, as for instance corresponding to nonconserved quantities, will in general differ from the corresponding moments (19) and (20) of f_{eq} [46].

4.1. System under Heat Flux and Flux Limiters. When we take into account the heat flux as the only dissipative source, the maximum entropy solution for the distribution function has the form

$$f_{\text{neq}} = A \exp\left(-\beta \frac{1}{2} m^* \mathbf{c}^2 - \left(\frac{1}{2} m^* \mathbf{c}^2 - \frac{5}{2\beta}\right) \gamma \cdot \mathbf{c}\right), \tag{21}$$

where β and γ are the L-Multipliers to be determined from the constraints expressed by (18) and (19). A is a normalization constant to be obtained from (16). The Lagrange multiplier corresponding to the velocity has been written as $-(5/2\beta)\gamma$, a relation which, as it is known, satisfies the restriction (17). Therefore, the term $(5/2\beta)\gamma \cdot \mathbf{c}$ follows from the requirement that the system is at rest; that is, $\langle \mathbf{c} \rangle = 0$ [47]; it behaves near equilibrium as $(5/2\beta)\mathbf{c} \approx ((5k_BT)/2)\mathbf{c}$.

By using the shorthand notation $\alpha = (1/2)m^*\beta$ and $B(\mathbf{c}) = ((1/2)m^*\mathbf{c}^2-(5/2\beta))$, we expand (21) up to the second-order in γ as

$$f_{\text{neq}} = A\exp\left(-\alpha\mathbf{c}^2\right)\left[1 - B(\mathbf{c})\,\gamma\cdot\mathbf{c} + \frac{1}{2}B^2(\mathbf{c})\left(\gamma\cdot\mathbf{c}\right)^2\right]. \tag{22}$$

Thus, if one restricts the expansion to the first-order in γ, one would obtain the Grad distribution function [10] corresponding to a steady heat flux without viscous pressure. Also this second-order expansion in f_{neq} (which corresponds to a fourth-order expansion in $\mathbf{c}$) is not exactly equal to the fourth-order expansion in the Grad's scheme as it will be commented below. Note finally that according to the usual definition of the entropy in terms of the velocity distribution function f

$$\rho s = -k_B\int f\ln(f)\,d\mathbf{c}. \tag{23}$$

We obtain, according to the restrictions imposed on f

$$ds = k_B\beta du + k_B\gamma\cdot d\mathbf{Q}. \tag{24}$$

Therefore, the entropy is no longer the local equilibrium entropy [1, 2], but it depends also on the heat flux $\mathbf{Q}$ (through the Lagrange multiplier γ). Later, we will include also the viscous pressure $\widehat{\theta}_{ij}$ as a further independent quantity.

Introduction of (22) into (10) yields up to second-order in L-Multipliers corresponding to heat flux

$$\widehat{\theta}_{ijk} = \frac{1}{5}\left(Q_i\delta_{jk} + Q_j\delta_{ik} + Q_k\delta_{ij}\right), \tag{25}$$

$$\widehat{\theta}_{ijrr} = \frac{5n}{2m^*\beta^2}\delta_{ij} + C_{2ij}(\beta,n)\,\gamma_i\gamma_j, \tag{26}$$

where $C_{2ij}(\beta,n) = A\int\exp(-\beta((m^*\mathbf{c}^2)/2))(B^2(\mathbf{c})/2)\mathbf{c}^2c_i^2c_j^2d\mathbf{c}$.

We conclude that relation (25) coincides with (11) and by eliminating the second-order correction term in the right-hand side of (26) we find the same result as (12). Indeed, we can justify the second-order correction appearing in (26) by using a generalized distribution function.

Introduction of (22) into (16)–(19) yields the following simplified expressions:

$$A = \frac{n}{m^*}\left(\frac{\alpha}{\pi}\right)^{3/2}\left[1 + \frac{5}{8}\frac{\gamma^2}{\alpha\beta^2}\right]^{-1}, \tag{27}$$

$$\frac{3}{2}nk_BT = \frac{3}{2}\frac{n}{\beta}\left[1 + \frac{15}{8}\frac{\gamma^2}{\alpha\beta^2}\right]\left[1 + \frac{5}{8}\frac{\gamma^2}{\alpha\beta^2}\right]^{-1}, \tag{28}$$

$$Q_i = -\frac{5}{2}\frac{n}{\beta^3m^*}\gamma_i\left[1 + \frac{5}{8}\frac{\gamma^2}{\alpha\beta^2}\right]^{-1}. \tag{29}$$

Then the relation between Q_i and γ_i can be written as

$$Q_i = -\frac{A'\gamma_i}{1 + B'\gamma_i^2}, \tag{30}$$

where A' and B' are functions of m^* and β and n are obtained directly from (28) and are written as $A' = (5/2)(n/\beta^3m^*)$ and $B' = (5/8)(1/\alpha\beta^2)$.

Thus, the expression (30) yields a flux limiter for the heat flux because of the relation

$$Q \le \frac{A'}{2\sqrt{B'}} = \left(\frac{5}{4}\right)^{1/2}\frac{n}{\sqrt{m^*}}(\beta)^{-(3/2)} \approx \left(\frac{5}{4}\right)^{1/2}\frac{n}{\sqrt{m^*}}(k_BT)^{3/2}. \tag{31}$$

This saturation behavior of heat flux is attained when $\gamma = \sqrt{1/B'}$. So as to have a more accurate expression of heat flux limiter, we should take higher-order terms in (22). Then, we conclude that within this approach, the heat flow is limited (31) which could described a flux-limited phenomena as mentioned below (13).

4.2. System Subject to Reduced Heat Flux and Viscous Pressure. The generalized distribution function obtained from IT which describes *a priori* the nonequilibrium steady states of a system subjected to heat flux Q and viscous pressure $\widehat{\theta}_{ij}$ takes the following form:

$$f_{\text{neq}} = A\exp\left(-\beta\frac{1}{2}m^*\mathbf{c}^2 - \left(\frac{1}{2}m^*\mathbf{c}^2 - \frac{5}{2\beta}\right)c_i\gamma_i - \Gamma_{\langle jk\rangle} : m^*c_{\langle jck\rangle}\right). \tag{32}$$

The expansion of this latter up to the second-order in the L-Multipliers, β, γ_i, and $\Gamma_{\langle jk\rangle}$ conjugated, respectively, to energy, heat flux, and viscous pressure, is written with a traceless symmetric tensor Λ as

$$f_{\text{neq}} = A\exp\left(-\alpha\mathbf{c}^2\right)\left[1 - B(\mathbf{c})(\gamma_ic_i) - m^*\left(\Gamma_{\langle jk\rangle} : \mathbf{\Lambda}\right)c_jc_k + \frac{\gamma_i^2}{2}B^2(\mathbf{c})\,c_i^2 + \frac{m^{*2}}{2}\left(\Gamma_{\langle jk\rangle} : \mathbf{\Lambda}\right)^2c_j^2c_k^2 + m^*\gamma_i\left(\Gamma_{\langle jk\rangle} : \mathbf{\Lambda}\right)B(\mathbf{c})\,c_ic_jc_k\right], \tag{33}$$

where A is the normalization constant.

Denoting that the condition imposed on the distribution function to give the *reduced heat flux* is written as

$$\int f_{\text{neq}}B(\mathbf{c})\,c_id\mathbf{k} = Q_i. \tag{34}$$

Hence, in order to incorporate nonlinear effects in the CR we propose up to the second-order in reduced heat flux Q_i and viscous pressure tensor $\widehat{\theta}_{\langle ij\rangle}$, the following expressions for the third- and fourth-order moments:

$$\begin{aligned}\widehat{\theta}_{ijk} = {} & \frac{1}{5}\left(Q_i\delta_{jk} + Q_j\delta_{ik} + Q_k\delta_{ij}\right)\\ & + \widehat{C}_1\left(Q_i\widehat{\theta}_{jk} + Q_j\widehat{\theta}_{ki} + Q_k\widehat{\theta}_{ij}\right)\\ & + \widehat{C}_2Q_l\left(\widehat{\theta}_{\langle li\rangle}\delta_{jk} + \widehat{\theta}_{\langle lj\rangle}\delta_{ki} + \widehat{\theta}_{\langle lk\rangle}\delta_{ij}\right),\end{aligned} \tag{35}$$

$$\begin{aligned}\widehat{\theta}_{ijrr} = {} & \left(\frac{5n}{2m^*\beta^2} + \widehat{C}_3Q^2\right)\delta_{ij} + \widehat{C}_4Q_iQ_j\\ & + \widehat{C}_5\widehat{\theta}_{\langle ij\rangle} + \widehat{C}_6\widehat{\theta}_{\langle il\rangle}\widehat{\theta}_{\langle jl\rangle},\end{aligned} \tag{36}$$

where $\widehat{C}_i$ $(i = 1,\ldots,6)$ are coefficients to be determined.

Note that all these coefficients are equal to zero in the simplified hypotheses which recover (11) and (12). In the following subsection, we will derive the coefficients $\widehat{C}_i$ $(i = 1,\ldots,6)$ within IT [17] by imposing some restrictions on Q_i, $\widehat{\theta}_{\langle ij\rangle}$, $\widehat{\theta}_{ijk}$, and $\widehat{\theta}_{ijrr}$.

4.2.1. Calculations of $\widehat{C}_i$. The *first two hypotheses* imposed over Q_i and $\widehat{\theta}_{jk}$ to evaluate $\widehat{C}_1$ (resp. $\widehat{C}_2$) are $Q_1 \neq 0$ and $\widehat{\theta}_{23} \neq 0$ (resp. $Q_1 \neq 0$ and $\widehat{\theta}_{11} \neq 0$), and, the other components of Q_i and $\widehat{\theta}_{ij}$ are equal to zero. After carrying out the calculations, (35) yields the expressions of $\widehat{C}_1$ and $\widehat{C}_2$ up to the second-order in L-Multipliers

$$\widehat{C}_1 = \frac{\widehat{\theta}_{123}}{Q_1\widehat{\theta}_{23}} = \frac{2}{5}\frac{m^*\beta}{n}\left[1 + \frac{5}{8}\frac{\gamma_1^2}{\alpha\beta^2} + \frac{1}{2}\frac{\left(\Gamma_{\langle 23\rangle} : \boldsymbol{\Lambda}\right)^2}{\beta^2}\right], \quad (37)$$

$$3\widehat{C}_1 + 2\widehat{C}_2 = \frac{5\widehat{\theta}_{111} - 3Q_1}{5Q_1\widehat{\theta}_{11}} = -\frac{69}{25}\frac{m^*\beta}{n}\left[\frac{\left(\Gamma_{\langle 11\rangle} : \boldsymbol{\Lambda}\right)}{\beta}\right]\Phi, \quad (38)$$

with

$$\Phi = \left(1 - \frac{\left(\Gamma_{\langle 11\rangle} : \boldsymbol{\Lambda}\right)}{\beta} + \frac{5}{8}\frac{\gamma_1^2}{\alpha^2\beta^2} + \frac{3}{2}\frac{\left(\Gamma_{\langle 11\rangle} : \boldsymbol{\Lambda}\right)^2}{\beta^2}\right) \times \left(\left(1 - \frac{27}{5}\frac{\left(\Gamma_{\langle 11\rangle} : \boldsymbol{\Lambda}\right)}{\beta}\right) \times \left(1 - 3\frac{\left(\Gamma_{\langle 11\rangle} : \boldsymbol{\Lambda}\right)}{\beta} + \frac{27}{8}\frac{\gamma_1^2}{\alpha^2\beta^2} + \frac{15}{2}\frac{\left(\Gamma_{\langle 11\rangle} : \boldsymbol{\Lambda}\right)^2}{\beta^2}\right)\right)^{-1}. \quad (39)$$

Note that we let $(\Gamma_{\langle 11\rangle} : \boldsymbol{\Lambda})/\beta$ term without simplification by β because it will play an important role as we will see later in this paper.

The *second two hypotheses* imposed over Q_i and $\widehat{\theta}_{ij}$ to evaluate $\widehat{C}_3$ (resp. $\widehat{C}_4$) are $Q_1 \neq 0$ (resp. $Q_1 \neq 0$ and $Q_2 \neq 0$), and, the other components of Q_i and $\widehat{\theta}_{ij}$ are equal to zero. After integration, (36) gives the following expressions for $\widehat{C}_3$ and $\widehat{C}_4$ up to second-order in the L-Multipliers

$$\widehat{C}_3 + \widehat{C}_4 = \frac{\widehat{\theta}_{11rr} - 5n/2m^*\beta^2}{Q_1^2} = \frac{166}{25}\frac{\beta}{n}\left[1 + \frac{5}{8}\frac{\gamma_1^2}{\alpha\beta^2}\right], \quad (40)$$

$$\widehat{C}_4 = \frac{\widehat{\theta}_{12rr}}{Q_1Q_2} = \frac{119}{25}\frac{\beta}{n}\left[1 + \frac{5}{8}\frac{\gamma_1^2 + \gamma_2^2}{\alpha\beta^2}\right]. \quad (41)$$

The *third two hypotheses* imposed over Q_i and $\widehat{\theta}_{ij}$ to evaluate $\widehat{C}_5$ (resp. $\widehat{C}_6$) are $\widehat{\theta}_{12} \neq 0$ (resp. $\widehat{\theta}_{11} \neq 0$), and, the other components of Q_i and $\widehat{\theta}_{ij}$ are equal to zero. Thus (36) yields the following expressions for $\widehat{C}_5$ and $\widehat{C}_6$ up to second-order in the L-Multipliers:

$$\widehat{C}_5 = \frac{\widehat{\theta}_{12rr}}{\widehat{\theta}_{\langle 12\rangle}} = \frac{7}{2}\frac{1}{\beta}, \quad (42)$$

$$\begin{aligned}\widehat{C}_6 &= \frac{\widehat{\theta}_{11rr} - 5n/2m^*\beta^2 - \widehat{C}_5\widehat{\theta}_{\langle 11\rangle}}{\widehat{\theta}^2_{\langle 11\rangle}} \\ &= -\frac{21}{4}\frac{m^*}{n}\left(\left[1 + \frac{8}{7}\frac{\left(\Gamma_{\langle 11\rangle} : \boldsymbol{\Lambda}\right)}{\beta} - \frac{75}{7}\frac{\left(\Gamma_{\langle 11\rangle} : \boldsymbol{\Lambda}\right)^2}{\beta^2}\right] \times \left[1 - \frac{\left(\Gamma_{\langle 11\rangle} : \boldsymbol{\Lambda}\right)}{\beta} + \frac{3}{2}\frac{\left(\Gamma_{\langle 11\rangle} : \boldsymbol{\Lambda}\right)^2}{\beta^2}\right]\right) \\ &\quad \times \left(\left[1 - 3\frac{\left(\Gamma_{\langle 11\rangle} : \boldsymbol{\Lambda}\right)}{\beta} + \frac{15}{2}\frac{\left(\Gamma_{\langle 11\rangle} : \boldsymbol{\Lambda}\right)^2}{\beta^2}\right]^2\right)^{-1}.\end{aligned} \quad (43)$$

Here, we should underline that the coefficients $\widehat{C}_i$ $(i = 1,\ldots,6)$ depend, in our case, not only on m^*, n, and T as in [22], but on the L-Multipliers β, γ_i, and $\Gamma_{\langle jk\rangle}$ by taking into account only the second-order in the expansion of the distribution function. Note that explicit expressions of the generalized distribution function corresponding to each hypothesis are given in Appendix A.

4.2.2. Comparison with Anile and Muscato [22]. In zero-order in L-Multipliers [21] or in first-order in the L-Multipliers with $(\Gamma_{\langle 11\rangle} : \Lambda)/\beta$ is equal to zero, the coefficients $\widehat{C}_i$ $(i = 1,\ldots,6)$, that is, (37)–(43), become $\widehat{C}_1 \approx (2/5)(m^*\beta/n)$, $\widehat{C}_2 \approx -(1/5)(m^*\beta/n)$, $\widehat{C}_3 \approx (47/25)(\beta/n)$, $\widehat{C}_4 \approx (119/25)(\beta/n)$, $\widehat{C}_5 \approx (7/2)(1/\beta)$, and $\widehat{C}_6 \approx -(21/4)(m^*/n)$. At this stage, the expressions for CR near equilibrium state take the forms

$$\widehat{\theta}_{\langle ijk\rangle} \approx \frac{2}{5}\frac{m^*\beta}{n}\left(Q_i\widehat{\theta}_{jk} + Q_j\widehat{\theta}_{ki} + Q_k\widehat{\theta}_{ij} - \frac{1}{2}Q_l\left(\widehat{\theta}_{\langle li\rangle}\delta_{jk} + \widehat{\theta}_{\langle lj\rangle}\delta_{ki} + \widehat{\theta}_{\langle lk\rangle}\delta_{ij}\right)\right), \quad (44)$$

$$\widehat{\theta}_{ijrr} \approx \left(\frac{5n}{2m^*\beta^2} + \frac{47}{25}\frac{\beta}{n}Q^2\right)\delta_{ij} + \frac{119}{25}\frac{\beta}{n}Q_iQ_j + \frac{7}{2}\frac{1}{\beta}\widehat{\theta}_{\langle ij\rangle} - \frac{21}{4}\frac{m^*}{n}\widehat{\theta}_{\langle il\rangle}\widehat{\theta}_{\langle jl\rangle}. \quad (45)$$

When we neglect the contribution of γ_i and $\Gamma_{\langle jk\rangle}$, we can recover easily (14) and (15) established by Anile and Muscato [22], with a slight suitable difference concerning the definitions of $\mathbf{Q}$, $\widehat{\theta}_{ij}$, $\widehat{\theta}_{ijk}$, and $\widehat{\theta}_{ijrr}$. In our opinion, the main difference is related to the use of different microscopic definitions of the heat flow and a vanishing expression of $(\Gamma_{\langle 11\rangle} : \boldsymbol{\Lambda})/\beta$. We will study later in this paper a nonvanishing expression of $(\Gamma_{\langle 11\rangle} : \boldsymbol{\Lambda})/\beta$ which will give us a solution to

TABLE 1: Estimates of the free parameter σ, (46), by using data extracted from [22].

Electron temperature: T (K)	275.27	281.77	348.27	449.46	1694.33	3281.89	5123.61
Electric field: $\mathbf{E}$ (V/cm)	1109	1449	5518	10768	41000	61230	80000
Free parameter: σ	2.594	2.534	2.050	1.589	0.421	0.218	0.139

overcome the slight differences among our expressions and those established by Anile and Muscato [22].

Otherwise to evaluate CR, Anile and Muscato [22] used for the heat-flux the expression $\mathbf{Q} = (1/2)m^*\mathbf{c}^2\mathbf{c}$ but in our case we have used $\mathbf{Q} = ((1/2)m^*\mathbf{c}^2 - (5/2\beta))\mathbf{c}$. Thus, the two definitions gave the same average value for $\mathbf{Q}$, as we have described above, but they yield different values for higher-order moments. The adopted definition in our calculation corresponds to what one calls *the reduced heat flow*, which it is used in fluctuation-dissipation theory where moments of $\mathbf{Q}$ play an important role.

Furthermore, if we simply compare (14) and (15) with those established in our approach (44) and (45), we can estimate the unknown free parameter σ. In fact, a definite value of this latter is provided by the maximum entropy approach or alternatively it may be determined by Monte Carlo simulations [1]. This is what we deal with in our approach based on IT. It is clear from (15) and (45) that the unknown parameter σ is connected to $\widehat{C}_3$ (resp. $\widehat{C}_4$). It is a nondimensional free parameter and a function of the lattice temperature T_L (300 K) and the electron temperature T as $\sigma \approx (57/25)(MT_L/nT)$ (resp. $\sigma \approx (119/50)(T_L/T)$) with $\beta \approx (1/k_BT)$ in the case of partial thermal equilibrium.

So as to compare our results with those obtained in Monte Carlo simulations, we plot in Figure 1 for the same values of T_L and T versus field $\mathbf{E}$. The data of these latters (see Table 1) are extracted from [22]. Thus, we choose the second expression of σ related to $\widehat{C}_4$ because the authors were silent about the expression of the parameter M used in expression (15)

$$\sigma \approx \frac{119}{50}\frac{T_L}{T}. \tag{46}$$

As shown in Figure 1, we find that free parameter σ which is obtained by using a fitting of Monte Carlo simulation data [22] becomes approximately constant at high field more than 60000 V/cm and $\sigma \le 0.139$.

In Figure 2, we plot the impact of electron temperature T on the nondimensional free parameter σ. It is clear from Figure 2 that at high electron temperature T is more than 3500 K and $\sigma \le 0.13$. This last σ-value is obtained without using any fitting. Hence, we can use the expression (46) to evaluate the value of the electric field which could be applied in future experiences dealing with electron transport in semiconductors and to compare with Monte Carlo simulation results, vice versa to have electron temperature T if we know electric field E.

4.2.3. Overcoming Slight Differences in CR with Anile and Muscato [22]. In order to overcome the slight difference between our results and those performed by Anile and Muscato [22], we have to take into account, in the first-order in L-Multipliers, a nonvanishing expression of $(\Gamma_{\langle 11\rangle} : \mathbf{\Lambda})/\beta$.

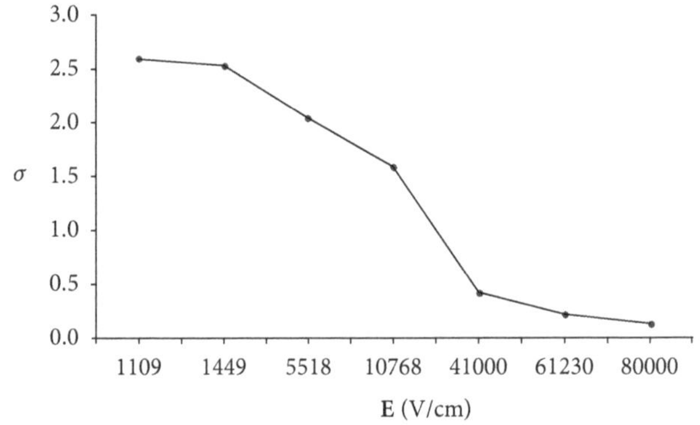

FIGURE 1: The free parameter σ from (46) versus electric field $\mathbf{E}$.

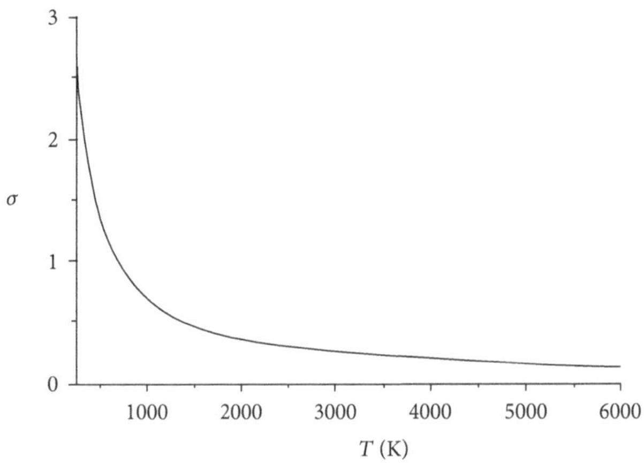

FIGURE 2: Determination of the free parameter σ from (46) by varying electron temperature T, without using a fit nor data of Monte Carlo simulation.

Denoting firstly by $\widehat{C}_i^A$ $(i = 1, \ldots, 6)$, the coefficients related to (14) and (15) which are written as $\widehat{C}_1^A \approx (2/15)(m^*\beta/n)$, $\widehat{C}_2^A \approx -(4/75)(m^*\beta/n)$, $\widehat{C}_3^A \approx ((\sigma/Mk_BT_L) - (2/5)(\beta/n))$, $\widehat{C}_4^A = (2\sigma/nk_BT_L)$, $\widehat{C}_5^A \approx (7/2)(1/\beta)$, and $\widehat{C}_6^A = (m^*/n)$ with $(1/k_BT) \approx \beta$. Thus, by subtracting from (38) three times (37) and adding (38) to (37), that is, (38) $-3\times$ (37) and (38) + (37), we obtain the coefficients $\widehat{C}_2$ and $\widehat{C}_1$ in the first-order in L-Multipliers with a nonvanishing $(\Gamma_{\langle 11\rangle} : \mathbf{\Lambda})/\beta$

$$\widehat{C}_2 = \widehat{C}_2^A\left[\frac{207}{8}\frac{(\Gamma_{\langle 11\rangle} : \mathbf{\Lambda})}{\beta} + \frac{45}{4}\right], \tag{47}$$

$$\widehat{C}_1 = -\frac{1}{5}\widehat{C}_1^A\left[\frac{207}{8}\frac{(\Gamma_{\langle 11\rangle} : \mathbf{\Lambda})}{\beta} + \frac{45}{4}\right] - \frac{m^*\beta}{n}\left[\frac{69}{100}\frac{(\Gamma_{\langle 11\rangle} : \mathbf{\Lambda})}{\beta} + \frac{2}{20}\right]. \tag{48}$$

We find from (47) that for $(\Gamma_{\langle 11\rangle} : \mathbf{\Lambda})/\beta = -(82/207)$ we have $\widehat{C}_2 = \widehat{C}_2^A$. This last equality into (48) yields $\widehat{C}_1 = \widehat{C}_1^A$. Thus, introducing the latter nonvanishing expression of $(\Gamma_{\langle 11\rangle} : \mathbf{\Lambda})/\beta$ into (43) without any trancation in L-Multipliers, it gives $\widehat{C}_6 = 0.857C_6^A$. Then, the closure relations (35) and (36) become

$$\begin{aligned}\widehat{\theta}_{\langle ijk\rangle} &\approx \frac{2}{5}\frac{m^*\beta}{n}\left(Q_i\widehat{\theta}_{jk} + Q_j\widehat{\theta}_{ki} + Q_k\widehat{\theta}_{ij}\right.\\&\qquad\left.-\frac{2}{15}Q_l\left(\widehat{\theta}_{\langle li\rangle}\delta_{jk} + \widehat{\theta}_{\langle lj\rangle}\delta_{ki} + \widehat{\theta}_{\langle lk\rangle}\delta_{ij}\right)\right),\\ \widehat{\theta}_{ijrr} &\approx \left(\frac{5n}{2m^*\beta^2} + \frac{47}{25}\frac{\beta}{n}Q^2\right)\delta_{ij} + \frac{119}{25}\frac{\beta}{n}Q_iQ_j + \frac{7}{2}\frac{1}{\beta}\widehat{\theta}_{\langle ij\rangle}\\&\quad + 0.857\frac{m^*}{n}\widehat{\theta}_{\langle il\rangle}\widehat{\theta}_{\langle jl\rangle}.\end{aligned} \tag{49}$$

By comparing (49) with (14) and (15), we conclude that we have overcome the silght difference arising at zero-order and at first-order in L-Multipliers where we have used a vanishing expression of $(\Gamma_{\langle 11\rangle} : \mathbf{\Lambda})/\beta$. Thus, the accuracy of our approach checked against the results of Anile and Muscato [22], involving the percent error between $\widehat{C}_i$ and $\widehat{C}_i^A$ is asserted such as $\widehat{C}_1/\widehat{C}_1^A = 100\%$, $\widehat{C}_2/\widehat{C}_2^A = 100\%$, $\widehat{C}_5/\widehat{C}_5^A = 100\%$, and $\widehat{C}_6/\widehat{C}_6^A = 85.7\%$. We recall, as we have processed above that $\widehat{C}_4$ and $\widehat{C}_4^A$ (resp. $\widehat{C}_3$ and $\widehat{C}_3^A$) allow us to find the nondimensional free parameter σ. It is clear now that taking into account higher-order in L-Multipliers will give better accurate results as it is seen from the calculation of coefficient $\widehat{C}_6$ determined without any truncation in L-Multipliers.

4.2.4. Closure Relations and Grad's Method [10]. The most well-known assumed form for the distribution function is the Grad type which is based on an expansion of the distribution function in Hermite polynomials [10]. This decouples the dependence of the closure coefficients and greatly simplifies the derivation. In fact, Grad considered both 13- and 20-moment closures, extensions to many moments have been considered by others [13, 48]. However, it is an unfortunate fact that members of the Grad hierarchy suffer from several problems: (i) the distribution function is not always positive; it is therefore not a properly defined probability density function and (ii) for modest departures from local equilibrium, it is possible for the resulting moment equations to become nonhyperbolic [49–51]. Here, we assert that Grad's method fall in nonequilibrium situation. Indeed, we expand the distribution function f_{neq} up to the fourth-order in the power of the velocity (this is the same order used in the precedent section). In the notation of Hermite polynomials where $H_i^{(n)}$ is a tensor of order n, Grad's distribution function writes as

$$f_G = f_{\text{eq}}\sum_{n=0}^{4}\frac{1}{n!}a_i^{(n)}H_i^{(n)}, \tag{50}$$

where the coefficient $a_i^{(n)}$ is also a tensor of order n.

The first four closure relations are written as

$$\widehat{\theta}_{ijk} = \sum_{n=1}^{4}\widehat{\theta}_{ijk}^n,\qquad \widehat{\theta}_{ijrr} = \sum_{n=1}^{4}\widehat{\theta}_{ijrr}^n. \tag{51}$$

In order to derive the coefficients $\widehat{C}_i$ appearing in (35) and (36), we should use the same conditions as in Section 4.2.1; that is, $Q_1 \neq 0$ and $\widehat{\theta}_{23} \neq 0$, and, the other components are equal to zero. Thus, up to the third-order Grad's distribution function takes the form

$$f_G = f_{\text{eq}}\left[1 + \frac{1}{2}\frac{p_{23}}{p}c_2c_3 + \frac{1}{6}\frac{Q_1c_1}{p\sqrt{RT}}\left(\mathbf{c}^2 - 5\right)\right]. \tag{52}$$

According to the last equation we observe that $\widehat{\theta}_{123} = 0$, which means that $\widehat{C}_1 = 0$. Grad's distribution function can be generalized by including the full third-order contributions

$$\begin{aligned}f_G = f_{\text{eq}}\Bigg[1 &+ \frac{\theta_{23}}{p}c_2c_3 + \frac{Q_1c_1}{p\sqrt{RT}}\left(c^2 - 5\right)\\&+\frac{\theta_{123}}{p\sqrt{RT}}c_1c_2c_3 + \cdots\Bigg],\end{aligned} \tag{53}$$

where $\widehat{\theta}_{123}$ is arbitrary.

Hence, if one imposes only conditions on p_{23} and on Q_1 but not on the additional third-order terms, we observe that $\widehat{C}_1$ is arbitrary too, since $\widehat{C}_1$ is given by $\widehat{C}_1 = \widehat{\theta}_{123}/(Q_1\widehat{\theta}_{23})$. We checked also the same method for the other coefficients $\widehat{C}_2, \widehat{C}_3, \ldots, \widehat{C}_6$ in (35) and (36). We concluded that at higher-order, if conditions are imposed only on heat flux and viscous pressure, Grad's method cannot give any restrictions in the closure relations (35) and (36). We can understand the difference between this model and our approach based on IT. Consequently, if we compare f_{neq} with f_G used to provid $\widehat{\theta}_{123}$, we will see clearly that in our approach $\widehat{\theta}_{123}$ is connected in a unique manner to γ_1 and $\Gamma_{\langle 23\rangle}$ but in Grad's method it is an independent coefficient. In fact, the distribution function of Grad assumes to $\widehat{\theta}_{123} = 0$ in the case, where only $Q_1 \neq 0$ and $\widehat{\theta}_{23} \neq 0$. Thus, this freedom of coefficients in Grad's method is present also for the others coefficients $\widehat{C}_2, \widehat{C}_3, \ldots, \widehat{C}_6$.

5. Nonlinear Closure Relations in Quantum Scheme

The present section is devoted to compare the aims and the results developed in Section 4 with those of the work of Trovato and Reggiani [23–26]. The authors proposed MEP in its local semiclassical and nonlocal quantum formulation as the basis of an unified physical approach. In fact, the local semiclassical case investigated MEP both in the usual kinetic theory of degenerate gases and in transport phenomena for hot carriers in nondegenerate semiconductors and mostly overviewed results obtained in the last decade.

The nonlocal quantum case mostly summarized recent results and presented a series of original findings. Otherwise, Romano [52] proposed quantum corrections to the hydrodynamical model of semiconductors based on MEP were obtained at $\hbar^2$ order with a Chapman-Enskog expansion in the high field approximation, modeling the $\hbar^2$ part of the collision term in a relaxation form. The author deduced limiting energy-transport and drift-diffusion models.

As constraints in QMEP approach, Trovato and Reggiani [24] have determined a closed quantum hydrodynamic (QHD) system for macroscopic variables. By showing that only a higher-order expansion of the distribution function can be fruitfully applied to describe transport phenomena under conditions extremely far from thermodynamic equilibrium in the presence of very high electric fields and associated gradients such as those occurring in submicron devices. Therefore, to describe accurately the transport phenomena in submicron semiconductor devices, they have considered a nonlinear expansion of the distribution function in terms of the first thirteen relevant macrovariables involved physical interpretation. For validation purposes, the approach was applied to many one-dimensional n^+nn^+ submicron Si structures by using different band structure models, different doping profiles, different lattice temperatures, and different applied biases. From the numerical results, it is shown that only a strongly nonlinear expansion provided a more accurate description of all moments [25]. We report below an abstract of those works which concern only nonlinear extended thermodynamics with thirteen moments and we comment briefly the main characteristics of this recent unified physical approach.

5.1. Unified Physical Approach due to Trovato and Reggiani [25]. In the local semiclassical case, Trovato and Reggiani [25] introduced BTE for a fermion or bose gas by developing a local theory with a temperature scheme to describe the state of a gas in which the moments of the distribution function $f(\mathbf{r},\mathbf{u},t)$ are taken as basic fields. Through a formal expansion around the local equilibrium configuration, they explained the extended thermodynamic theories of N moments and degree α, by determining a set of closed hydrodynamic systems for bosons and/or fermions, in the frame work of the usual gas dynamics. By introducing the kinetic field

$$\Psi\left\{n, mu_i, mu^2, mu_{\langle iuj\rangle}, mu^2u_i\right\}, \tag{54}$$

where $\mathbf{u}$ denotes velocity vector; they obtained the corresponding first thirteen macroscopic variables, that is, the masse density n, the moment density mu_i, the energy density mu^2, the traceless momentum flux density $mu_{\langle iuj\rangle}$, and the energy flux density mu^2u_i that admit a direct physical interpretation. In the same way, by defining the higher-order moments, the external field productions and collisional productions, they wrote the balance equations for usual first thirteen moments.

Thus, by considering the decomposition of a moment in its convective and central parts, they formally obtained the new set of variables $\{n, P, M_{\langle ij\rangle}, Q_i\}$, n being the numerical density, P the pressure, $M_{\langle ij\rangle}$ the stress deviator, and Q_i the heat flux. Analogously, they obtained the following constitutive equations for the third- and fourth-order moments tensors $\widehat{\theta}_{ijr}$ and $\widehat{\theta}_{illk}$ as

$$\begin{aligned}
\widehat{\theta}_{\langle ijk\rangle} &= M_{ijr} + 3M_{(\langle ij\rangle v_k)} \\
&\quad - \frac{2}{5}v_k\left(M_{\langle ri\rangle}\delta_{jk} + M_{\langle rj\rangle}\delta_{ki} + M_{\langle rk\rangle}\delta_{ij}\right) \\
&\quad + mnv_{(ivjvk)}, \\
\widehat{\theta}_{illk} &= M_{ijr} + 2v_rM_{\langle rik\rangle} + 7Pv_iv_k + \frac{28}{5}Q_{(ivk)} \\
&\quad + \frac{4}{5}\left(Q_rv_r\right)\delta_{ik} + v^2M_{\langle ik\rangle} + Pv^2\delta_{ik} \\
&\quad + 2v_r\left(M_{\langle ri\rangle}v_k + M_{\langle rk\rangle}v_i\right) + mnv^2v_iv_k,
\end{aligned} \tag{55}$$

where v_i is the mean velocity.

The central moments $M_{i_1\cdots i_nk}$ are expressed as

$$M_{i_1\cdots i_sk} = \int m\widetilde{u}_i\cdots\widetilde{u}_{i_s}\widetilde{u}_kf(\mathbf{r},\mathbf{u},t)\,d\mathbf{u}, \tag{56}$$

where $\widetilde{u} = u_i - v_i$ the peculiar velocity and $s = 0, 1, \ldots, M$ by construction with arbitrary values for the integer M.

The higher-order moments $\widehat{\theta}_{\langle ijk\rangle}$ and $\widehat{\theta}_{illk}$ (55) are determined by decomposing the distribution function and L-Multipliers into equilibrium and nonequilibrium parts where L-Multipliers are formulated versus bose and fermi integrals. In fact, by determining an analytic expression for the L-Multipliers versus n, P, Q_i, $M_{\langle ij\rangle}$ and the fermi and bose integral functions, Trovato and Reggiani [25] obtained an explicit distribution function up to the second-order of the local equilibrium. Thus, all the results obtained for the constitutive relations are determined in terms of the fermi and/or bose functions. Therefore, only by means of an efficient evaluation of these integrals, it is possible to analyse the gas behavior for various degeneracy levels.

5.2. Comparison with Trovato and Reggiani [25]. Within the QMEP Trovato and Reggiani [25] introduced a quantum entropy functional of the reduced density matrix; the principle of quantum maximum entropy is then asserted as fundamental principle of quantum statistical mechanics. Accordingly, they developed a comprehensive theoretical formalism to construct rigorously a closed QHD transport within a Wigner function approach [25]. Their approach is formulated both in thermodynamic equilibrium and nonequilibrium conditions, and the quantum contributions are obtained by only assuming that the L-Multipliers can be expanded in powers of $\hbar^2$, while the classical results are recovered in the limit $\hbar \rightarrow 0$. The authors asserted that the knowledge of the L-Multipliers involves the determination of all CR for the system, up to the first quantum correction [24]. Thus, to prove the validity of the QMEP results in the case of nonlinear extended thermodynamics with thirteen moments, Trovato and Reggiani compared the analytic results existing in the literature with some recent numerical methods for the evaluation of bose and fermi integrals.

Hence it is clear that the L-Multipliers play an important role in the conception dealing with CR in hydrodynamical models both in the relevant contributions of Trovato and Reggiani and in the present paper. In our case, we have assumed some restrictions on L-Multipliers where third- and fourth-order CR for the moments of the distribution function are expressed merely in terms of second-order L-Multipliers: (i) to recover the formulation of flux-limiters (30) and (ii) to overcome difficulties in comparing our results with those of Anile and Muscato [22]. In fact, we have used on the one hand a nonvanishing expression of $(\Gamma_{\langle 11\rangle} : \boldsymbol{\Lambda})/\beta$ at the first-order in L-Multipliers to find the accurate coefficients $\widehat{C}_1$ and $\widehat{C}_2$, and on the other hand the fall expression of $\widehat{C}_6$ without any truncation in L-Multipliers (49). Also, the fact that third- and fourth-order moments are related to the numerical values of bose and fermi integrals, in the unified physical approach due to Trovato and Reggiani, does not allow us to reach one of the main objectives of the present paper which is to find an analytical expression of the nondimensional free parameter σ (46).

6. Concluding Remarks

In this paper, we have studied within extended thermodynamics implications of higher-order moments, in particular second-order moments, so as to solve the closure problem presented in hydrodynamical models. This is done by means of a generalized distribution function which is obtained from maximum-entropy argument. Thus, we have compared the results of our approach with those obtained by both models based on the assumption of a drift-Maxwellian distribution function, and, on extended thermodynamics and entropy principle; then we have found a good agreement. Further, we have applied Grad's method to solve the same nonlinear CR. In the first part, we have restricted ourself using only heat flux as dissipative source. The CR found are the same as those obtained by a drift-Maxwellian assumption of the distribution function. We have been led to an expression for the heat flux describing the flux limited phenomena occurring in the submicron-electron devices. In the second part, we have considered a generalized distribution function for a nonequilibrium system under heat flux and viscous pressure in order to obtain constitutive equations of the third-order and the fourth-order moments. Indeed, we have found with a slight difference similar constitutive equations as those established by Anile and Muscato [22]. This slight difference which was published previously arised from the use of a vanishing first-order in L-Multipliers. Here, it was overcome by using a nonvanishing one.

Furthermore, we conclude that considerations proposed in this paper, which concern the definition of the reduced heat flux, allow us to express the nondimensional free parameter σ appearing in (15) as function of the lattice temperature T_L and the electron temperature T. In fact, the σ-parameter appearing in fourth-order moment tensor, for partial thermal equilibrium, is related both to the Monte Carlo simulation and to a fitting as described by Anile and coworkers. Finally, we have used Grad's method to calculate the CR for higher-order moments in a simple case. We have found that this method could not give any restrictions about higher-order moments. Also, we assert as Trovato and Reggiani [23–26] that the knowledge of the L-Multipliers involves the determination of all CR for the system.

In brief, we have constructed a generalized distribution function in the frame work of IT. This latter allowed us to reproduce the constitutive equations of the third-order and the fourth-order moments in terms of second-order L-Multipliers and to take into account the nonlinear effects in transport-hydrodynamical models of submicron-electron devices. We should not forgot that the generalized distribution function proposed in this paper is one of several contributions dealing with the extension of the canonical distribution function by adding to the equilibrium Hamiltonian an effective Hamiltonian due to the flow [1, 2, 53]. Thus, we have considered in (21) heat flux as a supplementary constraint, and in (32) both heat flux and viscous pressure as supplementary constraints in order to conceive the generalized distribution function used in our approach. At the end, for a full exploitation of the results mentioned in the present paper involving the nonlinear CR and the maximum-entropy should be tested against the Monte Carlo simulation incorporating second-order L-Multipliers.

Appendices

A. Explicit Expressions of f_{neq} Used in Section 4.1

The present appendix contains the explicit expressions of the generalized distribution function corresponding to each hypothesis proposed in our paper in order to derive the coefficients C_i, that is, (37), (38), (40), (41), (42), and (43).

For the *first tow hypotheses*, the distribution function is written, respectively,

$$\begin{aligned} f_{\text{neq}} = A\exp\left(-\alpha \mathbf{c}^2\right)\Big[& 1 - B(\mathbf{c})\gamma_1 c_1 - m^*\left(\Gamma_{\langle 23\rangle} : \boldsymbol{\Lambda}\right) c_2 c_3 \\ & + \frac{\gamma_1^2}{2} B^2(\mathbf{c}) c_1^2 + \frac{m^{*2}}{2}\left(\Gamma_{\langle 23\rangle} : \boldsymbol{\Lambda}\right)^2 c_2^2 c_3^2 \\ & + m^*\gamma_1\left(\Gamma_{\langle 23\rangle} : \boldsymbol{\Lambda}\right) B(\mathbf{c}) c_1 c_2 c_3 \Big], \\ f_{\text{neq}} = A\exp\left(-\alpha \mathbf{c}^2\right)\Big[& 1 - B(\mathbf{c})\gamma_1 c_1 - m^*\left(\Gamma_{\langle 11\rangle} : \Lambda\right) c_1^2 \\ & + \frac{\gamma_1^2}{2} B^2(\mathbf{c}) c_1^2 + \frac{m^{*2}}{2}\left(\Gamma_{\langle 11\rangle} : \boldsymbol{\Lambda}\right)^2 c_1^4 \\ & + m^*\gamma_1\left(\Gamma_{\langle 11\rangle} : \boldsymbol{\Lambda}\right) B(\mathbf{c}) c_1^3 \Big]. \end{aligned} \quad \text{(A.1)}$$

In the conditions of the *second two hypotheses*, the distribution function is written, respectively,

$$f_{\text{neq}} = A\exp\left(-\alpha\mathbf{c}^2\right)\Big[1 - B(\mathbf{c})\gamma_1 c_1 - m^*\left(\Gamma_{\langle 23\rangle} : \mathbf{\Lambda}\right)c_2c_3 + \frac{\gamma_1^2}{2}B^2(\mathbf{c})c_1^2 + \frac{m^{*2}}{2}\left(\Gamma_{\langle 23\rangle} : \mathbf{\Lambda}\right)^2 c_2^2c_3^2 + m^*\gamma_1\left(\Gamma_{\langle 23\rangle} : \mathbf{\Lambda}\right)B(\mathbf{c})c_1c_2c_3\Big],$$

$$f_{\text{neq}} = A\exp\left(-\alpha\mathbf{c}^2\right)\Big[1 - B(\mathbf{c})\gamma_1 c_1 - m^*\left(\Gamma_{\langle 11\rangle} : \mathbf{\Lambda}\right)c_1^2 + \frac{\gamma_1^2}{2}B^2(\mathbf{c})c_1^2 + \frac{m^{*2}}{2}\left(\Gamma_{\langle 11\rangle} : \mathbf{\Lambda}\right)^2 c_1^4 + m^*\gamma_1\left(\Gamma_{\langle 11\rangle} : \mathbf{\Lambda}\right)B(\mathbf{c})c_1^3\Big]. \tag{A.2}$$

The distribution functions related to the *third two hypotheses* are written, respectively,

$$f_{\text{neq}} = A\exp\left(-\alpha\mathbf{c}^2\right)\left[1 - B(\mathbf{c})\gamma_1 c_1 + \frac{\gamma_1^2}{2}B^2(\mathbf{c})c_1^2\right],$$

$$f_{\text{neq}} = A\exp\left(-\alpha\mathbf{c}^2\right)\Big[1 - B(\mathbf{c})\gamma_1 c_1 - B(\mathbf{c})\gamma_2 c_2 + \frac{\gamma_1^2}{2}B^2(\mathbf{c})c_1^2 + \frac{\gamma_2^2}{2}B^2(\mathbf{c})c_2^2\Big]. \tag{A.3}$$

B. Some Equations Used in Grad's Method

We underline here some equations used in Grad's method to deal with CR. Indeed, the first four Hermite polynomials $H_i^{(n)}$ appeared in the distribution function of Grad (50) [10] may be written as

$$\begin{aligned}
H^{(0)} &= 1,\\
H_i^{(1)} &= c_i,\\
H_{ij}^{(2)} &= c_ic_j - \delta_{ij},\\
H_{ijk}^{(3)} &= c_ic_jc_k - \left(c_i\delta_{jk} + c_j\delta_{ik} + c_k\delta_{ij}\right),\\
H_{ijkl}^{(4)} &= c_ic_jc_kc_l - \left(c_ic_j\delta_{kl} + c_ic_k\delta_{jl} + c_ic_l\delta_{jk} + c_jc_k\delta_{il} + c_jc_l\delta_{ik} + c_kc_l\delta_{ij}\right) + \left(\delta_{ij}\delta_{kl} + \delta_{ik}\delta_{jl} + \delta_{il}\delta_{jk}\right),
\end{aligned} \tag{B.1}$$

and the coefficient tensors $a_i^{(n)}$ are expressed in terms of the elementary moments as

$$\begin{aligned}
a^{(0)} &= 1,\\
a_i^{(1)} &= 0,\\
a_{ij}^{(2)} &= \frac{p_{ij}}{p},\\
a_{ijk}^{(3)} &= \frac{s_{ijk}}{p\sqrt{RT}},\\
a_{ijkl}^{(4)} &= \frac{M_{ijkl}}{pRT} - \frac{1}{p}\left(p_{ij}\delta_{kl} + p_{ik}\delta_{jl} + p_{il}\delta_{jk} + p_{jk}\delta_{il} + p_{jl}\delta_{ik} + p_{kl}\delta_{ij}\right) - \left(\delta_{ij}\delta_{kl} + \delta_{ik}\delta_{jl} + \delta_{il}\delta_{jk}\right),
\end{aligned} \tag{B.2}$$

where p_{mn}, s_{mnl}, and M_{ijkl} take the form

$$\begin{aligned}
p_{ij} &= \int c_ic_jf\,d\xi,\\
s_{ijk} &= \int c_ic_jc_kf\,d\xi,\\
M_{ijkl} &= \int c_ic_jc_kc_lf\,d\xi.
\end{aligned} \tag{B.3}$$

Then, the simplified expressions of the closure relations (51) are given by

$$\begin{aligned}
\widehat{\theta}_{ijk}^{(0)} &= \frac{Am^{*2}}{2}\int \exp\left(-\alpha\mathbf{c}^2\right)c_ic_jc_k\,d\mathbf{c} = 0,\\
\widehat{\theta}_{ijk}^{(1)} &= \frac{Am^{*2}}{2}\int c_ic_jc_k\exp\left(-\alpha\mathbf{c}^2\right)\left(0\,c_m\right)d\mathbf{c} = 0,\\
\widehat{\theta}_{ijk}^{(2)} &= \frac{Am^{*2}p^{mn}}{4p} : \int \exp\left(-\alpha\mathbf{c}^2\right)\left(c_mc_n - \delta_{mn}\right)c_ic_jc_kc_k\,d\mathbf{c} = 0,\\
\widehat{\theta}_{ijk}^{(3)} &= \frac{Am^{*2}s^{mnl}}{12p\sqrt{RT}} : \int \exp\left(-\alpha\mathbf{c}^2\right) \times \left(c_mc_nc_l - \left(c_m\delta_{nl} + c_n\delta_{ml} + c_l\delta_{mn}\right)\right)c_ic_jc_k\,d\mathbf{c},\\
\widehat{\theta}_{ijk}^{(4)} &= 0,\\
\widehat{\theta}_{ijrr}^{(0)} &= \frac{m^{*2}}{2}\int f_{\text{eq}}c_ic_j\mathbf{c}^2\,d\mathbf{c},\\
\widehat{\theta}_{ijrr}^{(1)} &= 0,\\
\widehat{\theta}_{ijrr}^{(2)} &= \frac{m^{*2}p^{mn}}{4p} : \int f_{\text{eq}}\left(c_mc_n - \delta_{mn}\right)c_ic_j\mathbf{c}^2\,d\mathbf{c},\\
\widehat{\theta}_{ijrr}^{(3)} &= 0,\\
\widehat{\theta}_{ijrr}^{(4)} &= \frac{m^{*2}}{2}\frac{1}{4!}a_{mnlr}^{(4)} : \int f_{\text{eq}}H_{mnlr}^{(4)}c_ic_j\mathbf{c}^2\,d\mathbf{c}.
\end{aligned} \tag{B.4}$$

C. Useful Integrals

This appendix contains some useful integrals that are needed to calculate equations presented in this paper. Let $F(\mathbf{c})$ be any scalar function of $\mathbf{c}$ such that the integrals appearing below converge, and let c_x and c_y be the components of the relative velocity $\mathbf{c}$ [1]. Then

$$\int F(\mathbf{c})\, c_x^2 d\mathbf{c} = \frac{1}{3}\int F(\mathbf{c})\, \mathbf{c}^2 d\mathbf{c},$$
$$\int F(\mathbf{c})\, c_x^4 d\mathbf{c} = \frac{1}{5}\int F(\mathbf{c})\, \mathbf{c}^4 d\mathbf{c}, \qquad (\mathrm{C.1})$$
$$\int F(\mathbf{c})\, c_x^2 c_y^2 d\mathbf{c} = \frac{1}{15}\int F(\mathbf{c})\, \mathbf{c}^4 d\mathbf{c}.$$

The following definite integrals are also useful:

$$\int_0^{+\infty} \exp\left(-\alpha \mathbf{c}^2\right) c^r d\mathbf{c} = \frac{\sqrt{\pi}}{2}\frac{1}{2}\frac{3}{2}\frac{5}{2}\cdots\frac{r-1}{2}\alpha^{-(r+1)/2}, \quad r \text{ even},$$

$$\int_0^{+\infty} \exp\left(-\alpha \mathbf{c}^2\right) c^r d\mathbf{c} = \frac{1}{2}\left(\frac{r-1}{2}\right)!\alpha^{-(r+1)/2}, \quad r \text{ odd}. \qquad (\mathrm{C.2})$$

Conflict of Interests

The author declares that there is no conflict of interests in this paper.

Acknowledgments

The author thanks Professor Jou, D. from Grup de Fsica Estatistica Departament de Fsica Universitat Autonóma de Barcelona (Spain), for his precious remarks, careful reading of the paper, and helpful discussions. The author thanks also Professor Trovato, M. from Dipartimento di Matematica Università di Catania (Italy) for giving his relevant contributions [23–26] and to Professor Bennai, M. for hosting this work at LPMC. A. Salhoumi's current address is Faculté des Sciences Juridiques Economiques et Sociales Ain Sebaa, Universitxé Hassan II-Mohammedia, Ain sebaa Morocco while his hosting address is Laboratoire de Physique de la Matière Condensée (LPMC), Faculté des Sciences Ben M'sik, Université Hassan II-Mohammedia, Morocco.

References

[1] D. Jou, J. Casas-Vazquez, and G. Lebon, *Extended Irreversible Thermodynamics*, Springer, Berlin, Germany, 2010.

[2] G. Lebon, D. Jou, and J. Casas-Vazquez, *Understanding Non-Equilibrium Thermodynamics*, Springer, Berlin, Germany, 2008.

[3] W. Hänsch, *The Drift-Diffusion Equation and Its Application in MOSFET Modeling*, Springer, Berlin, Germany, 1991.

[4] B. K. Blotekjaer, "Transport equations for electrons in two-valley semiconductors," *IEEE Transactions on Electron Devices*, vol. 17, no. 1, pp. 38–47, 1970.

[5] G. Baccarani and M. R. Wordeman, "An investigation on steady-state velocity overshoot in silicon," *Solid State Electronics*, vol. 29, pp. 970–977, 1982.

[6] W. Hänsch and M. Miura-Mattausch, "The hot-electron problem in small semiconductor devices," *Journal of Applied Physics*, vol. 60, no. 2, pp. 650–656, 1986.

[7] D. L. Woolard, H. Tian, R. J. Trew, M. A. Littlejohn, and K. W. Kim, "Hydrodynamic electron-transport model: nonparabolic corrections to the streaming terms," *Physical Review B*, vol. 44, no. 20, pp. 11119–11132, 1991.

[8] R. Thoma, A. Emunds, B. Meinerzhagen, H.-J. Peifer, and W. L. Engl, "Hydrodynamic equations for semiconductors with nonparabolic band structure," *IEEE Transactions on Electron Devices*, vol. 38, no. 6, pp. 1343–1353, 1991.

[9] A. M. Anile and S. Pennisi, "Thermodynamic derivation of the hydrodynamical model for charge transport in semiconductors," *Physical Review B*, vol. 46, no. 20, pp. 13186–13193, 1992.

[10] H. Grad, "Principles of the kinetic theory of gases in thermodynamics of gases," in *Handbuck der Physik*, S. Flügge, Ed., vol. 12, Springer, Berlin, Germany, 1958.

[11] Z. Banach and W. Larecki, "Nine-moment phonon hydrodynamics based on the modified Grad-type approach: formulation," *Journal of Physics A*, vol. 37, no. 41, pp. 9805–9829, 2004.

[12] I. Muller and T. Ruggeri, *Extended Thermodynamics*, Springer, New York, NY, USA, 1993.

[13] I. Muller and T. Ruggeri, *Rational Extended Thermodynamics*, Springer, New York, NY, USA, 1998.

[14] A. M. Anile and S. Pennisi, "Extended thermodynamics of the Blotekjaer hydrodynamical model for semiconductors," *Continuum Mechanics and Thermodynamics*, vol. 4, no. 3, pp. 187–197, 1992.

[15] A. M. Anile, S. Pennisi, and M. Muscato, "Extended thermodynamics of charge carrier transport in semiconductors," in *Nonlinear Hyperbolic Problems: Theoretical, Applied, and Computational Aspects*, A. Donato and F. Oliveri, Eds., vol. 43 of *Notes on Numerical Fluid Mechanics*, Vieweg, Braunschweig, Germany, 1993.

[16] S. R. de Groot and P. Mazur, *Nonequilibrium Thermodynamics*, Dover Publication, New York, NY, USA, 1985.

[17] R. M. Gray, *Entropy and Information Theory*, Springer, Berlin, Germany, 2010.

[18] T. Grasser, T. Tang, H. Kosina, and S. Selberherr, "A review of hydrodynamic and energy-transport models for semiconductor device simulation," *Proceedings of the IEEE*, vol. 91, no. 2, pp. 251–274, 2003.

[19] S. la Rosa, G. Mascali, and V. Romano, "Exact maximum entropy closure of the hydrodynamical model for sisemiconductors: the 8-moment case," *SIAM Journal on Applied Mathematics*, vol. 70, no. 3, pp. 710–734, 2009.

[20] J. G. McDonald and C. P. T. Groth, "Towards realizable hyperbolic moment closures for viscous heat-conducting gas ows based on a maximum-entropy distribution," *Continuum Mechanics and Thermodynamics*, vol. 25, no. 5, pp. 573–603, 2013.

[21] A. Salhoumi, M. Zakari, and Y. Boughaleb, "Nonlinear efiects in 3rd and 4th order moments of transporthydrodynamical models in semiconductors within extended thermodynamics," *The Moroccan Journal of Condensed Matter*, vol. 2, pp. 71–75, 1999.

[22] A. M. Anile and O. Muscato, "Extended thermodynamics tested beyond the linear regime: the case of electron transport in

silicon semiconductors," *Continuum Mechanics and Thermodynamics*, vol. 8, no. 3, pp. 131–142, 1996.

[23] M. Trovato and L. Reggiani, "Quantum hydrodynamic models from a maximum entropy principle," *Journal of Physics A*, vol. 43, no. 10, Article ID 102001, 2010.

[24] M. Trovato and L. Reggiani, "Quantum maximum entropy principle for a system of identical particles," *Physical Review E*, vol. 81, no. 2, Article ID 021119, 2010.

[25] M. Trovato and L. Reggiani, "Maximum entropy principle and hydrodynamic models in statistical mechanics," *Rivista del Nuovo Cimento*, vol. 35, pp. 99–266, 2012.

[26] M. Trovato and L. Reggiani, "Quantum maximum entropy principle for fractional exclusion statistics," *Physical Review Letters*, vol. 110, no. 2, Article ID 020404, 2013.

[27] E. Wigner, "On the quantum correction for thermodynamic equilibrium," *Physical Review*, vol. 40, no. 5, pp. 749–759, 1932.

[28] C. Kittel, *Quantum Theory of Solids*, John Wiley & Sons, New York, NY, USA, 1963.

[29] A. M. Anile and O. Muscato, "Improved hydrodynamical model for carrier transport in semiconductors," *Physical Review B*, vol. 51, no. 23, pp. 16728–16740, 1995.

[30] M. A. Stettler, M. A. Alam, and M. S. Lundstrom, "A critical examination of the assumptions underlying macroscopic transport equations for silicon devices," *IEEE Transactions on Electron Devices*, vol. 40, no. 4, pp. 733–740, 1993.

[31] A. M. Anile and V. Romano, "Covariant flux-limited diffusion theories," *Astrophysical Journal Letters*, vol. 386, no. 1, pp. 325–329, 1992.

[32] C. D. Levermore, "Relating Eddington factors to flux limiters," *Journal of Quantitative Spectroscopy and Radiative Transfer*, vol. 31, no. 2, pp. 149–160, 1984.

[33] A. R. Bell, R. G. Evans, and D. J. Nicholas, "Elecron energy transport in steep temperature gradients in laser-produced plasmas," *Physical Review Letters*, vol. 46, no. 4, pp. 243–246, 1981.

[34] E. M. Epperlein, G. J. Rickard, and A. R. Bell, "Two-dimensional nonlocal electron transport in laser-produced plasmas," *Physical Review Letters*, vol. 61, no. 21, pp. 2453–2456, 1988.

[35] M. Zakari, "A continued-fraction expansion for flux limiters," *Physica A*, vol. 240, no. 3-4, pp. 676–684, 1997.

[36] A. Salhoumi, M. Zakari, D. Jou, J. Casa-Vazquez, and Y. Boughaleb, "A brief glimpse of extended irreversible thermodynamics theory and some of its applications," *The Moroccan Journal of Condensed Matter*, vol. 1, pp. 62–70, 1998.

[37] A. Salhoumi, M. Zakari, and Y. Boughaleb, "Another contribution to calculate ballistic velocity by means of continued fraction expansion within extended irreversible thermodynamics," *Physics Letters A*, vol. 275, no. 5-6, pp. 486–492, 2000.

[38] F. Vazquez, "Size efiects on heat transport in small systems: dynamical phase transition from difiusive to ballistic regime," *Journal of Applied Physics*, vol. 105, pp. 64915–64919, 2009.

[39] A. M. Anile, W. Allegretto, and C. Ringhofer, *Mathematical Problems in Semiconductor Physics*, vol. 1823 of *Lecture Notes in Mathematics*, Springer, Berlin, Germany, 2003.

[40] A. M. Anile, A. Marrocco, V. Romano, and J. M. Sellier, "2D numerical simulation of the MEP energy-transport model with a mixed finite elements scheme," *Journal of Computational Electronics*, vol. 4, no. 3-4, pp. 231–259, 2005.

[41] V. Romano, "2D numerical simulation of the MEP energy-transport model with a finite difference scheme," *Journal of Computational Physics*, vol. 221, no. 2, pp. 439–468, 2007.

[42] M. Junk and V. Romano, "Maximum entropy moment system of the semiconductor Boltzmann equation using Kane's dispersion relation," *Continuum Mechanics and Thermodynamics*, vol. 17, no. 3, pp. 247–267, 2005.

[43] G. Mascali and V. Romano, "Numerical simulation of a hydrodynamic subband model for semiconductors based on the maximum entropy principle," *Il Nuovo Cimento C*, vol. 33, pp. 155–163, 2010.

[44] V. D. Camiola, G. Mascali, and V. Romano, "Numerical simulation of a double-gate MOSFET with a subband model for semiconductors based on the maximum entropy principle," *Continuum Mechanics and Thermodynamics*, vol. 24, pp. 417–436, 2012.

[45] G. Mascali and V. Romano, "A non parabolic hydrodynamical subband model for semiconductors based on the maximum entropy principle," *Mathematical and Computer Modelling*, vol. 55, no. 3-4, pp. 1003–1020, 2012.

[46] J. Casas-Vázquez and D. Jou, "Nonequilibrium temperature versus local-equilibrium temperature," *Physical Review E*, vol. 49, no. 2, pp. 1040–1048, 1994.

[47] R. Domínguez and D. Jou, "Thermodynamic pressure in nonequilibrium gases," *Physical Review E*, vol. 51, no. 1, pp. 158–163, 1995.

[48] H. Struchtrup, *Macroscopic Transport Equations for Rareffed Gas Flows*, Springer, Berlin, Germany, 2005.

[49] F. Brini, "Hyperbolicity region in extended thermodynamics with 14 moments," *Continuum Mechanics and Thermodynamics*, vol. 13, no. 1, pp. 1–8, 2001.

[50] M. Torrilhon, "Characteristic waves and dissipation in the 13-moment-case," *Continuum Mechanics and Thermodynamics*, vol. 12, no. 5, pp. 289–301, 2000.

[51] M. Torrilhon, "Hyperbolic moment equations in kinetic gas theory based on multi-variate Pearson-IV-distributions," *Communications in Computational Physics*, vol. 7, no. 4, pp. 639–673, 2010.

[52] V. Romano, "Quantum corrections to the semiclassical hydrodynamical model of semiconductors based on the maximum entropy principle," *Journal of Mathematical Physics*, vol. 48, no. 12, Article ID 123504, 2007.

[53] D. Jou, J. Casas-Vázquez, and M. Criado-Sancho, "Information theory and thermodynamics of polymer solutions under flow," *Physica A*, vol. 262, no. 1-2, pp. 69–75, 1999.

13

A Derivation of the Main Relations of Nonequilibrium Thermodynamics

Vladimir N. Pokrovskii

Moscow State University of Economics, Statistics and Informatics, Moscow 119501, Russia

Correspondence should be addressed to Vladimir N. Pokrovskii; vpok@comtv.ru

Academic Editors: C. D. Daub, P. Espeau, A. Ghoufi, and P. Trens

The principles of nonequilibrium thermodynamics are discussed, using the concept of internal variables that describe deviations of a thermodynamic system from the equilibrium state. While considering the first law of thermodynamics, work of internal variables is taken into account. It is shown that the requirement that the thermodynamic system cannot fulfil any work via internal variables is equivalent to the conventional formulation of the second law of thermodynamics. These statements, in line with the axioms introducing internal variables can be considered as basic principles of nonequilibrium thermodynamics. While considering stationary nonequilibrium situations close to equilibrium, it is shown that known linear parities between thermodynamic forces and fluxes and also the production of entropy, as a sum of products of thermodynamic forces and fluxes, are consequences of fundamental principles of thermodynamics.

1. Introduction

The modern nonequilibrium thermodynamics is formulated [1–3] as a generalisation of equilibrium thermodynamics, as adding some concepts and principles, in particular, the concepts of fluxes and thermodynamic forces, specific for non-equilibrium. Despite many different approaches the problem, reviewed recently by Muschik [4], the extension of equilibrium thermodynamics to non-equilibrium thermodynamics seems to need some justification. We are going to follow the approach [5, 6], which exploits additional variables, so-called *internal variables*,[1] to describe deviations of a state of thermodynamic system from equilibrium. It can be thought, that this approach allows one to explore the principles of non-equilibrium thermodynamics, providing some justification of the known linear relations and opens opportunities for non-linear generalizations. We have to note that formulation of the main principles of non-equilibrium thermodynamics in terms of internal variables is disputable, there is, at least, two explicit versions. In particular, one of the approaches [6] takes into consideration only distinctive internal variables, those that can be called [7] *complexity internal variables*. The other approach, which is followed in this paper, considers all quantities, which describe the deviation of the system from the equilibrium, to be internal variables. In this paper, we are trying to show advantages of our description [7] as compared with the alternative formulation [4, 6].

Section 2 begins with a description of a set of variables needed for the depiction of a non-equilibrium state of a thermodynamic system. Further, reproducing partly the previous paper [7], we pay a special attention to formulation of the first and second principles of thermodynamics in terms of internal variables. Dynamics of internal variables is discussed in Section 3. Section 4 is devoted to consideration of stationary states, while it is shown, that, in the areas close to equilibrium, the known linear relation between thermodynamic forces and fluxes and also expression for production of entropy, as a sum of products of fluxes and thermodynamic forces, follow the formulated general principles. Conclusion contains a discussion of the results.

2. The Main Concepts and Principles

Let us consider a thermodynamic system, *the equilibrium state* of which, on definition, is characterized by absolute temperature[2] T and *constitutive variables* $x_1, x_2, \ldots, x_n$. The specified constitutive variables determine thermodynamic system and define its borders. In the simplest case, a volume V, which contains a certain part of substance in a gas or liquid phase, is considered as a constitutive variable.

To define *a nonequilibrium state* of the same system, one needs, apart from temperature (in this case the concept of temperature ought to be discussed specially [6]) and constitutive variables, in an indefinite number of other variables $\xi_1, \xi_2, \ldots$. When a limited list of n constitutive variables is fixed, the other variables are called *internal variables*; in an equilibrium situation there are no internal variables or, otherwise, we can consider values of internal variables whatever their set is, are equal to zero, which corresponds to the agreement that the internal variables describe deviations of the system from the equilibrium.[3] The internal variables characterize local heterogeneity of system and, consequently, can be named as structural variables also. The thermodynamic system has no predetermined set of internal parameters; they arise, generally speaking, at any influence on the system and, consequently, it is possible to note also that internal variables in the system represent disappearing memory of the past interactions with the environment.

Thus, the state space of a system is defined as $(T, x_1, x_2, \ldots x_n; \xi_1, \xi_2 \ldots)$, where temperature T is a unique characteristic for both equilibrium and non-equilibrium, and any space deviations from this quantity are considered as internal variables. It is not the unique way of determining the system, but it is convenient to adopt the listed variables for the beginning. Under unchanging conditions, the thermodynamic system is trending to a unique steady macrostate, which is referred to as *the equilibrium state*. The trend of the system to the equilibrium state means a trend of all internal variables to zero values.

2.1. The First Principle of Thermodynamics. First of all, let us consider the balance of all influences on the system. Taking into account energy coming into the system with fluxes of heat and matter from the environment, ΔQ and ΔN_j, and work of the system[4] through constitutive variables (work, performed by the system, is considered positive), one can write for a change of total energy U of the system that

$$\Delta U = \Delta Q - \sum_{i=1}^{n} X_i \Delta x_i + \sum_{j=1}^{N} \mu_j \Delta N_j. \quad (1)$$

This is a conventional form of the balance of energy [4, 6].

In line with the total internal energy U, which includes possible potential energy of agitated internal variables, internal thermal energy E, which does not depend on internal variables, can be introduced [7]. In the equilibrium situations, total energy U coincides with internal energy E, but, in the non-equilibrium situations, these quantities are different.[5] Then, the law of conservation of energy (1) can be rewritten in the form

$$\Delta E = \Delta Q - \sum_{i=1}^{n} X_i \Delta x_i - \sum_{i} \Xi_i \Delta \xi_i + \sum_{j=1}^{N} \mu_j \Delta N_j, \quad (2)$$

where work of the system via internal variables is included. This form of *the first principle of thermodynamics* demonstrates, that the external influences (work via constitutive variables, heat energy and chemical energy of particles) contribute to a change of internal energy E and emerging of internal variables. It is important to take into account the processes of reallocation of energy due to the work of the internal variables in the explicit form.

2.2. The Second Principle of Thermodynamics. The work connected with the internal variables should be considered as an essential element of the description. Internal variables can be agitated by external influences, but the system cannot perform any positive work via internal variables, which means

$$\sum_{i} \Xi_i \Delta \xi_i \leq 0. \quad (3)$$

This statement, formulated earlier [7], introduces impossibility of the reversion of evolution of the thermodynamic system in time and can be considered as a formulation of *the second principle of thermodynamics*—the formulation, which is, as it will be shown below, equivalent to the formulation of the principle in terms of entropy (see Section 2.4).

2.3. The Introduction of Entropy. To define thermodynamic quantities as functions of state, that is, functions of the temperature and constitutive variables, one needs in the concept of entropy with help of which a change of internal energy E, which does not depend on internal variables on the definition, can be written as the total differential of variables of state:

$$dE = TdS - \sum_{i=1}^{n} X_i \, dx_i. \quad (4)$$

The relation is valid both for reversible and irreversible processes.

Entropy S can be defined here, on a comparison of relations (2) and (4), as a quantity, a change of which is determined by external influences and internal changes:

$$TdS = \Delta Q - \sum_{i} \Xi_i \Delta \xi_i + \sum_{j=1}^{K} \mu_j \Delta N_j. \quad (5)$$

Remarkably, when the external fluxes are absent, entropy, in virtue of (5), can be considered as a function of internal variables, whereas the change of entropy (5) can be considered as a total differential of the function. To calculate entropy S, in this case, one can consider process of changing of the system from a nonequilibrium state with the fixed values of variables $\xi_1, \xi_2, \ldots \xi_s$ to the equilibrium state. The result does not depend on the way of integration, and the difference between values of entropy in equilibrium and nonequilibrium states can be written symbolically as

$$S(T, \mathbf{x}, \boldsymbol{\xi}) - S(T, \mathbf{x}, 0) = -\frac{1}{T} \int_0^{\xi_1, \xi_2, \ldots \xi_s} \sum_{j} \Xi_j d\xi_j \leq 0. \quad (6)$$

Relation (6) can be considered as a definition of entropy of a nonequilibrium state of thermodynamic system. The definition is applicable to any systems, including open systems in situations far from equilibrium.

Note that relation (5) is a generalisation of Prigogine's relation, given by him for the systems with chemical reactions [1, Equation 3.52].[6]

2.4. Two Parts of the Entropy Variation. The total change of entropy in the system dS, defined by (5), can be split, due to Prigogine [1], into two components:

$$dS = d_eS + d_iS. \tag{7}$$

Change of Entropy due to the Fluxes. The component d_eS is connected with flows of heat and substances through the boundaries of the system:

$$d_eS = \frac{1}{T}\left(\Delta Q + \sum_{j=1}^{N}\mu_j\Delta N_j\right), \tag{8}$$

where N_j is a number of molecules of substance j in the system, and μ_j is a chemical potential. The quantity d_eS can be both positive (the flux of heat and/or substances into the system) and negative (the flux of heat and/or substances out of the system). In an *isolated* system, when there are neither heat fluxes nor matter fluxes between the system and environment, $d_eS = 0$.

The fluxes of heat and substances into the system or from the system appear at absence of balance between the system and environment. As characteristic quantities, real streams of heat and/or substances through borders of the system are fixed. The actual fluxes into thermodynamic system are specified by a problem under consideration, but, in any case, it is possible to define empirically a set of fluxes $J_1, J_2, \ldots, J_r$ which, it is supposed, are given.

It is convenient to write expression (8) for a change, due to external fluxes into the system, of entropy in a standardized form:

$$\frac{d_eS}{dt} = \frac{1}{T}\sum_{j=1}^{r} K_jJ_j, \tag{9}$$

where quantities K_j are characteristics of the system, that is, functions of temperature, constitutive variable, and also internal variables.

Production of Entropy. The internal part of the change of entropy d_iS is connected with some processes within the system:

$$d_iS = -\frac{1}{T}\sum_j \Xi_j\Delta\xi_j, \quad d_iS \geq 0. \tag{10}$$

That means, that some internal variables, which cannot be identified in advance and the number of which is not known, appear to be agitated during transition from one macroscopic state to another and then relax according to their internal laws. Under reversible processes, when the situation changes in such a slow way (characteristic time of process $t \gg$ characteristic time of relaxation τ_i), that all internal variables actually have their equilibrium values, internal production of entropy is equal to zero, $d_iS = 0$. By virtue of (3), the quantity d_iS can be only nonnegative, and this is a conventional expression of *the second law of thermodynamics.*

From (10), expression for *the production of entropy is as follows:*

$$\frac{d_iS}{dt} = -\frac{1}{T}\sum_j \Xi_j\frac{d\xi_j}{dt}, \quad \frac{d_iS}{dt} \geq 0. \tag{11}$$

Entropy increases, when relaxation of internal variables in the system occurs, so that the production of entropy is a manifestation of presence of internal variables, that is a manifestation of an internal complexity of the system.

3. Dynamics of Internal Variables

To move further, our speculations should be completed by equations, which allow to determine evolution of internal variables. Let us consider a system in a non-equilibrium state $(T, \mathbf{x}, \boldsymbol{\xi})$ and under external influences, which can be a change of constitutive variables and/or a change of temperature and concentration of substances in the environment. In the case of reversible processes, the external influences are considered to be weak, so that the changes of T and $\mathbf{x}$ follow external influences. In a general case, to describe the actual situation, one has to consider emerging and evolution of the internal variables, which are not influenced by the external forces directly. As an example, we can point to a situation, when temperature of the environment goes up and heat starts to get into the thermodynamic system. In this situation, gradients of temperature, that is, in our interpretation, internal variables are emerging and internal processes appear; the system trends to a new equilibrium, but the state of the system is determined by a play between external influences and internal processes. It is convenient to consider these contributions separately.

3.1. Preferable Values of Internal Variables. Under external influences, the system is found to be, generally speaking, in a non-equilibrium state, and for the description of the situation, one needs on introduction of some distinctive internal variables $\boldsymbol{\xi}$, as was explained, for example, by Kestin [8] and Muschik [9]. A change of entropy of the system $S(T, \mathbf{x}, \boldsymbol{\xi})$ occurs due to changing of arguments of the function and, apart from it, due to an incoming flux of entropy, that is, a contribution defined by expression (9). Similar to the case of isolated systems, when a maximum value of entropy defines the equilibrium state, extreme value of entropy under the presence of external influences defines special points in the state space. These points are defined by a relation obtained by equating total variation of entropy to zero:

$$\begin{gathered}\frac{\partial S}{\partial T}\frac{dT}{dt} + \frac{\partial S}{\partial \mathbf{x}}\frac{d\mathbf{x}}{dt} + \frac{\partial S}{\partial \boldsymbol{\xi}}\frac{d\boldsymbol{\xi}}{dt} \\ + \frac{1}{T}\sum_{j=1}^{r} K_j(T, \mathbf{x}, \boldsymbol{\xi}) J_j = 0.\end{gathered} \tag{12}$$

The incoming flux of entropy is different, generally speaking, from the flux of entropy, coming out of the environment.

The relation (12) at given temperature, constitutive variables, and external fluxes can be considered as an equation for unknown internal variables, which, in this case, determines *preferable values of internal variables* $\boldsymbol{\xi}^{\circ}$.

The rate of change of internal variables in a point of quasi-equilibrium can be replaced by rate of change of preferable variables (see (16)):

$$\left.\frac{d\boldsymbol{\xi}}{dt}\right|_{\boldsymbol{\xi}=\boldsymbol{\xi}^{\circ}} = \frac{d\boldsymbol{\xi}^{\circ}}{dt}. \tag{13}$$

This allows, using also definition of thermodynamic force, to write the equation for preferable values as

$$\Xi\left(\boldsymbol{\xi}^{\circ}\right)\frac{d\boldsymbol{\xi}^{\circ}}{dt} = T\left(\frac{\partial S}{\partial T}\frac{dT}{dt} + \frac{\partial S}{\partial \mathbf{x}}\frac{d\mathbf{x}}{dt}\right)_{\boldsymbol{\xi}=\boldsymbol{\xi}^{\circ}} + \sum_{j=1}^{r} K_j\left(T, \mathbf{x}, \boldsymbol{\xi}^{\circ}\right) J_j. \tag{14}$$

The preferable values of internal variables, corresponding to the actual values of other variables and external influence, are determined, apparently, by history of the application of external influences.

In general case, (14) cannot unambiguously define preferable values. However, the situation is not so hopeless, when considering special cases. If temperature and constitutive variables of the system do not change and external influences are given by constant fluxes of heat and substances into the system $J_1, J_2, \ldots, J_s$, it is possible to make some assumptions. We consider, that each flux corresponds to the only one of the internal variables, so that the number of distinctive preferable variables is equal to the number of fluxes, that is, $\xi_l^{\circ} \neq 0$ at $l = 1, 2, \ldots, s$ and $\xi_l^{\circ} = 0$ at $l = s+1, s+2, \ldots$. Taking also the arbitrariness in the fluxes into account, (14) can be reduced to a set of equations:

$$\frac{d\xi_i^{\circ}}{dt} = B_i J_i, \quad i = 1, 2, \ldots, s. \tag{15}$$

Values of factors B_i are settled by a choice of fluxes in respect to a choice of internal variables. Quantities B_i have only numerical values, when an "appropriate" choice of fluxes in the form of $J_j \sim \xi_j/\tau_j$, where τ_j is a time of relaxation, has been made. By a choice of the internal variables and the fluxes, it is possible to reduce values B_i to units, which is accepted further.

3.2. Evolution of Internal Variables. The actual values of the internal variables $\xi_1, \xi_2, \ldots$ differ from the preferable values. The deviations of internal variables from their current preferable values $\xi_i - \xi_i^{\circ}$, $(i = 1, 2, \ldots)$ determine the trending of the internal variables to a preferable values. The change of internal variables $\xi_1, \xi_2, \ldots$ is determined by internal laws of movement of particles of the thermodynamic system, so that an equation for a change of the internal variables can be written in the general form:

$$\frac{d\left(\xi_i - \xi_i^{\circ}\right)}{dt} = -R_{ij}\left(T, x_1, x_2, \ldots, x_n; \xi_1, \xi_2, \ldots; \xi_1^{\circ}, \xi_2^{\circ}, \ldots\right) \times \left(\xi_j - \xi_j^{\circ}\right), \quad i = 1, 2, \ldots. \tag{16}$$

This equation describes relaxation of internal variables to preferable values $\boldsymbol{\xi}^{\circ}$, which are defined by (14), in general form, or, for the simple situation, by (15). The right-hand side of (16) describes change of the variables, according to internal laws of the system, the external influences are presented via the preferable values. The sign "minus" is chosen for convenience: matrix R_{ij} in the situations close to equilibrium (at $\xi_i^{\circ} = 0$, $i = 1, 2, \ldots$) is positive definite in many simple cases.[7]

Let us notice, that the internal variables are depending generally on the space coordinates $\xi_i = \xi_i(t, x, y, z)$, so that the equations for change of internal variables can contain also terms (omitted here) for the description of processes of diffusion of internal variables. Moreover, the equation of evolution of internal variables should include random forces. Each internal variable can be presented as a sum of the regular (averaged in some way) and random components. Here the regular components of variables are considered only. Discussion of random components, also as well as fluctuations of thermodynamic variables, is omitted. For expansion of the description, it is possible to use the mathematical apparatus of stochastic nonlinear nonequilibrium thermodynamics described by Stratonovich [10].

4. The Stationary Nonequilibrium States

The situation is being simplified in a steady-state case, when a set of constant fluxes $J_1, J_2, \ldots, J_s$ is fixed. It is assumed that the number of fluxes corresponds to the number of complexity variables, which have distinctive constant values $\xi_1^{\circ}, \xi_2^{\circ}, \ldots, \xi_s^{\circ}$. Apart of it, an indefinite number of some other internal variables $\xi_{s+1}, \xi_{s+2}, \ldots$ can appear, and thermodynamic characteristics of the system are functions of the all internal variables. The non-equilibrium stationary states of thermodynamic system represent special interest, and it is remarkable that these states, also as equilibrium states of the thermodynamic systems, can be considered in a general way.

4.1. Dynamics of Internal Variables. To formulate the dynamic equation for internal variables for a stationary state, we take (16) with definition of the derivative of preferable variables (15), in which the quantities B_j is settled to be equal unities. This allows one to write the equation for dynamics of internal variables near a steady-state point:

$$\frac{d\xi_i}{dt} = J_i - R_{il}\left(\xi_l - \xi_l^{\circ}\right) - \frac{\partial R_{ik}}{\partial \xi_l}\left(\xi_k - \xi_k^{\circ}\right)\left(\xi_l - \xi_l^{\circ}\right) + \cdots, \quad i = 1, 2, \ldots, s, s+1, s+2, \ldots, \tag{17}$$

where a matrix R_{il} and its derivatives are fixed in a considered preferable point. The matrix R_{ij} depends on temperature T and constitutive variables $x_1, x_2, \ldots, x_n$ and assumed to be positive definite, that is to have positive eigenvalues.[8] Let us remind, that the preferable state is defined in such a way, that $\xi_l^\circ \neq 0$ at $l = 1, 2, \ldots, s$ and $\xi_l^\circ = 0$ at $l = s+1, s+2, \ldots$.

In stationary state, values of preferable variables are constant; however it is reached by balance of processes of relaxation and permanent excitation of internal variables by external fluxes. Considering the processes of relaxation and excitation in a stationary situation independent, instead of (17), one can write two equations, in linear approximation:

$$\frac{d\xi_i}{dt} = -R_{ij}\xi_j, \quad J_i = -R_{ij}\,\xi_j^\circ, \quad i = 1, 2, \ldots, s. \tag{18}$$

The rate of change of internal variables ξ_j depends on deviations of values of internal variables from the equilibrium values, but preferable values ξ_j°, which, on the assumption, are close to ξ_j, are determined by fluxes. An analysis of an empirical situation specifies the second set of equations from (18), thus, establishing the set of internal complexity variables.

4.2. Entropy Near a Stationary State. In situations close to equilibrium, an expansion of entropy into series with respect to internal variables contains no terms of the first order, in the simplest approach:

$$S(\boldsymbol{\xi}) = S(0) - \frac{1}{2}\sum_{i,j} S_{ij}\,\xi_i\xi_j + \cdots, \qquad S_{ij} = -\left(\frac{\partial^2 S}{\partial\xi_i\partial\xi_j}\right)_{T,\mathbf{x},\boldsymbol{\xi}=0}. \tag{19}$$

By virtue of (6), value of entropy of the system in a nonequilibrium state is less than value of entropy of the same system in the equilibrium state, so that one has to consider matrix S to be nonnegative determined. Components of the matrix S are functions of temperature and constitutive variables.

An expansion of entropy of the thermodynamic system near a preferable (stationary) point begins with linear terms:

$$\begin{aligned} S(\boldsymbol{\xi}) &= S(\boldsymbol{\xi}^\circ) - \sum_{j=1}^{s} S_j\left(\xi_j - \xi_j^\circ\right) \\ &\quad - \frac{1}{2}\sum_{i,j} S_{ij}\left(\xi_i - \xi_i^\circ\right)\left(\xi_j - \xi_j^\circ\right) + \cdots, \\ S_j &= -\left(\frac{\partial S}{\partial\xi_j}\right)_{T,\mathbf{x},\boldsymbol{\xi}^\circ}, \quad j = 1, 2, \ldots, s, \\ S_{ij} &= -\left(\frac{\partial^2 S}{\partial\xi_i\partial\xi_j}\right)_{T,\mathbf{x},\boldsymbol{\xi}^\circ}, \quad i, j = 1, 2, \ldots. \end{aligned} \tag{20}$$

It is taken into account that, apart from of the complexity internal variables $\xi_1, \xi_2, \ldots, \xi_s$, there is a set of internal variables with numbers $s+1, s+2, \ldots$, that describe deviations of the system from the stationary state. Matrixes S_j and S_{lk} are calculated not in equilibrium point, as in expression (19), but in the stationary point and depend, apart from temperature and constitutive variables, on the complexity internal variables.

As a function of the internal complexity variables, entropy has no features, which would allow to characterize the stationary state. Considering the variables of complexity to be fixed, expansion of entropy, as a function of all other possible internal variables, is reduced to a form:

$$S(\boldsymbol{\xi}) = S(\boldsymbol{\xi}^\circ) - \frac{1}{2}\sum_{i,j=s+1}^{\infty} S_{ij}\xi_i\xi_j + \cdots, \qquad S_{ij} = -\left(\frac{\partial^2 S}{\partial\xi_i\partial\xi_j}\right)_{T,\mathbf{x},\boldsymbol{\xi}^\circ} \quad i, j = s+1, s+2, \ldots. \tag{21}$$

It is assumed that a stationary state of thermodynamic system near the equilibrium state is steady, so that matrix $S_{lk}(\xi_1^\circ, \xi_2^\circ, \ldots, \xi_s^\circ)$ in expression (21) is positive definite; entropy, as a function of internal variables, in the fixed stationary state has a maximum. The properties of matrix S_{lk} in a stationary point, which is far from equilibrium, remain not certain.

Relation (21) allows us to write an expression for thermodynamic forces in a point near the stationary state:

$$\begin{aligned} \Xi_j(\boldsymbol{\xi}) &= -T\frac{\partial S}{\partial\xi_j} = \Xi_j(\boldsymbol{\xi}^\circ) \\ &\quad + TS_{jl}\left(\xi_l - \xi_l^\circ\right) + \cdots, \quad j = 1, 2, \ldots, s. \end{aligned} \tag{22}$$

The first terms of the expansion of the thermodynamic forces are constant. In the situations close to the equilibrium state, in the simplest approach, thermodynamic forces are connected linearly with internal variables:

$$\Xi_j = -T\frac{\partial S}{\partial\xi_j} = TS_{jk}\xi_k + \cdots. \tag{23}$$

Let us remember that, in the state of equilibrium, values of internal variables are considered to be zero, and thermodynamic forces Ξ_j disappear.

4.3. Production of Entropy. In a stationary nonequilibrium state, values of temperature and all constitutive variables of the thermodynamic system are constant. Values of thermodynamic functions of system, including entropy, also are constant; however, there is production of entropy inside the system and corresponding decrease in entropy of the system, due to fluxes of heat and/or substances, so that one can write

$$\frac{d_i S}{dt} = -\frac{d_e S}{dt}. \tag{24}$$

Equations (17) and (22) allow us to write an expansion of function of production of entropy (11) near a steady-state point:

$$\frac{d_iS}{dt} = -\frac{1}{T}\sum_{j=1}^{s}\Xi_i J_i + \frac{1}{T}\sum_{j,l=1}^{s}\left(-TS_{jl}J_j + \Xi_j R_{jl}\right)(\xi_l - \xi_l^\circ) + \sum_{j,k,l=1}^{s}\left(S_{jk}R_{jl} + \Xi_j\frac{\partial R_{jk}}{\partial \xi_l}\right)(\xi_k - \xi_k^\circ)(\xi_l - \xi_l^\circ) + \sum_{j,k,l=s+1}^{\infty} S_{jk}R_{jl}\xi_k\xi_l + \cdots . \tag{25}$$

Values of all matrixes are determined in the considered steady-state point $\xi_1^\circ, \xi_2^\circ, \ldots, \xi_s^\circ$.

In the simplest approximation, expression for production of entropy can be written as

$$\frac{d_iS}{dt} = -\frac{1}{T}\sum_{j=1}^{s}\Xi_j J_j. \tag{26}$$

This expression represents production of entropy in the conventional form as the sum of products of fluxes and thermodynamic forces. Equation (26) is considered as one of the basic statements of nonequilibrium thermodynamics [1–3]. The emerging of the sign "minus" in expression (26) is connected with the fact that signs of the fluxes are taken opposite to signs of the forces (the internal variables). One can note, that representation (26) is valid only for steady-state situations and for small deviations from equilibrium state.

Comparison of (9) and (26) for a stationary state, when relation (24) is valid, determines that quantity K_j in (9) coincides with thermodynamic force, $K_i = \Xi_i$, $i = 1, 2, \ldots, s$.

4.4. On the Criterion of Stability of Stationary States. Expression (25) shows that, in a steady-state point, production of entropy, as a function of the internal variables of complexity, has no peculiar points. However, if the steady-state point is fixed ($\xi_l = \xi_l^\circ$, $l = 1, 2, \ldots s$), expression (25) takes the form

$$\frac{d_iS}{dt} = -\frac{1}{T}\sum_{j=1}^{s}\Xi_i J_i + \sum_{j,k,l=s+1}^{\infty} S_{jk}R_{jl}\xi_k\xi_l + \cdots . \tag{27}$$

The behaviour of production of entropy in vicinity of a steady-state point is determined by terms of the second order with respect to all internal variables, excepting the complexity variables. The terms of the second order comprise a square form with the matrix that is a product of two matrixes R_{jk} and S_{lk}, which are calculated in the steady-state point and depend, apart from temperature and constitutive variables, on the complexity internal variables. In the equilibrium point and, assumingly, in the steady-state points near to the equilibrium state the matrixes are positive definite, so that production of entropy has a minimum, which confirms the validity of the Prigogine's principle of a minimum of production of entropy [1–3]. In the points, that are far from the equilibrium point, the matrixes R_{jk} and S_{lk} are not necessarily positive definite, so that stability of the system can be connected with a maximum of production of entropy, as it is stated by some investigators [11, 12].

4.5. The Relation between Fluxes and Thermodynamic Forces. Now, with help of (18) and (23), we can write expression (26) for production of entropy in other form as

$$\frac{1}{T}\sum_{j,k=1}^{s}\Xi_j R_{jk}\xi_k = -\sum_{j,k=1}^{s} S_{jk}\xi_k J_j. \tag{28}$$

The written equation, by virtue of the supposed arbitrariness and independence of internal variables, is followed by a relation between fluxes and thermodynamic forces:

$$\frac{1}{T}\sum_{j=1}^{s}\Xi_j R_{jk} = -\sum_{j=1}^{s} S_{jk}J_j. \tag{29}$$

This relation can be rewritten as

$$J_i = -L_{ik}\Xi_k, \qquad L_{ik} = \frac{1}{T}R_{il}S_{lk}^{-1}. \tag{30}$$

Considering linear approach, the components of matrixes R_{jk} and S_{lk} are constants, but in more general case it is necessary to consider them as functions of internal variables. By virtue of definition, the matrix L is positive definite.

The relation between fluxes and thermodynamic forces in linear approximation are fundamental relation of the nonequilibrium thermodynamics [1–3]. We have shown that these parities are consequence of the main principles of thermodynamics and valid under two conditions: first, deviations from an equilibrium state are small, and second, the state is stationary.

Let us note in addition that there is a statement [1–3] for the matrix L to be symmetric or antisymmetric:

$$L_{ij} = \pm L_{ji}. \tag{31}$$

For the proof of this statement, to which usually refer as to *Onsager principle*, one has to address the other principles and some assumptions considered in the following section.

4.6. Symmetry of Kinetic Coefficients. The proof of the symmetry of kinetic coefficients is based on the property of invariance of correlations of fluctuations of various quantities with respect to reversion of time, which is fair for equilibrium situations [1–3, 13]. Being interested in stationary states, it is possible to consider fluctuations of internal variables near their stationary values and to assume that time correlations of random deviations of internal variables also are invariant with respect to reversion of time. In other words, for correlations of various quantities, it is possible to write a relation

$$\left\langle (\xi_i - \xi_i^\circ)_t (\xi_k - \xi_k^\circ)_0 \right\rangle = \pm\left\langle (\xi_k - \xi_k^\circ)_t (\xi_i - \xi_i^\circ)_0 \right\rangle . \tag{32}$$

The "minus" sign arises in the case, when the internal variable itself changes the sign at the reversion of time.

Further, one can take advantage of the equation of evolution (16), which, with use of expression for thermodynamic force (22) and definition of a matrix of kinetic coefficients (30), can be presented in the form

$$\frac{d(\xi_i - \xi_i^\circ)}{dt} = -L_{il}\left(\Xi_l(\boldsymbol{\xi}) - \Xi_l(\boldsymbol{\xi}^\circ)\right). \tag{33}$$

This equation allows one, after differentiation of relation (32), to write down

$$\begin{aligned} &L_{il}\left\langle\left(\Xi_l(\boldsymbol{\xi}) - \Xi_l(\boldsymbol{\xi}^\circ)\right)(\xi_k - \xi_k^\circ)_0\right\rangle \\ &\quad = \pm L_{kl}\left\langle\left(\Xi_l(\boldsymbol{\xi}) - \Xi_l(\boldsymbol{\xi}^\circ)\right)(\xi_i - \xi_i^\circ)_0\right\rangle. \end{aligned} \tag{34}$$

By virtue of supposed independence of deviations of various variables from stationary values, the required relation (31) follows the equation written previously.

Expansion of the principle of invariance of correlations with respect to reversion of time for stationary situations allows proving the principle of symmetry of kinetic coefficients.

4.7. A Simple Example. Let us consider a portion of matter in a small volume $\Delta x \Delta y \Delta z$ as a thermodynamic system. In addition to the cases, when a stationary state is maintained by fluxes of heat and particles separately [7], let us consider the simultaneous action of these factors. We can assume that streams of heat and a substance are moving along the x-axis, so that

$$J_\mathrm{T} = \frac{1}{\Delta y \Delta z}\frac{\Delta Q}{\Delta t}, \qquad J_\mathrm{C} = \frac{1}{\Delta y \Delta z}\frac{\Delta m}{\Delta t}. \tag{35}$$

The fluxes of heat and substance are contributing independently in the entropy of the volume. The flux of heat, according to (8), determines the change of density of entropy in the volume:

$$\frac{d_\mathrm{heat} S}{dt} = \frac{1}{\Delta x}\left(\frac{J_\mathrm{T}}{T} - \frac{J_\mathrm{T}}{T + \Delta T}\right) \approx \frac{J_\mathrm{T}}{T^2}\nabla_x T. \tag{36}$$

In a similar way, one can define the change of density of entropy due to the flux of the substance:

$$\frac{d_\mathrm{subs} S}{dt} = \frac{J_\mathrm{C}}{T}\frac{\mu(c + \Delta c) - \mu(c)}{\Delta x} \approx \frac{J_\mathrm{C}}{T}\frac{\partial \mu}{\partial c}\nabla_x c. \tag{37}$$

The system is assumed to be characterised by chemical potential $\mu(c)$, which depends on concentration c.

It is convenient to introduce special symbols for the gradients of temperature and concentration, which can be considered to be internal complexity variables:

$$\xi_\mathrm{T} = \nabla_x T = \frac{\Delta T}{\Delta x}, \qquad \xi_\mathrm{C} = \nabla_x c = \frac{\Delta c}{\Delta x}. \tag{38}$$

Thus, the change of density of entropy, due to the external fluxes of heat and substance, can be written as follows:

$$\frac{d_e S}{dt} = \frac{J_\mathrm{T}}{T^2}\xi_\mathrm{T} + \frac{J_\mathrm{C}}{T}\frac{\partial \mu}{\partial c}\xi_\mathrm{C}. \tag{39}$$

The fluxes and gradients are connected with each other locally. It is well known [1–3], as an empirical fact, that, at simultaneous presence of the thermal and diffusion gradients, ξ_T and ξ_C cross-effects are observed; a gradient of temperature induces a flux of the substance and vice versa, a gradient of concentration induces a flux of heat, so that according to the last equation from (18), we record a linear parity connecting the fluxes and gradients for the considered system:

$$\begin{aligned} J_\mathrm{T} &= -R_\mathrm{TT}\xi_\mathrm{T} - R_\mathrm{TC}\xi_\mathrm{C}, \\ J_\mathrm{C} &= -R_\mathrm{CT}\xi_\mathrm{T} - R_\mathrm{CC}\xi_\mathrm{C}. \end{aligned} \tag{40}$$

These equations allow one to write the internal production of entropy, according to (26), in the form

$$\begin{aligned} \frac{d_i S}{dt} = \frac{1}{T}&\left[\left(R_\mathrm{TT}\xi_\mathrm{T} + R_\mathrm{TC}\xi_\mathrm{C}\right)\Xi_\mathrm{T}\right. \\ &\left. + \left(R_\mathrm{CT}\xi_\mathrm{T} + R_\mathrm{CC}\xi_\mathrm{C}\right)\Xi_\mathrm{C}\right]. \end{aligned} \tag{41}$$

In a steady-state situation, the thermodynamic state of the system does not change, so that (24) is valid, and taking relation (39) into account, one has another expression for production of entropy:

$$\frac{d_i S}{dt} = -\frac{J_\mathrm{T}}{T^2}\xi_\mathrm{T} - \frac{J_\mathrm{C}}{T}\frac{\partial \mu}{\partial c}\xi_\mathrm{C}. \tag{42}$$

Comparing (41) and (42), one finds the relations between fluxes and thermodynamic forces for a stationary case:

$$\begin{aligned} J_\mathrm{T} &= -T\left(R_\mathrm{TT}\Xi_\mathrm{T} + R_\mathrm{CT}\Xi_\mathrm{C}\right), \\ J_\mathrm{C} &= -\left(\frac{\partial \mu}{\partial c}\right)^{-1}\left(R_\mathrm{TC}\Xi_\mathrm{T} + R_\mathrm{CC}\Xi_\mathrm{C}\right). \end{aligned} \tag{43}$$

The condition of symmetry of kinetic coefficients (31) is followed by an equation:

$$\frac{1}{T}R_\mathrm{TC} = \frac{\partial \mu}{\partial c}R_\mathrm{CT}. \tag{44}$$

Entropy of the system is defined by formula (6), which, in considered case, takes the form

$$S - S_0 = -\frac{1}{T}\int_0^{\nabla_x T}\int_0^{\nabla_x c}\left(\Xi_\mathrm{T} d\xi_\mathrm{T} + \Xi_\mathrm{C} d\xi_\mathrm{C}\right). \tag{45}$$

The thermodynamic forces Ξ_T and Ξ_C, as functions of complexity variables, can be found from (40) and (43). After the calculation of the integrals, one gets a simple formula for entropy as a function of internal variables:

$$S - S_0 = -\frac{1}{2T}\left(\frac{1}{T}\left(\nabla_x T\right)^2 + \frac{\partial \mu}{\partial c}\left(\nabla_x c\right)^2\right). \tag{46}$$

It is remarkable that the condition of integrability and existence of entropy of the system appears to be identical to Onsager's relation, that is, equality of the nondiagonal components of a matrix in (43).

5. Conclusion

The usage of internal variables, playing a distinctive role in the description of nonequilibrium states of the system, allows us to develop a generalized view on the principle of non-equilibrium thermodynamics. In this paper, in line with the paper [7], the author aspired to show that the thermodynamics of nonequilibrium states could be formulated so consistently, as the thermodynamics of equilibrium states. At least, considering the situations close to equilibrium and using the simplest approach, we have obtained the fundamental relations of non-equilibrium thermodynamics. One can note the most essential features of our description.

(i) The reference to local equilibrium is not used. On the contrary, it is supposed, that local nonequilibrium exists. A set of internal variables $\xi_1, \xi_2, \ldots$ are introduced to describe deviation of the system from equilibrium.

(ii) While formulating the law of conservation of energy, the work connected with internal variables is considered, thus, one has to distinguish between the total internal energy $U(T, \mathbf{x}, \boldsymbol{\xi})$ and internal thermal energy $E(T, \mathbf{x})$ of the system. It allows to introduce entropy S for non-equilibrium states of thermodynamic system by (4), which keeps the meaning of entropy unchanged. As a consequence of this, temperature in any situation is defined through thermodynamic functions of the system as $T = (\partial E/\partial S)_{\mathbf{x}}$ or $T = (\partial U/\partial S)_{\mathbf{x},\boldsymbol{\xi}}$.

(iii) The second principle of thermodynamics may be formulated as impossibility of fulfilment of positive work by internal variables (Section 2.3). This formulation is equivalent to conventional formulations but excludes a flavour of mystique from concept of entropy.

(iv) Relation (5), that connects a change of entropy of a thermodynamic system with fluxes of heat and substance and also with change of internal variables, is valid for the cases of open systems and irreversible processes. It is a generalisation of Prigogine's relation, given by him for the systems with chemical reactions [1, Equation 3.52].

(v) The known linear relations of nonequilibrium thermodynamics (in particular, the parity between forces and the fluxes) follow from the general principles for the cases of steady-state processes running in the region of small deviations from equilibrium.

One can see that all known results of linear non-equilibrium thermodynamics follow the presented approach; however, the theory is open for nonlinear generalisations.

Endnotes

1. Perhaps, the first who used internal variables for the consecutive formulation of thermodynamics was Leontovich [5]. The first Russian edition of "Introduction to Thermodynamics," based on his prewar- and war-time lectures, has appeared, as a separate book, in 1950.

2. Temperature is usually measured in degrees, but, according to relationship with other physical quantities, it is energy. To omit the factor of transition from degrees to energy, it is convenient to agree that the absolute temperature T is measured in energy units as it was proposed by Landau and Lifshitz [13] and used here through the paper.

3. We do not include here into consideration the fluctuations in thermodynamic systems (local deviations of density, temperatures, concentration of substances, and other quantities from equilibrium values) which can be treated as huge number of uncontrollable internal variables.

4. In nonequilibrium situations, work of the system is not equal to the work of the environment on the system [14]. To balance the inner and outer forces, one needs, as it was shown by Gujrati [14], in some internal variables.

5. The alternative theory [4, 6] of non-equilibrium in terms of internal variables does not distinguish between the total and internal energies, thus, misinterpreting the role of internal variables.

6. The definition of entropy in the alternative theory [4, 6] of non-equilibrium in terms of internal variables does not include fluxes of substances (the last term in (5)), thus, failing to reproduce the Prigogine's result. Moreover, one needs in the omitted term to consider the processes of diffusion correctly, as was discussed earlier [7].

7. There are many important exclusions from this rule. A description of thermodynamic systems with oscillating chemical reactions, for example, requires matrix R_{ij} not to be positive definite.

8. It is known, that such combination (transformation) of variables can be chosen that the equation of dynamics (16) for these variables gets the form of the equation of a relaxation:

$$\frac{d\xi_i}{dt} = -\frac{1}{\tau_i}\xi_j, \quad i = 1, 2, \ldots, s, \tag{47}$$

where $\tau_i = \tau_i(T, x_1, x_2, \ldots, x_n)$ is a time of relaxation of corresponding variable. Changing under internal laws of the movement, the agitated internal variables trend to their unique equilibrium value, so that the system is moving to its equilibrium state.

Acknowledgment

The author is grateful to the anonymous reviewers of the paper for their helpful and constructive comments.

References

[1] I. Prigogine, *Introduction To Thermodynamics of Irreversible Processes*, Edited by C. C. Thomas, Sprinfild, Geneseo, Ill, USA, 1955.

[2] S. R. de Groot and P. Mazur, *Non-Equilibrium Thermodynamics*, NorthHolland, Amsterdam, The Netherlands, 1962.

[3] D. Kondepudi and I. Prigogine, *Modern Thermodynamics. From Heat Machines to Dissipative Structures*, John Wiley & Sons, Chichester, UK, 1999.

[4] W. Muschik, "Why so many "schools" of thermodynamics?" *Forschung im Ingenieurwesen*, vol. 71, no. 3-4, pp. 149–161, 2007.

[5] M. A. Leontovich, *Introduction to Thermodynamics. Statistical Physics*, Nauka, Glavnaja redaktzija fisiko-matematicheskoi literatury, Moscow, Russia, 1983.

[6] G. A. Maugin and W. Muschik, "Thermodynamics with internal variables Part I. general concepts," *Journal of Non-Equilibrium Thermodynamics*, vol. 19, no. 3, pp. 217–249, 1994.

[7] V. N. Pokrovski, "Extended thermodynamics in a discrete-system approach," *European Journal of Physics*, vol. 26, no. 5, pp. 769–781, 2005.

[8] J. Kestin, "the local-equilibrium approximation," *Journal of Non-Equilibrium*, vol. 18, no. 4, pp. 360–379, 1993.

[9] W. Muschik, "Comment to J. Kestin: internal variables in the local-equilibrium approximation," *Journal of Non-Equilibrium Thermodynamics*, vol. 18, pp. 380–388, 1993.

[10] R. L. Stratonovich, *Nonlinear Nonequilibrium Thermodynamics*, Springer, Heidelberg, Germany, 1994.

[11] A. Kleidon, "Nonequilibrium thermodynamics and maximum entropy production in the Earth system: applications and implications," *Naturwissenschaften*, vol. 96, no. 6, pp. 653–677, 2009.

[12] K. Hackl and F. D. Fischer, "On the relation between the principle of maximum dissipation and inelastic evolution given by dissipation potentials," *Proceedings of the Royal Society A*, vol. 464, no. 2089, pp. 117–132, 2008.

[13] L. D. Landau and E. M. Lifshitz, *Statistical Physics*, vol. 1, Pergamon Press, Oxford, UK, 3rd edition, 1986.

[14] P. D. Gujrati, "Nonequilibrium thermodynamics. Symmetric and unique formulation of the first law, statistical definition of heat and work, adiabati theorem and the fate of the Clausius inequality: a microscopi view," http://arxiv.org/abs/1206.0702.

Pippard Relations Close to the Melting Point in the Premelting Region of Hexadecane

Hamit Yurtseven,[1] Özcan Tilki,[1] and Mustafa Kurt[2]

[1] *Department of Physics, Middle East Technical University, 06531 Ankara, Turkey*
[2] *Department of Physics, Çanakkale 18 Mart University, 17100 Çanakkale, Turkey*

Correspondence should be addressed to Hamit Yurtseven, hamit@metu.edu.tr

Academic Editor: Dominique Richon

We examine here the validity of the Pippard relations close to the melting point in the premelting region of hexadecane. For this verification, we analyze the observed data for the thermal expansivity α_p obtained for various pressures at constant temperatures of 302 and 325 K in this system. By calculating the isothermal compressibility κ_T and the specific heat C_p in the same pressure region of the premelting region of hexadecane, we obtain that C_p varies linearly with α_p and also that α_p varies linearly with κ_T for this system. This indicat es that some molecular organic compounds, such as solid hexadecane studied here, can exhibit λ-phase transitions prior to melting which are expected to verify the Pippard relations.

1. Introduction

Phase behaviour of pure n-alkanes is important for industrial and scientific applications. The controlled precipitation of solid phases of these mixtures is widely applied in pharmaceutical, cosmetic, and food industries as a separation process, as also pointed out previously [1]. Those paraffin waxes are also used in latent thermal energy storage applications, and additionally for cool storage applications in regard to the phase transition processes a binary mixture of tetradecane and hexadecane is widely used [2].

In the literature, various sources of experimental data have been reported in regard to the thermodynamic properties of hexadecane. Some recent experimental reports are the following: T-X phase diagrams [1], entropy changes with molality at constant temperature [3], an anomalous heat effect during colloidal phase transitions [4], determination of the melting temperature (6.8°C) and a corresponding heat of fusion (121 kJ/kg) for a mixture of tetradecane (40%) and hexadecane (60%) [2], and also the crystallization temperatures of two binary mixtures: the tetradecane + hexadecane and tetradecane + pentadecane system up to 100 MPa [1].

Solid hexadecane among long-chain paraffin solids exhibits considerable precursor effects [5]. It has been shown that there exists an orientational disorder around the long axis of the molecule in solid hexadecane [6]. Experimentally, it has been demonstrated that this is a kinked chain-like disorder rather than a librational disorder [7, 8], as also indicated by Pruzan et al. [9]. This orientational disorder occurring in solid hexadecane gives rise to instability and melting of the solid phase. Instability and melting have been associated with a soft phonon of zero wave vector [10–13].

Solid hexadecane melts into the liquid phase as a second-order transition associated with a particular type of orientational disorder which depends on the symmetry [9]. As in many molecular organic compounds, solid hexadecane exhibits orientational disorder in a small range where the thermodynamic quantities such as the thermal expansivity α_p, isothermal compressibility κ_T, and the specific heat C_p show anomalous behaviour close to the melting point [5, 9].

The anomalous behaviour of α_p, κ_T, and C_p can be described by a power-law formula near the melting point in solid hexadecane. In the premelting region of hexadecane close to the melting point, we can assume that α_p, κ_T, and C_p exhibit similar critical behaviour associated with the same critical exponent. Buckingham and Fairbank [14] have indicated that C_p behaves like κ_T and α_p in the vicinity of the critical point. In the precursor zone of the solid hexadecane, C_p, α_p, and κ_T are expected to increase as approaching the melting point. This increase in the thermodynamic quantities considered is similar to the critical behaviour of λ-phase transitions. Pippard has established a linear variation of the specific heat C_p with the thermal expansivity α_p in the vicinity of the λ-phase transition in NH_4Cl [15]. The Pippard relations have been tested for a variety of materials exhibiting λ-phase transitions.

In this study, we also examine the Pippard relations in the premelting region of hexadecane near the melting point. For this examination, we calculate the isothermal compressibility κ_T and the specific heat C_p from our analysis of the thermal expansivity α_p which was measured experimentally at various pressures near the melting point for constant temperatures of 302 and 325 K in solid hexadecane [9]. By obtaining the values of the critical exponent γ from our analysis, we calculate κ_T and C_p as functions of pressure in the premelting region of hexadecane. We then establish a linear variation of C_p with α_p (first Pippard relation) and also a linear variation of α_p with κ_T (second Pippard relation) close to the melting point in this molecular organic compound.

In Section 2, we review the method and conditions about experimental study of Pruzan et al. [9]. In Section 3, our calculations and results are presented. In Sections 4 and 5, we present our discussion and conclusions, respectively.

2. Review of the Experimental Technique

Using a P-V-T method, the thermal expansivity α_p was observed by Pruzan et al. [9] at various pressures close to the melting point. Using a high-pressure calorimetric method, the heat of compression was measured. The measurement of heat, in open-system conditions, was conducted, which flows out of (into) the sample when it is compressed (decompressed) isothermally [9]. The thermal expansivity α_p was then determined according to the first Maxwell equation:

$$\alpha_p = -V^{-1}\left(\frac{\partial S}{\partial T}\right)_P, \qquad (1)$$

where V and S are the molar volume and entropy of the sample, respectively. This technique was referred to as the piezothermal method [9]. For the measurements, the pressure steps were about 20 MPa far from the melting point and they were 0.1–10 MPa close to the melting point [9]. For their measurements, Pruzan et al. [9] reported that their observations were made over different times, sometimes greater than 72 h, during which no change could be observed in the expansivity variation against pressure for hexadecane.

3. Calculations and Results

Near the melting point of solid hexadecane, the thermodynamic quantities such as the thermal expansivity α_p, isothermal compressibility κ_T, and the specific heat C_p can exhibit critical behaviour. This critical behaviour of the thermal expansivity can be expressed as

$$\alpha_p = A(P - P_m)^{-\gamma}, \qquad (2)$$

where γ is the critical exponent for α_p and A is the amplitude. Here, P_m denotes the melting pressure of solid hexadecane, which is close to the critical point [5]. Using the thermodynamic relation near the melting point:

$$\frac{\alpha_p}{\kappa_T} = \left(\frac{dP_m}{dT}\right), \qquad (3)$$

the isothermal compressibility κ_T can be expressed as a function of pressure

$$\kappa_T = \frac{A(P - P_m)^{-\gamma}}{(dP_m/dT)} \qquad (4)$$

by means of (2).

We can also obtain the pressure dependence of the specific heat C_p using the thermodynamic relation near the melting point as follows:

$$C_p = TV\alpha_p\left(\frac{dP_m}{dT}\right). \qquad (5)$$

In (5), the volume V is a function of pressure, which can be obtained as

$$V_T(P) = V_m \exp\left[\frac{A}{1-\gamma}\frac{1}{(dP_m/dT)}(P - P_m)^{1-\gamma}\right], \qquad (6)$$

where V_m denotes the melting volume of solid hexadecane. Equation (6) can be derived easily using the definition of the isothermal compressibility $\kappa_T = -(1/V)/(\partial V/\partial P)_T$. Thus, the pressure dependence of the molar volume $V_T(P)$ has been derived here, as given by (6), in the case when the isothermal compressibility κ_T is a function of pressure according to (4). We note here that when the isothermal compressibility κ_T is a function of the molar volume V, the pressure dependence of the molar volume $V_T(P)$ can also be derived. But, its functional form should not be the same as (6) since the volume dependence of the isothermal compressibility κ_T changes the pressure dependence of the molar volume in the premelting region close to the melting point of solid hexadecane. Finally, by means of (5), the pressure dependence of the specific heat C_p can be written as a power-law formula:

$$C_p = ATV\left(\frac{dP_m}{dT}\right)(P - P_m)^{-\gamma}. \qquad (7)$$

In the vicinity of the melting point of solid hexadecane, the specific heat C_p can be related to the thermal expansivity α_p linearly, according to the first Pippard relation:

$$C_p = TV\left(\frac{dP_m}{dT}\right)\alpha_p + T\left(\frac{dS}{dT}\right)_m. \qquad (8)$$

Similarly, the thermal expansivity α_p can be related linearly with the isothermal compressibility κ_T by the relation in the vicinity of the melting point:

$$\alpha_p = \left(\frac{dP_m}{dT}\right)\kappa_T + \frac{1}{V}\left(\frac{dV}{dT}\right)_m \qquad (9)$$

according to the second Pippard relation. Thus, linear plots of C_p against α_p (8) and α_p against κ_T (9) can be obtained near the melting point in solid hexadecane. From those linear plots, (8) and (9), we can extract the values of the slope dP_m/dT numerically for solid hexadecane close to the melting point.

We reanalyzed here the thermal expansivity α_p measured at various pressures in the premelting region close to the melting point for constant temperatures of 302 and 325 K in solid hexadecane due to Pruzan et al. who also analyzed their experimental data [9]. For our analysis, we followed the steps, as given below.

(1) We used the power-law formula (2) in the logarithmic form obtained as

$$\ln \alpha_p = \ln A - \gamma \ln(P - P_m) \tag{10}$$

for the premelting region of hexadecane. We then extracted the values of the critical exponent γ as 2.06 ± 0.05 and 1.97 ± 0.04 with the values of the amplitude A for constant temperatures of 302 and 325 K, respectively, as given in Table 1. Similarly, from their analysis of the thermal expansivity α_p, Pruzan et al. [9] have obtained the value of the critical exponent as $\gamma_m = 1.9$ in the melting zone of hexadecane. Figure 1 gives the thermal expansion α_p plotted as a function of $P - P_L$ in a log-log scale for constant temperatures of 302 K and 325 K. Since we used for our analysis the experimental data in the melting zone, which was obtained in the range $P - P_L < 50$ MPa where P_L is the liquid or melting (P_m) pressure [9], we were able to describe the critical behaviour of the thermal expansivity α_p in the premelting region of hexadecane according to (10). On the other hand, $P - P_L > 50$ MPa corresponds to the precursor zone (solid phase) for hexadecane [9]. These pressure ranges have also been reported in the work of Pruzan et al. [9] as $P_1 - P_L = 140$ MPa and $P_2 - P_L = 60$ MPa, where P_1 represents the beginning of the precursor range and P_2 is the pressure for a change of regime, which is closer to the critical pressure P_C for the solid hexadecane.

(2) Using those values of γ and A which we obtained (Table 1), we then calculated the isothermal compressibility κ_T as a function of pressure according to (4) where we used the experimental value of $dP_m/dT = 10$ MPa/K [9].

(3) The next step was to calculate the pressure dependence of the volume $V_T(P)$ in the premelting region of hexadecane near the melting point for constant temperatures of 302 and 325 K using (6). In this equation, the values of the melting volume V_m [16], as given in Table 1, were used with the experimental value of $dP_m/dT = 10$ MPa/K.

(4) Finally, the pressure dependence of the specific heat C_p was calculated by (7) for the premelting region of hexadecane near the melting point for those constant temperatures considered using $V_T(P)$ values (6) with the dP_m/dT value [5] given above.

Table 1: Values of γ and A extracted (Figure 1) for α_p using the experimental data [9] and the V_m values for the solid hexadecane at temperatures indicated.

T (K)	γ	A (MPa)$^\gamma$/K	$V_m \times 10^{-6}$ (m^3/mol)
302	2.06 ± 0.05	1.9837	294.93 at 303 K
325	1.97 ± 0.04	1.3624	300.13 at 323 K

Since we obtained the pressure dependence of the thermal expansivity α_p (2), isothermal compressibility κ_T (4), and the specific heat C_p (7) in the same pressure interval for constant temperatures of 302 and 325 K in the premelting region of hexadecane, we were then able to establish the Pippard relations ((8) and (9)) in this crystalline system. We plot the specific heat C_p against the thermal expansivity α_p in the premelting region of hexadecane near the melting point for constant temperatures of 302 K and 325 K in Figure 2.

We also examined the second Pippard relation (9) by α_p as a function of κ_T in the premelting region of hexadecane close to the melting point. Figure 3 gives our plot for constant temperatures of 302 and 325 K. From our plots of C_p against α_p (first Pippard relation) and α_p against κ_T (second Pippard relation), we extracted the value of the slope as $dP_m/dT = 10$ MPa/K. Since the experimental value of $dP_m/dT = 10$ MPa/K [5] was used as the initial value in (4) (step 2), (6) (step 3), and (7) (step 4) to calculate κ_T, $V_T(P)$, and C_P, respectively, as a function of pressure, we should expect the same slope value of 10 MPa/K from (8) as well.

4. Discussion

We examined here whether the specific heat C_p varies linearly with the thermal expansivity α_p (first Pippard relation) and also α_p varies linearly with the isothermal compressibility κ_T (second Pippard relation) in the vicinity of the melting point in the premelting region of hexadecane. As seen from our plots (Figures 2 and 3), we established the Pippard relations ((8) and (9)) with the slope $dP_m/dT = 10$ MPa/K which was extracted for this system. This indicates that those thermodynamic quantities, namely, C_p, α_p, and κ_T exhibit similar critical behaviour close to the melting point in the premelting region of hexadecane.

From our analysis of the measured thermal expansivity, we extracted the value of $\gamma = 2$ as the critical exponent for constant temperatures (302 K and 325 K), in the premelting region of hexadecane (Table 1). Our exponent value is very close to the value with the uncertainties (Table 1) of $\gamma = 1.9$ as obtained by Pruzan et al. [9]. They analyzed their experimental data according to the power-law relation (2) with the $P_m = P_C$. The critical pressure P_C at which α_p diverges due to the precursor transformation is always between P_2 and P_L [9]. For hexadecane, P_C is close to P_2, as indicated previously. Within the pressure ranges $P_1 - P_L = 140$ MPa and $P_2 - P_L = 60$ MPa, Pruzan et al. [9] obtained the values of $\gamma_S = 0.5$ in the solid phase, $\gamma_P = 0.7$ in the precursor zone, and $\gamma_m = 1.9$ in the melting zone, where the pressure

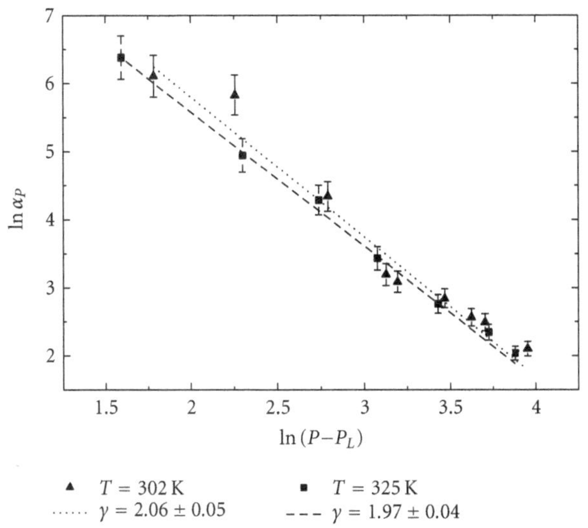

FIGURE 1: α_p versus P in a log-log scale close to P_m in the premelting region of hexadecane for T = 302 and 325 K (10) when $P_m = P_L$ [9]. The experimental data [9] with the uncertainties extracted (error bars) are also given here.

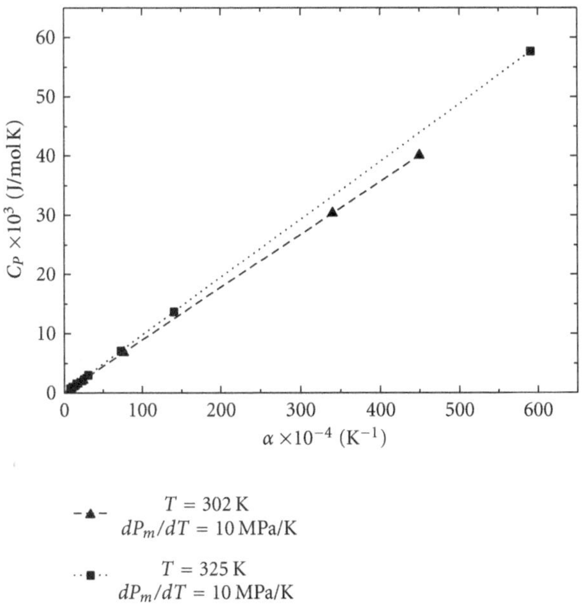

FIGURE 2: C_p versus α_p at various P close to P_m for T = 302 and 325 K in the premelting region of hexadecane (8). Observed α_p values [9] are shown here.

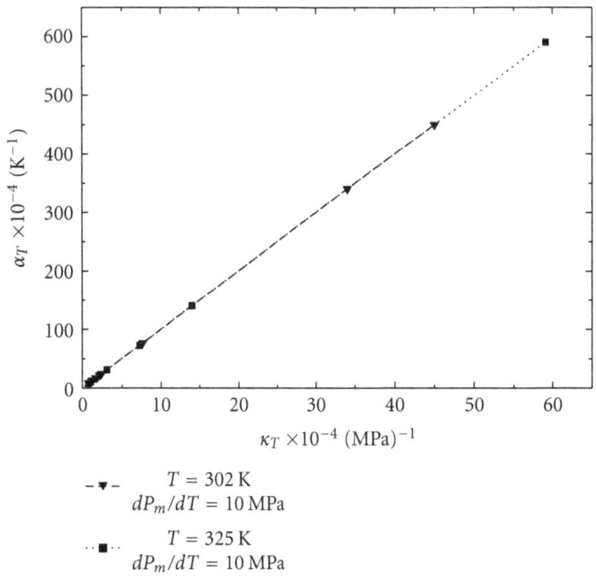

FIGURE 3: α_p versus κ_T close to the premelting region of hexadecane at various P for P_m, T = 302 and 325 K (9). Observed α_p values [9] are shown here.

range is greater than the melting pressure P_m. The exponent value of 1.9 or 2 from both analyses in the melting zone can be compared with an effective exponent greater than 1 for the critical behaviour of the thermal expansivity α_p due to surface melting model of Lipowsky and Speth [17, 18], as also pointed out previously [9]. It has been interpreted that the second regime (liquid phase) observed in molecular organic compounds might be ascribed to impurity effects for the critical exponents close to 2 [5]. Regarding the exponent value of 1.9 or 2, the melting zone can be shortened as the impurity decreases [19] close to the melting point in the premelting region of hexadecane. Our values of the critical exponent (γ_m = 2) for hexadecane can also be compared with those values for other organic compounds, which we obtained by Pruzan et al. [9]. Close to the melting point, they obtained the values of γ_m = 2 (cyclohexane), 1.8 (carbon tetrachloride), 1.6 (benzene), and 1.5 (butanediol), as given in Table 2 in [5]. Similarly, the γ_m values of 1.85 for hexamethylbenzene (HMB), 2 for paradichlorobenzene (PDCB), and the value of 1.7 for both paraozoxydianisol (PAA, C–N transition) and heptyloxyazoxybenzene (HOAOB, C–S transition) compounds (Table 2 in [9]) can also be compared with our exponent value of γ_m = 2 for hexadecane or γ_m = 1.9 as obtained by Pruzan et al. [9]. The organic compounds indicated above have similar chemical structure (C–H–· · · bonds) within the close melting points and pressure ranges [9]. This shows that within the pressure ranges studied all those organic compounds exhibit similar divergence behaviour of the thermal expansion α_p near the melting point according to a power-law formula (2). Thus, our exponent value is within expectations in regard to the experimental results for various organic compounds stated above.

Our γ values (Table 1) were used to calculate here the pressure dependence of volume $V_T(P)$ (6), isothermal compressibility κ_T (4), and the specific heat C_p (7), as indicated above. This calculation of C_p and κ_T then led us to plot C_p against α_p (Figure 2) and α_p against κ_T (Figure 3) with the zero values of the intercepts $T(dS/dT)_m$ and $(dV/dT)_m/V$, respectively, at the melting point in the premelting region of hexadecane. Those intercepts would have nonzero values according to the Pippard relations ((8) and (9)) if the temperature dependences of the entropy and volume are known experimentally or their functional forms can be derived for solid hexadecane. Values of the critical exponent γ (Table 1) and also the slope dP/dT were extracted from one source of experimental data [9] for hexadecane in the melting region. Regarding the anomalous behavior of α_p, κ_T, and C_p close to the melting point in the premelting region of hexadecane, the other sources of the experimental data rather than the measurements of the thermal expansivity α_p due to Pruzan et al. [9] can be used to examine our relations given here. Some new experimental measurements of the α_p, κ_T, and C_p at various pressures for constant temperatures can be performed in the premelting region near the melting point for hexadecane. When this experimental data is available in the literature which can be obtained to high accuracy, our analysis can then be reconsidered to examine the Pippard relations given here.

Regarding the uncertainties in the measurements of the thermal expansivity α_p for hexadecane, it was pointed out [9] that the uncertainty in the α_p data depends on the respective uncertainties in the heat of compression (1 or 2%) and in the pressure step. It was also pointed out [9] that far from the transition lines, pressure steps around 20 MPa caused an uncertainty in α_p of about 5%. Close to the melting lines, within the pressure steps of 0.1–10 MPa, the uncertainty in α_p increased, whereas for the measurements of the molar volume V the uncertainty was considerably small (~0.3%) [9].

Using the experimental data [9], we extracted that the uncertainties in α_p are about 5% close to the melting point, which may be higher than this value, as pointed out by Pruzan et al. [9]. We represented the error bars in Figure 1 for $T = 302$ K and 325 K, as stated above. This gave us the uncertainties in the critical exponent γ for the thermal expansivity α_p, which changes from 1.9 to 2.1 (Figure 1).

On the basis of the volume measured or calculated as a function of temperature in the premelting region of hexadecane, the temperature dependence of the thermal expansivity α_p, isothermal compressibility κ_T, and the specific heat C_p can be obtained, and the Pippard relations ((8) and (9)) can also be established at various temperatures near the melting point for constant pressures in this system.

5. Conclusions

We studied here the Pippard relations (C_p versus α_p and α_p versus κ_T) close to the melting point in the premelting region of hexadecane. Linear variations of C_p with α_p (first Pippard relation) and α_p with κ_T (second Pippard relation) were both obtained at various pressures for constant temperatures of 302 K and 325 K in this system.

The Pippard relations can be also be applied to the solid hexadecane in the premelting region using the temperature dependence of C_P, α_P, and κ_T for constant pressures. They can be verified for some other molecular organic compounds near the melting point when the experimental data for the thermodynamics quantities is available.

Nomenclature

α_p: Thermal expansivity
κ_T: Isothermal compressibility
C_P: Specific heat at constant pressure
S: Entropy
V_T: Volume at constant temperature
P_m: Melting pressure
γ: Critical exponent
A: Amplitude
V_m: Melting volume
P_L: Liquid pressure
P_C: Critical pressure
γ_s: Critical exponent in the solid phase
γ_P: Critical exponent in the precursor zone
γ_m: Critical exponent in the melting zone.

References

[1] M. Milhet, J. Pauly, J. A. P. Coutinho, M. Dirand, and J. L. Daridon, "Liquid-solid equilibria under high pressure of tetradecane + pentadecane and tetradecane + hexadecane binary systems," *Fluid Phase Equilibria*, vol. 235, no. 2, pp. 173–181, 2005.

[2] B. He, V. Martin, and F. Setterwall, "Phase transition temperature ranges and storage density of paraffin wax phase change materials," *Energy*, vol. 29, no. 11, pp. 1785–1804, 2004.

[3] Y. Hayami and G. H. Findenegg, "Surface crystallization and phase transitions of the adsorbed film of $F(CF_2)_{12}(CH_2)_{16}H$ at the surface of liquid hexadecane," *Langmuir*, vol. 13, no. 18, pp. 4865–4869, 1997.

[4] S. Roke, J. Buitenhuis, J. C. van Miltenburg, M. Bonn, and A. Van Blaaderen, "Interface-solvent effects during colloidal phase transitions," *Journal of Physics Condensed Matter*, vol. 17, no. 45, pp. S3469–S3479, 2005.

[5] A. R. Ubbelohde, *The Molten State of Matter*, John Wiley and Sons, Chichester, UK, 1978.

[6] H. L. Finke, M. E. Gross, G. Waddington, and H. M. Huffman, "Low-temperature thermal data for the nine normal paraffin hydrocarbons from octane to hexadecane," *Journal of the American Chemical Society*, vol. 76, no. 2, pp. 333–341, 1954.

[7] M. Maroncelli, P. Q. Song, H. L. Strauss, and R. G. Snyder, "Nonplanar conformers and the phase behavior of solid *n*-alkanes," *Journal of the American Chemical Society*, vol. 104, no. 23, pp. 6237–6247, 1982.

[8] D. L. Dorset, B. Moss, J. C. Wittman, and B. Lotz, "The premelt phase of *n*-alkanes crystallographic evidence for a kinked chain structure," *Proceedings of the National Academy of Sciences*, vol. 81, no. 6, pp. 1913–1917, 1984.

[9] P. Pruzan, D. H. Liebenberg, and R. L. Mills, "Experimental evidence for a second-order transformation prior to melting in ammonia, organic compounds and ice I," *Journal of Physics and Chemistry of Solids*, vol. 47, no. 10, pp. 949–961, 1986.

[10] B. J. Bulkin and F. T. Prochaska, "Vibrational spectra of liquid crystals. II. The Raman spectrum of *p*-Azoxyanisole in Crystal, Nematic, and Isotropic Phases, 10–100-cm^{-1} region," *Journal of Chemical Physics*, vol. 54, p. 635, 1971.

[11] R. Pynn and T. Riste, "Neutron scattering by a liquid crystal at temperatures close to its melting point," in *Anharmonic Lattices, Structural Transitions and Melting*, T. Riste, Ed., p. 363, Noordhoff, Leiden, The Netherlands, 1974.

[12] R. Pynn, "Calculations of neutron diffraction patterns obtained with a nematic liquid crystal," *Journal of Physics and Chemistry of Solids*, vol. 34, no. 4, pp. 735–747, 1973.

[13] P. B. Miller and J. D. Axe, "Internal strain and Raman-active vibrations in solids," *Physical Review*, vol. 163, no. 3, pp. 924–926, 1967.

[14] M. J. Buckingham and W. M. Fairbank, "The nature of the λ-transition in liquid Helium," in *Progress in Low Temperature Physics*, C. J. Gorter, Ed., p. 80, North-Holland, Amsterdam, The Netherlands, 1961.

[15] A. B. Pippard, *The Elements of Classical Thermodynamics*, Cambridge University Press, New York, NY, USA, 1957.

[16] J. Timmermans, *Physico-Chemical Constant of Pure Organic Compounds, Binary Systems in Concentrated Solutions*, vol. 4, Interscience Publishers, New York, NY, USA, 1960.

[17] R. Lipowsky, "Critical surface phenomena at first order bulk transitions," *Physical Review Letters*, vol. 49, pp. 1575–1578, 1982.

[18] R. Lipowsky and W. Speth, "Semi-infinite systems with first-order bulk transitions," *Physical Review B*, vol. 28, no. 7, pp. 3983–3993, 1983.

[19] G. Pilcher, "A simplified calorimeter for the precise determination of purity," *Analytica Chimica Acta*, vol. 17, pp. 144–160, 1957.

15

Second Law Analysis of a Gas-Liquid Absorption Film

Nejib Hidouri, Imen Chermiti, and Ammar Ben Brahim

University of Gabes, Engineers National School, Chemical and Process Engineering Department, Applied Thermodynamics Unit, Omar Ibn El Khattab Street, 6029 Gabès, Tunisia

Correspondence should be addressed to Nejib Hidouri; n_hidouri@yahoo.com

Academic Editor: Felix Sharipov

This paper reports an analytical study of the second law in the case of gas absorption into a laminar falling viscous incompressible liquid film. Velocity, temperature, and concentration profiles are determined and used for the entropy generation calculation. Irreversibilities due to heat transfer, fluid friction, and coupling effects between heat and mass transfer are derived. The obtained results show that entropy generation is mainly due to coupling effects between heat and mass transfer near the gas-liquid interface. Total irreversibility is minimum at the diffusion film thickness. On approaching the liquid film thickness, entropy generation is mainly due to viscous irreversibility.

1. Introduction

Gas-liquid (and vapor-liquid) flows occur in a wide variety of situations such as: power generation, chemical processing, and energy production facilities. Some practical applications include condensers and reboilers, gas-liquid reactors (used in the production of pharmaceuticals and specialty chemicals), wetted wall absorbers, power systems, core cooling of nuclear power plant, and falling film reactors. In addition to normal gravity applications, gas-liquid (and vapor-liquid) flows also occur in many planned space operations such as: power cycles, active thermal control systems [1], propulsion devices, and storage and transfer of cryogenic fluids.

Gas-liquid flows can be grouped into a number of different flow patterns. These flow patterns are based on the spatial and temporal distribution of the gas and liquid phases. Gravity is the major reason for such two-phase flows to exhibit a wide range of flow patterns with different characteristics.

Gas-liquid absorption is one of the most important separation processes, the gas-liquid contactors in which absorption takes place are typically spray, trayed, or packed systems involving falling films. Falling film flow has found wide applications in modern equipments, because it is widely used in modern process industries such as vertical condensers, film evaporators, distillation columns, and packed and absorption towers.

Various works are elaborated in order to study the coupled heat and mass transfer processes in gas or vapor absorption into a liquid film. The theoretical analysis of combined heat and mass transfer processes in the case of gas or vapor absorption into a laminar and/or turbulent liquid film as well as the temperature evolution and the concentration profiles in the film are developed [2–5]. Conlisk and Mao [6] considered the condensation of a two-component vapor mixture into a two-component liquid film on a falling film absorber in an absorption heat pump. Analytical expressions for temperature and mass fraction in the falling film have been determined. It is important in this situation to recall the experimental works of Miller and Keyhani [7], in which they presented experimental data of the local heat and mass transfer processes in aqueous LiBr absorption cycle. They studied heat and mass transfer processes and presented measurements of the temperature profile and the bulk concentration along the absorber length of the used test tube. Islam et al. [8] developed a linearized coupled model for heat and mass transfer in falling film absorbers. In order to test its accuracy, the developed model was compared with a nonlinear model and a numerical simulation. The used model provides analytical expressions of heat and mass transfer coefficients from the experimental data concerning vertical and horizontal tube absorbers. Killion and Garimella [9] reported an interesting review about the main used models to illustrate coupled heat and mass transfer phenomena during falling film absorption. Detailed expressions of the governing equations, the boundary conditions, and the methods of resolution used for modeling combined heat and mass transfer problems have been developed. Other extensive researches have been performed on gas absorption in liquid films with different flow regimes, geometries, and boundary conditions [10–12]. In many real cases, the heat effect is small and the absorption process may be considered isothermal [11, 13].

The improvement of industrial systems as well as the energy utilization in any fluid is one of the fundamental problems of the engineering processes, since improved systems will provide better material processing, energy conservation, and environmental effects. One of the used methods for predicting the performance of the engineering processes is the second law analysis. The second law of thermodynamics is applied to determine the irreversibilities in terms of the entropy generation. Entropy generation is the measure of the destruction of the available work of the system. In the present work, entropy generation analysis concerns gas absorption through a laminar falling liquid film. In the following paragraph, some researches about entropy generation analysis through a falling film will be presented.

The study of laminar falling viscous incompressible liquid film along an inclined porous heated plate has many significant applications in thermal engineering and industries, starting from petroleum drilling equipments to various industrial exchanger systems [16]. Adeyinka and Naterer [17] investigated the physical significance of entropy generation, and a resulting optimization correlation for a laminar film condensation on a flat plate was developed. Galović [18] studied entropy generation in evaporator/condenser types of heat exchangers. A model was developed to calculate the overall entropy generation of heat exchange for given values of heat transfer effectiveness and ratios of flow rates. Boulama et al. [19] derived an analytical expression of entropy generation rate in a fully developed, laminar, and time-independent simultaneous thermal and solute mixed convection for the case of a nonreacting binary gas mixture flowing upwards in a vertical parallel plate. Carrington and Sun [20] developed different expressions of entropy generation due to the coupling effects between heat and mass transfer in a multicomponent fluid. They identified restrictions on the applicability of expressions for entropy generation due to mass transfer and heat transfer. They used an ideal binary fluid to illustrate the analysis in the case of small pressure gradient in the diffusion flux direction. Magherbi et al. [21] reported a numerical study of entropy generation due to heat transfer, mass transfer, and fluid friction in steady state for laminar double diffusive convection in an inclined enclosure with heat and mass diffusive walls. They studied the influences of the inclination angle of the enclosure, the thermal Grashof number, and the buoyancy ratio on total entropy generation due to heat transfer, mass transfer, and fluid friction irreversibilities. Saouli and Aïboud-Saouli [22] investigated second law analysis of a laminar falling liquid film along an inclined heated plate. They considered the upper surface of the liquid film free and adiabatic and the lower wall fixed and impermeable. Results show that entropy generation number transversely decreases and increases for all values of group parameters. They found that the irreversibility ratio decreases in the transverse direction and increases as the group parameter increases. Makinde and Osalusi [23] analytically investigated entropy generation of a laminar falling viscous incompressible liquid film along an inclined porous heated plate. Velocity and temperature profiles are obtained and used to calculate entropy generation number. They found that, near the inclined porous heated plate, heat transfer irreversibility dominates, while near the liquid-free surface, viscous dissipation irreversibility dominates. Gorla and Pratt [24] analytically studied second law of a non-Newtonian laminar falling liquid film along an inclined heated plate, where the upper surface of the liquid film is considered free and adiabatic. Velocity and temperature profiles are calculated and used to compute the entropy generation number (N_s), the irreversibility ratio (ϕ), and the Bejan number (Be), for different values of the viscous dissipation parameter ($\mathrm{Br}\Omega^{-1}$), the viscosity index (n), and the dimensionless axial distance (X). Results show that heat transfer dominates the viscous irreversibility when $0 \leq \phi < 1$, and the fluid friction irreversibility dominates heat transfer irreversibility when $\phi > 1$. For $\phi = 1$, both heat transfer and fluid friction irreversibilities have equal contribution for entropy generation. Recently, Chermiti et al. [25] analytically investigated entropy generation of an isothermal absorption of carbon dioxide into a laminar viscous incompressible falling liquid film (H_2O). Entropy generation expression is developed, and the velocity profile as well as the absorbed gas concentration is determined and used for the calculation of the diffusive and the convective mass transfer fluxes, as well as the entropy generation. It was found that, from the gas-liquid interface towards the diffusion film thickness δ_d, entropy generation increases towards a maximum value due to diffusive mass transfer and then decreases towards a minimum value by convective mode. This minimum is obtained at the diffusion film thickness. For values of liquid film thickness greater than the diffusion film thickness (i.e., $x > \delta_d$), entropy generation linearly increases due to the fluid friction irreversibility. Near the diffusion film thickness, mass transfer irreversibility dominates. On approaching the liquid film thickness δ, viscous irreversibility dominates. The increase of the falling film length induces the decrease of entropy generation magnitude. Aïboud-Saouli et al. [26] presented the application of the second law to gravity-driven liquid film along an inclined heated plate in the presence of a transverse magnetic field and viscous dissipation effects. It was found that entropy generation increases with Hartman number, Brinkman number, and the dimensionless group. Irreversibility due to the magnetic field, the conduction heat in the transverse direction, and the fluid friction increases by the increase of Hartman number, Brinkman number, and the dimensionless group, respectively.

The main originality of the present work is the second law analysis for the case of gas absorption through a vertical laminar falling viscous incompressible liquid film, by the determination of different sources of irreversibility. The study is based on the analytical flow field solutions obtained for the velocity, the temperature, and the concentration profiles. The velocity distribution was determined by solving the momentum equation subject to an appropriate set of boundary conditions. The concentration profile was determined by solving the species conservation equation with appropriate boundary conditions, and the temperature distribution was obtained by using the energy conservation equation. In many previous works, authors are interested to determine the concentration and the temperature profiles only. In the case of vapor absorption [2, 3], a linear relation between the temperature and the concentration in equilibrium with vapor at constant pressure is adopted and makes the resolution of the system equations. In this study, velocity, temperature, and concentration fields are analytically determined and directly used to evaluate the entropy generation.

In the presence of simultaneous heat and mass transfer, four sources of irreversibility have been identified, which are due to coupling effects between heat and mass transfer by diffusion, due to heat transfer, due to coupling effects between heat and mass transfer by convection and fluid friction.

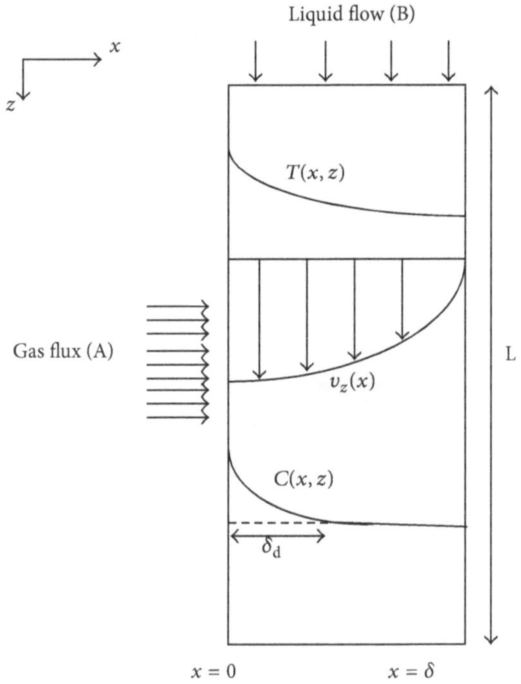

Figure 1: Description of gas absorption into a laminar falling liquid film: typical profiles of temperature, velocity, and concentration.

2. Mathematical Formulation

2.1. Physical Considerations. Generally, the fluid density depends on pressure and temperature. Thus, the presence of temperature variation or chemical species (thermal and concentration gradients) may change the fluid density which can result in thermal or solute natural convection. The system under consideration is illustrated in Figure 1. It consists of a laminar falling viscous incompressible liquid film (designated by the letter B), flowing down on a vertical plate under gravity. The gas, designated by the letter A, is absorbed without chemical reactions. In the present study, the assumptions employed in the problem formulation are

(i) physical absorption (no chemical reactions),

(ii) two dimensional flow in steady state regime,

(iii) falling laminar flow of a Newtonian incompressible viscous liquid film (water),

(iv) ammonia as absorbed gas,

(v) the diffusion film thickness is too small as compared to the falling film thickness (i.e., $\delta_d \ll \delta$), so that the thermophysical properties of the falling film are similar to those of the liquid,

(vi) the Boussinesq approximation is valid for assuming a linear variation of the fluid density with temperature and concentration,

(vii) the curvature of the liquid film is neglected, and the velocity only depends on the transverse direction (i.e., $v = v_z(x)$),

(ix) the absorption process takes place at constant pressure (i.e., $(\partial P/\partial z) \approx 0$),

(x) for the falling film (a thin film), the axial conduction terms are neglected as compared to the transverse terms (i.e., $(\partial^2 T/\partial z^2) \ll (\partial^2 T/\partial x^2)$). Further, for the thin film, small values of temperature difference ΔT are considered.

2.2. Velocity Profile. Under the Boussinesq approximation and for a small density variation, the fluid density linearly varies with temperature and species concentration such that

$$\rho = \rho_{\text{ref}}\left[1 - \beta_T\left(T - T_{\text{ref}}\right) - \beta_C\left(C - C_{\text{ref}}\right)\right]. \quad (1)$$

ρ_{ref} is the reference density measured at reference temperature (T_{ref}) and concentration (C_{ref}).

Thermal and solute expansion coefficients β_T and β_C are, respectively, given by

$$\beta_T = -\frac{1}{\rho_{\text{ref}}}\left(\frac{\partial \rho}{\partial T}\right), \qquad \beta_C = -\frac{1}{\rho_{\text{ref}}}\left(\frac{\partial \rho}{\partial C}\right). \quad (2)$$

The continuity and momentum equations are given by

$$\frac{\partial u}{\partial x} + \frac{\partial v}{\partial z} = 0,$$

$$\rho_{\text{ref}}\left(u\frac{\partial u}{\partial x} + v\frac{\partial u}{\partial z}\right) = -\frac{\partial P}{\partial x} + \eta_L\left(\frac{\partial^2 u}{\partial x^2} + \frac{\partial^2 u}{\partial z^2}\right), \quad (3)$$

$$\rho_{\text{ref}}\left(u\frac{\partial v}{\partial x} + v\frac{\partial v}{\partial z}\right) = -\frac{\partial P}{\partial z} + \eta_L\left(\frac{\partial^2 v}{\partial x^2} + \frac{\partial^2 v}{\partial z^2}\right) + f.$$

f describes the buoyancy forces due to thermal and concentration gradients. It is given by

$$f = g\int_{\rho_0}^{\rho_\delta} d\rho = g\left(\rho_\delta - \rho_0\right). \quad (4)$$

ρ_δ and ρ_0 are the fluid densities measured at the falling film thickness ($x = \delta$) and at the gas-liquid interface ($x = 0$), respectively. Since the flow takes place along a vertical plate, the velocity only depends on the diffusion film thickness (i.e., in transverse direction). That is, the transverse velocity $u \approx 0$ and $v = v_z(x)$ [27]. Further, the absorption process takes place at constant pressure. After rearrangement, the continuity and the momentum equations are, respectively, given by

$$\frac{\partial v}{\partial z} = 0, \quad (5)$$

$$\frac{\eta_L}{\rho_{\text{ref}}}\left(\frac{\partial^2 v}{\partial x^2}\right) + \frac{f}{\rho_{\text{ref}}} = 0. \quad (6)$$

By using (1) and (4), the buoyancy forces are given by

$$f = g\rho_{\text{ref}}\left[\left(\beta_T\left(T_0 - T_\delta\right) + \beta_C\left(C_0 - C_\delta\right)\right)\right]. \quad (7)$$

T_0 and T_δ are the gas-liquid interface and the falling film thickness temperatures, respectively. Thus, (6) can be written as follows:

$$\eta_L \frac{\partial^2 v}{\partial x^2} = g\rho_{\text{ref}}\left[\beta_T\left(T_\delta - T_o\right) + \beta_C\left(C_\delta - C_o\right)\right]. \quad (8)$$

This differential equation is the liquid motion equation. The parabolic profile of the velocity enables us to give the appropriate boundary conditions as follows:

$$\begin{aligned} &\text{at } x = 0, \quad v = v_{\max}, \\ &\text{at } x = \delta, \quad v = 0. \end{aligned} \quad (9)$$

The reference density is supposed to be equal to that of the liquid (i.e., $\rho_{\text{ref}} = \rho_L$). Thus, the solution of (8) is given by

$$\begin{aligned} v(x) &= v_{\max}\left(1 - \left(\frac{x}{\delta}\right)^2\right) \\ &= \frac{g\rho_L}{2\eta_L}\delta^2\left[\beta_T\left(T_o - T_\delta\right) + \beta_C\left(C_o - C_\delta\right)\right]\left(1 - \left(\frac{x}{\delta}\right)^2\right). \end{aligned} \quad (10)$$

For small density differences, thermal and solute expansion coefficients are given by [28, 29]

$$\beta_T = -\frac{1}{\rho_L}\left(\frac{\partial \rho}{\partial T}\right) \approx \frac{1}{\Delta T}, \qquad \beta_C = -\frac{1}{\rho_L}\left(\frac{\partial \rho}{\partial C}\right) \approx \frac{1}{\Delta C}, \quad (11)$$

where

$$\Delta T = T_0 - T_\delta, \qquad \Delta C = C_0 - C_\delta. \quad (12)$$

The maximum velocity is, therefore, given by

$$v_{\max} = \frac{g\rho_L}{\eta_L}\delta^2. \quad (13)$$

2.3. Concentration Profile. For the gas-liquid absorption system under consideration, the diffusing species is the absorbed gas, which is designated by the letter A (see Figure 1). Thus, the species conservation equation is given by

$$u\frac{\partial C_A}{\partial x} + v\frac{\partial C_A}{\partial z} - D_A\left(\frac{\partial^2 C_A}{\partial x^2} + \frac{\partial^2 C_A}{\partial z^2}\right) = 0. \quad (14)$$

As the Reynolds number of the system is smaller than 25 (laminar flow) [29], the above equation is, therefore, given by

$$v_{\max}\left(1 - \left(\frac{x}{\delta}\right)^2\right)\frac{\partial C_A}{\partial z} - D_A\frac{\partial^2 C_A}{\partial x^2} = 0. \quad (15)$$

For a small penetration distance of the gas into the falling liquid film and a short contact time (physical absorption), the film is supposed to move with a velocity equal to $v_{\max}$ [30, 31]. Finally, the required solution for the gas concentration through the falling film can be obtained by using the method of combination of variables (see details in [25]):

$$C_A = C_{A0}\left(1 - \operatorname{erf}\left(\frac{x}{\sqrt{4\left(D_A/v_{\max}\right)z}}\right)\right). \quad (16)$$

The transverse and axial molar fluxes J_{Ax} and J_{Az} of the gas are, respectively, given by

$$\begin{aligned} J_{Ax} &= -D_A\frac{\partial C_A}{\partial x}, \\ J_{Az} &= C_A v(x). \end{aligned} \quad (17)$$

2.4. Temperature Profile. Due to the strong dependence of the gas solubility on temperature, it is important to determine the temperature distribution through the falling liquid film. Indeed, the gas absorption heat is the most heat influencing the temperature profile and leading to a rise in the temperature of the liquid phase [32]. Energy conservation equation throughout the falling liquid film is given by

$$u\frac{\partial T}{\partial x} + v\frac{\partial T}{\partial z} = \frac{k_L}{\rho_L C_{p,L}}\left(\frac{\partial^2 T}{\partial x^2} + \frac{\partial^2 T}{\partial z^2}\right). \quad (18)$$

By taking into consideration the above-mentioned assumptions and by introducing the thermal diffusivity of the fluid given by $\alpha = k_L/(\rho_L \cdot C_{pL})$, the energy conservation equation is simply given by

$$\alpha\frac{\partial^2 T}{\partial x^2} = v_{\max}\left(1 - \left(\frac{x}{\delta}\right)^2\right)\frac{\partial T}{\partial z}. \quad (19)$$

The appropriate boundary conditions associated to this equation are

$$z = 0, \qquad T(x,0) = T_0, \tag{20}$$

$$x = 0, \qquad q^* = -k_L \frac{\partial T(0,z)}{\partial x}, \tag{21}$$

$$x = \delta, \qquad \frac{\partial T(\delta,z)}{\partial x} = 0. \tag{22}$$

q^* is the interfacial heat flux given by the heat of absorption.

Before proceeding with the solution of the temperature equation (19), it is convenient to rewrite the equations in a dimensionless form. Let us define the following dimensionless variables:

$$\theta(x,z) = \frac{T(x,z) - T_0}{\Delta T}, \qquad X = \frac{x}{\delta}, \qquad Z = \frac{\alpha z}{v_{\max}\delta^2}. \tag{23}$$

ΔT is the temperature difference given by $\Delta T = (q^*\delta)/k_L$.

Equation (19) with the new dimensionless variables becomes

$$\frac{\partial^2\theta}{\partial X^2} = \left(1 - X^2\right)\frac{\partial\theta}{\partial Z}. \tag{24}$$

The dimensionless boundary conditions are given by

$$Z = 0, \qquad (X,0) = 0, \tag{25}$$

$$X = 0, \qquad \frac{\partial\theta(0,Z)}{\partial X} = -1,$$
$$X = 1, \qquad \frac{\partial\theta(1,Z)}{\partial X} = 0. \tag{26}$$

To solve the dimensionless temperature equation (24), the best approach is to use the separation of variables method [33], so that the dimensionless temperature is given by

$$\theta(X,Z) = \theta_1(X) + \theta_2(Z). \tag{27}$$

The solution of (27) is the sum of two functions $\theta_1(X)$ and $\theta_2(Z)$ that depend only on X and Z, respectively. Consequently, the separated solutions into the energy equation give

$$\frac{\partial^2\theta_1(X)}{\partial X^2} = \left(1 - X^2\right)\frac{\partial\theta_2(Z)}{\partial Z}. \tag{28}$$

The two sides of the above equation are equal for any chosen values of X and Z, so both sides are equal to the same constant. In other words, the above equation yields a pair of separated differential equations given by

$$\frac{\partial^2\theta_1(X)}{\partial X^2} = \lambda_1\left(1 - X^2\right),$$
$$\frac{\partial\theta_2(Z)}{\partial Z} = \lambda_1. \tag{29}$$

The general solution of the dimensionless temperature is, therefore, given by

$$\theta(X,Z) = -\frac{X^4}{12}\lambda_1 + \frac{X^2}{2}\lambda_1 + X\lambda_2 + \lambda_1 Z + \lambda_3. \tag{30}$$

λ_1, λ_2, and λ_3 are constants of integration.

In order to determine the third constant λ_3, the mean bulk temperature should be introduced. It is given by

$$\theta_b(Z) = \int_0^1 \theta(X,Z)\,dX. \tag{31}$$

The first boundary condition given by (25) gives the mean bulk temperature $\theta_b(0) = 0$. By using this condition and the two other boundary conditions, the temperature profile is finally given by

$$T(x,z) = T_0 + \Delta T\left[-\frac{1}{8}\left(\frac{x}{\delta}\right)^4 + \frac{3}{4}\left(\frac{x}{\delta}\right)^2 - \left(\frac{x}{\delta}\right) + \frac{3}{2}\left(\frac{\alpha z}{v_{\max}\delta^2}\right) + \frac{11}{40}\right]. \tag{32}$$

3. Entropy Generation

The existence of thermal, velocity, and concentration gradients along the absorbing falling film yields a nonequilibrium state, which causes entropy generation. Under these conditions, the volumetric rate of entropy generation in the flow field with both heat and mass transfer and without body force and chemical reaction effects is given by [34]

$$\sigma_c = \vec{q}\cdot\overrightarrow{\text{grad}}\left(\frac{1}{T}\right) + \frac{\overline{\overline{\tau}} : \overline{\overline{\text{grad}}}(V)}{T} - \frac{1}{T}\sum_k \vec{J}_k\cdot\overrightarrow{\text{grad}}(\mu_k). \tag{33}$$

The first term of the right hand side of (33) is the entropy generation due to heat transfer, the second is due to fluid friction, and the third is due to mass transfer.

The energy flux $\vec{q}$ is the sum of the heat flux $\vec{q}'$, given by the Fourier law, and the enthalpy flux due to species diffusion. That is

$$\vec{q} = \vec{q}' - \sum_k h_k\vec{J}_k = -k_L\cdot\overrightarrow{\text{grad}}(T) - \sum_k h_k\vec{J}_k. \tag{34}$$

Rearrangement of (33) gives

$$\sigma_c = -\frac{1}{T^2}\left[\vec{q}'\cdot\overrightarrow{\text{grad}}(T)\right] + \frac{1}{T^2}\left[\left(\sum_k h_k\vec{J}_k\right)\cdot\overrightarrow{\text{grad}}(T)\right] + \frac{\overline{\overline{\tau}} : \overline{\overline{\text{grad}}}(\vec{V})}{T} - \frac{1}{T}\sum_k \vec{J}_k\cdot\overrightarrow{\text{grad}}(\mu_k). \tag{35}$$

The first term of the right-hand side of (35) is the irreversibility due to thermal gradients, the second is due to coupling effects between thermal gradients and mass diffusion, the third is due to fluid friction, and the fourth is due to mass diffusion.

The specific chemical potential of the gas (A) is given by

$$\mu_A(T,P_o) = C_{p,A}(T-T_o) - TC_{p,A}\text{Ln}\left(\frac{T}{T_o}\right) + h_{Ao}(T_o,P_o) - Ts_{Ao}(T_o,P_o). \quad (36)$$

P_o and T_o are the reference pressure and temperature, respectively. h_{Ao} and s_{Ao} are the enthalpy and entropy of the gas, at T_o and P_o.

The shear stress is simply given by [28]

$$\tau_{ij} = \eta_L\left(\frac{\partial v_i}{\partial x_j} + \frac{\partial v_j}{\partial x_i}\right). \quad (37)$$

Thus, the local entropy generation rate in a two-dimensional flow with a single diffusing species can, therefore, be written as follows:

$$\sigma_c = \frac{k_L}{T^2}\left[\left(\frac{\partial T}{\partial x}\right)^2 + \left(\frac{\partial T}{\partial z}\right)^2\right] + \frac{h_A}{T^2}\left[J_{Ax}\cdot\frac{\partial T}{\partial x} + J_{Az}\cdot\frac{\partial T}{\partial z}\right] + \frac{\eta_L}{T}\left(\frac{\partial v}{\partial x}\right)^2 - \frac{1}{T}\left(J_{Ax}\cdot\frac{\partial \mu_A}{\partial x} + J_{Az}\cdot\frac{\partial \mu_A}{\partial z}\right). \quad (38)$$

Expressions of specific enthalpy and entropy of the absorbed gas are, respectively, given by

$$h_A = C_{pA}(T-T_0) + h_{A0}, \qquad s_A = C_{pA}\text{Ln}\left(\frac{T}{T_0}\right) + s_{A0}. \quad (39)$$

By replacing the fluxes J_{Ax} and J_{Az} and the specific enthalpy and entropy given above by their expressions and by making the derivatives of the specific chemical potential, the expression of the local entropy generation rate for the case of gas absorption into a falling liquid film is, therefore, given by

$$\sigma_c = \frac{k_L}{T^2}\left[\left(\frac{\partial T}{\partial x}\right)^2 + \left(\frac{\partial T}{\partial z}\right)^2\right] - \left(\frac{D_A}{T^2}\right)\left(\frac{\partial C_A}{\partial x}\right)\left(\frac{\partial T}{\partial x}\right) \times \left[c_{pA}(T-T_0) + h_{A0} + T\left(c_{pA}\text{Ln}\left(\frac{T}{T_0}\right) + s_{A0}\right)\right] + \left(\frac{C_A v}{T^2}\right)\left(\frac{\partial T}{\partial z}\right) \times \left[c_{pA}(T-T_0) + h_{A0} + T\left(c_{pA}\text{Ln}\left(\frac{T}{T_0}\right) + s_{A0}\right)\right] + \frac{\eta_L}{T}\left(\frac{\partial v}{\partial x}\right)^2. \quad (40)$$

The first term of the right-hand side of (40) is the irreversibility due to heat transfer $\sigma_{c,\text{th}}$, the second and the third terms are due to the coupling effects between heat and mass transfer $\sigma_{c,m\text{-th}}$ (by diffusion and by convection, resp.), and the last one is due to the fluid friction $\sigma_{c,v}$ (viscous dissipation irreversibility).

Table 1: Gas and liquid physical properties [14, 15].

	Property	
Liquid	Inlet temperature, T_o (K)	298
	Reference pressure, P_o (atm)	1
	Dynamic viscosity, η_L (Pa · s)	$0,911 \cdot 10^{-3}$
(Water)	Mass density, ρ_L (kg · m^{-3})	1000
	Thermal conductivity, k_L (W · m^{-1} · K^{-1})	0,607
Gas	Diffusion coefficient, D_A (m^2 · s^{-1})	$1,9 \cdot 10^{-9}$
	Solubility in water at T_o, C_A (g · L^{-1})	540
(Ammonia)	Reference molar entropy, s_{Ao} (J · mol^{-1} · K^{-1})	192,77
	Specific heat, C_{pA} (J · kg^{-1} · K^{-1})	2129,45

4. Results and Discussions

Velocity and concentration are calculated by using (10) and (16), respectively. Temperature Equation (32) is solved for a given value of temperature difference ΔT. Values of x are selected in order to ensure a laminar flow of the liquid film. The absorption of ammonia into a falling water film is considered. All the following results are calculated using the physical properties of water and ammonia given in Table 1.

Figures 2–5 show the general behavior of irreversibilities due to thermal gradients, coupling effects between heat and mass transfer, fluid friction, and total entropy generation versus the transverse direction (i.e., from the gas-liquid interface to the falling liquid film thickness δ), for known values of the liquid film length and the temperature difference ΔT. The choice of the temperature difference ranging between 1°C and 5°C is based on an experimental study elaborated by Mahmoud [35] about heat and mass transfer accompanying ammonia vapor absorption into a solution of water-ammonia. The study shows that for a null fraction of ammonia in the solution and a penetration distance equal to 4×10^{-4} m, the solution temperature increases from 22°C to 29°C. Using this argument, it is important to consider the above-mentioned temperature difference values to examine the evolution of different sources of irreversibility in the system.

As it can be seen, Figure 2 shows that irreversibility due to thermal gradients $\sigma_{c,\text{th}}$ is at its maximum value at the gas-liquid interface, where the gas absorption process begins. For any fixed value of the temperature difference ΔT, entropy generation due to thermal gradients exponentially decreases as x increases; since the temperature is a fractional polynomial function of the transverse coordinate x, its derivative gives another polynomial function with lower degree (i.e., the irreversibility due to thermal gradients). Consequently, thermal irreversibility mainly happens near the gas-liquid interface, then it decreases and tends towards a minimum constant value on approaching the falling film thickness δ, where the absorption process is practically absent.

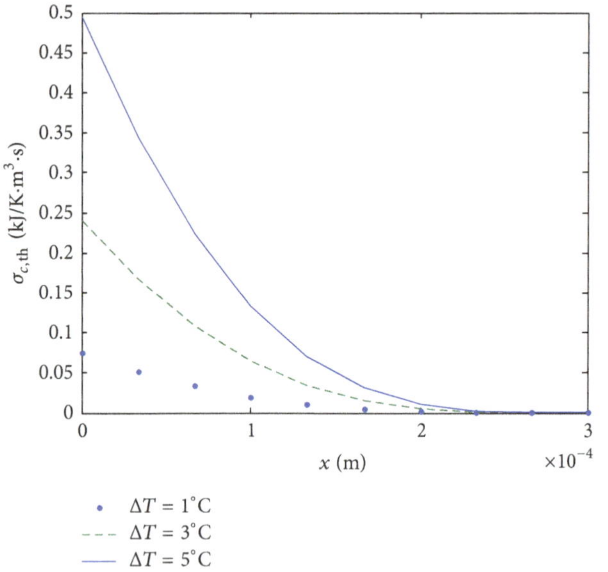

FIGURE 2: Entropy generation due to thermal gradients, $\sigma_{c,\text{th}}$ versus x and ΔT.

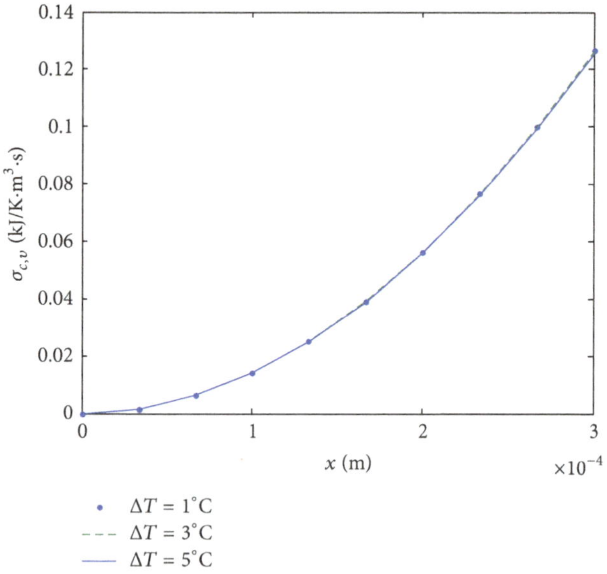

FIGURE 4: Entropy generation due to fluid friction, $\sigma_{c,v}$ versus x and ΔT.

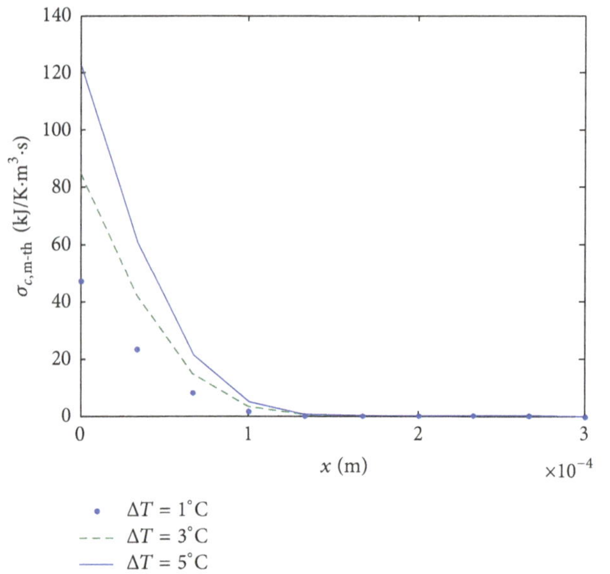

FIGURE 3: Entropy generation due to heat and mass transfer, $\sigma_{c,m\text{-th}}$ versus x and ΔT.

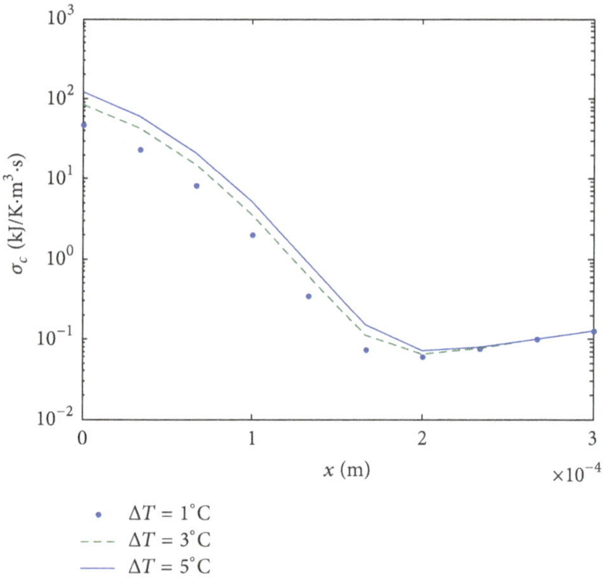

FIGURE 5: Total entropy generation, σ_c versus x and ΔT.

For a fixed value of the transverse distance x, it can be observed that as the temperature gradients increase, entropy generation due to thermal gradients also increases.

Figure 3 illustrates entropy generation due to the coupling effects between heat and mass transfer, $\sigma_{c,m\text{-th}}$. It can be seen that near the gas-liquid interface, entropy generation is at its maximum value, since the absorption process mainly takes place close to the gas-liquid interface, which is accompanied by the heat of absorption. Thus, entropy generation due to the coupling effects between heat and mass transfer is similar to the concentration profile, which is maximum at the gas-liquid interface (i.e., at $x = 0$) as found by Chermiti et al. [25]. The increase of the transverse distance (i.e., $x > 0$) induces the decrease of $\sigma_{c,m\text{-th}}$, to become null near the falling film thickness (i.e., $x = \delta$), for all the studied values of ΔT. This is due to the fact that when the transverse distance increases, the absorption process and the fluid motion decrease. Consequently, irreversibilities due coupling effects between heat and mass transfer by diffusion and convection decrease. At the film thickness ($x = \delta$), the absorption process is insignificant (no concentration gradients).

It could be noticed that, at lower values of temperature gradient ΔT, the gas absorption proceeds at a faster rate due to the availability of larger driving potentials for mass and heat transfer.

Throughout the diffusion film thickness region, the decrease of the temperature difference induces the decrease of both irreversibilities due to heat transfer (Figure 2) and coupling effects between heat and mass transfer (Figure 3). This is attributed to the decrease of temperature and concentration as well as velocity gradients in this region.

Figure 4 shows entropy generation due to fluid friction, $\sigma_{c,v}$ along the liquid film. As shown, the analytically obtained viscous dissipation irreversibility is null near the gas-liquid interface, since velocity is at its maximum value at this region. As a consequence, the friction effects contribution is insignificant near the interface.

On approaching the film thickness, viscous dissipation irreversibility significantly increases and it is maximum at the falling film thickness ($x = \delta$). The fluid friction irreversibility exhibits a parabolic profile. This behavior is expected, since its expression includes the term (x^2) via the velocity expression (see (40)). The fluid friction irreversibility magnitude slightly increases with the decrease of the temperature difference ΔT as it is inversely proportional to the temperature.

As a consequence, the viscous dissipation irreversibility through the falling liquid film increases as the transverse distance increases and, inversely, as the temperature difference decreases.

Total entropy generation σ_c behavior is illustrated in Figure 5. As it can be seen, entropy generation is at its maximum value near the gas-liquid interface for all the studies values of ΔT. This is attributed, on one hand, to the dominant effect of coupled heat and mass transfer irreversibility induced by the absorption process and, on the other hand, to the thermal irreversibility. As mentioned, at the interface region, irreversibility due to fluid friction is insignificant (Figure 3). Further, the increase of the temperature difference induces the increase of total entropy generation magnitude.

On increasing the transverse distance, total entropy generation significantly decreases towards a minimum value for all the studied values of the temperature difference. Entropy generation is minimum at the diffusion film thickness δ_d. The decrease of entropy generation from the gas-liquid interface to the diffusion film thickness is due to the decrease of both convective and diffusive irreversibilities, since the gas diffusion occurs at the beginning of the falling film thickness and the mass transfer by convection decreases while approaching the diffusion film thickness. At this point, the influence of both thermal and coupled effects between heat and mass transfer (by diffusion and convection) irreversibilities on total entropy generation becomes insignificant. For $x > \delta_d$, total entropy generation increases towards an asymptotic value, which is mainly due to the viscous dissipation irreversibility. This result is consistent with the findings of Chermiti et al. [25]. For $x > \delta_d$, entropy generation slightly increases with the decrease of the temperature difference as explained above.

5. Conclusions

The second law of thermodynamics is analytically investigated for the case of gas absorption through a laminar falling film. Under some physical assumptions, velocity, temperature, and gas concentration equations are derived and used for the entropy generation calculation concerning the absorption of ammonia in a water falling film. The analysis shows that entropy generation is due to heat transfer, coupling effects between heat and mass transfer (by convection and diffusion), and fluid friction. Entropy generation due to fluid friction is insignificant at the gas-liquid interface, where thermal and coupled effects between heat and mass transfer irreversibilities dominate. At the interface ($x = 0$), total irreversibility magnitude is maximum, whereas it is minimum at the diffusion film thickness ($x = \delta_d$). On approaching the liquid film thickness ($x = \delta_d$), entropy generation is mainly due to fluid friction effects. At the diffusion film region ($0 < x \leq \delta_d$), entropy generation magnitude increases with the increase of the temperature difference of the falling film; this behavior is reversed for values of the transverse distance greater than the diffusion film thickness ($\delta_d < x \leq \delta$).

Nomenclature

C_A: Molar concentration of gas (mol·m^{-3})
C_{pA}: Isobaric specific heat of gas (J·kg^{-1}·K^{-1})
D_A: Molecular diffusivity coefficient (m^2·s^{-1})
erf: Error function
q: Energy flux (J·s^{-1}·m^2)
J_A: Mass transfer flux of gas (mol·m^{-2}·s^{-1})
h_A: Specific enthalpy of gas (J/mol)
k_L: Liquid thermal conductivity (W·m^{-1}·°C^{-1})
L: Falling film length (m)
P: Pressure (Pa)
s_A: Specific entropy of gas (J·kg^{-1}·K^{-1})
T: Temperature (K)
u, v: Falling film velocity components (m·s^{-1})
x, z: Rectangular coordinates (m)
X, Z: Dimensionless coordinates (—).

Greek Symbols

α: Thermal diffusivity of liquid (m^2·s^{-1})
β_c: Solute expansion coefficient (m^3·mol^{-1})
β_T: Thermal expansion coefficient (K^{-1})
η_L: Dynamic viscosity (Pa·s)
δ: Falling film thickness (m)
δ_d: Diffusion film thickness (m)
ΔT: Temperature difference through the liquid film (°C)
μ_A: Specific chemical potential of gas (J·mol^{-1})
ρ: Fluid mass density (kg·m^{-3})
τ: Stress tensor (N·m^{-2})
θ: Dimensionless temperature (—)
λ: Integration constant (—)
σ_c: Entropy generation rate per unit volume (J·m^{-3}·s^{-1}·K^{-1}).

References

[1] A. A. M. Delil, "Research issues on two-phase loops for space applications," in *Proceedings of the Symposium on Space Flight Mechanics*, Institute of Space and Astronautical Science, Sagamihara, Japan, 2000.

[2] G. Grossman, "Analysis of interdiffusion in film absorption," *International Journal of Heat and Mass Transfer*, vol. 30, no. 1, pp. 205–208, 1987.

[3] G. Grossman and M. T. Heath, "Simultaneous heat and mass transfer in absorption of gases in turbulent liquid films," *International Journal of Heat and Mass Transfer*, vol. 27, no. 12, pp. 2365–2376, 1984.

[4] L. P. Kholpanov and E. Y. Kenig, "Coupled mass and heat transfer in a multicomponent turbulent falling liquid film," *International Journal of Heat and Mass Transfer*, vol. 36, no. 14, pp. 3647–3657, 1993.

[5] B. J. C. van der Wekken and R. H. Wassenaar, "Simultaneous heat and mass transfer accompanying absorption in laminar flow over a cooled wall," *International Journal of Refrigeration*, vol. 11, no. 2, pp. 70–77, 1988.

[6] A. T. Conlisk and J. Mao, "Nonisothermal absorption on a horizontal cylindrical tube—1. The film flow," *Chemical Engineering Science*, vol. 51, no. 8, pp. 1275–1285, 1996.

[7] W. A. Miller and M. Keyhani, "Experimental analysis of local heat and mass transfer data for vertical falling film absorption," Tech. Rep., Japan Aerospace Exploration Agency (JAXA), Tokyo, Japan, 1999, Fluid Mechanics and Thermodynamics.

[8] R. Islam, N. E. Wijeysundera, and J. C. Ho, "Simplified models for coupled heat and mass transfer in falling-film absorbers," *International Journal of Heat and Mass Transfer*, vol. 47, no. 2, pp. 395–406, 2004.

[9] J. D. Killion and S. Garimella, "A critical review of models of coupled heat and mass transfer in falling-film absorption," *International Journal of Refrigeration*, vol. 24, no. 8, pp. 755–797, 2001.

[10] S. M. Yih and R. C. Seagrave, "Mass transfer in laminar falling liquid films with accompanying heat transfer and interfacial shear," *International Journal of Heat and Mass Transfer*, vol. 23, no. 6, pp. 749–758, 1980.

[11] G. Grossman, "Simultaneous heat and mass transfer in film absorption under laminar flow," *International Journal of Heat and Mass Transfer*, vol. 26, no. 3, pp. 357–371, 1983.

[12] M. Danish, R. K. Sharma, and S. Ali, "Gas absorption with first order chemical reaction in a laminar falling film over a reacting solid wall," *Applied Mathematical Modelling*, vol. 32, no. 6, pp. 901–929, 2008.

[13] H. L. Goff, A. Ramadane, and P. L. Goff, "Modélisation des transferts couples de matière et de chaleur dans l'absorption gaz-liquide en film ruisselant laminaire," *International Journal of Heat and Mass Transfer*, vol. 28, no. 11, pp. 2005–2017, 1985.

[14] K. Othmer, *Encyclopedia of Chemical Technology*, Springer, New York, NY, USA, 4th edition, 1993.

[15] C. L. Yaws, *Handbook of Thermal Conductivity*, Gulf Publishing Corporation, Houston, Tex, USA, 1997.

[16] O. D. Makinde, "Exothermic explosions in a slab: a case study of series summation technique," *International Communications in Heat and Mass Transfer*, vol. 31, no. 8, pp. 1227–1231, 2004.

[17] O. B. Adeyinka and G. F. Naterer, "Optimization correlation for entropy production and energy availability in film condensation," *International Communications in Heat and Mass Transfer*, vol. 31, no. 4, pp. 513–524, 2004.

[18] A. Galović, "Non-dimensional entropy analysis of condenser and/or evaporator type heat exchangers," *International Journal of Heat Exchangers*, vol. 5, no. 2, pp. 337–346, 2004.

[19] K. G. Boulama, N. Galanis, and J. Orfi, "Entropy generation in a binary gas mixture in the presence of thermal and solutal mixed convection," *International Journal of Thermal Sciences*, vol. 45, no. 1, pp. 51–59, 2006.

[20] C. G. Carrington and Z. F. Sun, "Second law analysis of combined heat and mass transfer phenomena," *International Journal of Heat and Mass Transfer*, vol. 34, no. 11, pp. 2767–2773, 1991.

[21] M. Magherbi, H. Abbassi, N. Hidouri, and A. Ben Brahim, "Second law analysis in convective heat and mass transfer," *Entropy*, vol. 8, no. 1, pp. 1–17, 2006.

[22] S. Saouli and S. Aïboud-Saouli, "Second law analysis of laminar falling liquid film along an inclined heated plate," *International Communications in Heat and Mass Transfer*, vol. 31, no. 6, pp. 879–886, 2004.

[23] O. D. Makinde and E. Osalusi, "Entropy generation in a liquid film falling along an inclined porous heated plate," *Mechanics Research Communications*, vol. 33, no. 5, pp. 692–698, 2006.

[24] R. S. R. Gorla and D. M. Pratt, "Second law analysis of a non-Newtonian laminar falling liquid film along an inclined heated plate," *Entropy*, vol. 9, no. 1, pp. 30–41, 2007.

[25] I. Chermiti, N. Hidouri, and A. Ben Brahim, "Entropy generation in gas absorption into a falling liquid film," *Mechanics Research Communications*, vol. 38, no. 8, pp. 586–593, 2011.

[26] S. Aïboud-Saouli, S. Saouli, N. Settou, and N. Meza, "Thermodynamic analysis of gravity-driven liquid film along an inclined heated plate with hydromagnetic and viscous dissipation effects," *Entropy*, vol. 8, no. 4, pp. 188–199, 2006.

[27] S. J. Lee and W. K. Lee, "Physical absorption of oxygen in laminar falling films of dilute polymeric aqueous solutions," *Korean Journal of Chemical Engineering*, vol. 1, no. 1, pp. 51–57, 1984.

[28] P. A. Nikrityuk, *Computational Thermo-Fluid Dynamics: In Material Sciences and Engineering*, John Wiley & Sons, New York, NY, USA, 2011.

[29] N. Gabbiye, *Modeling and simulation of interface mass transfer with chemical reaction [M.S. thesis of Sciences in Chemical Engineering]*, Faculty of Technology, Department of Chemical Engineering, Addis Ababa University, 2004.

[30] R. B. Bird, W. E. Stewart, and E. N. Lightfoot, *Transport Phenomena*, John Wiley & Sons, New York, NY, USA, 2nd edition, 2002.

[31] R. K. Sharma, M. Danish, and S. Ali, "Solution of physical gas absorption in falling liquid film by alternate analytical methods," *Indian Chemical Engineering Journal*, vol. 48, no. 3, pp. 190–193, 2006.

[32] J. R. Bourne, U. V. Stockar, and G. C. Coggan, "Gas absorption with heat effects. I. A new computational method," *Industrial and Engineering Chemistry Process Design and Development* , vol. 13, no. 2, pp. 115–123, 1974.

[33] J. Y. San, W. M. Worek, and Z. Lavan, "Entropy generation in combined heat and mass transfer," *International Journal of Heat and Mass Transfer*, vol. 30, no. 7, pp. 1359–1369, 1987.

[34] J. O. Hirschfelder, C. F. Curtiss, and R. B. Bird, *The Molecular Theory of Gases and Liquids*, John Wiley & Sons, New York, NY, USA, 1954.

[35] I. Mahmoud, *Mass and Heat Transfer during Absorption of Ammonia Vapor into Ammonia-Water Mixture*, Department of Energy and Materials Science, Graduate School of Science and Engineering, Saga University, Saga, Japan, 2004.

16

MHD Stagnation-Point Flow of Casson Fluid and Heat Transfer over a Stretching Sheet with Thermal Radiation

Krishnendu Bhattacharyya

Department of Mathematics, The University of Burdwan, Burdwan, West Bengal 713104, India

Correspondence should be addressed to Krishnendu Bhattacharyya; krish.math@yahoo.com

Academic Editor: Felix Sharipov

The two-dimensional magnetohydrodynamic (MHD) stagnation-point flow of electrically conducting non-Newtonian Casson fluid and heat transfer towards a stretching sheet have been considered. The effect of thermal radiation is also investigated. Implementing similarity transformations, the governing momentum, and energy equations are transformed to self-similar nonlinear ODEs and numerical computations are performed to solve those. The investigation reveals many important aspects of flow and heat transfer. If velocity ratio parameter (B) and magnetic parameter (M) increase, then the velocity boundary layer thickness becomes thinner. On the other hand, for Casson fluid it is found that the velocity boundary layer thickness is larger compared to that of Newtonian fluid. The magnitude of wall skin-friction coefficient reduces with Casson parameter (β). The velocity ratio parameter, Casson parameter, and magnetic parameter also have major effects on temperature distribution. The heat transfer rate is enhanced with increasing values of velocity ratio parameter. The rate of heat transfer is enhanced with increasing magnetic parameter M for $B > 1$ and it decreases with M for $B < 1$. Moreover, the presence of thermal radiation reduces temperature and thermal boundary layer thickness.

1. Introduction

In fluid dynamics the effects of external magnetic field on magnetohydrodynamic (MHD) flow over a stretching sheet are very important due to its applications in many engineering problems, such as glass manufacturing, geophysics, paper production, and purification of crude oil. The flow due to stretching of a flat surface was first investigated by Crane [1]. Pavlov [2] studied the effect of external magnetic field on the MHD flow over a stretching sheet. Andersson [3] discussed the MHD flow of viscous fluid on a stretching sheet and Mukhopadhyay et al. [4] presented the MHD flow and heat transfer over a stretching sheet with variable fluid viscosity. On the other hand, Fang and Zhang [5] reported the exact solution of MHD flow due to a shrinking sheet with wall mass suction. Bhattacharyya and Layek [6] showed the behavior of solute distribution in MHD boundary layer flow past a stretching sheet. Furthermore, many vital properties of MHD flow over stretching sheet were explored in various articles [7–12] in the literature. Several important investigations on the flow due to stretching/shrinking sheet are available in the literature [13–16].

Chiam [17] investigated the stagnation-point flow towards a stretching sheet with the stretching velocity of the plate being equal to the straining velocity of the stagnation-point flow and found no boundary layer structure near the sheet. Mahapatra and Gupta [18] reconsidered the stagnation-point flow problem towards a stretching sheet taking different stretching and straining velocities and they observed two different kinds of boundary layer near the sheet depending on the ratio of the stretching and straining constants. The detailed discussion on the stagnation-point flow over stretching/shrinking sheet can be found in the works of Mahapatra and Gupta [19], Nazar et al. [20], Layek et al. [21], Nadeem et al. [22], Bhattacharyya [23–25], Bhattacharyya et al. [26–28], Bhattacharyya and Vajravelu [29], and Van Gorder et al. [30].

Many fluids used in industries show non-Newtonian behaviour, so the modern-day researchers are more interested in those industrial non-Newtonian fluids and their dynamics. A single constitutive equation is not enough to cover all properties of such non-Newtonian fluids and hence many non-Newtonian fluid models [31–34] have been proposed to clarify all physical behaviours. Casson fluid is one of the types of such non-Newtonian fluids, which behaves like an elastic solid, and for this fluid, a yield shear stress exists in the constitutive equation. Fredrickson [35] investigated the steady flow of a Casson fluid in a tube. The unsteady boundary layer flow and heat transfer of a Casson fluid over a moving flat plate with a parallel free stream were studied by Mustafa et al. [36] and they solved the problem analytically using homotopy analysis method (HAM). Bhattacharyya et al. [37, 38] reported the exact solution for boundary layer flow of Casson fluid over a permeable stretching/shrinking sheet with and without external magnetic field. The important characteristics of the flows of various non-Newtonian fluids over a stretching/shrinking sheet can be found in the articles [39–46].

Motivated by the previously mentioned investigations on flow of non-Newtonian fluids due to a stretching sheet and its vast applications in many industries, in the present paper, the steady two-dimensional MHD stagnation-point flow of electrically conducting non-Newtonian Casson fluid and heat transfer past a stretching sheet in presence of thermal radiation effect are investigated. Using similarity transformations, the governing equations are transformed. The converted self-similar ordinary differential equations are solved by shooting method. The numerical results are plotted in some figures to see the effects of physical parameters on the flow and heat transfer.

2. Mathematical Analysis of the Flow

Consider the steady two-dimensional incompressible flow of electrically conducting Casson fluid bounded by a stretching sheet at $y = 0$, with the flow being confined in $y > 0$. It is also assumed that the rheological equation of state for an isotropic and incompressible flow of a Casson fluid can be written as [37, 47]

$$\tau_{ij} = \begin{cases} \left(\mu_B + \dfrac{p_y}{\sqrt{2\pi}}\right) 2e_{ij}, & \pi > \pi_c, \\ \left(\mu_B + \dfrac{p_y}{\sqrt{2\pi_c}}\right) 2e_{ij}, & \pi < \pi_c, \end{cases} \tag{1}$$

where μ_B is plastic dynamic viscosity of the non-Newtonian fluid, p_y is the yield stress of fluid, π is the product of the component of deformation rate with itself, namely, $\pi = e_{ij}e_{ij}$, e_{ij} is the (i, j)th component of the deformation rate, and π_c is critical value of π based on non-Newtonian model.

Under the previous conditions, the MHD boundary layer equations for steady stagnation-point flow can be written as

$$\frac{\partial u}{\partial x} + \frac{\partial v}{\partial y} = 0, \tag{2}$$

$$u\frac{\partial u}{\partial x} + v\frac{\partial u}{\partial y} = U_s\frac{dU_s}{dx} + \upsilon\left(1 + \frac{1}{\beta}\right)\frac{\partial^2 u}{\partial y^2} - \frac{\sigma H_0^2}{\rho}\left(u - U_s\right), \tag{3}$$

where u and v are the velocity components in x and y directions, respectively, $U_s = ax$ is the straining velocity of the stagnation-point flow with $a\,(>0)$ being the straining constant, υ is the kinematic fluid viscosity, ρ is the fluid density, $\beta = \mu_B\sqrt{2\pi_c}/p_y$ is the non-Newtonian or Casson parameter, σ is the electrical conductivity of the fluid, and H_0 is the strength of magnetic field applied in the y direction, with the induced magnetic field being neglected.

The boundary conditions for the velocity components are

$$u = U_w, \quad v = 0 \quad \text{at } y = 0, \qquad u \longrightarrow U_s \quad \text{as } y \longrightarrow \infty, \tag{4}$$

where $U_w = cx$ is stretching velocity of the sheet with $c\,(>0)$ being the stretching constant.

The stream function Ψ is introduced as

$$u = \frac{\partial \Psi}{\partial y}, \qquad v = -\frac{\partial \Psi}{\partial x}. \tag{5}$$

For relations of (5), (2) is satisfied automatically and (3) takes the following form:

$$\frac{\partial \Psi}{\partial y}\frac{\partial^2 \Psi}{\partial x \partial y} - \frac{\partial \Psi}{\partial x}\frac{\partial^2 \Psi}{\partial y^2} = U_s\frac{dU_s}{dx} + \upsilon\left(1 + \frac{1}{\beta}\right)\frac{\partial^3 \Psi}{\partial y^3} - \frac{\sigma H_0^2}{\rho}\left(\frac{\partial \Psi}{\partial y} - U_s\right). \tag{6}$$

Also, boundary conditions in (4) reduce to

$$\frac{\partial \Psi}{\partial y} = U_w, \quad \frac{\partial \Psi}{\partial x} = 0, \quad \text{at } y = 0, \qquad \frac{\partial \Psi}{\partial y} \longrightarrow U_s \quad \text{as } y \longrightarrow \infty. \tag{7}$$

Now, the dimensionless variable for the stream function is implemented as

$$\Psi = \sqrt{c\upsilon}\,x f(\eta), \tag{8}$$

where the similarity variable η is given by $\eta = y\sqrt{c/\upsilon}$.

Using relation (8) and similarity variable, (6) finally takes the following self-similar form:

$$\left(1 + \frac{1}{\beta}\right) f''' + ff'' - f'^2 - M\left(f' - B\right) + B^2 = 0, \tag{9}$$

where primes denote differentiation with respect to η, $M = \sigma H_0^2/\rho c$ is the magnetic parameter, and $B = a/c$ is the velocity ratio parameter.

The boundary conditions reduce to

$$f(\eta) = 0, \quad f'(\eta) = 1 \quad \text{at } \eta = 0, \qquad f'(\eta) \longrightarrow B \quad \text{as } \eta \longrightarrow \infty. \tag{10}$$

3. Analysis of Heat Transfer

For the temperature distribution in the flow field with thermal radiation, the governing energy equation can be written as

$$u\frac{\partial T}{\partial x} + v\frac{\partial T}{\partial y} = \frac{\kappa}{\rho c_p}\frac{\partial^2 T}{\partial y^2} - \frac{1}{\rho c_p}\frac{\partial q_r}{\partial y}, \tag{11}$$

where T is the temperature, κ is the thermal conductivity, c_p is the specific heat, and q_r is the radiative heat flux.

The appropriate boundary conditions are

$$T = T_w \quad \text{at } y = 0, \qquad T \longrightarrow T_\infty \quad \text{as } y \longrightarrow \infty, \tag{12}$$

where T_w is the constant temperature at the sheet and T_∞ is the free stream temperature assumed to be constant.

Using the Rosseland approximation for radiation [48], $q_r = -(4\sigma^*/3k_1)\partial T^4/\partial y$ is obtained, where σ^* is the Stefan-Boltzmann constant and k_1 is the absorption coefficient. We presume that the temperature variation within the flow is such that T^4 may be expanded in a Taylor's series. Expanding T^4 about T_∞ and neglecting higher-order terms we get $T^4 = 4T_\infty^3 T - 3T_\infty^4$.

Now (11) reduces to

$$u\frac{\partial T}{\partial x} + v\frac{\partial T}{\partial y} = \frac{\kappa}{\rho c_p}\frac{\partial^2 T}{\partial y^2} + \frac{16\sigma T_\infty^3}{3k_1\rho c_p}\frac{\partial^2 T}{\partial y^2}. \tag{13}$$

Next, the dimensionless temperature θ is introduced as

$$\theta(\eta) = \frac{T - T_\infty}{T_w - T_\infty}. \tag{14}$$

Using (8), (14), and the similarity variable, (13) reduces to

$$(3R + 4)\theta'' + 3R\mathrm{Pr} f\theta' = 0, \tag{15}$$

where primes denote differentiation with respect to η, $\mathrm{Pr} = c_p\mu/\kappa$ is the Prandtl number, and $R = \kappa^* k_1/4\sigma T_\infty^3$ is the thermal radiation parameter.

The boundary conditions for θ are obtained from (12) as

$$\theta(\eta) = 1 \quad \text{at } \eta = 0, \qquad \theta(\eta) \longrightarrow 0 \quad \text{as } \eta \longrightarrow \infty. \tag{16}$$

4. Skin Friction Coefficient and Nusselt Number

The physical quantities of interest are the wall skin friction coefficient C_f and the local Nusselt number Nu_x, which are defined as

$$C_f = \frac{\tau_w}{\rho U_w^2(x)}, \qquad \mathrm{Nu}_x = \frac{x q_w}{\alpha(T_w - T_\infty)}, \tag{17}$$

where τ_w is the shear stress or skin friction along the stretching sheet and q_w is the heat flux from the sheet and those are defined as

$$\tau_w = \left(\mu_B + \frac{p_y}{\sqrt{2\pi}}\right)\left(\frac{\partial u}{\partial y}\right)_{y=0}, \qquad q_w = \alpha\left(\frac{\partial T}{\partial y}\right)_{y=0}. \tag{18}$$

Thus, we get the wall skin friction coefficient C_f and the local Nusselt number Nu_x as follows:

$$C_f\sqrt{\mathrm{Re}_x} = \left(1 + \frac{1}{\beta}\right)f''(0), \qquad \frac{\mathrm{Nu}_x}{\sqrt{\mathrm{Re}_x}} = -\theta'(0), \tag{19}$$

where $\mathrm{Re}_x = U_w x/v$ is the local Reynolds number.

5. Numerical Method for Solution

Equations (9) and (15) along with boundary conditions (10) and (16) are solved using shooting method [49–51] by converting them to an initial value problem. In this method, it is necessary to choose a suitable finite value of $\eta \to \infty$, say η_∞. The following system is set

$$f' = p, \qquad p' = q, \qquad q' = \frac{\{p^2 + M(p - B) - fq - B^2\}}{(1 + 1/\beta)}, \qquad \theta' = z, \qquad z' = \frac{3R\mathrm{Pr} fz}{(3R + 4)} \tag{20}$$

with the boundary conditions

$$f(0) = 0, \qquad p(0) = 1, \qquad \theta(0) = 1. \tag{21}$$

In order to integrate (20) with (21) as an initial value problem, the values of $q(0)$, that is, $f''(0)$ and $z(0)$, that is, $\theta'(0)$, are required, but no such values are given in the boundary conditions. The suitable guess values for $f''(0)$ and $\theta'(0)$ are chosen and then integration is carried out. Then, the calculated values for $f'$and θ at $\eta_\infty = 15$ (say) are compared with the given boundary conditions $f'(15) = B$ and $\theta(15) = 0$ and the estimated values, $f''(0)$ and $\theta'(0)$, are adjusted to give a better approximation for the solution. We take the series of values for $f''(0)$ and $\theta'(0)$ and apply the fourth-order classical Runge-Kutta method with step-size $\Delta\eta = 0.01$. The previous procedure is repeated until we get the asymptotically converged results within a tolerance level of 10^{-5}.

6. Results and Discussion

The abovementioned numerical scheme is carried out for various values of physical parameters, namely, the velocity ratio parameter (B), the magnetic parameter (M), the Casson parameter (β), the Prandtl number (Pr), and the thermal radiation parameter (R) to obtain the effects of those parameters on dimensionless velocity and temperature distributions. The obtained computational results are presented graphically in Figures 1–12 and the variations in velocity and temperature are discussed.

Firstly, a comparison of the obtained results with previously published data is performed. The values of wall skin-friction coefficient $f''(0)$ for Newtonian fluid case ($\beta = \infty$)

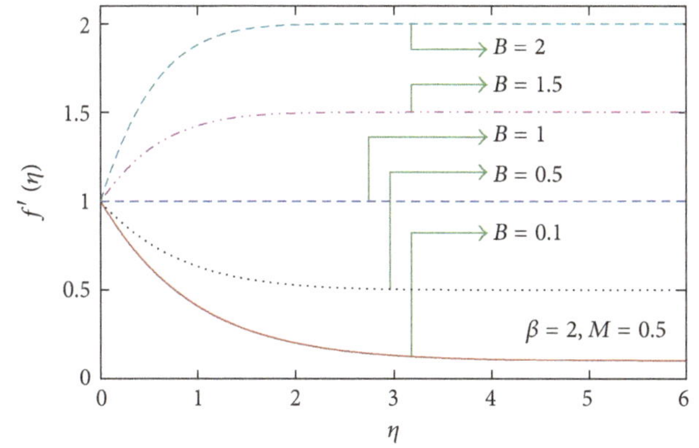

FIGURE 1: Velocity profiles $f(\eta)$ for several values of B.

TABLE 1: Values of $f''(0)$ for several values of B with $M = 0$ and $\beta = \infty$ (Newtonian fluid case without magnetic field).

B	Mahapatra and Gupta [19]	Nazar et al. [20]	Present study
0.1	−0.9694	−0.9694	−0.969386
0.2	−0.9181	−0.9181	−0.918107
0.5	−0.6673	−0.6673	−0.667263
2.0	2.0175	2.0176	2.017503
3.0	4.7293	4.7296	4.729284

TABLE 2: Values of η_δ and $\eta_{\delta T}$ for several values of B, β, M, Pr, and R.

$\beta = 2$, $M = 0.5$, Pr $= 1$, $R = 1$	$B \rightarrow$		0.1	0.5	1	1.5	2.0
	η_δ		5.76	4.81	—	2.59	2.48
	$\eta_{\delta T}$		10.78	6.59	5.04	4.25	3.75
$M = 0.5$, Pr $= 1$, $R = 1$	$\beta \rightarrow$		0.5	1	2	5	∞
	η_δ	$B = 0.1$	8.14	6.65	5.76	5.15	4.70
		$B = 2$	3.50	2.86	2.48	2.22	2.03
	$\eta_{\delta T}$	$B = 0.1$	9.45	10.24	10.78	11.17	11.48
		$B = 2$	3.82	3.78	3.75	3.73	3.72
$\beta = 2$, Pr $= 1$, $R = 1$	$M \rightarrow$		0	0.5	1	2	
	η_δ	$B = 0.1$	6.49	5.76	5.20	4.43	
		$B = 2$	2.53	2.48	2.43	2.33	
	$\eta_{\delta T}$	$B = 0.1$	10.17	10.78	11.21	11.75	
		$B = 2$	3.76	3.75	3.74	3.73	
$\beta = 2$, $M = 0.5$, $R = 1$	Pr $\rightarrow$		0.5	1	2		
	$\eta_{\delta T}$	$B = 0.1$	14.46	10.78	6.69		
		$B = 2$	5.22	3.75	2.70		
$\beta = 2$, $M = 0.5$, Pr $= 1$	$R \rightarrow$		1	2	5		
	$\eta_{\delta T}$	$B = 0.1$	10.78	8.58	7.09		
		$B = 2$	3.75	3.20	2.81		

in the absence of external magnetic field for different values of velocity ratio parameter (B) are compared with those obtained by Mahapatra and Gupta [19], Nazar et al. [20] in Table 1 in order to verify the validity of the numerical scheme used and those are found in excellent agreement.

The velocity boundary layer thickness (δ) and thermal boundary layer thickness (δ_T) are, respectively, described by the equations $\delta = \eta_\delta \sqrt{\upsilon/c}$ and $\delta_T = \eta_{\delta T}\sqrt{\upsilon/c}$. The dimensionless boundary layer thicknesses η_δ and $\eta_{\delta T}$ are defined as the values of η (nondimensional distance from the surface) at which the difference of dimensionless velocity $f'(\eta)$ and the parameter B has been reduced to 0.001 and the dimensionless temperature $\theta(\eta)$ has been decayed to 0.001, respectively. The velocity and thermal boundary layer thicknesses for various parametric values are given in Table 2. The velocity boundary layer thickness decreases with increasing values B (both for $B > 1$ and $B < 1$) and also the thermal boundary layer thickness decreases with increasing B. So, when the straining velocity rate increases compared to that of stretching velocity rate, then both the boundary layer thicknesses reduce. Actually, downward vorticity due to straining velocity causes the reduction of boundary layer thickness. Similar to velocity ratio parameter, the increase of Casson parameter β also makes the velocity boundary layer thickness thinner. So, the velocity boundary layer thickness for Casson fluid is larger than that of Newtonian fluid. Thus, the plasticity of the fluid causes the increment of the velocity boundary layer thickness. On the other hand, for $B = 0.1$ (<1), the thermal boundary layer thickness increases with increasing values of Casson parameter, but for $B = 2$ (>1) the thermal boundary layer thickness decreases with Casson parameter. Furthermore, due to magnetic field, the velocity boundary layer thickness reduces in all cases. But the thermal boundary layer thickness reduces (increases) for $B = 2$ ($B = 0.1$) with stronger magnetic field. Finally, for the Prandtl number and for radiation parameter, the thermal boundary layer thickness decreases, which is the same as that of Newtonian fluid case.

The velocity and temperature profiles for various values of velocity ratio parameter B are plotted in Figures 1 and 2, respectively. Depending on the velocity ratio parameter, two different kinds of boundary layers are obtained as described by Mahapatra and Gupta [18] for Newtonian fluid. In the first kind, the velocity of fluid inside the boundary layer decreases from the surface towards the edge of the layer (for $B < 1$) and in the second kind the fluid velocity increases from the surface towards the edge (for $B > 1$). Those characters can be seen from velocity profiles in Figure 1. Also, it is important to note that if $B = 1$ ($a = c$), that is, the stretching velocity and the straining velocity are equal, then there is no boundary layer of Casson fluid flow near the sheet, which is similar to that of Chiam's [17] observation for Newtonian fluid. From Figure 2, it is seen that in all cases thermal boundary layer is formed and the temperature at a point decreases with B.

The effects of Casson parameter β on the velocity and temperature fields are depicted in Figures 3 and 4. It is worthwhile to note that the velocity increases with the increase in values of β for $B = 2$ and it decreases with β for $B = 0.1$. Consequently, the velocity boundary layer thickness reduces for both values of B. Due to the increase of Casson parameter β, the yield stress p_y falls and consequently

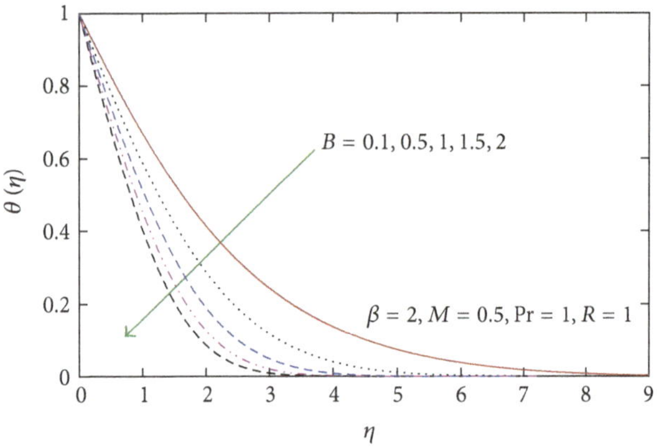

FIGURE 2: Temperature profiles $\theta(\eta)$ for several values of B.

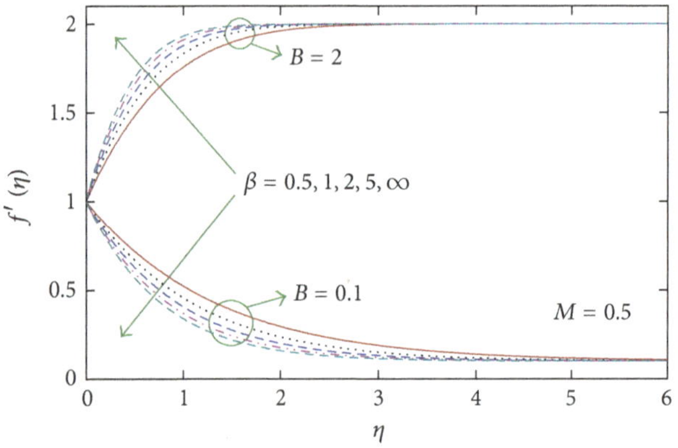

FIGURE 3: Velocity profiles $f(\eta)$ for several values of β.

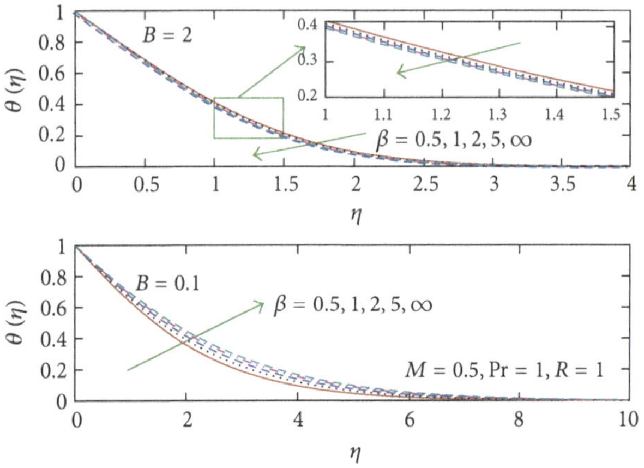

FIGURE 4: Temperature profiles $\theta(\eta)$ for several values of β.

velocity boundary layer thickness decreases. The influences of Casson parameter on the temperature profiles are different in two cases, $B = 2$ and $B = 0.1$. Temperature at a point decreases with increasing β for $B = 2$ and increases with increasing β for $B = 0.1$.

In Figures 5 and 6, the velocity and temperature profiles are presented for several values of magnetic parameter M. Similar to that of Casson parameter, due to the increase of

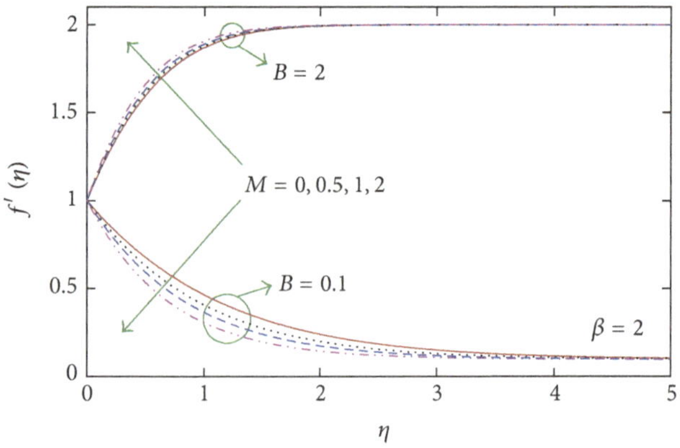

FIGURE 5: Velocity profiles $f(\eta)$ for several values of M.

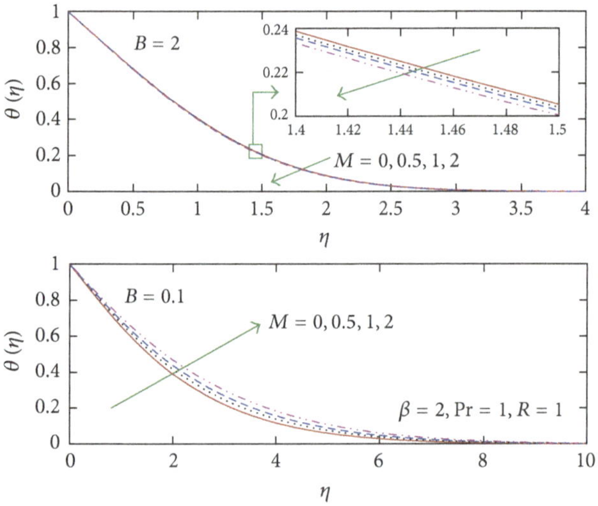

FIGURE 6: Temperature profiles $\theta(\eta)$ for several values of M.

magnetic parameter the dimensionless velocity at fixed η increases for $B = 2$ and for $B = 0.1$ the velocity decreases. Consequently, for both types of boundary layers, the thickness decreases. The Lorentz force induced by the dual actions of electric and magnetic fields reduces the velocity boundary layer thickness by opposing the transport phenomenon. Also, for $B = 2$, the temperature decreases with M and increases with M for $B = 0.1$.

The dimensionless temperature profiles $\theta(\eta)$ for several values of Prandtl Number Pr and thermal radiation parameter R are exhibited in Figures 7 and 8, respectively, for two values of B. In both cases ($B = 0.1$ and 2), the temperature decreases with increasing values of Prandtl number and radiation parameter and the thermal boundary layer thickness becomes smaller in all cases. Actually, the rate of heat transfer is enhanced with Prandtl Number and radiation parameter and this causes the reduction of thermal boundary layer thickness.

The physical quantities, the wall skin friction coefficient C_f, and the local Nusselt number Nu_x, which have immense engineering applications, are proportional to the values of $(1 + 1/\beta)f''(0)$ and $-\theta'(0)$, respectively. The values of (1 +

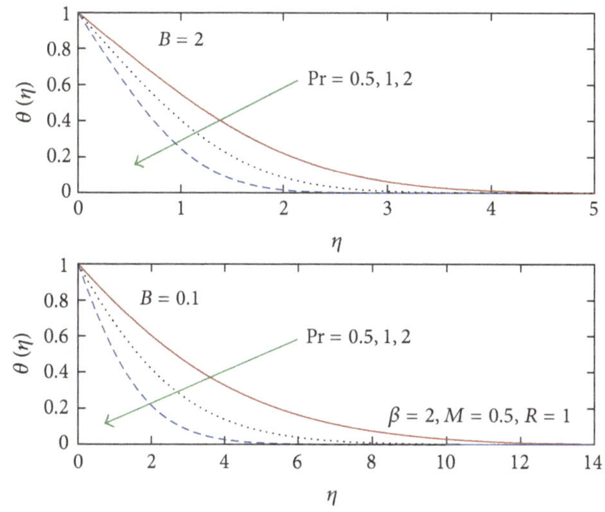

FIGURE 7: Temperature profiles $\theta(\eta)$ for several values of Pr.

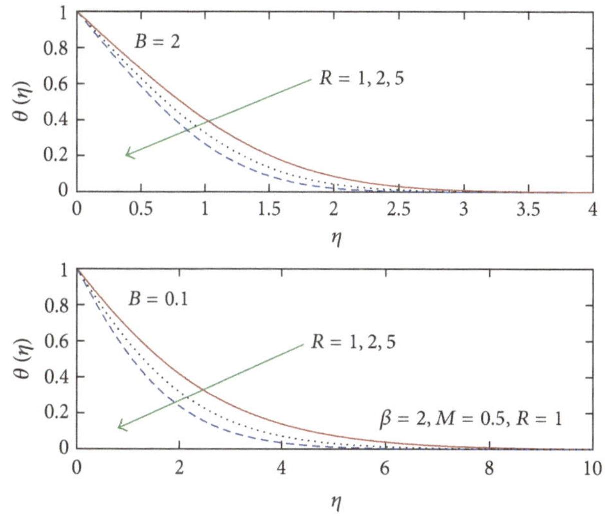

FIGURE 8: Temperature profiles $\theta(\eta)$ for several values of R.

$1/\beta)f''(0)$ and $-\theta'(0)$ against the magnetic parameter M are plotted in Figures 9 and 10 for different values of B. From the figures, it is observed that the magnitude of wall skin friction coefficient decreases with increasing values of velocity ratio parameter B when $B < 0.1$, whereas for $B > 0.1$ the magnitude of skin-friction increases with B. The local Nusselt number (Figure 10) increases with B; that is, the heat transfer rate is enhanced with B. Due to higher values of Casson parameter β, the magnitude of $(1 + 1/\beta)f''(0)$ decreases (Figure 11) for both values of B (for $B > 1$ as well as for $B < 1$). On the other hand, the value of $-\theta'(0)$, that is, the heat transfer (Figure 12), increases with β for $B = 2$ (>1) and decreases with β when $B = 0.1$ (<1). Finally, from those figures (Figures 9–12), it can be noticed that the wall skin friction coefficient always becomes larger when the external magnetic field is stronger and the rate of heat transfer is enhanced (reduced) with increasing magnetic parameter M for $B > 1$ ($B < 1$).

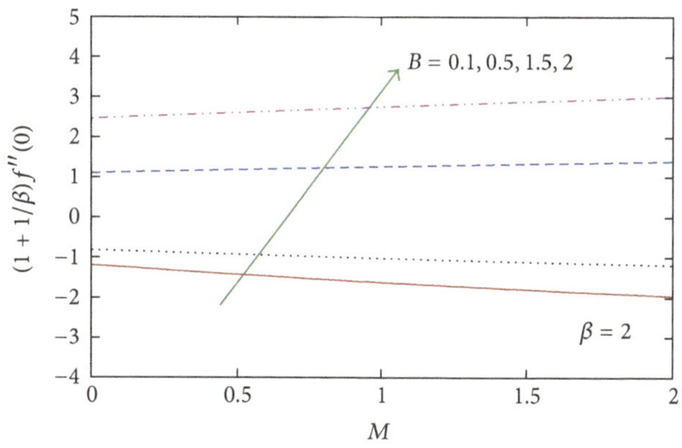

FIGURE 9: The values of $(1+1/\beta)f''(0)$ for several values of B and M.

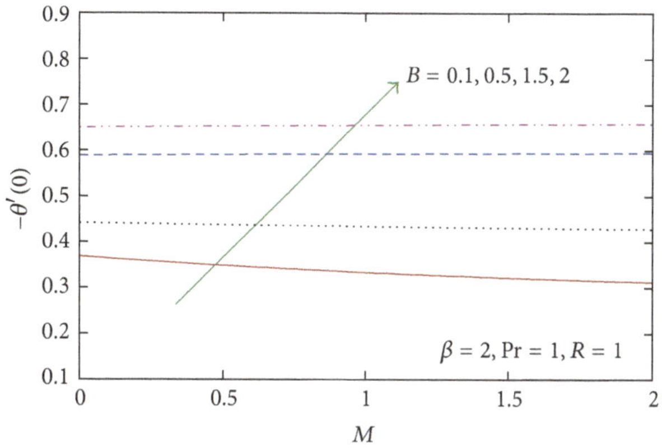

FIGURE 10: The values of $-\theta'(0)$ for several values of B and M.

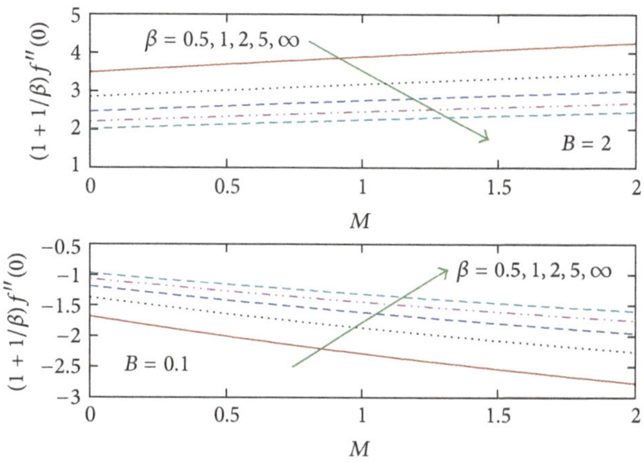

FIGURE 11: The values of $(1 + 1/\beta)f''(0)$ for several values of β and M.

7. Conclusions

The MHD stagnation-point flow of Casson fluid and heat transfer over a stretching sheet are investigated taking into consideration the thermal radiation effect. Using similarity transformations, the governing equations are transformed to self-similar ordinary differential equations which are then

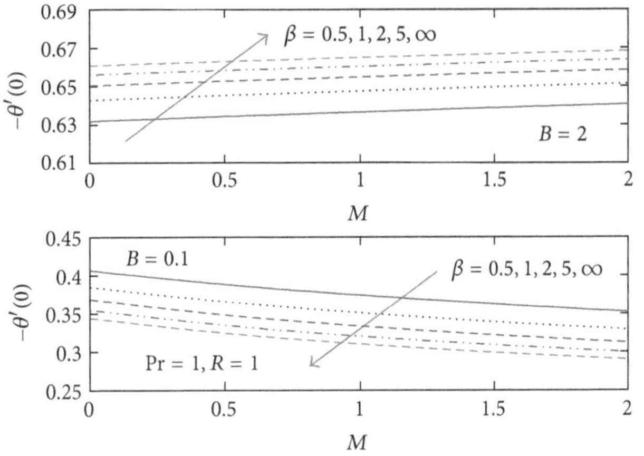

FIGURE 12: The values of $-\theta'(0)$ for several values of β and M.

solved using shooting method. From the study, the following remarks can be summarized.

(a) The velocity boundary layer thickness reduces with velocity ratio parameter and magnetic parameter.

(b) The velocity boundary layer thickness for Casson fluid is larger than that of Newtonian fluid.

(c) For Casson fluid, that is, for decrease of Casson parameter, the thermal boundary layer thickness decreases for $B = 0.1$ (<1) and, in contrast, for $B = 2$ (>1) the thickness increases.

(d) Due to thermal radiation, the temperature inside the boundary layer decreases.

(e) The magnitude of wall skin-friction coefficient decreases with Casson parameter β.

Nomenclature

a: Straining constant
B: Velocity ratio parameter
c: Stretching constant
C_f: Wall skin friction coefficient
c_p: Specific heat
f: Dimensionless stream function
f': Dimensionless velocity
H_0: Strength of magnetic field applied in the y direction
k_1: Absorption coefficient
M: Magnetic parameter
Nu_x: Local Nusselt number
Pr: Prandtl number
p: A variable
p_y: Yield stress of fluid
q: A variable
q_r: Radiative heat flux
q_w: Heat flux from the sheet
R: Thermal radiation parameter
Re_x: Local Reynolds number
T: Temperature
T_w: Constant temperature at the sheet
T_∞: Free stream temperature
U_s: Straining velocity of the stagnation-point flow
U_w: Stretching velocity of the sheet
u: Velocity component in x direction
v: Velocity component in y direction
x: Distance along the sheet
y: Distance perpendicular to the sheet
z: A variable.

Greek Symbols

β: Non-Newtonian/Casson parameter
δ: Velocity boundary layer thickness
δ_T: Thermal boundary layer thickness
η: Similarity variable
η_∞: Finite value of η
η_δ: Dimensionless velocity boundary layer thickness
$\eta_{\delta T}$: Dimensionless thermal boundary layer thickness
κ: Thermal conductivity
μ_B: Plastic dynamic viscosity of the non-Newtonian fluid
π: Product of the component of deformation rate with itself
π_c: Critical value of π
υ: Kinematic fluid viscosity
ρ: Fluid density
Ψ: Stream function
σ: Electrical conductivity of the fluid
σ^*: Stefan-Boltzmann constant
τ_w: Shear stress
θ: Dimensionless temperature.

Conflict of Interests

The author declares that there is no conflict of interests regarding the publication of this paper.

Acknowledgments

The author gratefully acknowledges the financial support of the *National Board for Higher Mathematics* (NBHM), Department of Atomic Energy, Government of India, for pursuing this work. The author also thanks the referees for their valuable comments and suggestions which helped a lot in the improvement of the quality of the paper.

References

[1] L. J. Crane, "Flow past a stretching plate," *Zeitschrift für Angewandte Mathematik und Physik*, vol. 21, no. 4, pp. 645–647, 1970.

[2] K. B. Pavlov, "Magnetohydrodynamic flow of an incompressible viscous fluid caused by the deformation of a plane surface," *Magnetohydrodynamics*, vol. 10, pp. 146–148, 1974.

[3] H. I. Andersson, "MHD flow of a viscoelastic fluid past a stretching surface," *Acta Mechanica*, vol. 95, no. 1-4, pp. 227–230, 1992.

[4] S. Mukhopadhyay, G. C. Layek, and S. A. Samad, "Study of MHD boundary layer flow over a heated stretching sheet with variable viscosity," *International Journal of Heat and Mass Transfer*, vol. 48, no. 21-22, pp. 4460–4466, 2005.

[5] T. Fang and J. Zhang, "Closed-form exact solutions of MHD viscous flow over a shrinking sheet," *Communications in Nonlinear Science and Numerical Simulation*, vol. 14, no. 7, pp. 2853–2857, 2009.

[6] K. Bhattacharyya and G. C. Layek, "Chemically reactive solute distribution in MHD boundary layer flow over a permeable stretching sheet with suction or blowing," *Chemical Engineering Communications*, vol. 197, no. 12, pp. 1527–1540, 2010.

[7] K. Bhattacharyya, "Effects of radiation and heat source/sink on unsteady MHD boundary layer flow and heat transfer over a shrinking sheet with suction/injection," *Frontiers of Chemical Engineering in China*, vol. 5, no. 3, pp. 376–384, 2011.

[8] K. Bhattacharyya, "Effects of heat source/sink on mhd flow and heat transfer over a shrinking sheet with mass suction," *Chemical Engineering Research Bulletin*, vol. 15, no. 1, pp. 12–17, 2011.

[9] K. Bhattacharyya and I. Pop, "MHD Boundary layer flow due to an exponentially shrinking sheet," *Magnetohydrodynamics*, vol. 47, pp. 337–344, 2011.

[10] H. Tabaei, M. A. Moghimi, A. Kimiaeifar, and M. A. Moghimi, "Homotopy analysis and differential quadrature solution of the problem of free-convective magnetohydrodynamic flow over a stretching sheet with the Hall effect and mass transfer taken into account," *Journal of Applied Mechanics and Technical Physics*, vol. 52, no. 4, pp. 624–636, 2011.

[11] R. Kandasamy, I. Muhaimin, and G. Kamachi, "Scaling group transformation for the effect of temperature-dependent nanofluid viscosity on an mhd boundary layer past a porous stretching surface," *Journal of Applied Mechanics and Technical Physics*, vol. 52, no. 6, pp. 931–940, 2011.

[12] A. M. Salem and R. Fathy, "Effects of variable properties on MHD heat and mass transfer flow near a stagnation point towards a stretching sheet in a porous medium with thermal radiation," *Chinese Physics B*, vol. 21, Article ID 054701, 2012.

[13] S. Mukhopadhyay and R. S. R. Gorla, "Effects of partial slip on boundary layer flow past a permeable exponential stretching sheet in presence of thermal radiation," *Heat and Mass Transfer*, vol. 48, pp. 1773–1781, 2012.

[14] K. Bhattacharyya and G. C. Layek, "Slip effect on diffusion of chemically reactive species in boundary layer flow over a vertical stretching sheet with suction or blowing," *Chemical Engineering Communications*, vol. 198, no. 11, pp. 1354–1365, 2011.

[15] K. Bhattacharyya, "Boundary layer flow and heat transfer over an exponentially shrinking sheet," *Chinese Physics Letters*, vol. 28, no. 7, Article ID 074701, 2011.

[16] I. C. Mandal and S. Mukhopadhyay, "Heat transfer analysis for fluid flow over an exponentially stretching porous sheet with surface heat flux in porous medium," *Ain Shams Engineering Journal*, vol. 4, pp. 103–110, 2013.

[17] T. C. Chiam, "Stagnation-point flow towards a stretching plate," *Journal of the Physical Society of Japan*, vol. 63, no. 6, pp. 2443–2444, 1994.

[18] T. R. Mahapatra and A. S. Gupta, "Magnetohydrodynamic stagnation-point flow towards a stretching sheet," *Acta Mechanica*, vol. 152, no. 1-4, pp. 191–196, 2001.

[19] T. R. Mahapatra and A. S. Gupta, "Heat transfer in stagnation-point flow towards a stretching sheet," *Heat and Mass Transfer*, vol. 38, no. 6, pp. 517–521, 2002.

[20] R. Nazar, N. Amin, D. Filip, and I. Pop, "Unsteady boundary layer flow in the region of the stagnation point on a stretching sheet," *International Journal of Engineering Science*, vol. 42, no. 11-12, pp. 1241–1253, 2004.

[21] G. C. Layek, S. Mukhopadhyay, and S. A. Samad, "Heat and mass transfer analysis for boundary layer stagnation-point flow towards a heated porous stretching sheet with heat absorption/generation and suction/blowing," *International Communications in Heat and Mass Transfer*, vol. 34, no. 3, pp. 347–356, 2007.

[22] S. Nadeem, A. Hussain, and M. Khan, "HAM solutions for boundary layer flow in the region of the stagnation point towards a stretching sheet," *Communications in Nonlinear Science and Numerical Simulation*, vol. 15, no. 3, pp. 475–481, 2010.

[23] K. Bhattacharyya, "Dual solutions in boundary layer stagnation-point flow and mass transfer with chemical reaction past a stretching/shrinking sheet," *International Communications in Heat and Mass Transfer*, vol. 38, no. 7, pp. 917–922, 2011.

[24] K. Bhattacharyya, "Dual solutions in unsteady stagnation-point flow over a shrinking sheet," *Chinese Physics Letters*, vol. 28, no. 8, Article ID 084702, 2011.

[25] K. Bhattacharyya, "Heat transfer in unsteady boundary layer stagnation-point flow towards a shrinking sheet," *Ain Shams Engineering Journal*, vol. 4, pp. 259–264, 2013.

[26] K. Bhattacharyya, S. Mukhopadhyay, and G. C. Layek, "Reactive solute transfer in magnetohydrodynamic boundary layer stagnation-point flow over a stretching sheet with suction/blowing," *Chemical Engineering Communications*, vol. 199, no. 3, pp. 368–383, 2012.

[27] K. Bhattacharyya, M. G. Arif, and W. A. Pramanik, "MHD boundary layer stagnation-point flow and mass transfer over a permeable shrinking sheet with suction/blowing and chemical reaction," *Acta Technica*, vol. 57, pp. 1–15, 2012.

[28] K. Bhattacharyya, S. Mukhopadhyay, and G. C. Layek, "Slip effects on an unsteady boundary layer stagnation-point flow and heat transfer towards a stretching sheet," *Chinese Physics Letters*, vol. 28, no. 9, Article ID 094702, 2011.

[29] K. Bhattacharyya and K. Vajravelu, "Stagnation-point flow and heat transfer over an exponentially shrinking sheet," *Communications in Nonlinear Science and Numerical Simulation*, vol. 17, no. 7, pp. 2728–2734, 2012.

[30] R. A. Van Gorder, K. Vajravelu, and I. Pop, "Hydromagnetic stagnation point flow of a viscous fluid over a stretching or shrinking sheet," *Meccanica*, vol. 47, no. 1, pp. 31–50, 2012.

[31] W. L. Wilkinson, "The drainage of a maxwell liquid down a vertical plate," *The Chemical Engineering Journal*, vol. 1, no. 3, pp. 255–257, 1970.

[32] K. R. Rajagopal, "Viscometric flows of third grade fluids," *Mechanics Research Communications*, vol. 7, no. 1, pp. 21–25, 1980.

[33] K. R. Rajagopal, T. Y. Na, and A. S. Gupta, "Flow of a viscoelastic fluid over a stretching sheet," *Rheologica Acta*, vol. 23, no. 2, pp. 213–215, 1984.

[34] C. Dorier and J. Tichy, "Behavior of a bingham-like viscous fluid in lubrication flows," *Journal of Non-Newtonian Fluid Mechanics*, vol. 45, no. 3, pp. 291–310, 1992.

[35] A. G. Fredrickson, *Principles and Applications of Rheology*, Prentice-Hall, Englewood Cliffs, NJ, USA, 1964.

[36] M. Mustafa, T. Hayat, I. Pop, and A. Aziz, "Unsteady boundary layer flow of a Casson fluid due to an impulsively started moving flat plate," *Heat Transfer*, vol. 40, no. 6, pp. 563–576, 2011.

[37] K. Bhattacharyya, T. Hayat, and A. Alsaedi, "Exact solution for boundary layer flow of Casson fluid over a permeable stretching/shrinking sheet," *Zeitschrift für Angewandte Mathematik und Mechanik*, 2013.

[38] K. Bhattacharyya, T. Hayat, and A. Alsaedi, "Analytic solution for magnetohydrodynamic boundary layer flow of Casson fluid over a stretching/shrinking sheet with wall mass transfer," *Chinese Physics B*, vol. 22, Article ID 024702, 2013.

[39] T. Hayat, Z. Abbas, and N. Ali, "MHD flow and mass transfer of a upper-convected Maxwell fluid past a porous shrinking sheet with chemical reaction species," *Physics Letters Section A*, vol. 372, no. 26, pp. 4698–4704, 2008.

[40] T. Hayat, S. Iram, T. Javed, and S. Asghar, "Shrinking flow of second grade fluid in a rotating frame: an analytic solution," *Communications in Nonlinear Science and Numerical Simulation*, vol. 15, no. 10, pp. 2932–2941, 2010.

[41] K.-L. Hsiao, "Viscoelastic fluid over a stretching sheet with electromagnetic effects and nonuniform heat source/sink," *Mathematical Problems in Engineering*, vol. 2010, Article ID 740943, 14 pages, 2010.

[42] B. Sahoo and S. Poncet, "Flow and heat transfer of a third grade fluid past an exponentially stretching sheet with partial slip boundary condition," *International Journal of Heat and Mass Transfer*, vol. 54, no. 23-24, pp. 5010–5019, 2011.

[43] K. Bhattacharyya, S. Mukhopadhyay, G. C. Layek, and I. Pop, "Effects of thermal radiation on micropolar fluid flow and heat transfer over a porous shrinking sheet," *International Journal of Heat and Mass Transfer*, vol. 55, no. 11-12, pp. 2945–2952, 2012.

[44] S. Mukhopadhyay, "Casson fluid flow and heat transfer over a nonlinearly stretching," *Chinese Physics B*, vol. 22, Article ID 074701, 2013.

[45] S. Mukhopadhyay, P. R. De, and G. C. Layek, "Heat transfer characteristics for the Maxwell fluid flow past an unsteady stretching permeable surface embedded in a porous medium with thermal radiation," *Journal of Applied Mechanics and Technical Physics*, vol. 54, pp. 385–396, 2013.

[46] T. Javed, N. Ali, Z. Abbas, and M. Sajid, "Flow of an Eyring-Powell non-Newtonian fluid over a stretching sheet," *Chemical Engineering Communications*, vol. 200, pp. 327–336, 2013.

[47] M. Nakamura and T. Sawada, "Numerical study on the flow of a non-Newtonian fluid through an axisymmetric stenosis," *Journal of Biomechanical Engineering*, vol. 110, no. 2, pp. 137–143, 1988.

[48] M. Q. Brewster, *Thermal Radiative Transfer Properties*, John Wiley and Sons, 1972.

[49] K. Bhattacharyya, S. Mukhopadhyay, and G. C. Layek, "MHD boundary layer slip flow and heat transfer over a flat plate," *Chinese Physics Letters*, vol. 28, no. 2, Article ID 024701, 2011.

[50] S. Mukhopadhyay, K. Bhattacharyya, and G. C. Layek, "Steady boundary layer flow and heat transfer over a porous moving plate in presence of thermal radiation," *International Journal of Heat and Mass Transfer*, vol. 54, no. 13-14, pp. 2751–2757, 2011.

[51] K. Bhattacharyya, "Dual solutions in boundary layer stagnation-point flow and mass transfer with chemical reaction past a stretching/shrinking sheet," *International Communications in Heat and Mass Transfer*, vol. 38, no. 7, pp. 917–922, 2011.

17

A New Cryogenic Air Separation Process with Flash Separator

Zeinab A. M. Khalel, Ali A. Rabah, and Taj Alasfia M. Barakat

Department of Chemical Engineering, Faculty of Engineering, University of Khartoum, P.O. Box 321, Khartoum 11111, Sudan

Correspondence should be addressed to Ali A. Rabah; rabahss@hotmail.com

Academic Editors: R. R. Burnette, T. Chen, P. Espeau, and P. Trens

A new cryogenic air separation process with flash separator is developed. A flash separator is added to the conventional double-column cryogenic air separation process. The flash separator is used to replace the turbine required to recover a portion of the energy in the double-column air separation process. The flash separator served dual purposes of throttling and separation. Both the conventional and the new processes are simulated using Aspen Plus version 11.1 the model air flow rate and compositions are taken as 50000 Nm^3/h of air at standard conditions of 1 atm and 25°C and feed composition of 79.1% N_2 and 20.9% O_2. The new process decreases the energy consumption and increases the productivity.

1. Introduction

There are mainly three methods of air separation, cryogenic distillation, pressure swing absorption, and membrane separation. Although the latter two methods have become more competitive, cryogenic distillation remained the dominant choice for mass production of high-purity O_2. The first cryogenic distillation unit with single-column was built in 1902 and shortly followed by double-column process. The double column consists of compressors, low- and high-pressure columns, and two sets of heat exchangers. The low- and high-pressure columns are thermally interconnected [1].

Although the double-column process is widely used, its high energy consumption remains challenging. Therefore in the recent past, the double column has been subjected to many energy-efficient modifications such as changing the operation condition of the low-pressure column from low to moderate pressure [2]. This modification resulted in smaller size of the column and 10% energy savings. The modification has also the advantage of argon recovery (up to 5% greater argon recovery over the conventional). Another modification is the addition of a heat pump to a side rectifier which is thermally linked to the two columns. The heat pump enhances the separation by providing a supplementary crude argon condensing duty [3]. Reference [4] did an exergy analysis of the double-column air separation process. It found that more than half of the exergy loss takes place in the liquefaction unit and almost one-third in the air compression unit, and minor exergy losses are taking place in the distillation unit and the main heat exchangers. Reference [5] has changed the design of the double column to single column using self-heat recuperation technology (i.e., heat from the top vapor stream is recuperated and exchanged with heat in the bottom liquid and feed streams). The author's modifications have improved the energy consumption by around 36% of the conventional process.

This work is aimed to develop an energy-efficient air separation process. Flash separator has the advantage of working as throttling device on one hand and vapor liquid separation unit on the other hand. It is introduced into the conventional double-column air separation process to replace the turbine needed to expand the air prior to injection to lower-pressure distillation column.

2. Process Description

Figure 1 shows schematic presentation of the conventional double-column cryogenic air separation process. It consists of two compressors (C1 and C2), a turbine (T), two thermally linked distillation columns (low-pressure column (LPC) and high-pressure column (HPC)), and two sets of heat exchangers (HX1, HX2). The air is compressed from atmospheric pressure to 6 atm in C1. The compressed air is then split into

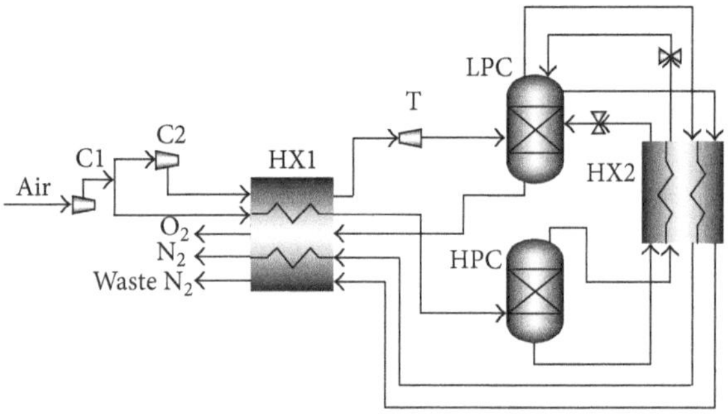

FIGURE 1: Conventional double-column cryogenic air separation process.

two streams; one stream for HPC (hereafter is called HPC split) and the other stream for LPC (hereafter is called LPC split). The LPC split is compressed in C2 further to 7.5 atm, cooled in HX1, throttled in T to 1.3 atm, and finally fed to LPC. The HPC split is cooled in HX1 and then fed to HPC. The products of LPC are pure N_2 (top), pure O_2 (bottom), and waste N_2 (side stream). The products of HPC are pure N_2 (top) and air-enriched oxygen (bottom). The pure N_2 of HPC is working as reflux to the LPC and the air-enriched oxygen of HPC is an extra feed to the LPC. The two columns are thus thermally linked by exchanging the latent heat of vaporization via condenser of the HPC and reboiler of the LPC. The products of the LPC are liquid pure oxygen from the bottom, gas pure nitrogen from the top, and side product of waste nitrogen; all these products are heated in the two sets of heat exchangers prior to be sent as final products.

Figure 2 shows schematic presentation of the new cryogenic air separation process with flash separator. The air is compressed in C3 from atmospheric pressure to 4 atm instead of 6 atm as in the conventional process then split into two streams. The HPC split is compressed in C4 further to 6 atm and subsequently cooled in HX1 and then injected to HPC. Hence this stream is maintained at the same conditions as these in the conventional process. The LPC split is cooled in HX1 and then throttled in a flash separator (F) to 1.3 atm. The bottom product of the flash separator is air-enriched O_2 and the top product is a waste N_2. The waste N_2 is a final product (mixed with the waste nitrogen from the LPC). The bottom product of the flash (air-riched O_2) is fed to the LPC. The rest of the process remains the same as in the conventional process.

Clearly the flash separator replaces the turbine of the conventional double column. Although the turbine is used to recover part of the work introduced into the compressors, the other point to be noticed in this new design is that the LPC split is compressed to 4 atm only rather than 7.5 atm as in the conventional process.

3. Simulation

Both the conventional and the new cryogenic air separation processes are simulated using Aspen Plus version 11.1 with the Peng-Robinson equation of state as a fluid package [6].

TABLE 1: Model air specification.

Parameter	Value	Unit
Flow rate	50000	Nm^3/h
Pressure	1	atm
Temperature	298	°K
N_2	79.1	Mole %
O_2	20.9	Mole %

TABLE 2: Design and operating conditions of distillation columns.

Parameter	Column	
	LPC	HPC
No. of stages	20	15
Reboiler duty	2141 kW	—
Condenser duty	—	2146 kW

The following assumptions are considered in this work.

(1) The model air is assumed to gain 5 K in the processes of drying and cleaning prior to the inlet of the main compressor.

(2) The pressure drop across the heat exchangers is taken as 0.1 atm.

(3) The pressure drop along the HPC and the LPC is 0.1 atm.

(4) The minimum temperature approach is 1.5 K for the HX1 and 2.5 K for HX2.

(5) 5 kW is lost during the thermal integration between the two columns.

The model air flow rate and composition are shown in Table 1.

The compressors C1, C3, and C4 are simulated as multistage compressors (two stages), while C2 is simulated as single stage because the pressure difference is only 1.5 atm. The cooling duty for multistage compressor is neglected as cooling water is assumed to be taken from large reservoir like the electrical energy consumed in pumps. The pumps are considered as auxiliary devices and common in both conventional and new processes. The distillation columns in both processes are simulated by a block known as RadFrac. The two sets of heat exchangers (HX1, HX2) are simulated by block called MHeatX.

4. Results and Discussion

The core of any cryogenic air separation process is distillation unit; here and as mentioned before the HPC split specification remains the same in the conventional and new processes, so the value of the condenser duty of the HPC from the first run is input as the value of the reboiler duty of the LPC. These values are shown in Table 2. Also the number of stages of the HPC and the LPC is adjusted using the composition profile; the number of stages is taken when there is no much difference in the product purities with the increased number of states; the number of stages is given in Table 2.

TABLE 3: Specification of product streams.

Product	Conventional process			New process		
	Flow rate	O_2	N_2	Flow rate	O_2	N_2
O_2 (g)	8038	99.99	0.01	9189	99.98	0.02
N_2 (g)	10602	0.13	99.87	11178	0.13	99.87
Waste N_2 (g)	31360	7.65	92.35	29633	4.21	95.79

The flow rate in Nm^3/h and composition in mole %

TABLE 4: Specification of energy consumption units.

Block	Conventional process			New process		
	Pressure (atm)		Work (kW)	Pressure (atm)		Work (kW)
	In	Out		In	Out	
C1	1	6	4390	—	—	—
C2	6	7.5	79	—	—	—
C3	—	—	—	1	4	3296
C4	—	—	—	4	6	748
Turbine/flash	7.4	1.3	−81	3.9	1.3	—
Net energy consumption			4388 kW			4044 kW
Specific energy consumption			0.55 kW·h/Nm^3 O_2			0.44 kW·h/Nm^3 O_2

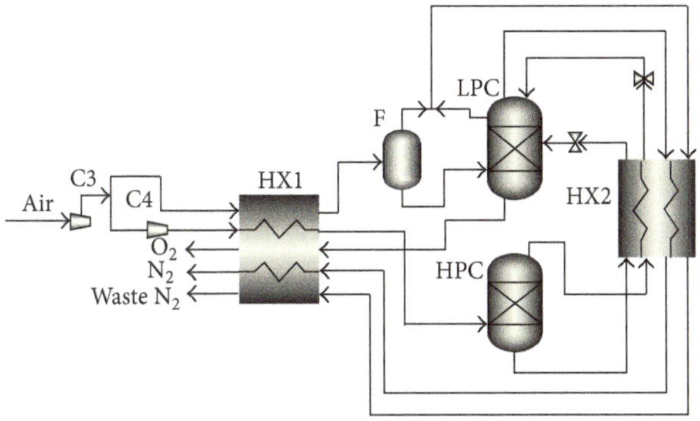

FIGURE 2: New cryogenic air separation process with flash separator.

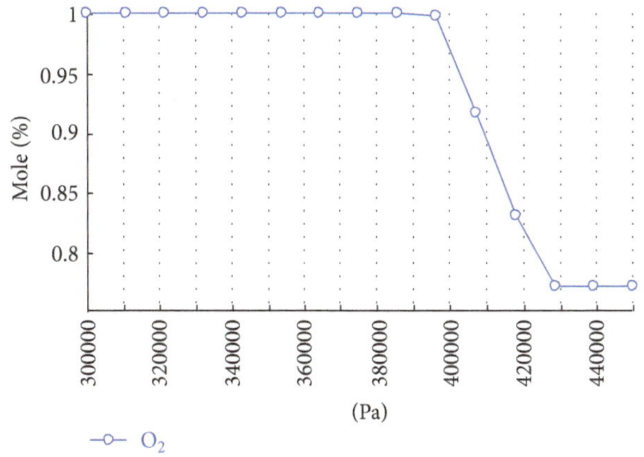

FIGURE 3: The outlet pressure of C3 versus product oxygen purity.

Table 3 shows the simulation result of products flow rates and composition. It can be seen that there is no significant difference in the product purity of both process, but in the new process there is an increase in the flow rate of oxygen by 14.3% and nitrogen by 5.4%, while the flow rate of the waste nitrogen (undesirable product) decreased by 5.5%.

This is attributed to the change in LPC feed. In the new process the feed to LPC is reduced by the early separation of the waste N_2 in the flash separator. Hence the feed is less in quantity and enrich in O_2 than that in the conventional process. This change in the feed quantity and composition is expected to have either influence on the LPC unit specification or the quality of the product. Since the LPC column specifications are maintained similar to those in the conventional process, the feed change is reflected in the products quantity.

Also it is important to mention that in the conventional process the outlet pressure of C2 controlled the pressure drop across the turbine and then the temperature drop across it, and it also affects the amount of recovered energy; on the other hand the outlet pressure of C3 in the new process manipulates the liquid fraction of air to the flash separator and, subsequently, the quantity and composition of the LPC feed from the bottom of the flash separator "the air-enriched oxygen", which is directly reflected on the oxygen flow rate and purity. The effect of the outlet pressure of C3 on product oxygen purity and flow rate is shown in Figures 3 and 4, respectively.

The main consumption of energy in both processes is in the compressors, while other utilities energy consumption like pumps is excluded, because it is almost the same in the new and the conventional processes.

The summary of the energy consumption in the two processes is shown in Table 4. It can be seen that portion

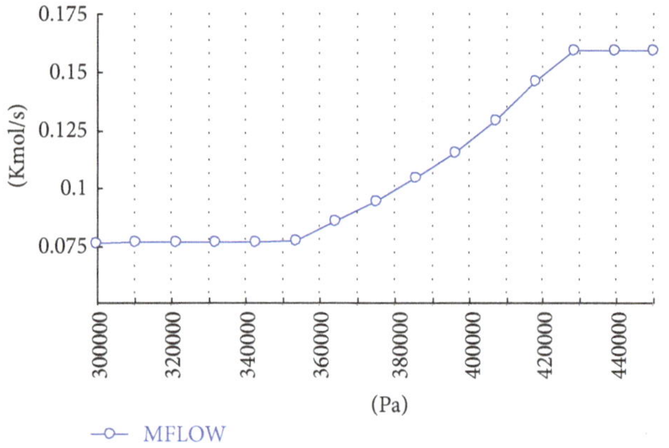

FIGURE 4: The outlet pressure of C3 versus product oxygen flow rate.

of the energy consumed in C1 and C2 of the conventional process is recovered by the turbine; however still the net energy consumption of the new process is lower due to the decrease in the compression duty.

For comparison between two processes, the specific energy consumption (SEC) may be a more meaningful measure than the net energy consumption. The SEC is defined as the net energy consumption per Nm^3/h of the product, that is, Nm^3/h of O_2. Hence SEC for the conventional process and new process is 0.55 kW·h/Nm^3 O_2 and 0.44 kW·h/Nm^3 O_2, respectively, as shown in Table 4. This is equivalent to an enhancement in SEC by about 20%; this decrease is because the new process produces more oxygen and consumes less energy than the conventional process.

5. Conclusion

A new cryogenic air separation process with flash separator is developed and simulated using Aspen Plus software. In the new process the turbine of conventional double column process is replaced by a flash separator. From thermodynamic point of view the new design is better than the conventional design in two aspects: (1) the SEC is lower than that of the conventional by 20% and (2) the pure O_2 produced is increased by 14.3% and pure N_2 is increased by 5.4%. On the other hand flash separator is more reliable than the turbine as it contains no moving parts.

References

[1] F. G. Kerry, *Industrial Gas Handbook: Gas Separation and Purification*, CRC & Taylor & Francis, New York, NY, USA, 2006.

[2] H. Cheung, "Moderate-pressure cryogenic air separation process," *Gas Separation and Purification*, vol. 5, no. 1, pp. 25–28, 1991.

[3] R. Agrawal, D. W. Woodward, and T. F. Yee, "Argon production from air distillation: use of a heat pump in a ternary distillation with a side rectifier," *Gas Separation and Purification*, vol. 8, no. 1, pp. 37–43, 1994.

[4] R. L. Cornelissen and G. G. Hirs, "Exergy analysis of cryogenic air separation," *Energy Conversion and Management*, vol. 39, no. 16–18, pp. 1821–1826, 1998.

[5] Y. Kansha, A. Kishimoto, T. Nakagawa, and A. Tsutsumi, "A novel cryogenic air separation process based on self-heat recuperation," *Separation and Purification Technology*, vol. 77, no. 3, pp. 389–396, 2011.

[6] Aspen Technology, *Aspen Plus User Guide*, Aspen Technology, Cambridge, Mass, USA, 2000.

18

Thermodynamic and Interfacial Properties of DTABr/CTABr Mixed Surfactant Systems in Ethanolamine/Water Mixtures: A Conductometry Study

Olaseni Segun Esan,[1] Medinat Olubunmi Osundiya,[2] Christopher Olumuyiwa Aboluwoye,[1] Owoyomi Olanrewaju,[3] and Jide Ige[3]

[1] *Chemistry Department, Adekunle Ajasin University, Akungba-Akoko 001, Nigeria*
[2] *Chemistry Department, Lagos State University, Ojo Apapa, Lagos 0001, Nigeria*
[3] *Department of Chemistry, Obafemi Awolowo University, Ile Ife 432, Nigeria*

Correspondence should be addressed to Olaseni Segun Esan; lincolnolaseni@yahoo.com

Academic Editors: M. Appell and C. Pierleoni

Mixed-micelle formation in the binary mixtures of dodecyltrimethylammonium bromide (DTABr) and cetyltrimethylammonium bromide (CTABr) surfactants in water-ethanolamine mixed solvent systems has been studied by conductometric method in the temperature range of 298.1 to 313.1 K at 5 K intervals. It was observed that the presence of ethanolamine forced the formation of mixed micelle to lower total surfactant concentration than in water only. The synergistic interaction was quantitatively investigated using the theoretical models of Clint and Rubingh. The interaction parameter β_{12} was negative at all the mole fractions of DTABr in the surfactant mixtures indicating a strong synergistic interaction, with the presence of ethanolamine in the solvent system resulting in a more enhanced synergism in micelle formation than in water only. The free energy of micellization ΔG_M values was more negative in water-ethanolamine mixed solvent system than in pure water indicating more spontaneity in mixed micelle formation in the presence of ethanolamine than in pure water.

1. Introduction

Mixed micelles are widely recognized and employed in numerous fields ranging from biological systems to technical applications such as in detergency, cosmetic, pharmaceuticals, enhanced oil recovery, and as flotation agents. This is because they provide a direct and convenient approach to improve physical stability as well as better chemical and surface active properties over the individual surfactant; hence little quantity is required for any application [1–4]. The physicochemical properties of these systems are dependent on environmental conditions such as the presence of additives, with different additives behaving differently in mixed surfactant systems.

In a given medium (usually, water or binary mixtures of either two nonaqueous solvents or nonaqueous solvent with water), amphiphilic molecules self-aggregate together to minimize the unfavorable interaction with the solvent (medium) and form different types of aggregates (depending on their properties and the other physicochemical conditions which are experienced by them). Hydrophobic or, more generally, solvophobic interactions play an important role in raising the above situation and, therefore, several studies have been made by altering the medium (solvent) properties either by the incorporation of additives [5–10] or by mixing with other solvents [11–19].

The effects of different kinds of additives including glycols, hydrazine, adn formamide on the micellization of single surfactants have been widely studied [20–23]. These solvents are known to have high cohesive energies and considerable hydrogen bonding ability which favor aggregation of surfactant monomers to form micelle. Notwithstanding the recognition of the importance of "water structure" in the micelle formation in aqueous surfactant solutions, no significant attempt seems to have been made to investigate the role of "structuredness" which may exist in many polar organic solvents due to the presence of one or more potential hydrogen bonding centers in their molecules. Addition of alcohols and amine has been reported to have a drastic effect on the micellization of surfactants resulting in a significant change in the critical micelle concentration [22]. Ethanolamine combines the properties of amine and alcohol; it exhibits the unique capability of undergoing reactions common to both groups [23]. It is therefore worthwhile to consider the effect of an organic molecule that combines the properties of alcohol and amine as a single entity on the mixed micelle formation. A literature survey shows that considerable efforts have been made on the study of mixed surfactant systems [24–26] to characterize micelles and their formation in water.

In this work, results of the conductometric study of the physicochemical properties such as micellar composition, counterion binding, interaction parameter, thermodynamic parameters of mixed micelle formation of mixed cationic surfactants in ethanolamine-water binary system using dodecyltrimethyl ammonium bromide (DTABr) and cetyltrimethyl ammonium bromide (CTABr) are presented. The results have been analyzed in light of the existing theories of Clint's for ideal mixing and Rubingh model for nonideal mixing [2, 27]. It may be mentioned that although comprehensive studies on binary mixtures of cationic surfactants are reported in the literature, fundamental investigations on the effect of ethanolamine on micellization have not been mentioned.

2. Experimental

2.1. Materials. Cetyltrimethylammoniumbromide (CTABr) and dodecytrimethylammonium bromide (DTABr) ethanolamine were purchased from Sigma-Aldrich (USA). All the products were used without further purification. All solutions were prepared in doubly distilled water and the experiment was performed between 298.1 and 313.1 K.

2.2. Electrical Conductivity Measurements. The conductivity measurements were made on a DDS-307 conductivity meter (Shanghai Precision Scientific Instrument Co., Ltd.). The conductivity (κ) measurements were made as a function of total surfactant concentration at different mole fraction of DTABr in the DTABr-CTABr mixtures. The mixed solutions were prepared by diluting the concentrated stock solutions and were kept for at least 30 min for equilibration before measuring the conductivity. The conductivity meter was calibrated with different concentrations of solutions of potassium chloride KCl (Merck, purity > 99%). Electrodes were inserted in double walled glass cell containing the solution. The cell constant of the cell used was $1\,\mathrm{cm}^{-1}$. The measurement of conductivity was carried out with an absolute accuracy up to ±3%. The conductivity measurements were made at different temperatures from 298.1 to 313 K for the mixed surfactant/ethanolamine/water system.

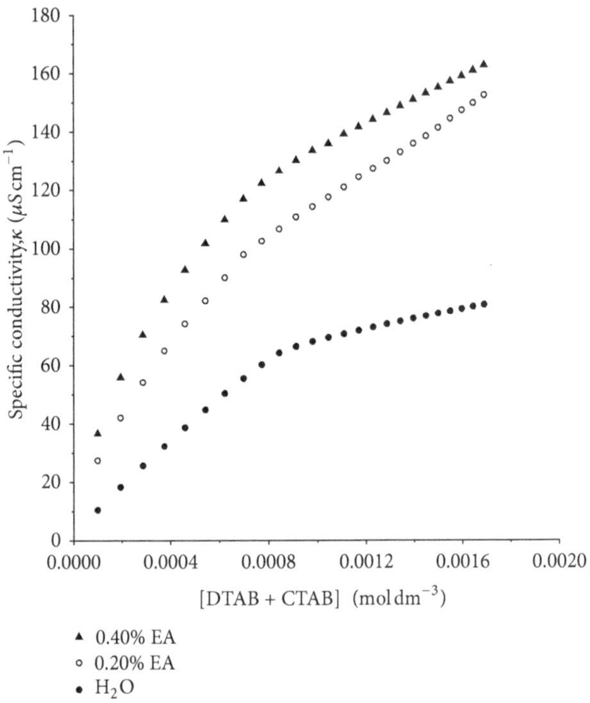

FIGURE 1: Plots of specific conductivity against concentration of DTABr/CTABr in ethanolamine + water mixtures at 298.1 K.

3. Results and Discussion

3.1. Critical Micelle Concentration (CMC). The determination of the CMC and the degree of counter-ion dissociation (binding) for each solvent composition and at different temperatures was carried out from the conductivity-surfactant concentration curve. A number of these plots are shown in Figure 1.

As shown in Figure 1, it is well known that the location of the break point appearing in the plot of specific conductivity versus concentration gives the CMC, while the ratio of the slopes of the plot above and below the CMC gives the degree of dissociation [28]. However, as the concentration of ethanolamine and temperature increased, there appeared a curvature around CMC making the precise determination of CMC difficult. Similar observations are reported in the literature and in most cases have been addressed by either carrying out a nonlinear fit of the raw conductivity-surfactant data or by the analysis of the plots of differential conductivity. This latter approach is adopted in this work, which is based on the assumption that the first derivative of the specific conductivity-concentration curve is well-described by the Boltzmann-type decreasing sigmoid given by [29]

$$\frac{d\kappa}{dc} = A_1 + \frac{A_1 - A_2}{1 + e^{(c-\mathrm{cmc})/d}}, \qquad (1)$$

where κ is the specific conductivity, c is the total concentration of mixed DTABr-CTABr, A_1 and A_2 are the upper and lower limit of the sigmoid, respectively, the CMC value is the center of the sigmoid, and d is the time constant, which is directly related to the independent variable range, where

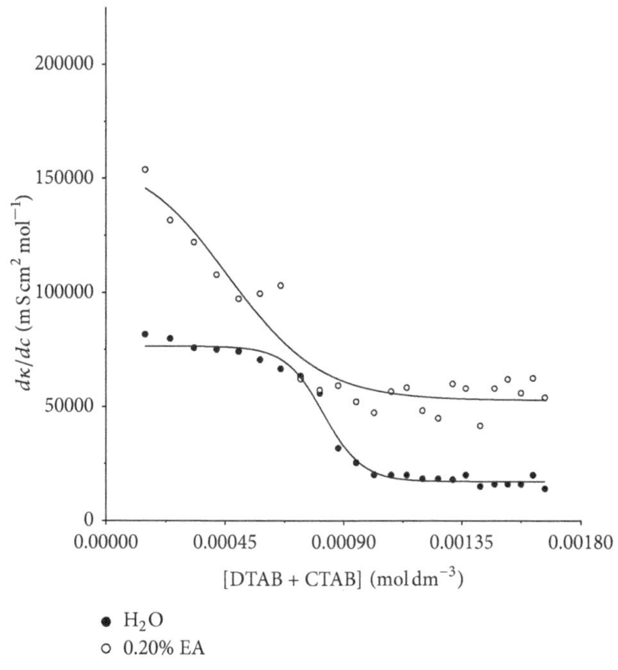

FIGURE 2: Plot of the differential conductivity versus the average concentration of DTABr/CTABr binary system in ethanolamine + water mixture at 298.1 k.

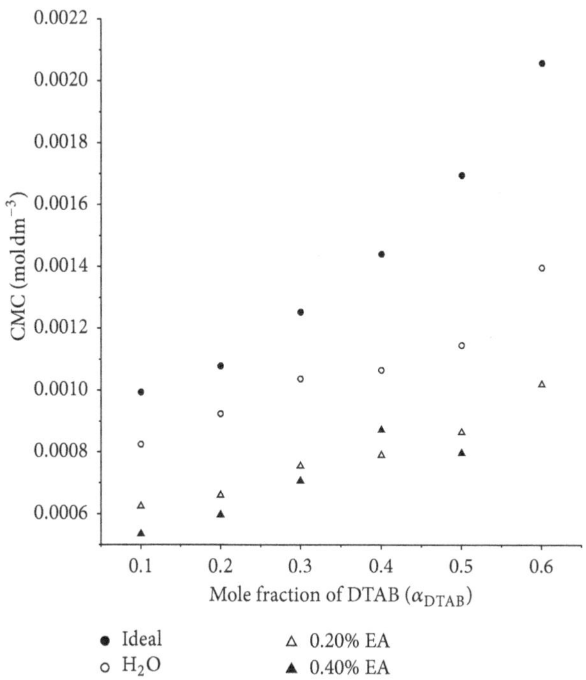

FIGURE 3: The critical micelle concentration of DTABr/CTABr binary system in ethanolamine + water mixture as a function of the mole fraction of DTABr in the mixture.

the sudden change of the dependent variable occurs. Typical plots of differential conductivity versus average surfactant concentration are shown in Figure 2.

3.2. Effects of Ethanolamine on Micellar Behaviour in DTABr-CTABr Mixtures. The critical micelle concentration of the surfactant mixtures at different mole fraction of DTABr in the presence of ethanolamine was found to be significantly lower than the values in pure aqueous system as can be seen in Tables 1, 2, and 3. Also, the values decreased with increase in ethanolamine concentrations from 0.2% to 0.4% at any given mole fraction of DTABr, though the CMC increases along this trend as depicted by Figure 3. This observation can be explained in terms of the reinforcement of water structure through extensive intermolecular hydrogen bond formation due to the presence of the amino nitrogen atom; this of course increases the hydrophobic effect favouring micellization. Hence, a concomitant decrease in CMC was observed in ethanolamine and this increases with increase in concentration of ethanolamine in the mixed solvent systems.

3.3. Surfactant-Surfactant Mixed Micelle in the Absence and in the Presence of Ethanolamine. Like amine additive, addition of ethanolamine may affect micellization of mixed ionic surfactant in two ways. First, the nitrogen atoms bearing lone pair of electron tend to form hydrogen bond with water and break down the "ice-berg" structure formed around the monomer. Secondly, the hydrophobic alkyl portion of the ethanolamine has the tendency to incorporate in micelles, which may also replace some of the water molecules at the surface. This replacement increases the distance between head group and decreases the charge density of the micellar surface. These effects contribute to the micellization of mixed DTABr-CTABr surfactant in water-ethanolamine medium with concomitants decrease in CMC.

Mixed systems comprising of DTABr and CTABr in water-ethanolamine undergo several physico-chemical changes due to surfactant-surfactant and surfactant-solvent interactions resulting in enhanced micellar properties as compared to pure aqueous system. To investigate the mixing behavior between DTABr and CTABr in this mixed solvent system, the Rubingh's regular solution theory was employed [2]. This model is based on regular solution approach for the treatment of a nonideal mixing, and due to its simplicity, it has been mostly used. The nonideality is introduced with the inclusion of the activity coefficient, f_i, in (2); that is,

$$\frac{1}{\text{cmc}^*} = \sum \frac{\alpha_i}{f_i \text{cmc}_i}. \quad (2)$$

In the case of ideal behavior, $f_{\text{DTAB}} = f_{\text{CTABr}} = 1$ and (2) can be reduced to the form (for binary surfactant systems)

$$\frac{1}{\text{cmc}^*} = \frac{\alpha_1}{c_{\text{DTABr}}} + \frac{(1-\alpha_1)}{c_{\text{CTABr}}}, \quad (3)$$

as proposed by Clint's [27] for ideal mixed micelle.

TABLE 1: The degree of counterion binding, thermodynamic function, micellar composition, interaction parameters, and critical micelle concentration of the DTABr + CTABr in aqueous medium.

Temp (K)	α_{DTABr}	β	ΔG^E (J/mol)	ΔH_M (J/mol)	ΔS_M (J/mol·K)	ΔG_M (J/mol)	χ_1	β_{12}	f_{DTABr}	f_{CTABr}	cmc* (mM)
298.1	0.1	0.79	−1081.9	−1081.9	3.17	−2010.9	0.124	−4.02	0.046	0.940	0.824
	0.2	0.68	−893.1	−893.1	3.27	−1867.5	0.134	−3.11	0.097	0.946	0.923
	0.3	0.67	−818.2	−818.2	3.46	−1851.0	0.147	−2.64	0.146	0.944	1.003
	0.4	0.59	−1187.4	−1187.4	4.15	−2427.3	0.200	−2.99	0.147	0.887	1.102
	0.5	0.73	−1402.9	−1402.9	4.57	−2767.0	0.239	−3.12	0.166	0.837	1.150
	0.6	0.65	−1282.1	−1282.1	4.74	−2695.3	0.253	−2.71	0.225	0.836	1.401
303.1	0.1	0.75	−714.4	−714.4	2.52	−1479.1	0.090	−3.45	0.058	0.972	0.871
	0.2	0.68	−909.6	−909.6	3.39	−1937.8	0.142	−2.96	0.112	0.942	0.939
	0.3	0.65	−751.9	−751.9	3.34	−1766.0	0.139	−2.50	0.157	0.953	1.053
	0.4	0.72	−898.7	−898.7	3.83	−2061.4	0.173	−2.49	0.183	0.928	1.151
	0.5	0.70	−1299.9	−1299.9	4.48	−2659.3	0.230	−2.91	0.177	0.857	1.181
	0.6	0.69	−1243.8	−1243.8	4.70	−2668.7	0.253	−2.61	0.237	0.846	1.420
308.1	0.1	0.74	−686.0	−686.0	2.52	−1463.4	0.090	−3.25	0.067	0.973	0.885
	0.2	0.68	−595.6	−595.6	2.76	−1444.5	0.103	−2.51	0.131	0.973	0.983
	0.3	0.63	−415.1	−415.1	2.71	−1250.1	0.100	−1.79	0.234	0.982	1.135
	0.4	0.73	−761.1	−761.1	3.64	−1882.2	0.159	−2.22	0.207	0.945	1.193
	0.5	0.71	−1302.8	−1302.8	4.48	−2684.5	0.230	−2.87	0.182	0.859	1.217
	0.6	0.68	−1117.5	−1117.5	4.60	−2533.8	0.242	−2.38	0.254	0.870	1.484
313.1	0.1	0.74	−217.9	−217.9	1.47	−678.1	0.043	−2.01	0.154	0.996	0.954
	0.2	0.68	−331.8	−331.6	2.16	−1009.4	0.073	−1.89	0.196	0.990	1.033
	0.3	0.64	−325.9	−325.9	2.47	−1100.8	0.088	−1.56	0.272	0.998	1.158
	0.4	0.75	−743.7	−743.7	3.59	−1869.9	0.156	−2.17	0.213	0.948	1.201
	0.5	0.73	−698.1	−698.1	4.28	−2041.1	0.212	−1.61	0.368	0.931	1.267
	0.6	0.72	−1116.6	−1116.6	4.58	−2551.4	0.240	−2.35	0.257	0.873	1.493

*Critical micellar concentration.

Jana and Moulik [26] also proposed that the composition of the mixed micelle and the interaction parameter (β_{12}) can be calculated by employing the following equation:

$$\chi_{DTABr}^2 \ln\left(\frac{\alpha_{DTABr}\text{cmc}^*}{\chi_{DTABr}\cdot \text{cmc}_{DTABr}}\right) = \left(1-\chi_{DTABr}\right)^2 \ln\left(\frac{\left(1-\alpha_{DTABr}\right)\cdot \text{cmc}^*}{\left(1-\chi_{DTABr}\right)\cdot \text{cmc}_{CTABr}}\right), \quad (4)$$

$$\beta_{12} = \frac{\ln\left(\alpha_{DTABr}\text{cmc}^*/\chi_{DTABr}\text{cmc}_{DTABr}\right)}{\left(1-\chi_{DTABr}\right)^2},$$

where in the binary solution

$$f_{DTABr} = \exp\left(\beta\left(1-\chi_i\right)^2\right), \quad f_{CTABr} = \exp\left(\beta\chi_i^2\right). \quad (5)$$

The value of χ_1, β_{12}, f_{DTABr}, and f_{CTABr} for all the system are presented in Tables 1 to 3. It can be seen from Tables 1 to 3 that the β_{12} values are not constant throughout the concentration range, but they are all negative indicating synergism in mixed DTABr-CTABr system. The larger the negative value of β_{12}, the stronger the attractive interaction between the two different surfactant molecules, and the greater the synergism in mixed micelle formation between them [30]. As proposed by Maeda [31], apart from electrostatic interactions, the chain/chain interactions play a major role in the formation of mixed micelle especially for the nonsimilar chain length. Since the hydrocarbon chains of these surfactants are different from each other, a nonideal behavior is expected. Figure 1 represents the plot of experimental CMC and the ideal CMC* versus the mole fraction of DTABr. The experimental CMC value deviates from the ideal CMC* value in the whole mixing range indicating a strong synergism exists in the mixed micelle formation process. It was observed that the experimental data of mixed CMC value are always lower than the ideal CMCs obtained from Clint's equation. Similar interaction was reported earlier by several authors [32].

The activity coefficient f_i shows effect and contribution of individual component in the mixed micelle. The f_{DTABr} values of the DTABr which increased linearly with temperature are obviously very low, indicating that the DTABr in the mixed micelle is very far to its standard state.

As depicted in Table 1, the micellar mole fraction (χ_{DTAB}) is lesser and increases linearly in aqueous medium compared with what is obtained in ethanolamine-water mixture, as shown in Tables 2 and 3 with increasing mole fraction of DTABr. This signifies less surface activity of DTABr rather than micellization in the mixed state while in aqueous medium rather than in ethanolamine medium.

TABLE 2: The degree of counterion binding, thermodynamic function, micellar composition, interaction parameters, and critical micelle concentration of the DTABr + CTABr in 0.2% ethanolamine.

Temp (K)	α_{DTABr}	β	ΔG^E (J/mol)	ΔH_M (J/mol)	ΔS_M (J/mol·K)	ΔG_M (J/mol)	χ_1	β_{12}	f_{DTABr}	f_{CTABr}	cmc* (mM)
298.1	0.1	0.67	−2406.7	−2406.7	4.18	−3625.4	0.202	−6.03	0.021	0.783	0.625
	0.2	0.66	−2333.7	−2333.7	4.45	−3660.0	0.227	−5.37	0.040	0.759	0.667
	0.3	0.56	−2091.9	−2091.9	4.58	−3448.8	0.237	−4.67	0.066	0.769	0.755
	0.4	0.54	−2257.3	−2257.3	4.82	−3694.6	0.267	−4.65	0.081	0.718	0.789
	0.5	0.53	−2333.4	−2333.4	5.02	−3830.2	0.292	−4.55	0.102	0.678	0.865
	0.6	0.50	−2262.3	−2262.3	5.17	−3802.6	0.313	−4.25	0.134	0.660	1.019
303.1	0.1	0.75	−2769.9	−2769.9	4.34	−4084.9	0.216	−6.49	0.019	0.739	0.579
	0.2	0.71	−2607.5	−2607.5	4.56	−3988.3	0.237	−5.72	0.036	0.725	0.618
	0.3	0.63	−2577.3	−2577.3	4.75	−4017.9	0.259	−5.33	0.053	0.699	0.667
	0.4	0.71	−2423.0	−2423.0	4.87	−3899.5	0.273	−4.85	0.077	0.697	0.761
	0.5	0.57	−2480.9	−2480.9	5.06	−4013.3	0.297	−4.72	0.097	0.659	0.835
	0.6	0.53	−2296.2	−2296.2	5.17	−3861.9	0.313	−4.23	0.135	0.660	1.021
308.1	0.1	0.73	−3124.3	−3124.3	4.47	−4500.8	0.228	−6.92	0.016	0.697	0.537
	0.2	0.66	−2973.0	−2973.0	4.69	−4416.6	0.251	−6.17	0.031	0.677	0.571
	0.3	0.64	−2760.1	−2760.1	4.80	−4240.4	0.265	−5.54	0.051	0.678	0.641
	0.4	0.62	−3011.7	−3011.7	5.05	−4565.9	0.295	−5.65	0.060	0.611	0.646
	0.5	0.58	−2757.3	−2757.3	5.12	−4335.5	0.306	−5.07	0.087	0.621	0.775
	0.6	0.56	−2488.2	−2488.2	5.21	−4092.4	0.319	−4.47	0.126	0.634	0.971
313.1	0.1	0.70	−2911.6	−2911.6	4.36	−4276.6	0.218	−6.56	0.018	0.732	0.572
	0.2	0.66	−2766.9	−2766.9	4.59	−4203.6	0.241	−5.82	0.035	0.714	0.607
	0.3	0.59	−2700.0	−2700.0	4.77	−4192.6	0.260	−5.39	0.052	0.694	0.660
	0.4	0.61	−2732.6	−2732.6	4.95	−4282.3	0.283	−5.18	0.069	0.661	0.712
	0.5	0.53	−2716.9	−2716.9	5.09	−4313.6	0.303	−4.92	0.091	0.635	0.796
	0.6	–	–	–	–	–	–	–	–	–	–

*Critical micellar concentration.

3.4. Thermodynamic of Micellization. The regular solution theory (RST) was used to compute the thermodynamic function of mixing which assumes that the excess entropy of mixing is zero. According to this theory, the excess free energy, excess enthalpy, and enthalpy of micellization, ΔH_M are given by

$$G^E = H^E = \Delta H_M = RT\chi_{DTABr} \ln f_{DTABr} + (1 - \chi_{DTABr}) \ln f_{CTABr}. \quad (6)$$

The excess free energy of micellization represents the deviation from the ideal behavior ($G^E = \Delta G_M - \Delta G_M^{ideal}$). For an ideal mixing, the free energy of micellization can be given by

$$\Delta G_M^{ideal} = RT\left(\chi_{DTABr} \ln \chi_{DTABr} + (1 - \chi_{DTABr}) \ln (1 - \chi_{DTABr})\right). \quad (7)$$

The nonideal free energy of micellization is given by

$$\Delta G_M = RT\left(\chi_{DTABr} \ln (\chi_{DTABr} \cdot f_{DTABr}) + (1 - \chi_{DTABr}) \ln ((1 - \chi_{DTABr}) \cdot f_{CTABr})\right). \quad (8)$$

Using the values of enthalpy and free energy of micellization, entropy of micellization can be obtained from

$$\Delta S_M = \frac{\Delta H_M - \Delta G_M}{T}. \quad (9)$$

As reported in Tables 1 to 3, the free energy of micellization (ΔG_M) is all negative and becomes more negative as ethanolamine concentration in the mixed solvent increases over the temperature range of study, suggesting that micellization process occurs less spontaneous in aqueous medium compared with the mixed solvent.

The negative value of ΔH_M implies micellization of DTABr-CTABr in water, and ethanolamine-water mixture is an exothermic process and the tendency becomes more negative with enhanced temperature. Indication from this result shows that the total energy change including translational energy loss by individual and the heat released by the interaction among hydrocarbon bonds is less than the energy absorbed when the iceberg structure of water was destroyed during micellization.

In Tables 1 to 3, the values of ΔS_M are all positive as temperature increases, indicating that the degree of disorderliness increases due to micellization and the entropy change is favorable to the formation of mixed micelle. In aqueous medium, as shown in Table 1, ΔH_M is much less than $T\Delta S_M$, indicating that the enthalpy change contribution to the Gibb's energy of micellization is smaller. This confirmed that the micellization process is entropy driven, while in ethanolamine medium, $-T\Delta S_M$ is less than ΔH_M, indicating that the process is enthalpy driven in the mixed micelle.

TABLE 3: The degree of counterion binding, thermodynamic function, micellar composition, interaction parameters, and critical micelle concentration of the DTABr + CTABr in 0.4% ethanolamine.

Temp (K)	α_{DTABr}	β	ΔG^E (J/mol)	ΔH_M (J/mol)	ΔS_M (J/mol·K)	ΔG_M (J/mol)	χ_1	β_{12}	f_{DTABr}	f_{CTABr}	cmc* (mM)
298.1	0.1	0.78	−3040.6	−3040.6	4.47	−4374.2	0.229	−4.95	0.016	0.694	0.535
	0.2	0.72	−2716.1	−2716.1	4.62	−4094.0	0.244	−5.94	0.033	0.702	0.596
	0.3	0.71	−2329.3	−2329.3	4.66	−3720.3	0.249	−5.03	0.058	0.732	0.706
	0.4	0.63	−1917.7	−1917.7	4.64	−3308.1	0.247	−4.14	0.097	0.774	0.797
	0.5	0.55	−2586.2	−2586.2	5.09	−4106.4	0.303	−4.94	0.091	0.635	0.872
	0.6	—	—	—	—	—	—	—	—	—	—
303.1	0.1	0.67	−3194.3	−3194.3	4.51	−4562.3	0.233	−7.09	0.015	0.680	0.520
	0.2	0.65	−3233.4	−3233.4	4.79	−4685.5	0.263	−6.62	0.027	0.632	0.524
	0.3	0.62	−2513.2	−2513.2	4.72	−3946.2	0.256	−5.24	0.054	0.709	0.679
	0.4	0.58	−2525.1	−2525.1	4.90	−4012.8	0.277	−5.00	0.073	0.681	0.738
	0.5	0.57	−2490.2	−2490.2	5.05	−4023.5	0.297	−4.73	0.097	0.658	0.833
	0.6	—	—	—	—	—	—	—	—	—	—
308.1	0.1	0.71	−3733.6	−3733.6	4.63	−5160.9	0.245	−7.87	0.011	0.623	0.451
	0.2	0.69	−3616.1	−3616.1	4.88	−5121.7	0.275	−7.09	0.024	0.586	0.478
	0.3	0.57	−2879.2	−2879.2	4.84	−4372.2	0.269	−5.71	0.048	0.660	0.620
	0.4	0.54	−3006.2	−3006.2	5.04	−4559.9	0.295	−5.64	0.061	0.611	0.647
	0.5	0.51	−2864.9	−2864.9	5.14	−4451.6	0.310	−5.22	0.083	0.604	0.750
	0.6	—	—	—	—	—	—	—	—	—	—
313.1	0.1	0.75	−3627.1	−3627.1	4.63	−5075.6	0.245	−7.54	0.014	0.636	0.481
	0.2	0.64	−3114.6	−3114.6	4.72	−4591.7	0.255	−6.30	0.030	0.664	0.559
	0.3	0.63	−2764.0	−2764.0	4.79	−4263.8	0.263	−5.48	0.051	0.685	0.660
	0.4	0.61	−2237.9	−2237.9	4.76	−3730.5	0.260	−4.47	0.087	0.739	0.711
	0.5	—	—	—	—	—	—	—	—	—	—
	0.6	—	—	—	—	—	—	—	—	—	—

*Critical micellar concentration.

4. Conclusion

Mixed micellization of DTABr and CTABr was investigated in aqueous and ethanolamine-water mixed solvent systems, and the results of the study allow us to conclude the following.

(1) The observed CMC values of DTABr-CTABr mixture obtained by Rubingh model are lower than those predicted from Clint's model, indicating the presence of nonideality in the mixed micelle formation.

(2) It was observed that changing the mole fraction (α_{DTAB}) over the temperature of studies, the CMC increases, signifying the dependence of CMC of mixed DTABr-CTABr on temperature.

(3) The higher cohesive energy of ethanolamine in relation to water justifies the higher aggregation of DTABr-CTABr, as the concentration of ethanolamine in the solvent system increases.

(4) The negative value of excess free energy (G^E) and interaction parameter (β_{12}) obtained for the binary combination of DTABr and CTABr at different temperature indicate the synergistic behavior and also signify evidence of significant degree of interaction in mixed-micelle formation, which increases as the amount of ethanolamine in the water-ethanolamine mixed solvent system increases.

(5) The negative value of ΔH_M and positive value of ΔS_M over the temperature of studies both in aqueous and ethanolamine medium signify exothermic process and increase degree of disorderliness indicating that entropy change is favorable to the formation of mixed micelle. The ΔG_M values become more negative in the mixed solvent systems than in pure aqueous system indicating that the micellization process is more spontaneous in the mixed solvent system than in the pure aqueous system.

References

[1] M. Akram, N. Azum, A. Z. Naqvi, and K.-U. Kabir-Ud-Din, "Conductometric and fluorimetric investigations on the properties of mixed micelles of two cationic gemini surfactants," *Journal of Chemical and Engineering Data*, vol. 55, no. 11, pp. 4746–4751, 2010.

[2] H. D. Holland and D. Rubingh, "Chapter 1: an overview," in *Mixed Surfactant System*, H. D. Holland and D. Rubingh, Eds., vol. 501 of *ACS Symposium Series*, pp. 2–30, ACS, Washington, DC, USA, 1992.

[3] T. Matsuda, Y. Asoh, M. Villeneuve, H. Matsubara, T. Takiue, and M. Aratono, "Nonideal mixing of dodecyltrimethyl-ammonium halides and nonionic surfactant in adsorbed films and micelles," *Colloid and Polymer Science*, vol. 282, no. 4, pp. 324–329, 2004.

[4] M. Manabe, M. Funamoto, F. Kohgami, H. Kawamura, and H. Katsuura, "Critical composition of ionic-nonionic mixed micelles for counterion condensation," *Colloid and Polymer Science*, vol. 281, no. 3, pp. 239–245, 2003.

[5] D. Myres, *Surfactant Science and Technology*, VCH Publishers, Weinheim, Germany, 1988.

[6] P. K. Misra, B. K. Mishra, and G. B. Behera, "Micellization of ionic surfactants in tetrahydrofuran-water and acetonitrile-water mixed-solvent systems," *Colloids and Surfaces*, vol. 57, no. 1, pp. 1–10, 1991.

[7] M. S. Bakshi, G. Kaur, and G. Kaur, "Hexadecyltrimethylammonium bromide + tetradecyltrimethylammonium bromide mixed micelles in aqueous glycol oligomers," *Journal of Macromolecular Science A*, vol. 36, no. 5-6, pp. 697–718, 1999.

[8] S. Kumar, D. Sharma, and K.-U. Kabir-ud-Din, "Cloud point phenomenon in anionic surfactant + quaternary bromide systems and its variation with additives," *Langmuir*, vol. 16, no. 17, pp. 6821–6824, 2000.

[9] K.-U. Kabir-ud-Din, A. Z. Naqvi, A. B. Khan, M. D. A. Al-Ahmadi, and M. Akram, "Energetics of drug-additive systems at the cloud point," *Journal of Chemical and Engineering Data*, vol. 54, no. 2, pp. 387–391, 2009.

[10] P. R. Desai, N. J. Jain, R. K. Sharma, and P. Bahadur, "Effect of additives on the micellization of PEO/PPO/PEO block copolymer F127 in aqueous solution," *Colloids and Surfaces A*, vol. 178, no. 1–3, pp. 57–69, 2001.

[11] R. Jha and J. C. Ahluwalia, "Thermodynamics of micellization of Triton X-100 in aqueous ethylene glycol solutions," *Journal of Physical Chemistry*, vol. 95, no. 20, pp. 7782–7784, 1991.

[12] K. Mukherjee, D. C. Mukherjee, and S. P. Moulik, "Thermodynamics of micellization of aerosol OT in binary mixtures of water, formamide, ethylene glycol, and dioxane," *Journal of Physical Chemistry*, vol. 98, no. 17, pp. 4713–4718, 1994.

[13] K. Gracie, D. Turner, and R. Palepu, "Thermodynamic properties of micellization of sodium dodecyl sulfate in binary mixtures of ethylene glycol with water," *Canadian Journal of Chemistry*, vol. 74, no. 9, pp. 1616–1625, 1996.

[14] C. C. Ruiz, "Thermodynamics of micellization of tetradecyltrimethylammonium bromide in ethylene glycol-water binary mixtures," *Colloid and Polymer Science*, vol. 277, no. 7, pp. 701–707, 1999.

[15] C. Carnero Ruiz, J. A. Molina-Bolívar, J. Aguiar, G. MacIsaac, S. Moroze, and R. Palepu, "Thermodynamic and structural studies of Triton X-100 micelles in ethylene glycol-water mixed solvents," *Langmuir*, vol. 17, no. 22, pp. 6831–6840, 2001.

[16] C. Carnero Ruiz, J. A. Molina-Bolívar, J. Aguiar, G. MacIsaac, S. Moroze, and R. Palepu, "Effect of ethylene glycol on the thermodynamic and micellar properties of Tween 20," *Colloid and Polymer Science*, vol. 281, no. 6, pp. 531–541, 2003.

[17] M. D. M. Graciani, M. Muñoz, A. Rodríguez, and M. L. Moyá, "Water-N,N-dimethylformamide alkyltrimethylammonium bromide micellar solutions: Thermodynamic, structural, and kinetic studies," *Langmuir*, vol. 21, no. 8, pp. 3303–3310, 2005.

[18] A. Rodríguez, M. Del Mar Graciani, and M. L. Moya, "Effects of addition of polar organic solvents on micellization," *Langmuir*, vol. 24, no. 22, pp. 12785–12792, 2008.

[19] A. Shrivastava and K. K. Ghosh, "Micellization of cetyl triphenyl phosphonium bromide surfactant in binary aqueous solvents," *Journal of Surfactants and Detergents*, vol. 11, no. 4, pp. 287–292, 2008.

[20] R. Jha and J. C. Ahluwalia, "Thermodynamics of micellization of Triton X-100 in aqueous ethylene glycol solutions," *Journal of Physical Chemistry*, vol. 95, no. 20, pp. 7782–7784, 1991.

[21] C. Carnero Ruiz, J. A. Molina-Bolívar, J. Aguiar, G. MacIsaac, S. Moroze, and R. Palepu, "Thermodynamic and structural studies of Triton X-100 micelles in ethylene glycol-water mixed solvents," *Langmuir*, vol. 17, no. 22, pp. 6831–6840, 2001.

[22] B.-Y. Jiang, J. Du, S.-Q. Cheng, Q. Wang, and X.-C. Zeng, "Effects of amine additives on critical micelle concentration of ionic surfactants," *Journal of Dispersion Science and Technology*, vol. 24, no. 6, pp. 755–760, 2003.

[23] Y. Zang, X. Yongshen, and Q. Shouji, "Synthesis and properties of mono or double long-chain alkanolamine surfactant," *Journal of Surfactant and Detergent*, vol. 16, no. 6, pp. 841–848, 2013.

[24] P. K. Jana and S. P. Moulik, "Interaction of bile salts with hexadecyltrimethylammonium bromide and sodium dodecyl sulfate," *Journal of Physical Chemistry*, vol. 95, no. 23, pp. 9525–9532, 1991.

[25] M. J. Rosen, "Chapter 11: molecular interactions and synergism in mixtures of two surfactants," in *Surfactants and Interfacial Phenomena*, Wiley, New York, NY, USA, 1989.

[26] P. K. Jana and S. P. Moulik, "Employment of a useful liquid membrane electrode system to characterise the micelles of bile salts and other detergents in pure and mixed states," *Colloid & Polymer Science*, vol. 272, no. 7, pp. 837–845, 1994.

[27] J. H. Clint, "Mixed micelle theory as an aid to surfactant formulation," in *The Structure, Dynamics and Equilimium Properties of Colloidal Systems*, D. M. Bloor, Ed., p. 7184, Kluwer Academic, London, UK, 1990.

[28] J. Aguiar, J. A. Molina-Bolívar, J. M. Peula-García, and C. Carnero Ruiz, "Thermodynamics and micellar properties of tetradecyltrimethylammonium bromide in formamide-water mixtures," *Journal of Colloid and Interface Science*, vol. 255, no. 2, pp. 382–390, 2002.

[29] K.-U. Kabir-Ud-Din and P. A. Koya, "Effect of acetonitrile on the micellization and thermodynamic parameters of tetradecyltrimethylammonium bromide: conductometric and fluorimetric studies," *Journal of Molecular Liquids*, vol. 158, no. 2, pp. 111–116, 2011.

[30] L.-S. Hao, Y.-T. Deng, L.-S. Zhou, H. Ye, Y.-Q. Nan, and P. Hu, "Mixed micellization and the dissociated margules model for cationic/anionic surfactant systems," *Journal of Physical Chemistry B*, vol. 116, no. 17, pp. 5213–5225, 2012.

[31] H. Maeda, "A simple thermodynamic analysis of the stability of ionic/nonionic mixed micelles," *Journal of Colloid and Interface Science*, vol. 172, no. 1, pp. 98–105, 1995.

[32] S. A. Moore, K. M. Glenn, A. M. MacDonald, and R. M. Palepu, "Micellar and associated thermodynamic properties of binary mixtures of alkyl triphenyl phosphonium bromides in ethylene glycol and water mixtures," *Colloid and Polymer Science*, vol. 285, no. 5, pp. 543–552, 2007.

Second Law Analysis for Third-Grade Fluid with Variable Properties

Samuel O. Adesanya

Department of Mathematical Sciences, Redeemer's University, Redemption City, PMB 2011, Sagamu, Ogun State, Nigeria

Correspondence should be addressed to Samuel O. Adesanya; adesanyaolumide@yahoo.com

Academic Editor: Pedro Jorge Martins Coelho

This paper investigates the entropy generation in a third-grade fluid flow with variable properties through a channel. Approximate solutions to the nonlinear boundary-value problem are obtained using Adomian decomposition method (ADM). Variation of important parameters on the fluid velocity, temperature distribution, entropy generation and irreversibility ratio are presented graphically and discussed.

1. Introduction

Over the past few decades, there has been a tremendous increase in the study of heat transfer to viscous fluids due to its numerous applications in many industrial and engineering processes. As postulated in the second law of thermodynamics, the study could give insight into the thermal performance of the system by improving energy that is available for work [1, 2].

In the class of third-grade fluid, quite a lot has been done in recent times on entropy generation by assuming a constant thermal conductivity. For instance, Pakdemirli and Yilbas investigated the entropy generation in the flow of third-grade fluid through a pipe with constant viscosity in [3] while the temperature dependent viscosity has been investigated in [4] by using Vogel model. We refer interested readers to [5–14] for more interesting results on third-grade fluid flow. Moreover, Makinde and Aziz [15] presented the second law analysis of a pressure-driven temperature dependent fluid flow with asymmetry at the walls. Kahraman and Yürüsoy [16] examined the entropy generation due to non-Newtonian fluid flow in an annular pipe with relative rotation using a third-grade fluid model while Chauhan and Kumar [17] presented a non-Newtonian third-grade fluid flow in an annulus partially filled by a porous medium of very small permeability and many more results on entropy generation in literature.

Surprisingly, in spite of the enormous amount of work done on the entropy generation, it is observed that not much has been done on the exergy analysis of third-grade fluid flow with variable thermal conductivity and internal heat generation. As shown by Hayat et al. [18], these variations can significantly affect the flow field. Hence a more accurate result could be obtained by taking these variations into consideration.

Therefore, the specific objective of this paper is to investigate the entropy generation in the flow of third-grade fluid channel flow with temperature dependent properties. The problem under discussion is strongly nonlinear boundary valued problem. Approximate solution will be obtained using Adomian decomposition method as presented by Siddiqui et al. [19] for the third-grade fluid flow. Therefore, in the absence of internal heat generation and neglecting variations with temperature, the result obtained in [19] will be fully recovered. The seminumerical method has also been used to obtain approximate solution to different fluid flow problems in [20–22]. To the best of the author's knowledge, the study reported here on entropy generation in a heat generating third-grade fluid flow with variable properties has not been undertaken in literature.

The rest of the paper is organized as follows. In the following section, the mathematical analysis of the flow is presented. Section 3 describes the method of obtaining the approximate solution of the nonlinear problem. In Section 4,

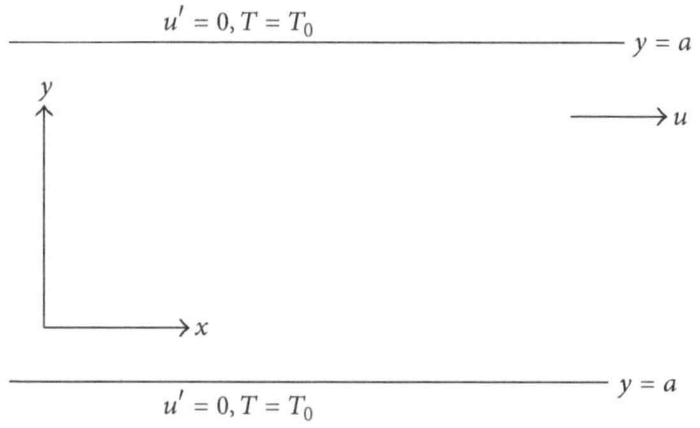

FIGURE 1: Channel geometry.

results are presented and discussed before the paper is concluded in Section 5.

2. Mathematical Analysis

Consider the steady flow of a third-grade fluid with variable viscosity and thermal conductivity through infinite parallel plates of distance $2a$ apart as shown in Figure 1.

The fluid is assumed to be reactive and internal heat generation is assumed to be a linear function of temperature. Then the governing equations for the fully developed flow can be written as [15, 18]

$$0 = -\frac{dP}{dx} + \frac{d}{dy'}\left(\mu'\frac{du'}{dy'}\right) + 6\beta_3\frac{d^2u'}{dy'^2}\left(\frac{du'}{dy'}\right)^2,$$
$$0 = \frac{d}{dy'}\left(k'\frac{dT}{dy'}\right) + \mu'\left(\frac{du'}{dy'}\right)^2 + 2\beta_3\left(\frac{du'}{dy'}\right)^4 + Q_0\left(T - T_0\right), \quad (1)$$

and the entropy generation is given as

$$E_G = \frac{k'}{T_0^2}\left(\frac{dT}{dy'}\right)^2 + \frac{1}{T_0}\left(\frac{du'}{dy'}\right)^2\left(\mu' + 2\beta_3\left(\frac{du'}{dy'}\right)^2\right) \quad (2)$$

with appropriate boundary conditions

$$u' = 0, \quad T = T_0, \quad \text{on } y' = a,$$
$$\frac{du'}{dy'} = 0 = \frac{dT}{dy'}, \quad \text{on } y' = 0; \quad (3)$$

the temperature dependent viscosity and thermal conductivity take the form

$$\mu' = \mu - \eta_0\left(T - T_0\right),$$
$$k' = k + \eta_1\left(T - T_0\right), \quad (4)$$

where u' is the fluid velocity, T is the fluid temperature, P is the fluid pressure, a is the channel half width, β_3 is the material coefficient, k is the thermal conductivity, μ' is the dynamic viscosity, $\eta_{0,1}$ are the viscosity and thermal conductivity variation parameters, Q_0 is the heat generated internally, T_0 and T_1 are referenced fluid temperature and E_G is the entropy generation.

Introducing the following dimensionless parameters and variables:

$$y = \frac{y'}{a}, \quad u = \frac{u'}{U}, \quad \gamma = \frac{\beta_3 U^2}{a^2\mu},$$
$$\theta = \frac{T - T_0}{T_1 - T_0}, \quad \alpha = \frac{\eta_0\left(T_1 - T_0\right)}{\mu}, \quad \Omega = \frac{T_1 - T_0}{T_0},$$
$$G = -\frac{a^2}{\mu U}\frac{dP}{dx}, \quad \delta = \frac{Q_0 a^2}{k}, \quad \text{Br} = \frac{\mu U^2}{k_0\left(T_1 - T_0\right)}, \quad (5)$$
$$\lambda = \frac{\eta_1\left(T_1 - T_0\right)}{k}, \quad N_s = \frac{T_0^2 a^2 E_G}{k\left(T_1 - T_0\right)^2},$$

we get the following dimensionless equations together with appropriate boundary conditions:

$$\frac{d^2u}{dy^2} = \alpha\left[\theta\frac{d^2u}{dy^2} + \frac{du}{dy}\frac{d\theta}{dy}\right] - 6\gamma\frac{d^2u}{dy^2}\left(\frac{du}{dy}\right)^2 - G;$$
$$u\left(\pm 1\right) = 0,$$
$$\frac{d^2\theta}{dy^2} = -\lambda\left[\left(\frac{d\theta}{dy}\right)^2 + \theta\frac{d^2\theta}{dy^2}\right] \quad (6)$$
$$- \text{Br}\left(\frac{du}{dy}\right)^2\left\{\left(1 - \alpha\theta\right) + 2\gamma\left(\frac{du}{dy}\right)^2\right\} + \delta\theta;$$
$$\theta\left(\pm 1\right) = 0,$$

$$N_s = \left(1 + \lambda\theta\right)\left(\frac{d\theta}{dy}\right)^2 + \frac{\text{Br}}{\Omega}\left(\frac{du}{dy}\right)^2\left\{\left(1 - \alpha\theta\right) + 2\gamma\left(\frac{du}{dy}\right)^2\right\}. \quad (7)$$

If we define

$$N_1 = \left(1 + \lambda\theta\right)\left(\frac{d\theta}{dy}\right)^2,$$
$$N_2 = \frac{\text{Br}}{\Omega}\left(\frac{du}{dy}\right)^2\left\{\left(1 - \alpha\theta\right) + 2\gamma\left(\frac{du}{dy}\right)^2\right\}, \quad (8)$$

then the irreversibility ratio denoted by Bejan number (Be) can be written as

$$\text{Be} = \frac{N_1}{N_s} = \frac{N_1}{N_1 + N_2} = \frac{1}{1 + \Phi}, \quad \Phi = \frac{N_2}{N_1}, \quad (9)$$

here u is the dimensionless velocity, α is the viscosity variation parameter, γ is the third-grade material effect, G is the pressure gradient, θ is the dimensionless fluid temperature, λ is the thermal conductivity variation parameter, Br is Brinkman number, δ is the internal heat generation parameter, Ω is the parameter that measures the temperature difference between the two heat reservoirs, N_s is the dimensionless entropy generation rate, and U represents the characteristic velocity.

Table 1: Convergence result for $\gamma = 0.2, \delta = 1, \mathrm{Br} = 0.4, \alpha = 0.5 = \lambda, y = 0.1$.

n	u_n	$\sum_{n=0}^{m} u_n$	θ_n	$\sum_{n=0}^{m} \theta_n$
0	0.445252	0.445252	0.0507153	0.0507153
1	−0.000116788	0.445135	−0.000256831	0.0504585
2	-1.96106×10^{-6}	0.445134	6.56661×10^{-6}	0.0504650
3	-2.58151×10^{-8}	0.445133	-1.93224×10^{-7}	0.0504648
4	1.74273×10^{-10}	0.445133	5.79716×10^{-9}	0.0504648

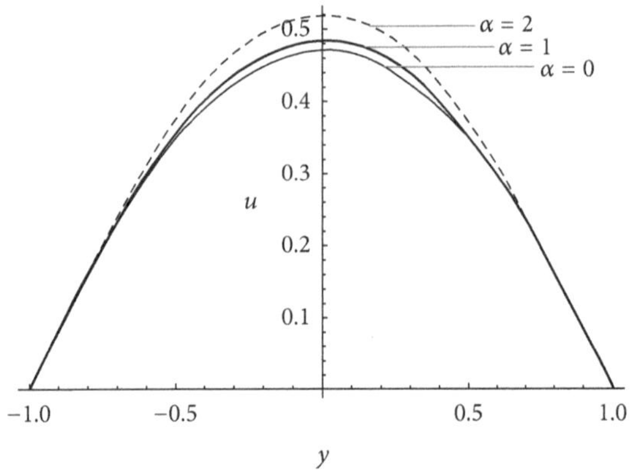

Figure 2: Effect of viscosity variation parameter on velocity profile.

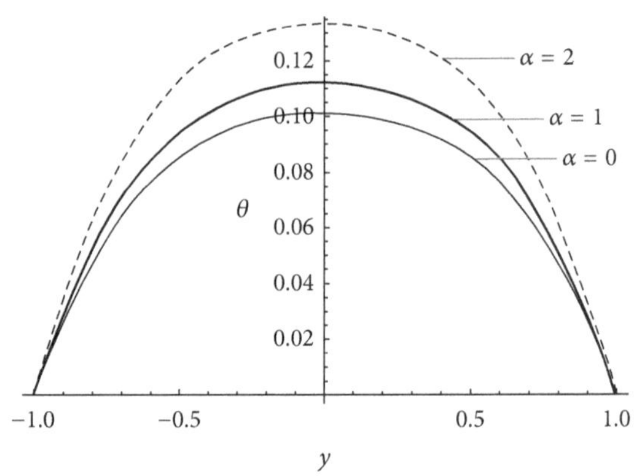

Figure 3: Effect of viscosity variation parameter on temperature profile.

3. Adomian Decomposition Method of Solution

The integral form of (6) can be written as

$$u(y) = a_0 + \iint_0^y \left\{ \alpha \left[\theta \frac{d^2u}{dY^2} + \frac{du}{dY}\frac{d\theta}{dY} \right] - 6\gamma \frac{d^2u}{dY^2}\left(\frac{du}{dY}\right)^2 - G \right\} dY\, dY,$$

$$\theta(y) = b_0 - \iint_0^y \left\{ \lambda \left[\left(\frac{d\theta}{dY}\right)^2 + \theta \frac{d^2\theta}{dY^2} \right] + \mathrm{Br}\left(\frac{du}{dY}\right)^2 \times \left\{ (1 - \alpha\theta) + 2\gamma \left(\frac{du}{dY}\right)^2 \right\} + \delta\theta \right\} dY\, dY, \quad (10)$$

where the constants a_0 and b_0 are to be determined later; the standard Adomian decomposition method [19–21] assumes series solutions in the form

$$u = \sum_{n=0}^{\infty} u_n, \qquad \theta = \sum_{n=0}^{\infty} \theta_n; \quad (11)$$

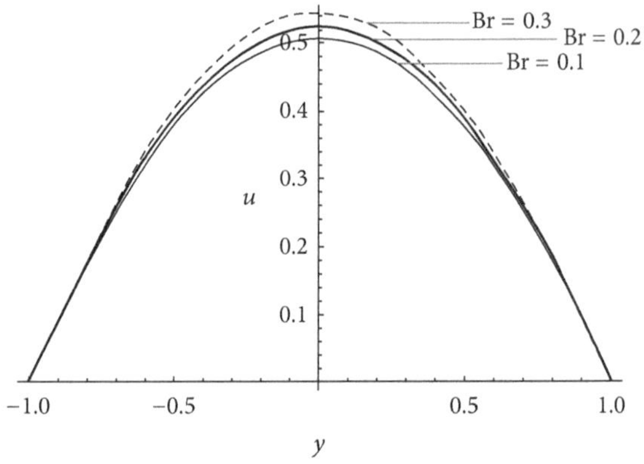

Figure 4: Effect of Brinkman number on velocity profile.

substituting (11) into the integral equations (10) yields the following recursive algorithm:

$$\theta_0(y) = b_0,$$

$$u_0(y) = a_0 - \iint_0^y (G)\, dY\, dY,$$

$$u_{n+1}(y) = \iint_0^y \{\alpha A_n - 6\gamma B_n\}\, dY\, dY, \quad n \geq 1,$$

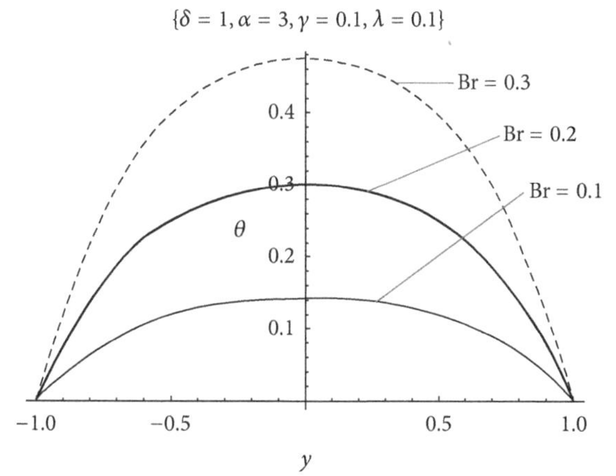

FIGURE 5: Effect of Brinkman number on temperature profile.

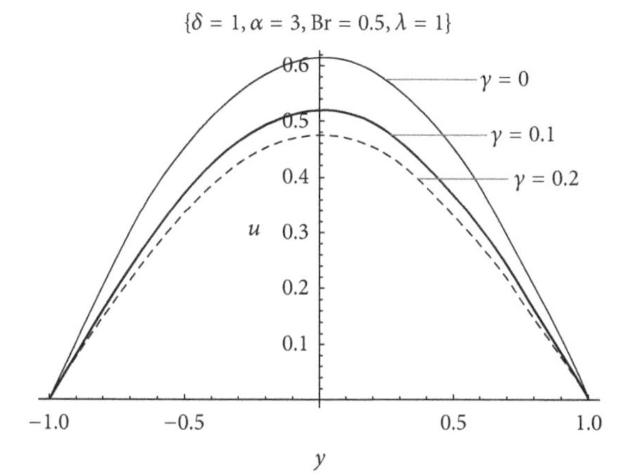

FIGURE 6: Effect of non-Newtonian material on velocity profile.

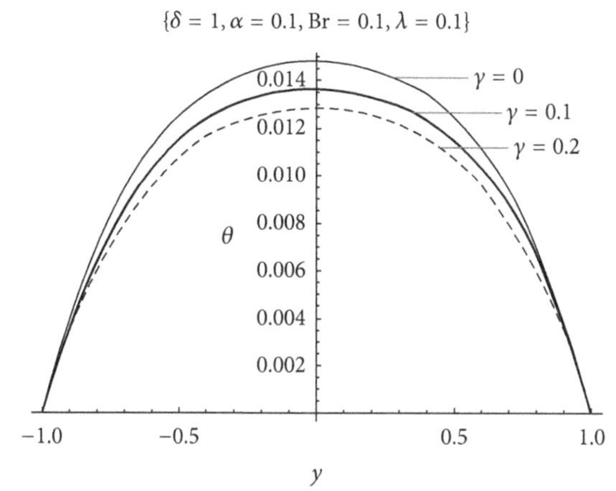

FIGURE 7: Effect of non-Newtonian material on temperature profile.

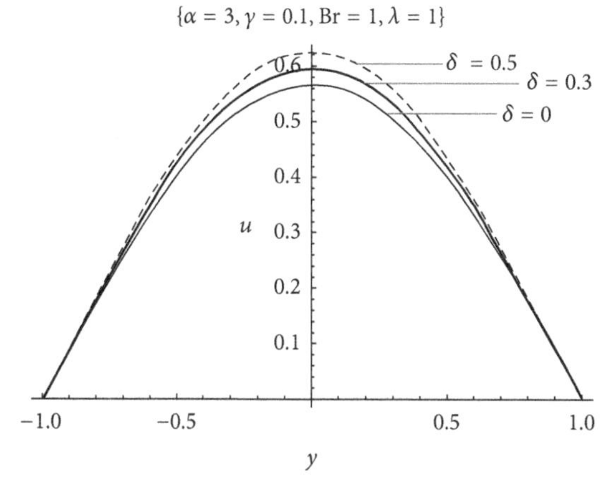

FIGURE 8: Effect of internal heat generation parameter on velocity profile.

$$\theta_{n+1}(y) = -\iint_0^y [\lambda C_n + \mathrm{Br} F_n \{(1-\alpha\theta_n) + 2\gamma F_n\} + \delta\theta_n]\, dY\, dY; \quad n \geq 1; \tag{12}$$

the approximate solution is given by the partial sum

$$\theta = \sum_{n=0}^{m} \theta_n, \qquad u = \sum_{n=0}^{m} u_n, \tag{13}$$

where m represent the truncation point. The nonlinear terms in (12) represented by

$$A_n = \frac{du_n}{dY}\frac{d\theta_n}{dY} + \theta_n \frac{d^2 u_n}{dY^2}, \qquad B_n = \frac{d^2 u_n}{dY^2}\left(\frac{du_n}{dY}\right)^2,$$
$$C_n = \left(\frac{d\theta_n}{dY}\right)^2 + \theta_n \frac{d^2\theta_n}{dY^2}, \qquad F_n = \left(\frac{du_n}{dY}\right)^2 \tag{14}$$

are decomposed into Adomian polynomials as follows:

$$\begin{aligned}
A_0 &= \frac{d\theta_0}{dY}\frac{du_0}{dY} + \theta_0 \frac{d^2 u_0}{dY^2}, \\
A_1 &= \frac{d\theta_0}{dY}\frac{du_1}{dY} + \frac{d\theta_1}{dY}\frac{du_0}{dY} + \theta_1 \frac{d^2 u_0}{dY^2} + \theta_0 \frac{d^2 u_1}{dY^2}, \\
A_2 &= \frac{d\theta_0}{dY}\frac{du_2}{dY} + \frac{d\theta_1}{dY}\frac{du_1}{dY} + \frac{d\theta_2}{dY}\frac{du_0}{dY} + \theta_2 \frac{d^2 u_0}{dY^2} \\
&\quad + \theta_1 \frac{d^2 u_1}{dY^2} + \theta_0 \frac{d^2 u_2}{dY^2},
\end{aligned} \tag{15}$$

similarly,

$$\begin{aligned}
B_0 &= \frac{d^2 u_0}{dY^2}\left(\frac{du_0}{dY}\right)^2, \\
B_1 &= 2\frac{d^2 u_0}{dY^2}\left(\frac{du_0}{dY}\right)\left(\frac{du_1}{dY}\right) + \frac{d^2 u_1}{dY^2}\left(\frac{du_0}{dY}\right)^2, \\
B_2 &= \frac{d^2 u_0}{dY^2}\left(\frac{du_1}{dY}\right)^2 + 2\frac{d^2 u_0}{dY^2}\left(\frac{du_0}{dY}\right)\left(\frac{du_2}{dY}\right) \\
&\quad + 2\frac{d^2 u_1}{dY^2}\left(\frac{du_0}{dY}\right)\left(\frac{du_1}{dY}\right) + \frac{d^2 u_2}{dY^2}\left(\frac{du_0}{dY}\right)^2,
\end{aligned} \tag{16}$$

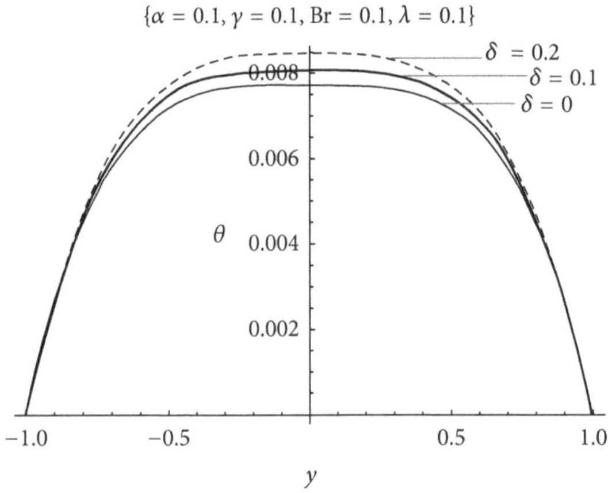

FIGURE 9: Effect of internal heat generation parameter on temperature profile.

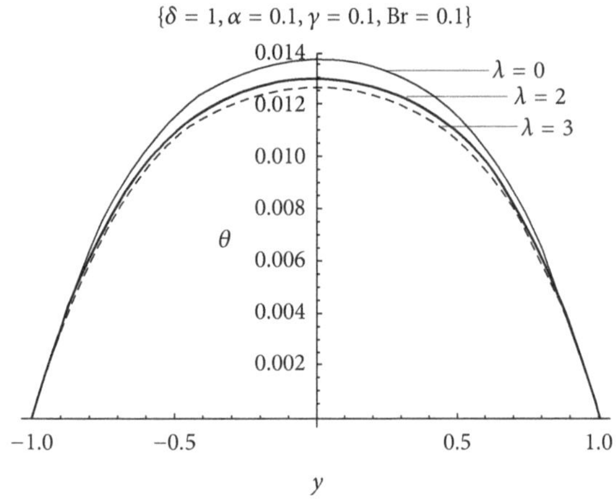

FIGURE 11: Effect of thermal conductivity variation parameter on temperature profile.

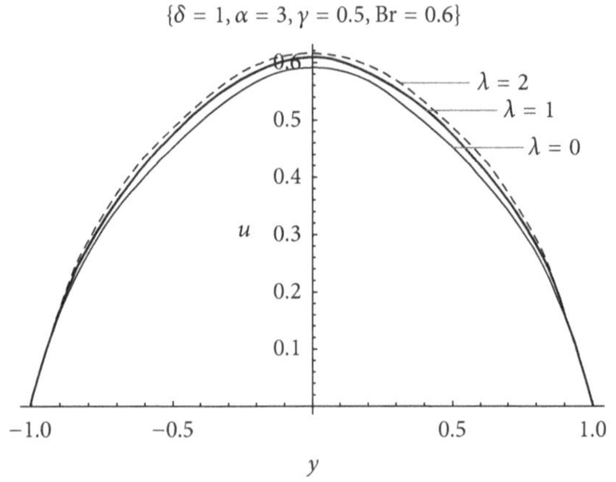

FIGURE 10: Effect of thermal conductivity variation parameter on velocity profile.

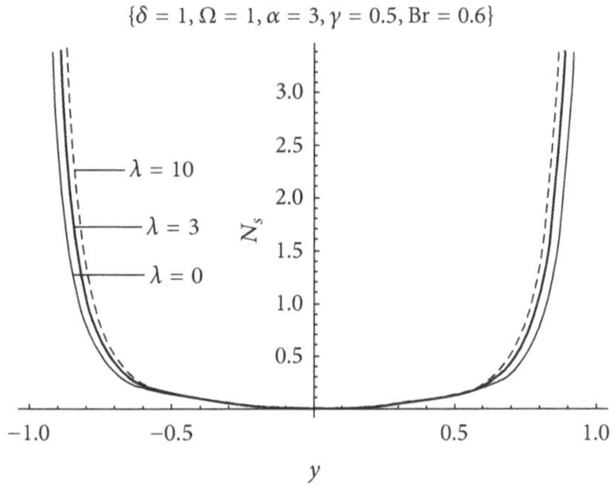

FIGURE 12: Effect of thermal conductivity variation parameter on entropy generation rate.

with

$$F_0 = \left(\frac{du_0}{dY}\right)^2,$$

$$F_1 = 2\frac{du_0}{dY}\frac{du_1}{dY}, \quad (17)$$

$$F_2 = 2\frac{du_0}{dY}\frac{du_2}{dY} + \left(\frac{du_1}{dY}\right)^2, \ldots,$$

$$C_0 = \left(\frac{d\theta_0}{dY}\right)^2 + \theta\frac{d^2\theta_0}{dY^2},$$

$$C_1 = 2\frac{d\theta_0}{dY}\frac{d\theta_1}{dY} + \theta_0\frac{d^2\theta_1}{dY^2} + \theta_1\frac{d^2\theta_0}{dY^2},$$

$$C_2 = 2\frac{d\theta_0}{dY}\frac{d\theta_2}{dY} + \left(\frac{d\theta_1}{dY}\right)^2 + \theta_2\frac{d^2\theta_0}{dY^2} + \theta_1\frac{d^2\theta_1}{dY^2} + \theta_0\frac{d^2\theta_2}{dY^2}, \ldots. \quad (18)$$

4. Results and Discussion

The major objective of this work is to investigate the effects of temperature dependent viscosity and thermal conductivity on heat generating third-grade fluid flow through a channel with isothermal temperature. Approximate solutions to the nonlinear boundary-value problem are obtained in the form of a rapidly convergent Adomian series solution. Table 1 confirmed that series solution is convergent and reliable. Figure 2 shows the velocity profile for variations in the viscosity variation parameter. As observed from the graph, an increase in the viscosity variation parameter enhances the flow. This is physically true due to the melting effect on the fluid viscosity. Similarly, Figure 3 shows that an increase in the viscosity variation parameter increases the temperature profile due to rise in the temperature difference within the channel. Figures 4 and 5 show the velocity and temperature profiles, respectively, for variation in Brinkman number. From the plots it is observed that an increase in the Brinkman number leads to an increase in both the fluid velocity and temperature due to rise in the kinetic energy

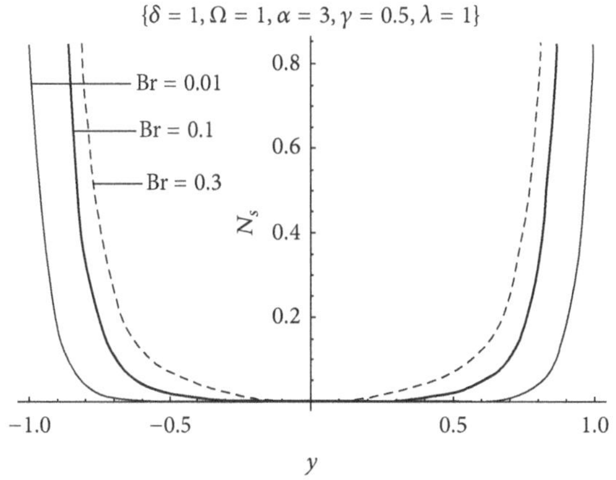

FIGURE 13: Effect of Brinkman number on entropy generation rate.

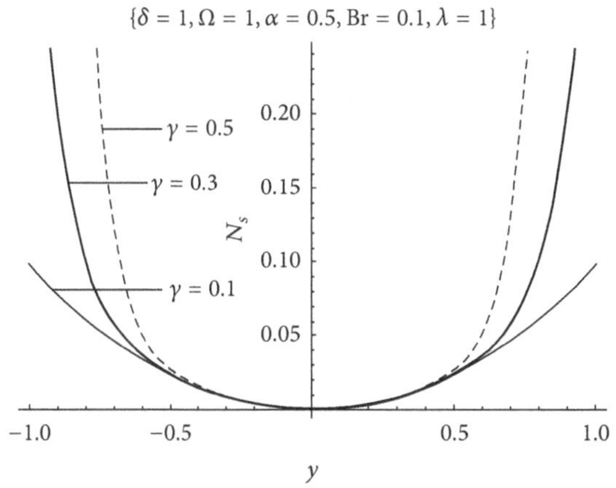

FIGURE 14: Effect of third-grade material effect on entropy generation rate.

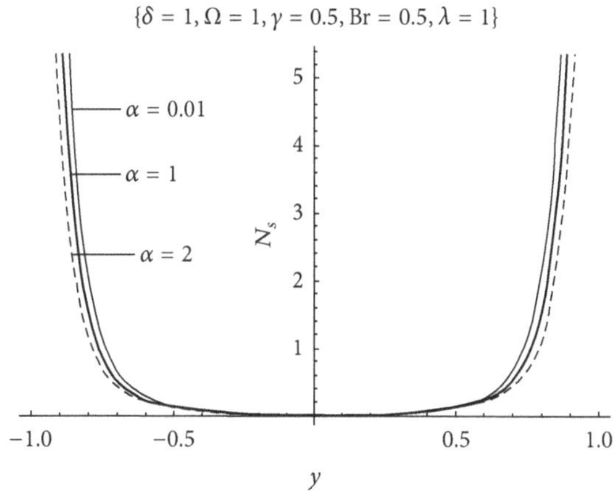

FIGURE 15: Effect of viscosity variation parameter on entropy generation rate.

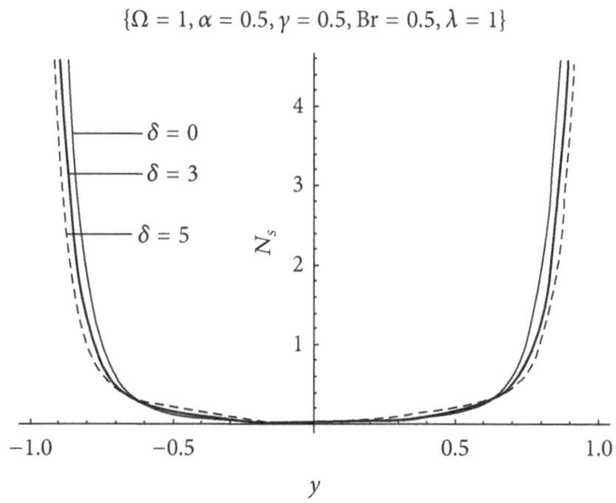

FIGURE 16: Effect of internal heat generation parameter on entropy generation rate.

within the moving fluid. Moreover, Figures 6 and 7 represent the effect of third-grade material effect on the velocity and temperature profiles; the result shows that an increase in the non-Newtonian material effect decreases both the velocity and temperature profiles due to fluid thickening. Figure 8 depicts the velocity profile for variations in internal heat generation parameter. The result shows that an increase in heat generation parameter increases the flow velocity due to rise in the concentration of the reacting fluid. However, as observed in Figure 9, an increase in the internal heat generation increases the temperature distribution within the channel. Figures 10 and 11 display the effect of variations in thermal conductivity on both the velocity and temperature distributions within the channel. As observed from Figure 10, an increase in the thermal conductivity variation parameter increases the velocity distribution within the channel. On the other hand, an increase in the thermal conductivity variation parameter implies a decrease in the thermal conductivity of the fluid. This ultimately decreases the fluid temperature as observed in Figure 11. As observed from the plots, Figures 12, 13, and 14 show that, in an increase in thermal conductivity variation parameter, Brinkman number and third-grade material effect enhance the entropy generation rate within the channel, while increase in viscosity variation parameter decreases the entropy generation rate within the channel as seen in Figure 15. In Figure 16, rise in the internal heat generation parameter is observed to increase entropy generation in the center line of the channel while entropy generation decreases at the walls. Moreover, Figures 17 and 18 shows that as the internal heat generation and Brinkman number increases heat transfer dominates over the fluid viscosity within the channel. Figure 19 represents the effect of viscosity variation parameter on the irreversibility ratio; from the plot it is observed that as the viscosity variation parameter increases, heat transfer dominates over the fluid viscosity at the walls whereas at the centerline of the channel fluid viscosity is observed to dominate over heat transfer. Finally, Figures 20 and 21 show that rise in increase in both third-grade material effect and thermal conductivity variation parameter shows that fluid viscosity dominates over the heat transfer within the channel.

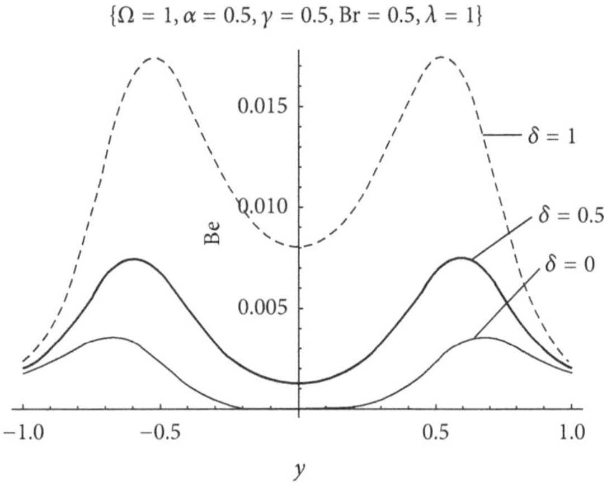

FIGURE 17: Effect of internal heat generation on the irreversibility ratio.

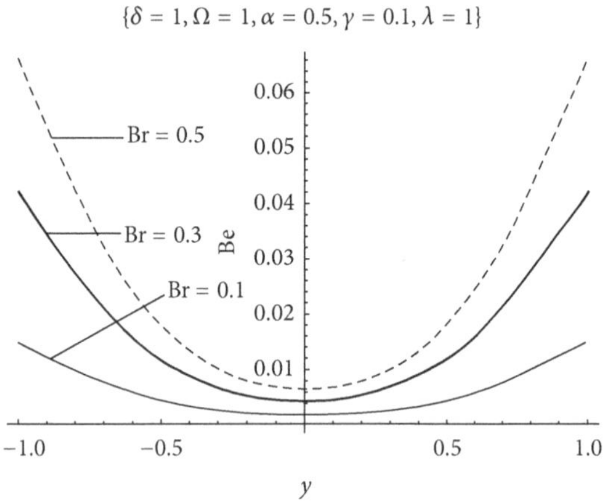

FIGURE 18: Effect of Brinkman number on the irreversibility ratio.

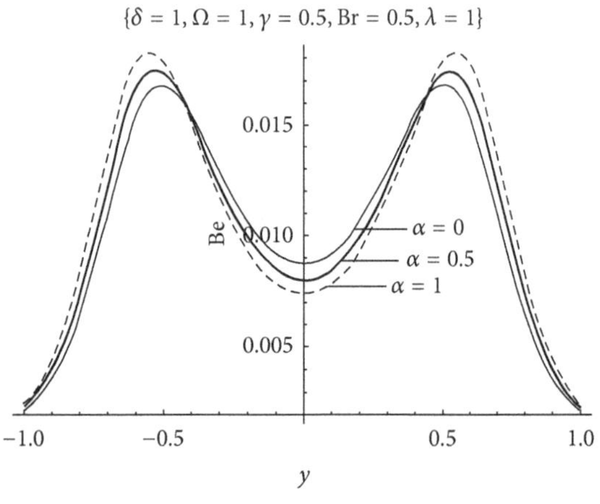

FIGURE 19: Effect of viscosity variation parameter on the irreversibility ratio.

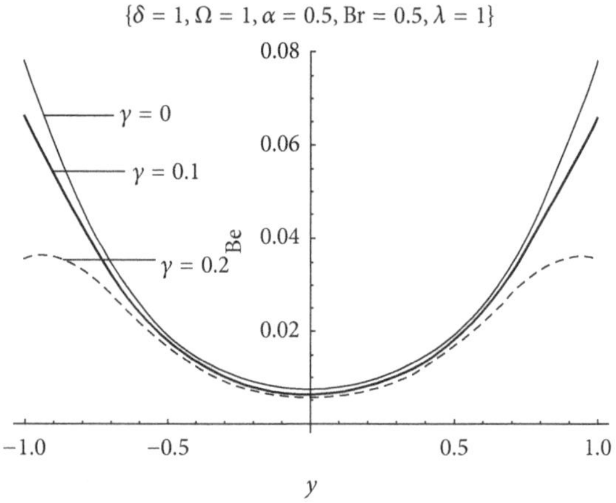

FIGURE 20: Effect of third-grade material effect on irreversibility ratio.

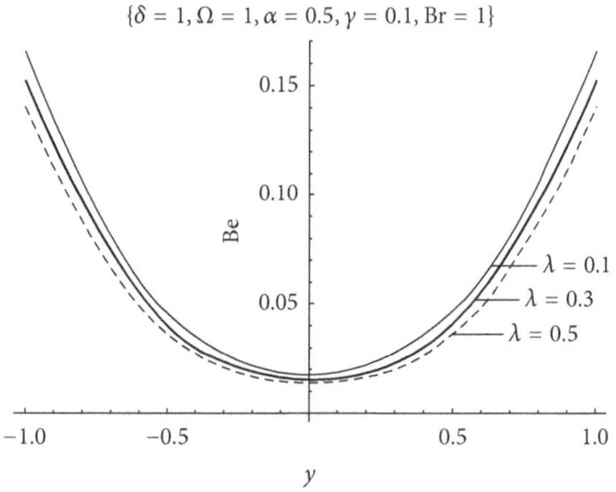

FIGURE 21: Effect of thermal conductivity variation parameter on irreversibility ratio.

5. Conclusion

In this paper, the entropy generation analysis for third-grade fluid flow with variable properties through a channel with uniform wall temperature is studied. In the limiting case, there is perfect agreement between the present result and that obtained in [19] when $\lambda = \delta = \alpha = 0$. Summarily, the main contributions to knowledge from the computation are as follows.

(1) Rise in the viscosity variation parameter is observed to decrease the entropy generation rate within the channel. Moreover, in the middle of the channel, fluid viscosity is observed to dominate over heat transfer. A reverse behaviour is observed at the channel walls.

(2) An increase in the fluid thermal conductivity enhances entropy generation rate. Interestingly, it supports the dominance of fluid viscosity over the heat transfer within the channel.

(3) An increase in the internal heat generation parameter increases the entropy generation rate in the centreline of the channel only.

One major area of future research is the flow behavior for large values of the thermal conductivity parameter. To achieve this further analysis will be required to determine the inherent thermal criticality.

Conflict of Interests

The author declares that there is no conflict of interests regarding the publication of this paper.

References

[1] S. O. Adesanya and O. D. Makinde, "Entropy generation in couple stress fluid flow through porous channel with fluid slippage," *International Journal of Exergy*. In press.

[2] S. O. Adesanya and O. D. Makinde, "Effects of couple stresses on entropy generation rate in a porous channel with convective heating," *Computational and Applied Mathematics*.

[3] M. Pakdemirli and B. S. Yilbas, "Entropy generation in a pipe due to non-Newtonian fluid flow: constant viscosity case," *Sadhana*, vol. 31, no. 1, pp. 21–29, 2006.

[4] M. Pakdemirli and B. S. Yilbas, "Entropy generation for pipe flow of a third grade fluid with Vogel model viscosity," *International Journal of Non-Linear Mechanics*, vol. 41, no. 3, pp. 432–437, 2006.

[5] R. Ellahi, "A study on the convergence of series solution of non-Newtonian third grade fluid with variable viscosity: by means of homotopy analysis method," *Advances in Mathematical Physics*, vol. 2012, Article ID 634925, 11 pages, 2012.

[6] R. Ellahi, T. Hayat, and F. M. Mahomed, "Generalized couette flow of a third-grade fluid with slip: the exact solutions," *Zeitschrift fur Naturforschung A*, vol. 65, no. 12, pp. 1071–1076, 2010.

[7] R. Ellahi and A. Riaz, "Analytical solution for MHD flow in a third-grade fluid with variable viscosity," *Mathematical and Computer Modelling*, vol. 52, no. 9-10, pp. 1783–1793, 2010.

[8] R. Ellahi, T. Hayat, F. M. Mahomed, and S. Asghar, "Effects of slip on the non-linear flows of a third grade fluid," *Nonlinear Analysis: Real World Applications*, vol. 11, no. 1, pp. 139–146, 2010.

[9] S. Abbasbandy, T. Hayat, F. M. Mahomed, and R. Ellahi, "On comparison of exact and series solutions for thin film flow of a third-grade fluid," *International Journal for Numerical Methods in Fluids*, vol. 61, no. 9, pp. 987–994, 2009.

[10] R. Ellahi, P. D. Ariel, T. Hayat, and S. Asghar, "Effect of heat transfer on a third grade fluid in a flat channel," *International Journal for Numerical Methods in Fluids*, vol. 63, no. 7, pp. 847–859, 2010.

[11] T. Hayat, R. Ellahi, and F. M. Mahomed, "The analytical solutions for magnetohydrodynamic flow of a third order fluid in a porous medium," *Zeitschrift fur Naturforschung A Journal of Physical Sciences*, vol. 64, no. 9-10, pp. 531–539, 2009.

[12] T. Hayat, R. Ellahi, and F. M. Mahomed, "Exact solutions for thin film flow of a third grade fluid down an inclined plane," *Chaos, Solitons & Fractals*, vol. 38, no. 5, pp. 1336–1341, 2008.

[13] R. Ellahi and S. Afzal, "Effects of variable viscosity in a third grade fluid with porous medium: an analytical solution," *Communications in Nonlinear Science and Numerical Simulation*, vol. 14, no. 5, pp. 2056–2072, 2009.

[14] R. Ellahi, "The thermodynamics, stability, applications and techniques of differential type: a review," *Reviews in Theoretical Science*, vol. 2, pp. 116–123, 2014.

[15] O. D. Makinde and A. Aziz, "Second law analysis for a variable viscosity plane Poiseuille flow with asymmetric convective cooling," *Computers & Mathematics with Applications*, vol. 60, no. 11, pp. 3012–3019, 2010.

[16] A. Kahraman and M. Yürüsoy, "Entropy generation due to non-newtonian fluid flow in annular pipe with relative rotation: constant viscosity case," *Journal of Theoretical and Applied Mechanics*, vol. 46, no. 1, pp. 69–83, 2008.

[17] D. S. Chauhan and V. Kumar, "Entropy analysis for third-grade fluid flow with temperature-dependent viscosity in annulus partially filled with porous medium," *Theoretical and Applied Mechanics*, vol. 40, no. 3, pp. 441–464, 2013.

[18] T. Hayat, S. A. Shehzad, and S. Asghar, "MHD flow, thixotropic fluid, thermal radiation, variable thermal conductivity and thermal radiation," *Walailak Journal of Science & Technology*, vol. 10, no. 1, pp. 29–42, 2013.

[19] A. M. Siddiqui, M. Hameed, B. M. Siddiqui, and Q. K. Ghori, "Use of Adomian decomposition method in the study of parallel plate flow of a third grade fluid," *Communications in Nonlinear Science and Numerical Simulation*, vol. 15, no. 9, pp. 2388–2399, 2010.

[20] S. O. Adesanya, "Linear stability analysis of a plane-Poiseuille hydromagnetic flow using Adomian decomposition method," *UPB Scientific Bulletin A*, vol. 75, no. 2, pp. 99–106, 2013.

[21] S. O. Adesanya, "Steady magnetohydrodynamic visco-elastic heat generating /absorbing slip flow through a porous medium with radiation effect," *International Journal of Heat and Technology*, vol. 30, no. 1, pp. 69–74, 2012.

[22] S. O. Adesanya and O. D. Makinde, "Heat transfer to magnetohydrodynamic non-newtonian couple stress pulsatile flow between two parallel porous plates," *Zeitschrift fur Naturforschung A: Journal of Physical Sciences*, vol. 67, no. 10-11, pp. 647–656, 2012.

20

Thermodynamical Analysis of the Flow and Heat Transfer over a Static and a Moving Wedge

Adnan Saeed Butt and Asif Ali

Department of Mathematics, Quaid-I-Azam University, Islamabad 45320, Pakistan

Correspondence should be addressed to Adnan Saeed Butt; adnansaeedbutt85@gmail.com

Academic Editors: G. L. Aranovich, A. Chagovetz, and M. Sanati

The first and second law characteristics of fluid flow and heat transfer over a static and a moving wedge are investigated. With the help of suitable similarity transformations, the governing boundary layer equations for the velocity and temperature fields are transformed into ordinary differential equations and are solved numerically. The velocity and the temperature profiles are obtained for various parameters and are utilized to compute the entropy generation number Ns and the Bejan number Be. The effects of various physical parameters on the entropy generation number and the Bejan number are depicted through graphs and are discussed qualitatively. It is observed that the entropy production rate is less in case of wedge moving in the opposite direction to flow as compared to static wedge.

1. Introduction

Newton's second law of motion and laws of thermodynamics are the fundamental principles on which all the flow and heat transfer systems are built today. All other laws play a supportive role. First law of thermodynamics gives information about the energy of the system quantitatively. On the other hand, second law of thermodynamics states that all real life processes are irreversible. The irreversibility of the processes is measured by entropy generation. The pioneer work on entropy generation in flow systems was done by Bejan [1, 2]. He showed that the engineering design of a thermal system can be improved by minimizing the entropy generation. Later on, a lot of research has been done by many other investigators by taking different geometrical configurations related to the thermally designed systems and fluid flow processes. Sahin [3] analyzed the entropy effects in viscous fluid flow in a circular duct with the isothermal boundary condition. The first and second law characteristics of fluid flow and heat transfer in a channel having two parallel plates with a finite gap between them were analytically investigated by Mahmud and Fraser [4]. Yilbas et al. [5] discussed the entropy production effects in a semiblocked pipe by taking into consideration the swirling effects. The influence of magnetic field on entropy generation in laminar forced flow past a horizontal flat plate was examined by Al-Odat et al. [6]. Selamet and Arpaci [7] investigated the entropy effects in boundary layer flows. Makinde and Osalusi [8] studied entropy effects in a liquid film falling along an inclined heated porous flat plate. Makinde [9] extended his work and analyzed the irreversibility effects in non-Newtonian liquid film falling under the influence of gravity force. The entropy generation effects in the slip MHD flow in a single rotating disk were studied by Arikoglu et al. [10]. Makinde [11] discussed the irreversibility effects in variable viscosity channel flow with convective cooling at the walls. Second law analysis for a gravity driven variable viscosity liquid film along an inclined heated plate with convective cooling was made by Makinde [12]. Révèillere and Baytaş [13] studied entropy generation in laminar boundary layer flow over an isothermal permeable flat surface. Makinde and Bég [14] used perturbation technique together with Hermite-Pade approximation to examine the entropy generation in a reactive hydromagnetic channel flow. The first and second law characteristics of fluid flow in a porous medium over a permeable stretching surface were investigated by Tamayol et al. [15]. The thermodynamical analysis of MHD boundary layer flow over a flat plate with variable viscosity was done by Makinde [16]. He considered that the thermal radiation

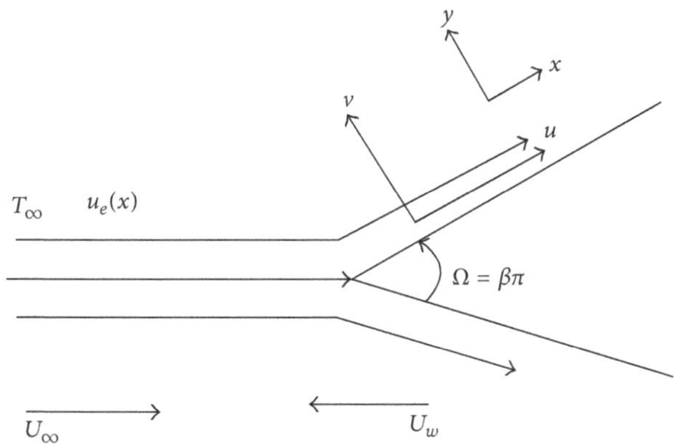

FIGURE 1: Schematic diagram of the problem.

is also present there, and the surface is convectively heated. It was found that the variable viscosity results in more entropy production. Makinde and Aziz [17] discussed the entropy generation effects in a long hollow cylinder with time dependent internal heating and convective heating at both inner and outer surfaces. Analytical solution of the problem was obtained, and a thermal stability criterion was also examined. Chen et al. [18] made use of differential transform method to study the entropy effects in the mixed convective flow in a vertical channel. Butt et al. [19] examined the viscoelastic effects on the entropy generation in fluid flow and heat transfer through a porous medium over a permeable stretching surface. The study of entropy production in MHD slip flow over a vertical surface with convective boundary was done by Butt et al. [20]. Munawar and Ali [21] examined the entropy generation in fluid flow and heat transfer over an oscillating stretching cylinder. The effects of magnetic field on entropy generation in flow over a radially stretching sheet were investigated by Butt and Ali [22].

2. Mathematical Formulation of the Problem

Consider a steady state two-dimensional boundary layer flow of a laminar incompressible fluid past a static and a moving wedge as shown in the Figure 1. The coordinate system is chosen such that the x-coordinate is parallel to the surface of the wedge and the y-coordinate is normal to it. The properties of the fluid are assumed to be constant. It is assumed that the velocity of the free stream (inviscid flow) is $u_e(x) = U_\infty x^m$ and that of the moving wedge is $u_w(x) = U_w x^m$, where U_∞, U_w, and m are constants with $0 \le m \le 1$. Using the boundary layer approximations, the governing equations for the steady flow and heat transfer over a wedge are given by

$$\frac{\partial u}{\partial x} + \frac{\partial v}{\partial y} = 0, \quad (1)$$

$$u\frac{\partial u}{\partial x} + v\frac{\partial u}{\partial y} = \nu\frac{\partial^2 u}{\partial y^2} + u_e\frac{du_e}{dx}, \quad (2)$$

$$u\frac{\partial T}{\partial x} + v\frac{\partial T}{\partial y} = \alpha\frac{\partial^2 T}{\partial y^2} + \frac{\nu}{\rho c_p}\left(\frac{\partial u}{\partial y}\right)^2. \quad (3)$$

The corresponding boundary conditions for the flow are

(i) static wedge:

$$u = 0, \quad v = 0 \quad \text{at } y = 0, \qquad u = u_e(x) \quad \text{at } y \longrightarrow \infty, \quad (4)$$

(ii) moving wedge:

$$u = u_w(x), \quad v = 0 \quad \text{at } y = 0, \qquad u = u_e(x) \quad \text{at } y \longrightarrow \infty. \quad (5)$$

The thermal boundary conditions are

$$T = T_w \quad \text{at } y = 0, \qquad T = T_\infty \quad \text{at } y \longrightarrow \infty. \quad (6)$$

Here u and v are the velocity components along the x and y-axes, respectively, ν is the kinematic viscosity, α is the thermal diffusivity, ρ is the density of the fluid, c_p is the specific heat of the fluid at a constant pressure, T is the temperature of the fluid, and T_w is the wall temperature, T_∞ is the temperature far away from the surface.

Introducing the similarity transformation, we have

$$\psi(x,y) = \sqrt{\frac{2\nu x u_e(x)}{m+1}}\, f(\eta), \qquad \eta(x,y) = \sqrt{\frac{(m+1)\,u_e(x)}{2\nu x}}\, y, \quad (7)$$

where ψ is the stream function defined such that $u = \partial\psi/\partial y$, $v = -\partial\psi/\partial x$ satisfy (1) identically. The similarity transformation for temperature is

$$\theta = \frac{T - T_\infty}{T_w - T_\infty}. \quad (8)$$

Substituting (7) and (8) into (1)–(6), we have

$$f''' + ff'' + \frac{2m}{m+1}\left(1 - f'^2\right) = 0, \qquad \theta'' + \Pr f\theta' + \Pr\mathrm{Ec} f''^2 = 0. \quad (9)$$

The boundary conditions for flow are

(i) static wedge:

$$f(0) = 0, \quad f'(0) = 0, \quad f'(\infty) = 1, \quad (10)$$

(ii) moving wedge:

$$f(0) = 0, \quad f'(0) = \lambda, \quad f'(\infty) = 1. \quad (11)$$

The boundary conditions for temperature in nondimensional form are

$$\theta(0) = 1, \qquad \theta(\infty) = 0, \quad (12)$$

where prime denotes the differentiation with respect to η, $\mathrm{Pr} = \mu c_p/k$ is the Prandtl number, $\mathrm{Ec} = U^2/c_p(T_w - T_\infty)$ is the Eckert number, and the parameters λ and β are defined as

$$\lambda = \frac{U_w}{U_\infty}, \qquad \beta = \frac{2m}{m+1}. \tag{13}$$

Here λ is the moving wedge parameter where $\lambda > 0$ corresponds to the motion of wedge in the direction of free stream and $\lambda < 0$ corresponds to a moving wedge in the opposite direction of the free stream velocity. The case $\lambda = 0$ corresponds to a static wedge.

3. Entropy Generation

According to Bejan [2, 23], the entropy generation per unit volume is defined as

$$S_G = \frac{k}{T^2}\left[\left(\frac{\partial T}{\partial x}\right)^2 + \left(\frac{\partial T}{\partial y}\right)^2\right] + \frac{\mu}{T}\left(\frac{\partial u}{\partial y}\right)^2. \tag{14}$$

The first term in (14) is the entropy generation due to heat transfer, and the second term is the entropy generation due to viscous dissipation. The dimensionless form of the entropy generation is given by

$$\begin{aligned}\mathrm{Ns} &= \frac{S_G}{S_o} \\ &= \left(\left(\frac{m-1}{2}\right)^2\eta^2 + \left(\frac{m+1}{2}\right)\mathrm{Re}_x\right)\theta'^2 \\ &\quad + \frac{\mathrm{Br}}{\Omega}\left(\frac{m+1}{2}\right)\mathrm{Re}_x f''^2,\end{aligned} \tag{15}$$

where $S_o = k(T_w - T_\infty)^2/T_w{}^2x^2$ is the characteristic entropy generation rate, $\Omega = (T_w - T_\infty)/T_w$ is the dimensionless temperature difference, $\mathrm{Re}_x = Ux/\nu$ is the local Reynolds number, and $\mathrm{Br} = \mathrm{PrEc}$ is the Brinkman number.

An alternative irreversibility distribution parameter is the Bejan number, which gives an idea whether the fluid friction irreversibility dominates over heat transfer irreversibility or the heat transfer irreversibility dominates over fluid friction rreversibility. It is simply the ratio of entropy generation due to heat transfer to the total entropy generation:

$$\mathrm{Be} = \frac{\text{Entropy generation due to heat transfer}}{\text{Total entropy generation}}. \tag{16}$$

When $\mathrm{Be} \gg 0.5$, the irreversibility due to heat transfer dominates, whereas when $\mathrm{Be} \ll 0.5$, the irreversibility due to viscous effects dominates. When $\mathrm{Be} = 0.5$, the heat transfer and the fluid friction irreversibilities are equal.

4. Results and Discussion

The nonlinear differential equation (9) with boundary conditions (10)–(12) is solved using the shooting method with Runge-Kutta algorithm. A step size of 0.001 is taken to obtain the numerical solution, and the accuracy goal is kept equal to 10^{-7}. The effects of different physical parameters on velocity and temperature profiles are analyzed by plotting graphs. Moreover, the graphs of the entropy generation number and the Bejan number are presented to examine the entropy production in the considered problem.

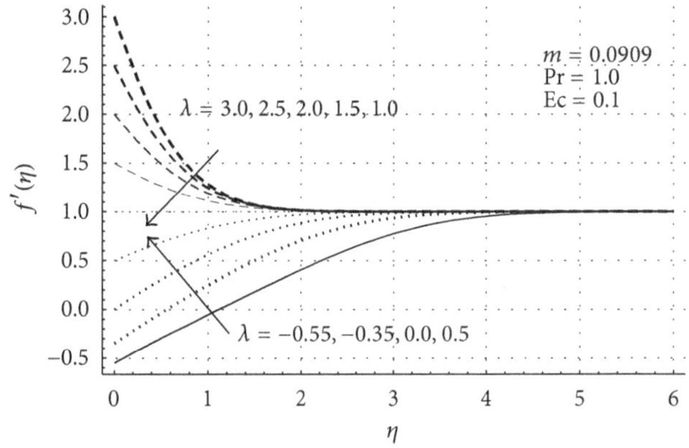

FIGURE 2: Effects of moving wedge parameter λ on $f'(\eta)$.

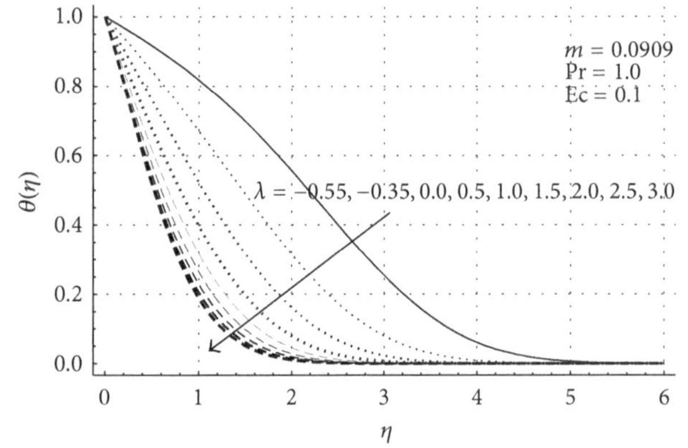

FIGURE 3: Effects of moving wedge parameter λ on $\theta(\eta)$.

Figure 2 illustrates that for nonpositive values of λ; that is, when the wedge is moving in the opposite direction of stream velocity, the momentum boundary layer thickness is more than the case when both wedge and stream are moving in the same direction $\lambda \geq 0$. Moreover $\lambda = 1$ corresponds to the case when the fluid and the wedge move with same velocities. In this situation, there is no shear stress at the surface. Figure 3 depicts the effects of moving wedge parameter λ on temperature profile. It is observed that the thermal boundary layer thickness decreases all values of λ. The influence of Prandtl number Pr on temperature profile is presented in Figure 4. It is observed that with an increase in the Prandtl number, the thermal diffusion rate decreases which ultimately results in decreasing of the thermal boundary layer thickness. The effects of Eckert number Ec on temperature profile are shown in Figure 5. An increase in thermal boundary layer is noticed with an increase in the value of Ec.

The variations in entropy generation number Ns are plotted against η in Figures 6, 7, 8, and 9 for different

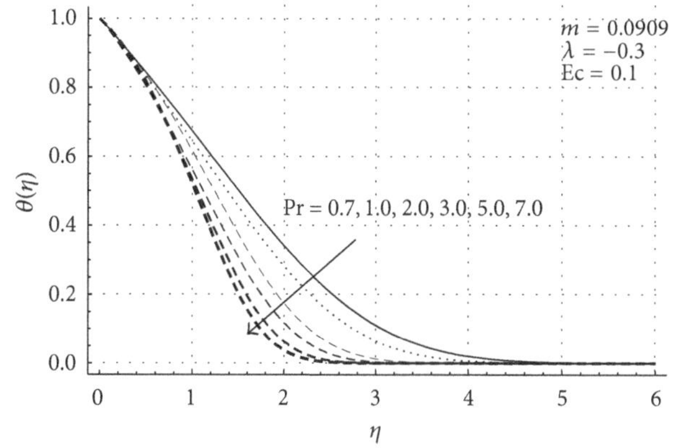

FIGURE 4: Effects of Prandtl number on temperature profile $\theta(\eta)$.

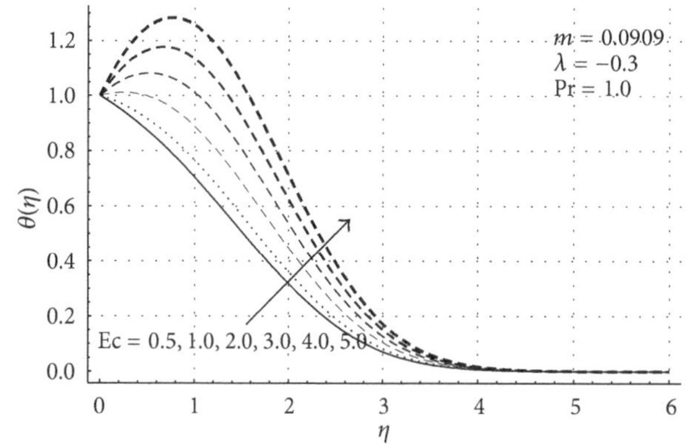

FIGURE 5: Effects of Eckert number on temperature profile $\theta(\eta)$.

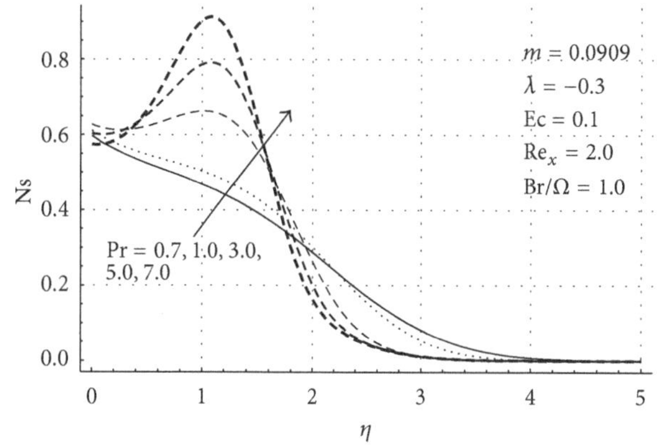

FIGURE 6: Effects of Prandtl number Pr on entropy generation number Ns.

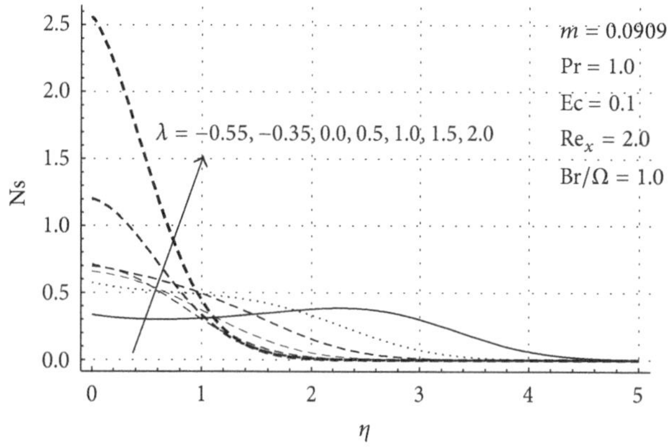

FIGURE 7: Effects of moving wedge parameter λ on entropy generation number Ns.

physical parameters. Figure 6 shows the effects of Prandtl number Pr on entropy profile. For a particular value of Pr entropy generation increases and attains its peak value within the boundary layer region. Then it decreases sharply and becomes zero in the free stream flow region. Also it is noteworthy that the irreversibility effects are observed within a small region of boundary layer for higher values of Prandtl number as compared to the small Prandtl number. The effects of moving wedge parameter λ on Ns are shown in Figure 7. It is noticed from the figure that near the surface entropy production increases by increasing λ and an opposite behavior is observed far away from the surface. Moreover, in the case when wedge and free stream are moving oppositely, that is, $\lambda < 0$, the entropy effects can be observed for large distance within the boundary layer region. The effects of the group parameter Br/Ω on entropy generation number Ns are shown in Figure 8. Increasing the value of Br/Ω results in more entropy production. It is also observed that for a particular value of Br/Ω, the entropy generation increases with an increase in η and attains its maximum value at $\eta = 1.7$ (roughly) and then decreases and approaches zero as the free stream region is reached. The influence of the local Reynolds number Re_x on Ns is depicted from Figure 9. It can be seen that the entropy production increases as Re_x increases.

The Bejan number Be is plotted in Figures 10–13 for different ranges of various parameters. Figure 10 depicts that, for small values of the Prandtl number Pr, the irreversibility effects due to viscous dissipation are dominant near the wedge surface, and afterward heat transfer irreversibility becomes prominent. However, for large values of Pr, the scenario is quite different. It is noticed that, at the surface, irreversibility due to viscous dissipation is dominant, and then the heat transfer irreversibility becomes strong in the boundary layer region by increasing Pr. In the main stream region entropy production due to viscous dissipation effects again become strong as Pr increases. Figure 11 shows the behavior of the Bejan number for various values of moving wedge parameter λ. It is quite evident that, by increasing λ in the range $-0.55 \leq \lambda \leq 1$ the entropy effects due to heat transfer become strong. For $\lambda > 1$, an opposite trend is noticed, and the entropy due to fluid friction is dominant. This means that irreversibility due to viscous effects remains strong when wedge moves in the opposite direction to the free stream velocity or when the velocity of the wedge is greater than the velocity of free stream flow. Obviously, this conduct is due to large frictional forces generated near the surface in both cases. Figure 12 illustrates that the fluid friction irreversibility becomes dominant by increasing the group parameter Br/Ω at the surface. It is quite obvious that this behavior is due to the increasing effects of viscous

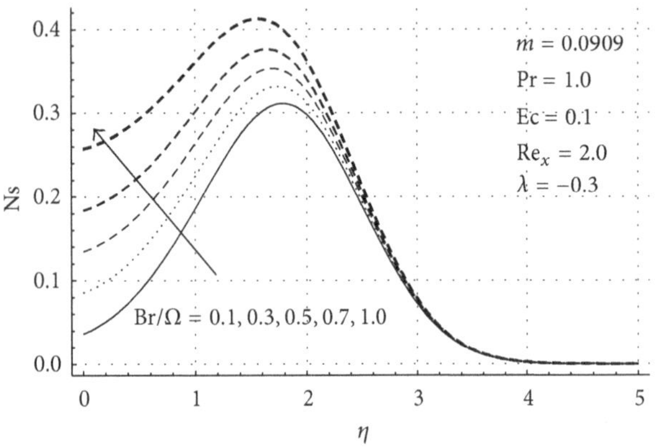

FIGURE 8: Effects of group parameter Br/Ω on entropy generation number Ns.

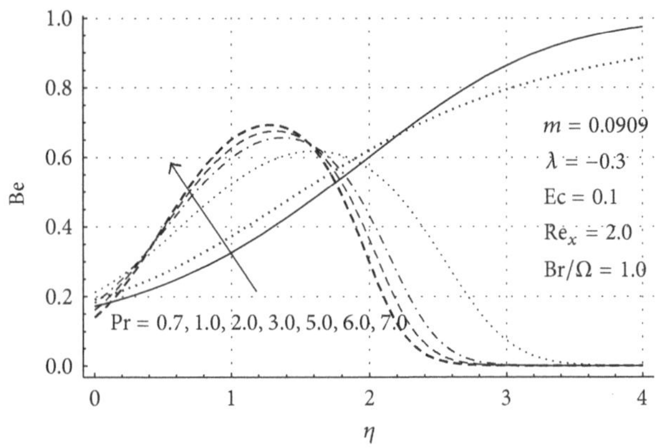

FIGURE 10: Effects of Prandtl number Pr on the Bejan number Be.

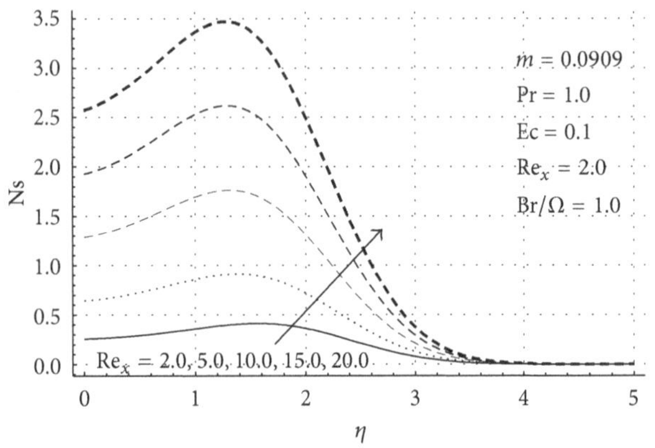

FIGURE 9: Effects of local Reynolds number Re_x on entropy generation number Ns.

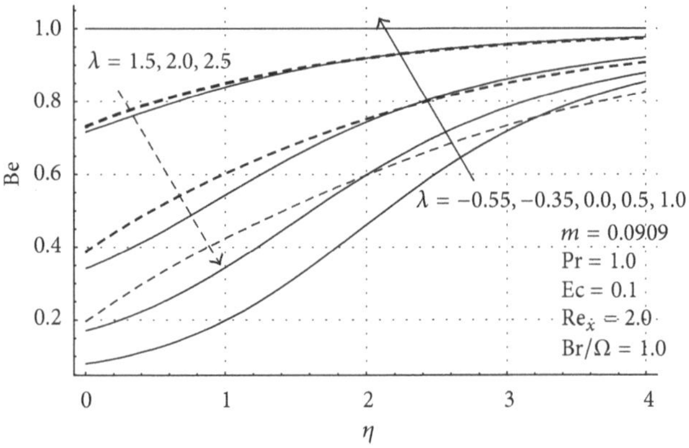

FIGURE 11: Effects of moving wedge parameter λ on the Bejan number Be.

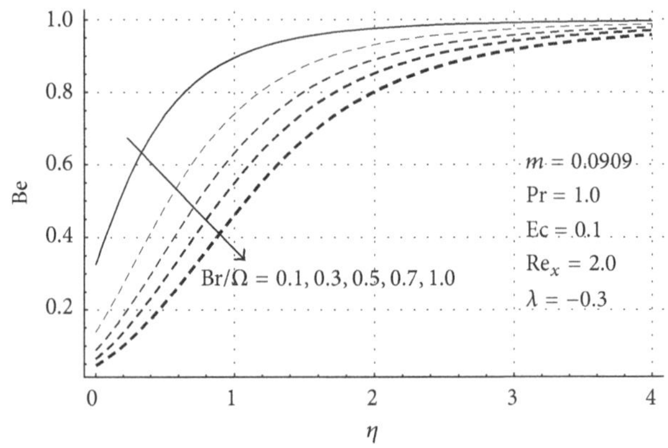

FIGURE 12: Effects of group parameter Br/Ω on the Bejan number Be.

dissipation in the flow. In Figure 13, the influence of local Reynolds number Re_x on Bejan number is presented. The heat transfer irreversibility becomes weak as Re_x increases. However, these effects are not significant near the surface.

5. Conclusion

The second law (of thermodynamics) characteristics of viscous incompressible fluid flow over a static and moving wedge are investigated. The velocity and the temperature fields are used to obtain the expression for the entropy generation. In order to make a comparison whether the heat transfer irreversibility dominates over the viscous dissipation irreversibility or vice versa, the Bejan number is plotted for different physical parameter, and the results are discussed briefly. It is noticed that when the wedge and fluid are moving in opposite directions ($\lambda < 0$), the momentum boundary layer is thicker than the case when the fluid and the wedge are moving in same direction. However, the thermal boundary layer thickness decreases for both moving and static wedge. It is concluded that the fluids with large Pr have small thermal boundary layers. Moreover, the thermal boundary layer thickens by increasing the Eckert number Ec. The effects of Prandtl number Pr on entropy generation number are increasing within the boundary layer region. However, for large values of Pr, the irreversibility effects are observed in a small region. The entropy generation increases as λ increases near the surface and decreases far away from the surface. Moreover, it can be seen that the entropy generation is minimum when wedge moves in the opposite direction of free stream velocity. The presence of the group parameter Br/Ω and the local Reynolds number Re_x is to increase

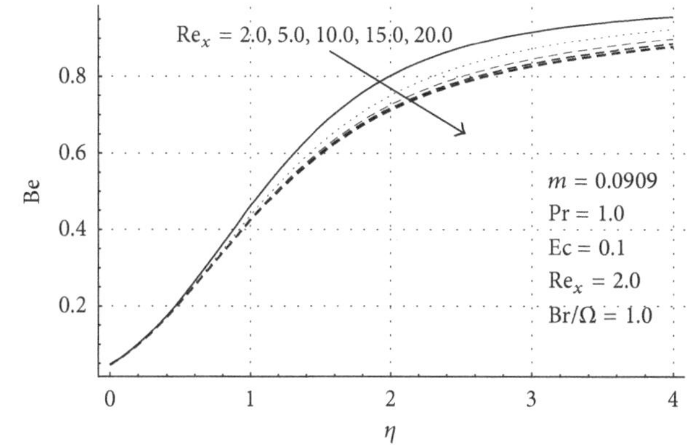

FIGURE 13: Effects of local Reynolds number Re_x on the Bejan number Be.

entropy production. It is observed from the analysis that fluid friction irreversibility dominates when the wedge is moving in the opposite direction of free stream velocity or when the wedge is moving quickly as compared to the free stream velocity. The fluid friction irreversibility effects become stronger as the group parameter Br/Ω and the local Reynolds number Re_x increase.

References

[1] A. Bejan, "A study of entropy generation in fundamental convective heat transfer," *Journal of Heat Transfer*, vol. 101, no. 4, pp. 718–725, 1979.

[2] A. Bejan, *Entropy Generation Through Heat and Fluid Flow*, John Wiley and Sons, Toronto, Canada, 1994.

[3] A. Z. Sahin, "Second law analysis of laminar viscous flow through a duct subjected to constant wall temperature," *Journal of Heat Transfer*, vol. 120, pp. 76–83, 1998.

[4] S. Mahmud and R. A. Fraser, "Thermodynamic analysis of flow and heat transfer inside channel with two parallel plates," *Exergy*, vol. 2, pp. 140–146, 2002.

[5] B. S. Yilbas, S. Z. Shuja, and M. O. Budair, "Second law analysis of a swirling flow in a circular duct with restriction," *International Journal of Heat and Mass Transfer*, vol. 42, no. 21, pp. 4027–4041, 1999.

[6] M. Q. Al-Odat, R. A. Damseh, and M. A. Al-Nimr, "Effect of magnetic field on entropy generation due to laminar forced convection past a horizontal flat plate," *Entropy*, vol. 6, no. 3, pp. 293–303, 2004.

[7] A. Selamet and V. S. Arpaci, "Entropy production in boundary layers," *Journal of Thermophysics and Heat Transfer*, vol. 4, no. 3, pp. 404–407, 1990.

[8] O. D. Makinde and E. Osalusi, "Entropy generation in a liquid film falling along an inclined porous heated plate," *Mechanics Research Communications*, vol. 33, no. 5, pp. 692–698, 2006.

[9] O. D. Makinde, "Irreversibility analysis for a gravity driven non-Newtonian liquid film along an inclined isothermal plate," *Physica Scripta*, vol. 74, no. 6, article 007, pp. 642–645, 2006.

[10] A. Arikoglu, I. Ozkol, and G. Komurgoz, "Effect of slip on entropy generation in a single rotating disk in MHD flow," *Applied Energy*, vol. 85, no. 12, pp. 1225–1236, 2008.

[11] O. D. Makinde, "Irreversibility analysis of variable viscosity channel flow with convective cooling at the walls," *Canadian Journal of Physics*, vol. 86, no. 2, pp. 383–389, 2008.

[12] O. D. Makinde, "Thermodynamic second law analysis for a gravity-driven variable viscosity liquid film along an inclined heated plate with convective cooling," *Journal of Mechanical Science and Technology*, vol. 24, no. 4, pp. 899–908, 2010.

[13] A. Réveillère and A. C. Baytaş, "Minimum entropy generation for laminar boundary layer flow over a permeable plate," *International Journal of Exergy*, vol. 7, no. 2, pp. 164–177, 2010.

[14] O. D. Makinde and O. A. Bég, "On inherent irreversibility in a reactive hydromagnetic channel flow," *Journal of Thermal Science*, vol. 19, no. 1, pp. 72–79, 2010.

[15] A. Tamayol, K. Hooman, and M. Bahrami, "Thermal analysis of flow in a porous medium over a permeable stretching wall," *Transport in Porous Media*, vol. 85, no. 3, pp. 661–676, 2010.

[16] O. D. Makinde, "Second law analysis for variable viscosity hydromagnetic boundary layer flow with thermal radiation and Newtonian heating," *Entropy*, vol. 13, no. 8, pp. 1446–1464, 2011.

[17] O. D. Makinde and A. Aziz, "Analysis of entropy generation and thermal stability in a long hollow cylinder with asymmetry convective cooling," *Heat and Mass Transfer*, vol. 47, no. 11, pp. 1407–1415, 2011.

[18] C.-K. Chen, H.-Y. Lai, and C.-C. Liu, "Numerical analysis of entropy generation in mixed convection flow with viscous dissipation effects in vertical channel," *International Communications in Heat and Mass Transfer*, vol. 38, no. 3, pp. 285–290, 2011.

[19] A. S. Butt, S. Munawar, A. Ali, and A. Mehmood, "Effect of viscoelasticity on entropy generation in a porous medium over a stretching plate," *World Applied Sciences Journal*, vol. 17, no. 4, pp. 516–523, 2012.

[20] A. S. Butt, S. Munawar, A. Ali, and A. Mehmood, "Entropy generation in hydrodynamic slip flow over a vertical plate with convective boundary," *Journal of Mechanical Science and Technology*, vol. 26, no. 9, pp. 2977–2984, 2012.

[21] A. S. Butt and A. Ali, "Effects of magnetic field on entropy generation in flow and heat transfer due to radially stretching surface," *Chinese Physics Letters*, vol. 30, no. 2, pp. 02704–02708, 2012.

[22] S. Munawar and M. A. Ali, "Thermal analysis of the flow over an oscillatory stretching cylinder," *Physica Scripta*, vol. 86, Article ID 065401, 2012.

[23] L. C. Woods, *Thermodynamics of Fluid Systems*, Oxford University Press, Oxford, UK, 1975.

21

Equilibrium Molecular Interactions in Pure Gases

Boris I. Sedunov

The Information Systems and Computer Technologies Department, Russian New University, Radio St. 22, Moscow 105005, Russia

Correspondence should be addressed to Boris I. Sedunov, tropic1@ya.ru

Academic Editor: Juan Ortega

The equilibrium molecular interactions in pure real gases are investigated based on the chemical thermodynamics principles. The parallels between clusters in real gases and chemical compounds in equilibrium media have been used to improve understanding of the real gas structure. A new approach to the equilibrium constants for the cluster fractions and new methods to compute them and their significant parameters from the experimental thermophysical data are developed. These methods have been applied to some real gases, such as Argon and Water vapors and gaseous Alkanes. It is shown that the four-particle clusters make a noticeable contribution in the thermophysical properties of the equilibrium Water vapor. It is shown also that the effective bond energy for dimers in Alkanes linearly grows with the number of carbon atoms in the molecule.

1. Introduction

Modern technologies benefit from the utilization of extra pure and highly uniform materials and substances and precise knowledge of their thermophysical properties. For computer-aided design and control of technological processes, precise models of thermophysical properties and equations of state for pure gases are very important because the real gases are widely used both as initial and intermediate materials in modern technologies. The models for pure gases can be used to build more complex models for mixtures of different gases. This work has a goal to show no obvious relations between equilibrium molecular interaction parameters in pure real gases and to find useful expressions to predict equilibrium properties of pure gases in changing external conditions. The work is based on a phenomenological approach that, nevertheless, provides a vision of the real gas structure on a microscopic level.

The paper tries to answer the following questions, arising in courses of molecular and chemical physics.

(i) What are the equilibrium constants and bond energies for real gas clusters considered as the low bond energy analogues of chemical compounds?

(ii) How to define the bond energies for clusters that in contrast to a wide separation of quantum levels of compounds have a dense spacing of multiple energetic levels?

(iii) How to build the bridge between the virial equations of state based on variables that are sums of partial quantities for cluster fractions and the chemical equilibrium law, where arguments for concentrations of chemical compounds are the partial concentrations of reacting components [1]?

(iv) How to transform the Sackur-Tetrode equation for entropy [2, 3] of ideal atomic gases for real gases with account of the molecular interactions and internal movements?

The paper intends to show that clusters in equilibrium real gases obey to laws that differ from the well-known laws for chemical compounds equilibrium [1]. One of ways to understand the properties of real gases is their comparison with chemical media, where dimerized compounds, such as Alkali metal [4, 5] or Nitrogen Dioxide [1, 6] dimers, may be formed in equilibrium conditions from their monomers. While the concentrations of chemical compounds and their equilibrium constants may be found from spectroscopic measurements, the thermal movement shadows and erodes the spectral lines of clusters. There are many papers on spectroscopic investigation of molecular clusters in molecular beams, but not in equilibrium gases. There are many works on molecular dynamics simulation of gaseous clusters, which do not build the bridge between the molecular interaction parameters and the thermophysical properties of real gases. This work develops the reverse methods to estimate the bond parameters of clusters in pure real gases from thermophysical data.

There is a productive way to the clusters' equilibrium constants and molecular interaction parameters, based on the processing of known thermophysical data. But found so the cluster fractions' apparent equilibrium constants responsible for the Pressure-Density-Temperature relations (PDT equilibrium constants) are essentially different from the potential-energy (PE-) based equilibrium constants in real gases, while this difference for chemical compounds may be negligible. The paper strives to explain these facts.

This paper deals with the extension of the thermodynamics to real gaseous systems. All conclusions of the developed here theory are verified with the utilization of precise experimental data from the NIST (the National Institute of Standards and Technology USA) databases [7, 8]. Now, to provide the computer-aided design and quality control for many technological processes, comprehensive databases of thermophysical properties for every used substance are accumulated and regularly updated. In modern science and technology, their creation goes on constantly with a steady growth of the data abundance and precision. Constant improvement of comprehensive and precise databases of the NIST [7, 8] is the best example of this kind of scientific activity. The systematization of experimental data, from one side, provides practical application, testing, and correction of the existing theoretical models and methods but, from the other side, opens possibilities for further development of the theoretical basis. This paper shows how these possibilities work.

2. The Equilibrium Structure of Pure Gases

2.1. Clusters and Monomers in Real Gases. Real gas consists of basic particles (atoms or molecules), interacting through their interaction potential [1, 9–11]. The main part of gas consists of monomers that are the free moving basic particles performing elastic interactions with other particles without falling in bound states. Due to the attraction part of the interaction potential, some basic particles collect in temporary stable agglomerations with different numbers of particles, Figure 1. These agglomerations are named clusters [12–15] that move in a gas-like free particles and sometimes collide with other clusters or monomers. In spite of the plurality of clusters existing in a real gas, all equilibrium thermophysical properties of real gases are estimated for a fixed number of basic particles, in this paper, for a mole. Due to constantly changing interactions between basic particles in a real gas, it is difficult to define correctly the monomers and clusters from the microscopic point of view.

In this paper, the thermophysical properties of the monomer and cluster fractions are found by processing the known thermophysical properties for all real gas. The results of this processing are not obvious; they reflect both elastic and no elastic interactions between basic particles.

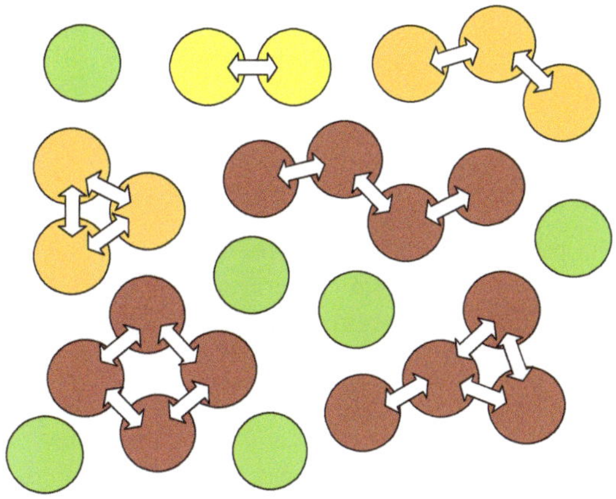

Figure 1: Schematic view of the real gas structure: the basic particles of the real gas (balls of all colors) in forms of monomers (green balls) and clusters (other colors). Yellow balls: dimer; orange balls: trimers with open chain and closed isomer configurations; red balls: tetramers (quatromers) with open chain, closed, and partly closed structures; white arrows: intermolecular bonds.

Figure 1 presents the schematic view of the real gas structure that is used here only for interpretation of the computer-aided analysis of experimental data, but not for microscopic computations of the real gas properties.

Figure 1 does not show all types of the 4-particle isomers and does not show larger clusters at all. Not to overload the picture, there is not shown the tetragonal configuration of the tetramer with 6 bonds between particles. But the figure clearly shows that different isomers at the same number of particles in a cluster can possess different numbers of bonds. The number of bonds characterizes the cluster's bond energy, and the isomer configuration determines the cluster's entropic factor, connected with the number of states in this configuration. Two of these factors are responsible for prevailing of some types of isomer structures at a given temperature and for a change of the prevailing types with a change of temperature. It will be shown how interaction parameters found by phenomenological analysis of the experimental data point out on the prevailing types of isomers.

The collisions in a gas can be divided on elastic and no elastic types. After an elastic collision, the numbers of basic particles in colliding clusters and their isomer configurations stay not changed. Figure 1 shows only the bonds between basic particles resulting from the no elastic collisions but does not show the elastic forces between particles that influence significantly the thermophysical properties of real gases. Clusters at any type of collisions, even at the no elastic, conserve their common momentum. Two colliding monomers perform only the elastic collision and after this collision fly apart again as monomers. To form the bound state of a dimer, the two monomers at their meeting should collide also with the third particle (monomer or cluster) to give off the excess of their common kinetic energy.

Generally speaking, the clusters are the products of the no elastic collisions, in which the cluster's bound state is formed by transferring the excess kinetic energy to some third particle. After the no elastic collision, the colliding particles fall into potential wells created by the attraction part of molecular interactions, Figure 2.

Figure 2 shows principal differences between molecular interactions in clusters and chemical compounds.

(i) The potential well for particles in a cluster is not as deep as in a chemical compound that shortens the clusters' lifetimes by many orders of magnitude as compared to chemical compounds.

(ii) The average width of the potential well in a cluster is much larger than in a chemical compound that lowers distances between energetic levels in clusters by many orders of magnitude as compared to chemical compounds.

(iii) The average numbers of bound states in clusters are many orders of magnitude larger than the numbers of bound states in chemical compounds.

(iv) The potential well around one gaseous particle may attract several particles, but the chemical potential well may disappear for the third particle after formation of the two-particle compound.

Particles in clusters and compounds do not occupy the position at the minimum of their potential. In gases, the thermal movement lifts the particles to higher levels, but, in compounds, their ground level stays over the potential minimum due to quantum laws. We should develop the method of finding the average energy of clusters in their bound states as an equivalent to the Schrodinger equation for the ground level of chemical compounds.

2.2. The Equilibrium Structure of Real Gases. The real gas consists of cluster fractions with a fixed number n of basic particles and with a definite isomer structure of bonds. The equilibrium structure of a real gas can be expressed through the cluster fractions' populations. The cluster fractions keep dynamic equilibrium of their populations through basic particles exchange. Unlike crystals, where structure means the configuration of the elementary cell, regularly repeated in space [16], the equilibrium structure of gases is disordered but statistically stable and determining properties of real gases. Due to differences between the isomers' energy and entropy factors, some isomer configurations are prevailing at a given temperature. The change of prevailing types of isomers is responsible for the temperature dependences of the clusters' molecular interaction parameters that may be found from the equilibrium thermophysical properties of gases.

The bonds between basic particles in real gases are much weaker than the chemical bonds, how it is reflected by the Figure 2. Their values in most cases are comparable with the energy of thermal movement; therefore, the relaxation time to establish the cluster fractions populations is much lower than the typical times for experimental determination of thermophysical properties of gases. But this rule may be not valid for overcooled gases and for the sound speed or the complex dielectric permeability measurements.

In this work, the main variable for the computer-aided analysis of thermophysical properties is the molar density of the monomer fraction [17–19], D_m. The values of the monomer fraction density can be found from the isothermal pressure dependence of the total density D.

2.3. Monomer Fraction Density and the Chemical Potential. Due to a fast time to establish equilibrium in the cluster fractions populations, the Gibbs energy for a mole of basic particles (the chemical potential), G, does not depend on their aggregate state: the molar Gibbs energy of basic particles in each cluster fraction, G_n, and in the monomer fraction, G_m, is the same and is equal to the total Gibbs energy of the gas: $G_n = G_m = G$ [17, 18]. This statement is correct only for equilibrium systems. In a no equilibrium situation, the basic particles' Gibbs energies for different cluster fractions may split, like the Fermi quasilevels for electrons and holes in semiconductors.

For atomic ideal gases, the molar Gibbs energy $G = -RT\ \ln(V/V_q)$, where V is the molar volume of gas and V_q is the molar quantum volume of atoms [20]. In molecular gases, we should add the part of the molar Gibbs energy responsible for the internal molecular movements: rotations of molecules and vibrations of their atoms, G_0 [18].

The monomer fraction in a gas behaves as an ideal gas with a molar volume $V_m = 1/D_m$; therefore, the molar Gibbs energy G of basic particles is [18]:

$$G = G_m = G_0 + RT\ \ln\left(D_m V_q\right), \qquad (1)$$

where V_q is the molar quantum volume for basic particles that is proportional to the third degree of the thermal de Broglie wavelength:

$$V_q = \frac{h^3 N_A^4}{(2\pi\ MRT)^{3/2}}, \qquad (2)$$

where h is the Plank's constant, M is molecular weight of basic particles in kg/mol, N_A is Avogadro number, and R is universal gas constant. The compact form of (1) combines in V_q [20] the entropy dependence on atomic weight, temperature, and Plank constant.

Equation (1) is the thermodynamics-based definition of the monomer fraction density. This equation reminds the well-known equation for fugacity f : $G = B(T) + RT\ln(f)$, suggested by Lewis [21] to solve the problems of no ideality of real gases. The monomer fraction density is more evident: it has a clear statistical and mechanical sense. Its introduction may help in a better understanding of properties of widely used variables: fugacity and activity.

In this theory, G_0, like $B(T)$ in the theory of fugacity [21], is supposed to be independent on the density and aggregate state of basic particles. All possible changes in the internal movements' part of the Gibbs energy are accounted for by the pressure dependence of D_m.

G corresponds to the thermodynamics equation [1, 9, 10]:

$$\left(\frac{\partial G}{\partial P}\right)\Bigg|_T = V, \qquad (3)$$

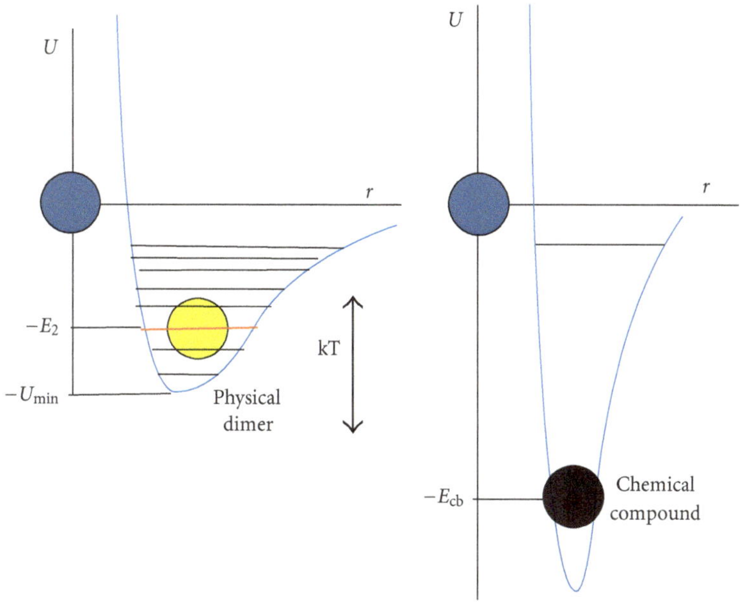

FIGURE 2: Comparison of a cluster with a chemical compound. Blue lines: molecular interaction potentials $U(r)$; narrow lines: energetic levels densely spaced in the cluster and widely separated in the chemical compound. E_2: the cluster's effective bond energy, comparable with kT; E_{cb}: the chemical bond energy, greatly exceeding the kT value.

that together with (1) leads to the differential equation [17, 18] for D_m:

$$\left(\frac{\partial D_m}{\partial P}\right)\bigg|_T = \frac{D_m}{RTD}. \tag{4}$$

The numeric integration of (4) should start from such a small pressure P_1, at which the difference between the real density D_1 and the ideal density $D_{p1} = P_1/RT$ is small. Then, the initial condition for numeric integration of (1) may be $D_{m1} = 2D_{p1} - D_1$. Equation (4) is valid for all fluid phases: gaseous, liquid, and supercritical. Figure 3 shows the D_m pressure dependence in comparison with the total density dependence at subcritical and supercritical temperatures in Nitrogen.

The figure shows that the share of monomers in the total density falls with growing pressure, but still remains large both in liquid and supercritical media. The integration over the gas-liquid phase transition boundary does not put any problem, because the monomer fraction density does not change its value at the phase transition. When the monomer fraction density is found, it becomes possible to use it as an argument for other thermodynamics functions, such as the entropy or potential energy, and to use it as an argument in series expansions of different values for properties of real gases.

2.4. Generalization of the Sackur-Tetrode Equation for Real Gases. The theory of Sackur [2] and Tetrode [3] for entropy of ideal atomic gases had been created in 1912 at the very beginning of the quantum theory development and when the discussion around the third law of thermodynamics still had not been finished. In spite of its applicability only to ideal atomic gases, it made a significant contribution both in thermal and quantum physics. Belloni wrote [22] that Fermi came to the famous Fermi-Dirac statistics trying to find his own way to the Sackur-Tetrode equation. The theory was confirmed experimentally by calculation of the entropy for dilute Neon gas taking into account the third law of thermodynamics. The measurements started from the near zero Kelvin temperature and included solid and liquid phases.

According to this theory, entropy of an ideal atomic gas with a molar volume V equals to

$$S = 2.5R + R\ln\left(\frac{V}{V_q}\right), \tag{5}$$

where V_q is the molar quantum volume given by (2).

Equation (5) with account of (1) and $G = H - TS$ can be transformed to

$$G - G_0 = RT\ \ln\left(D_m V_q\right) = H - H_0 - T(S - S_0), \tag{6}$$

where H_0 and S_0 are the enthalpy and entropy of the internal molecular movements: rotations and vibrations. Thus, we come to the Generalized Sackur-Tetrode equation [17–19]:

$$S = S_0 + \frac{H - H_0}{T} - R\ln\left(D_m V_q\right). \tag{7}$$

To test the developed here theory, we compared the experimental values for entropy of Water in gaseous and liquid phases with computed by (7) model values; D_m was found by integration of (4). All experimental data have been taken from the NIST database [7].

Figure 4 shows the pressure dependence of the Water entropy in gaseous and liquid phases at a temperature 350 K and the multiplied by factor 10^9 deviation of the model

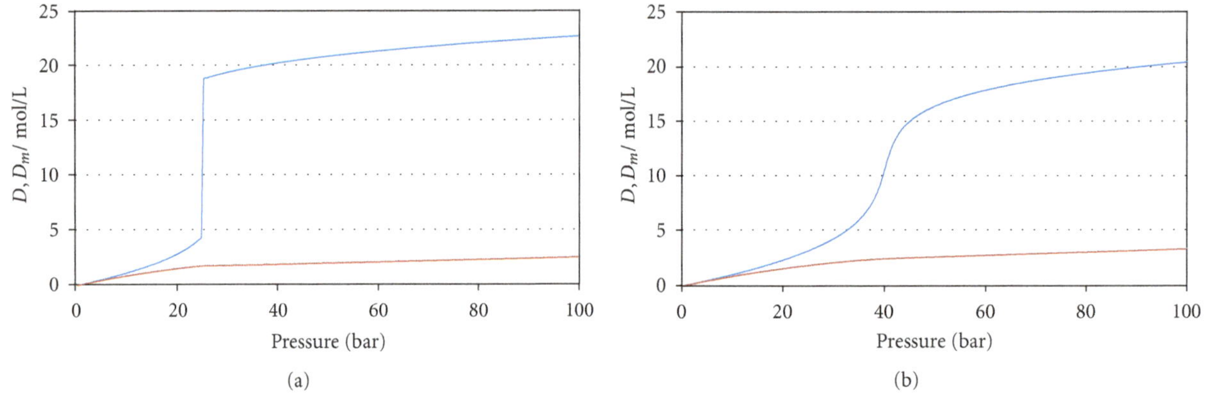

FIGURE 3: Density (blue lines) and monomer fraction density (red lines) versus pressure in gaseous, liquid, and supercritical Nitrogen at (a) subcritical temperature, $T = 120$ K; (b) supercritical temperature, $T = 130$ K.

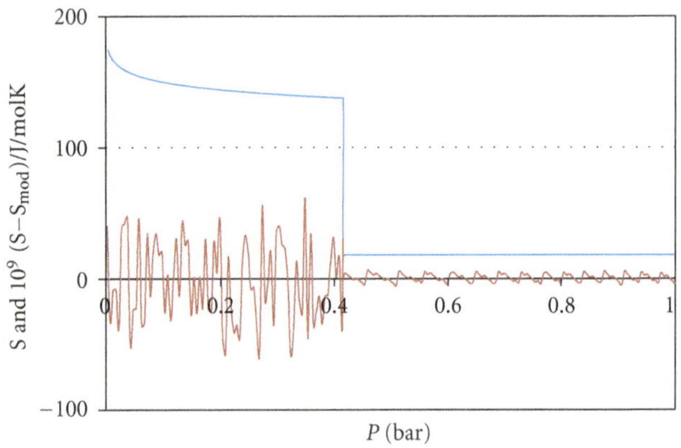

FIGURE 4: Confirmation of a high precision of the entropy computation by the generalized Sackur-Tetrode equation for Water in the gaseous and liquid phases at temperature 350 K. Blue line: experimental and computed values for entropy; red line: multiplied by factor 10^9 deviations of the computed values from the experimental ones.

values from the experimental data. For H_0, the result of extrapolation to zero pressure of the experimental values for $H(P_i)$ was taken. For $S_0 - R\ln(V_q)$ value, the averaged difference between experimental values for S and computed values for $(H - H_0)/T - R\ln(D_m)$ was taken. So this test estimates only the spread of computations. Nevertheless, the level of spread is nine orders of magnitude lower than the total entropy change that tells about correctness of the theory and about the high precision of computations. This high precision of computations results from remarkable features of the NIST database [7]:

(i) the experimental data have been interpolated with a very high precision;

(ii) the data for different thermophysical properties have been mutually correlated.

This test had been applied to a number of pure gases in different conditions, including gaseous, liquid, and supercritical phases, and the results confirm the high precision of computations of the $D_m(P, T)$ and $S(P, T)$ values. On this basis, the D_m values in a couple with the NIST tables [7] permit to find the cluster fractions equilibrium constants and molecular interaction parameters for pure real gases with a good precision.

3. Monomer Fraction Density as a New Variable for Thermophysical Functions

3.1. The Partial Potential Energies for Cluster Fractions in Gases. The potential energy $U(T, P)$ of a real gas can be defined as the difference between the molar internal energy $E(T, P)$ of a real gas and its limit $E_0(T)$ at zero pressure:

$$U(T,P) = E(T,P) - E_0(T). \tag{8}$$

The $E_0(T)$ corresponds to the internal energy for a mole of an ideal gas. In terms of clusters, the potential energy of real gases is equal to a sum of partial potential energies of the cluster fractions with account of the partial kinetic energies change with a change of density.

The contribution ΔU_{uv} of chemical compounds, such as the alkali metal [4, 5] or the Nitrogen Dioxide [1, 6] dimers, with their chemical bond energies E_{cb}, molar densities D_d, and equilibrium constants C_d, to the total equilibrium potential energy of a unit volume is equal to

$$\Delta U_{\mathrm{uv}} = -\ E_{\mathrm{cb}}\ D_d = -\ E_{\mathrm{cb}} C_d D_m^2. \tag{9}$$

The molar densities of dimers D_d and monomers D_m for chemical compounds can be found from the spectroscopy data. And, on this basis, it is possible to find the equilibrium constants C_d [4, 6]. But partial densities for clusters in real gases cannot be found in this way, and we have to use the potential energy to derive them. Instead of the molar potential energy U, we expand the potential energy UD of a unit volume by the monomer fraction density. This expansion does not separate the contributions of different isomers with the same number n of particles in the cluster fraction. In the result, we find the clusters' PE equilibrium constants as the thermodynamically averaged values for all possible isomers at a constant number n.

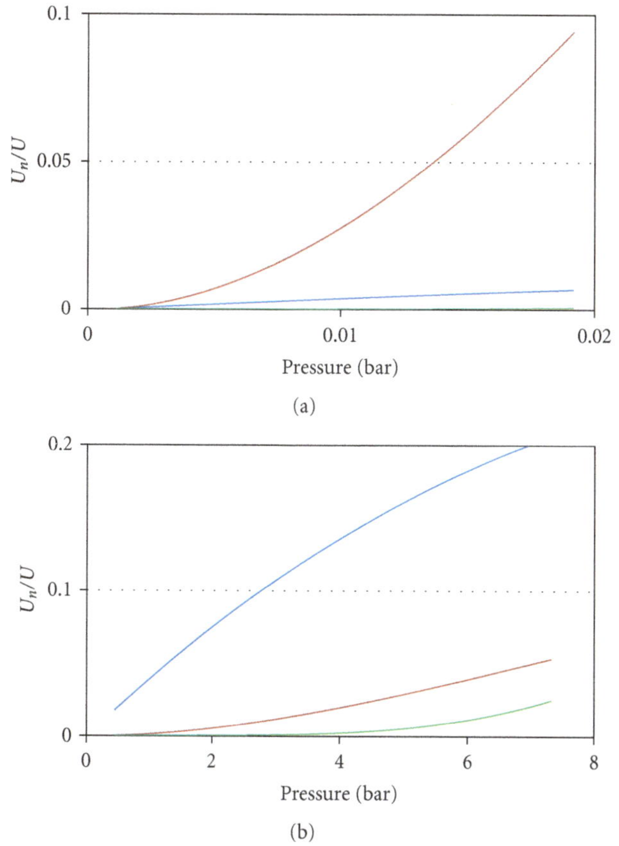

FIGURE 5: Shares of the partial potential energies U_n in the total potential energy U of the Water vapor. Blue lines: $n = 3$; red lines: $n = 4$; green lines: residuals; (a) $T = 290$ K; (b) $T = 440$ K.

We assume for the product UD an equation:

$$UD = -\sum (E_n C_{\text{un}} D_m^n). \tag{10}$$

To find the coefficients $K_{\text{un}} = C_{\text{un}} E_n$, at $K_{u1} = 0$, we expand the value UD/D_m^2 by D_m to the second exponent of D_m. It gives contributions to potential energy of small cluster fractions with a number of particles in a cluster up to four. A special method for this series expansion was developed to take into account the higher-order residual terms of the series expansion. The maximal density was limited by the criterion: the residual term should be small as compared to the found terms of expansion at this density. For example, the density for Argon gas was limited at the 0.3 mol/L level.

The mathematical aspects of this method stay too far from the paper on physical aspects of molecular interactions in real gases and may be described separately. The typical results of the series expansion for the Water vapor potential energy are shown at the Figure 5. The Figure shows the pressure dependences for shares of the trimers' and tetramers' fractions in the total potential energy at low and high temperatures. For comparison the share of the residual terms has been also plotted.

It is seen that, at $T = 290$ K, the share of tetramers is much larger than the trimers' share and reaches almost 10%. The residuals' share at this temperature is much lower than the trimers' share. At $T = 440$ K, the share of trimers is much larger than the tetramers' share and exceeds 20%. The residuals' share at this temperature is twice lower than the tetramers' share but quickly grows when pressure approaches its saturation value. Small values of residuals confirm the correctness of the expansion method in the temperature, range 273.16–440 K for pressures up to the saturation level. For a larger temperature the upper limit of pressure should be lower than its saturation value.

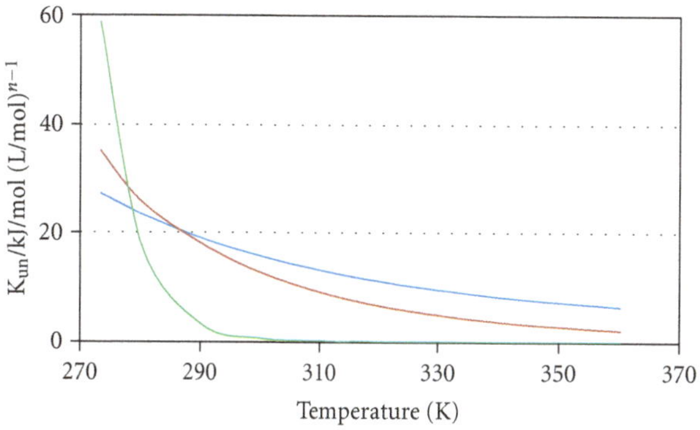

FIGURE 6: The series expansion coefficients for unit volume potential energy of the Water vapor. Blue line: K_{u2} (T); red line: K_{u3} $(T)/10$; green line: $K_{u4}/10^6$.

The resulting series expansion coefficients for the Water vapor are shown at Figure 6.

Figure 6 shows that near the triple point the potential energy of the Water vapor contains a large contribution from the 4-particle cluster fraction. For this computation, the data from the NIST database [7] have been downloaded with a precision of 10 digits. The database [7] contains the thermodynamically averaged experimental data. Therefore, all statistical processing of the experimental data had been performed at the database preparation. The correctness of this method has been confirmed by the correlation of different thermophysical values: partial energies, pressures, densities, activation energies, clusters' bond energies, and bond volumes, Joule-Thomson coefficients in wide ranges of T and P.

3.2. Activation Energies for Cluster Fractions. The temperature dependence of the clusters' PE equilibrium constants C_{un} shows that they correspond to real physical clusters, being analogues to chemical compounds. Figure 7 confirms the Boltzmann law for the temperature dependences of the expansion coefficients K_{un} for partial potential energies in rather wide temperature ranges for the Water vapor and Carbon Monoxide gas.

Figure 7 shows that it is possible to find the activation energies for rather wide temperature ranges that are given by tangents of slopes of the $\ln(K_{un})$ versus $1/T$ dependence. Figure 7(b) gives three values for activation energies E_n of the cluster fractions in the Carbon Monoxide gas. It is seen that the E_3/E_2 ratio is around 2.5 that signals about a complex binding of the third molecule to the dimer with an average number of bonds more than one. But the value for the E_4/E_2

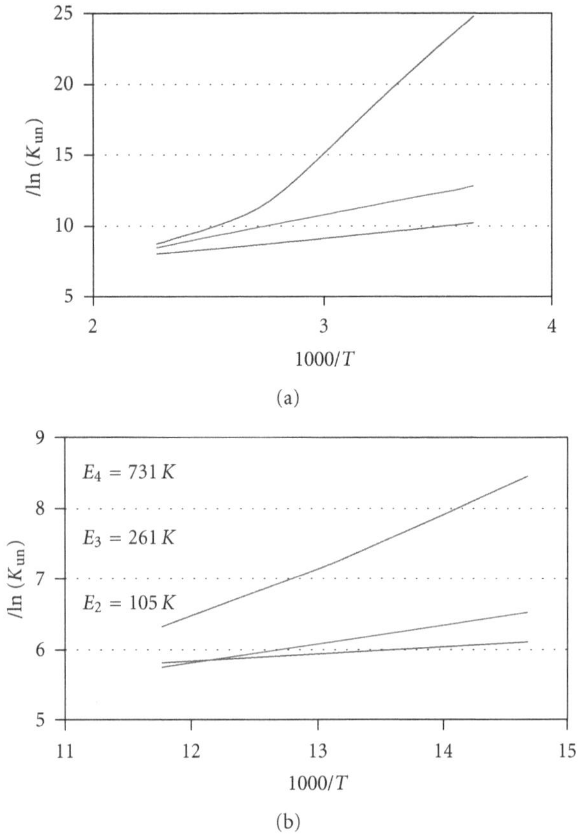

Figure 7: Linear dependences of the $\ln(K_{un})$ on the reverse temperature $1000/T$ for the Water vapor and Carbon Monoxide, corresponding to the Boltzmann law: (a) H_2O at T between 273.16 and 440 K; (b) CO at T between 68.16 and 85 K.

ratio near 7 signals about a tightly bonded structure of the tetramers in the CO gas near the triple point.

A similar picture is in the Water vapor at temperatures lower than 360 K, Figure 7(a). The E_3/E_2 ratio is around 2 that signal about an open chain structure of trimers with two Hydrogen bonds, but the E_4/E_2 ratio near 8 signals about a tightly bonded, solid-like, tetramers structure in the H_2O vapor near the triple point.

The diagram for the Water vapor reflects a very interesting phenomenon: the transformation of tightly bonded tetramers in the open chain tetramers at the temperature range 360–380 K. It is not a phase transition in tetramers: it is a substitution of one predominant isomer type by another, with smaller bond energy, but much larger number of bound states. This substitution is not abrupt but takes almost 20 K for the temperature zone, where different isomers coexist in a mixed isomer fraction. So the phenomenological approach may give some vision of the microscopic structure of clusters. It could be useful to proceed this research in the mutually coordinated directions: from the molecular dynamics simulations and from the thermophysical properties analysis.

3.3. Equilibrium Constants for Physical Clusters. To find the PE equilibrium constants C_{un} from the power expansions of the potential energy, we should first find the clusters' effective bond energies E_n. We can start from the found values of the activation energies E_n. In the temperature range, where E_n is constant,

$$C_{un}(T) = \frac{K_{un}}{E_n}. \tag{11}$$

With found clusters' PE equilibrium constants, we can find the cluster fraction's effective volume of the attraction zone V_n from the expression for C_{un} (T):

$$C_{un}(T) = V_n \exp\left(\frac{E_n}{T}\right). \tag{12}$$

The applicability of the Boltzmann law to clusters in the corresponding temperature range may be confirmed by the constant value of the found V_n.

The cluster fractions' equilibrium constants found from the potential energy, PE equilibrium constants, stay close to the chemical compounds' equilibrium constants in the form of their temperature, dependences. These equilibrium constants correspond to real physical clusters. But, at high temperatures there may be some difference between the spectroscopy and potential energy-based equilibrium constants. The spectroscopy data for compounds' concentrations are not sensitive to the repulsions of the compounds' components, while the potential energy of a cluster may feel these repulsions. The method of accounting for the repulsions in the PE equilibrium constant has to be developed.

3.4. The System of Equations for Pressure and Density in Real Gases. Having found the isothermal row of data $D_{mi}(T,P_i)$, we can reverse this dependence and expand the pressure by D_m:

$$P = RT\sum\left(C_n D_m^n\right). \tag{13}$$

The right side of (13) looks like a sum of partial pressures for n-particle cluster fractions. Therefore, the expansion coefficients $C_n(T)$ can be considered as apparent equilibrium constants for partial pressures of n-particle cluster fractions, at $C_1 = 1$.

Applying (4) to the expansion (13), we come [17, 18] to the series expansion for the total density D by powers of the D_m:

$$D = \sum\left(nC_n D_m^n\right). \tag{14}$$

This equation again looks like the sum of partial densities of basic particles in the n-particle cluster fractions. We came to this series expansion not from microscopic considerations, but on the thermodynamics basis. It confirms our decision to consider $C_n(T)$ as the apparent PDT equilibrium constants in spite of their strange temperature dependences, as it will be shown later.

3.5. Chemical Equilibrium for Apparent Cluster Fractions. From (14), we see that the apparent n-particle cluster fraction's molar density D_n, found from the PDT relations, is equal to

$$D_n = C_n D_m^n. \quad (15)$$

This result corresponds to the chemical equilibrium theory, but, for apparent cluster densities. The apparent n-particle cluster fraction's molar density D_n is proportional to the nth power of the monomer fraction density D_m. It should be stressed again that we did not use the microscopic considerations to come to this expression.

So the knowledge of the $D(D_m, T)$ and $P(D_m, T)$ dependences can be used for computation of the apparent PDT equilibrium constants C_n by the joint series expansion of the isothermal $D(D_m, T)$ and $P(D_m, T)$ dependences. In this approach, it is impossible to determine the partial contributions of isomers; therefore, C_n reflects a joint contribution of all isomers and mainly of the predominant types of isomers, existing at a given temperature in the n-particle cluster fraction.

It should be specially noted that the (13) and (14) are the series expansions of P and D by the monomer fraction density D_m. In this respect, they differ [23] from the well known virial expansions [1, 9–11, 24] of P by the total density or D by the total pressure.

At small pressures, the difference between D and D_m is small that makes possible to use D instead of D_m in the series expansion (13). At these conditions, only the contributions of dimers in the total density and pressure are essential. For this reason, the second PDT equilibrium constant C_2 correlates with the second virial coefficient B, namely, $C_2 = -B$. It is well known [24] that, at the Boyle temperature T_B, the second virial coefficient B changes its sign. Therefore, the apparent dimers' PDT equilibrium constant C_2 and apparent dimers' fraction density become negative at $T > T_B$.

4. Parameters of Molecular Interactions in Pure Real Gases

The physical clusters can be characterized by their effective parameters: the bond energy E_n and the attraction zone volume V_n. The bond energies for any n may be measured in Joule per mole. After dividing the bond energy values by the universal gas constant R, we receive their values in K. In this paper, the bond energies are estimated in Kelvin just to compare them with the used values of temperature T, because the E_n/T ratio is very important for the analysis of equilibrium constants. But, for V_n, the units of measurement depend on n: for $n = 2$, it is the molar volume (L/mol); for $n > 2$, the dimension of V_n is $(\text{L/mol})^{n-1}$.

4.1. The Clusters' Effective Bond Energy. In some papers, the maximal depth $U_{\min}$ of the potential well in dimers, shown on the Figure 2(a), is considered as the bond energy, but this value never corresponds to the estimated activation energy for the dimer fraction density. For the reason that the energetic levels in the cluster's bound state are spaced very densely and the gaps between them are much smaller than kT, the bond energy E_n is a thermodynamic average of localized energy levels. To distinguish it from the chemical bond with clearly defined quantum energetic level, it may be named as the effective bond energy. This value corresponds to the estimated activation energy of the cluster fraction density. The $E_2/U_{\min}$ ratio depends on the form of the interaction potential $U(r)$.

4.2. The Clusters' Effective Volume of the Attraction Zone. The effective volume of the dimer's attraction zone V_2 corresponds to the average number of localized states in the attraction zone $N_{b2} = V_2/V_q$. This value for the number of bound states in dimers reflects the smooth dependence of the attraction energy on the interparticle distance in conditions of densely spaced energetic levels. It strongly depends on temperature due to the temperature dependence of V_q and differs from the well-known chemical compound's number of states that is the quantum degeneration degree of the compound's ground level and does not depend on temperature.

For larger clusters, the average number of bound states $N_{bn} = V_n/(V_q)^{n-1}$, where V_n is the effective volume of the cluster's attraction zone with its dimension equal to $3(n-1)$. For all pure gases, there exist the relatively large ranges of temperature, in which both E_n and V_n are almost constant. These constants can be considered as the fixed parameters of the molecular interactions in pure gases. Besides, for the pair molecular interaction the third parameter, the excluded volume, reflecting the average volume of the repulsion zone, may be found. The question about the excluded volumes for clusters is not so clear and waits for its decision.

4.3. The PDT Equilibrium Constants Determination. To remove the large regular part from the $D(D_m, T)$ and to pay more attention to the clusters contributions in it, we expand by the D_m the function $(D - D_p)/D_m^2$. Here, $D_p = P/RT$ is the density of ideal gas at pressure P. The coefficients of this expansion are $C_2(T), 2C_3(T), 3C_4(T), \ldots$ This type of serial expansion can be named as a joint series expansion of the system of (13) and (14), because, during the expansion, we use both equations of this system. This procedure is repeated for different temperatures, and, thus, we receive the temperature dependences of the $C_n(T)$. Figure 8 shows the results of the $C_n(T)$ computation for Water vapor, $n = 2, 3, 4$.

The presented on Figure 8 results of the computerized analysis of experimental thermophysical data from the NIST database [7] for the Water vapor show that the apparent equilibrium constant for 4-particle cluster fraction is very large and its temperature dependence is very steep. It confirms the made earlier conclusion about a large value of the effective bond energy E_4 for tetramers in the Water vapors. But, now, it is impossible to find the activation energy because the apparent PDT equilibrium constants do not correspond to the Boltzmann law.

Figure 9 shows that the apparent PDT equilibrium constant is lower than the PE equilibrium constant on almost a constant value.

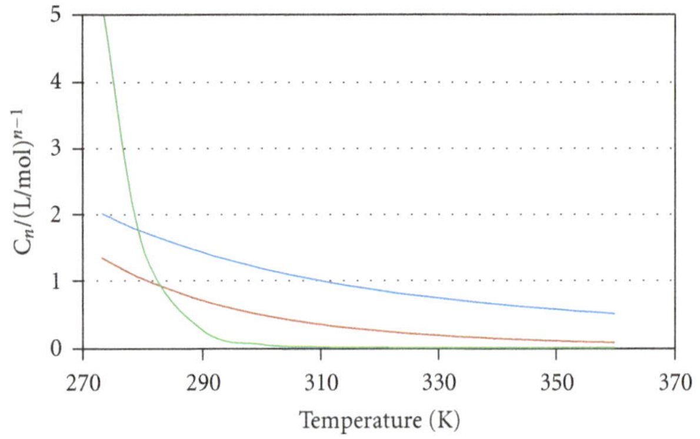

FIGURE 8: The apparent PDT equilibrium constants for the Water vapor at temperatures between the triple and normal boiling points. Blue line: $C_2(T)$; red line: $C_3(T)/10$; green line: $C_4(T)/10^5$.

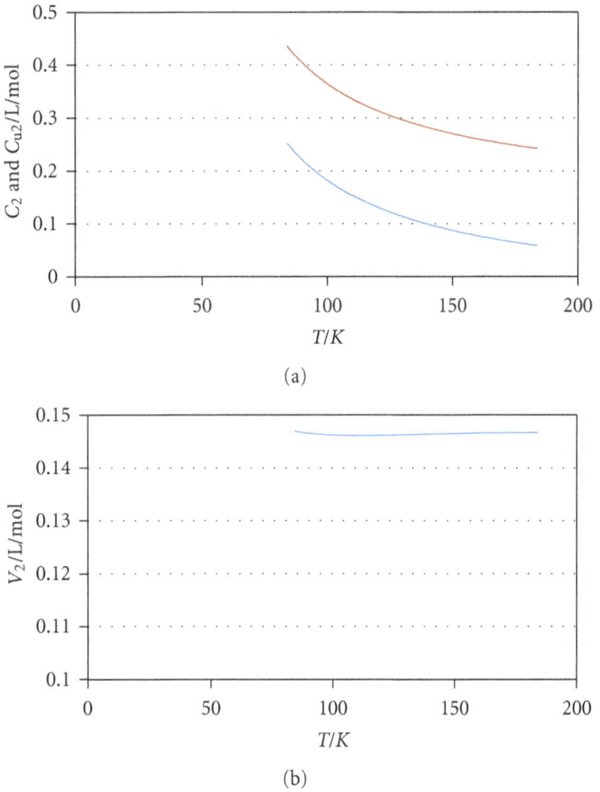

FIGURE 9: Results of computation of the equilibrium constants and the effective volume of the attraction zone for dimers in Argon. (a) Blue line: the apparent PDT equilibrium constant; red line: the PE equilibrium constant. (b) The effective volume of the attraction zone.

To come to the effective bond energies for different cluster fractions, we should use both the apparent PDT equilibrium constants and the series expansion coefficients for potential energy.

4.4. A New Method for Estimation of the Clusters' Effective Bond Energies. The comparison of the found earlier PE

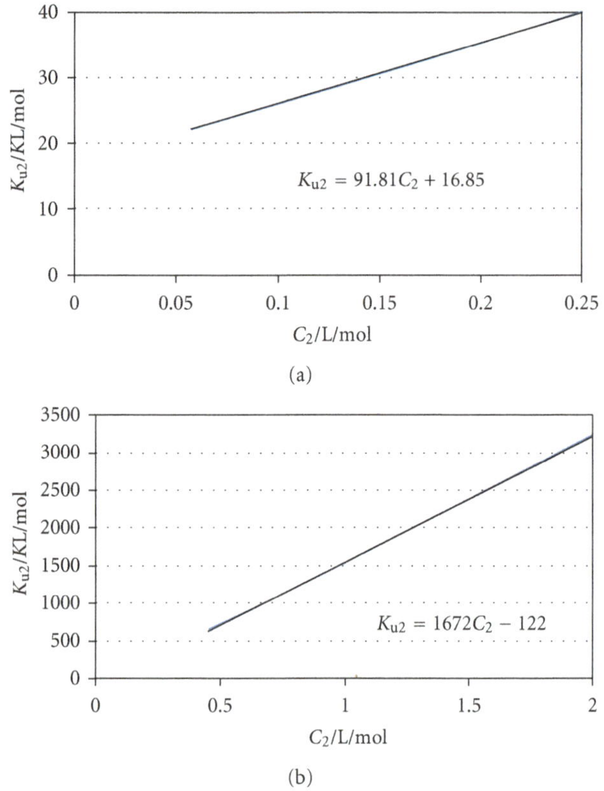

FIGURE 10: The correlations between the second-order series expansion coefficients of the potential energy UD for a unit volume and the difference between the real and ideal densities $(D - D_p)$: (a) for the gaseous Argon in the temperature range between the triple point T_{tr} and T_{tr} + 100 K; (b) for the Water vapor in the temperature range between the triple point and the normal boiling point.

equilibrium constants C_{un} with the PDT equilibrium constants shows that C_n contains in addition to the equilibrium constants C_{un} corresponding to real physical clusters the virtual part C_{vn}:

$$C_n = V_n \exp\left(\frac{E_n}{T}\right) + C_{vn}. \tag{16}$$

The analysis of data for many gases shows that the C_{vn} changes very slightly if compared to the changes of the C_n or C_{un} with temperature. In some temperature ranges, it is almost constant. Therefore, the diagram of the $K_{un} = C_{un}E_n$ versus C_n is close to linear with a tangent of slope equal to E_n.

To find from the found $K_{un}(T)$ dependences the effective bond energies E_n and the PE equilibrium constants, we plot the K_{un} versus C_n. The resulting curve can be named as the correlation curve between the expansion coefficients K_{un} and C_n. Figure 10 presents an example of such a plot for dimers in Argon and Water vapors.

Figure 10 shows that the correlation curves for dimers in the Argon and Water vapors in the temperature range between the triple point and the triple point + 100 K are close to linear. Their tangents of slope give us the averaged for the said temperature ranges, the dimers' effective bond energies E_2. For these diagrams potential energy was divided by the gas constant R; therefore, dimension of the K_{u2} is (K L/mol) and the average tangent of slope gives the effective bond energy in K. From Figure 10, it is seen that the effective bond energy for dimers in the Argon gas is equal to 91.81 K and, in the Water vapor, it is equal to 1672 K. This procedure can be repeated for other n; thus, we come to the cluster fractions' effective bond energies. For the Water vapor, the results are shown in the Table 1. The difference of the E_2 values in the Table 1 and Figure 10(b) results from small nonlinearities of the used dependences.

4.5. Estimation of the Bond Volumes. For the Argon gas, the found values of V_2 are very close to a constant value equal to 0.146 L/mol. An additional confirmation gives the activation energy found from the $\ln(C_{u2})$ plot versus $1/T$ that is close to the found value for $E_2 = 91.81$ K. The effective bond energies determination errors result from the nonlinearities of the correlation curves shown on the Figure 10. For dimers in Argon and Water vapor, they do not exceed 2% in the temperature range under investigation. So we come to the same results for effective bond energies as from the estimations of the activation energy and from the correlation between the PDT and PE expansion coefficients.

For the Water vapor, the found values for the effective volumes V_n of the attraction zones and effective bond energies E_n in clusters are shown in the Table 1.

5. The Results of the Molecular Interaction Parameters Computation for Some Gases

5.1. The Effective Bond Parameters for the Water Vapor. In addition to results for the Water vapor, shown on the Figure 7(a), we present here the values of the effective bond energies and effective volumes of the attraction zones for the temperature range between the triple point and the normal boiling point, see Table 1. The normal boiling point for Water stays near the center of the temperature range, where the exchange of the predominant types of tetramers takes place. So the table gives parameters for the low temperature range, where tetramers have rigid solid-like structure with a large value of the bond energy and with a very small value for the effective volume of the attraction zone.

The shown E_n/E_2 ratios permit to estimate the type of the predominant isomers. The trimers' ratio being near 2 tells about the open chain trimers' structure with two Hydrogen bonds, but the tetramers' ratio near 8 signals about the closed structure with eight Hydrogen bonds. The number of Hydrogen bonds may be smaller, if every bond becomes stronger in the tightly bound cluster. But the conclusion about the solid-like crystalline structure of tetramers in the Water vapor near its triple point is correct at any case. An additional confirmation of this idea is provided by the V_n/V_2^{n-1} ratio. In trimers, it is around 3 and tells about the possible number of diverse open chain configurations with a large freedom of molecular movements. But, in tetramers, it is very small that signals about the rigid crystalline structure of clusters in this range of temperatures.

These results have been achieved after a large work. But the conclusion about an essential role of tetramers in the Water vapor near the triple point may be seen directly from the density dependence of the Joule-Thomson coefficient. This coefficient may give very useful information about molecular interactions in gases.

Table 1: The effective bond parameters for the Water vapor clusters at T between the triple and normal boiling points.

Parameters for n	2	3	4
Bond energy E_n, K	1690	3370	13500
E_n/E_2 ratio		1.99	7.99
Bond volume $V_n(\text{L/mol})^{n-1}$	3.9 10^{-3}	4.7 10^{-5}	1.4 10^{-16}
V_n/V_2^{n-1} ratio		3.1	2.4 10^{-9}

5.2. The Joule-Thomson Coefficient for the Water Vapor. The shown on Figures 6 and 8 large values of equilibrium constants for tetramers in the Water vapor are reflected in the Joule-Thomson coefficient μ density dependence. According to Jame Wisniak [25], "the Joule-Thomson coefficient may be calculated from a knowledge of the PVT relationship of the fluid (equation of state) and the specific heat at constant pressure":

$$\mu = \frac{1}{C_p}\left(T\left(\frac{\partial V}{\partial T}\right)\Big|_P - V\right). \tag{17}$$

It is useful to remove large regular part from V by introduction of the interaction volume $V_{\text{int}} = RT/P - V$. After that, (17) may be rewritten as

$$\mu C_p = V_{\text{int}} - T\left(\frac{\partial V_{\text{int}}}{\partial T}\right)\Big|_P. \tag{18}$$

The interaction volume is proportional to the used in our series expansion value $(D - D_p)$:

$$V_{\text{int}} = \frac{D - D_p}{D\, D_p}. \tag{19}$$

In the Water vapor, the difference between D, D_p and D_m is very small. Therefore, it is possible to change the denominator of (19) on the D_m^2. Then, the V_{int} becomes just the value $(D - D_p)/D_m^2$ that was used in the series expansion to find the PDT equilibrium constants. According to the system (13) and (14),

$$\frac{D - D_p}{D_m^2} = C_2 + 2C_3 D_m + 3C_4\, D_m^2 + \cdots. \tag{20}$$

Therefore, we can transform (18) in the series expansion for the μC_p:

$$\mu C_p = C_2 - \frac{TdC_2}{dT} + 2\left(C_3 - \frac{TdC_3}{dT}\right)D_m + 3\left(C_4 - \frac{TdC_4}{dT}\right)D_m^2. \tag{21}$$

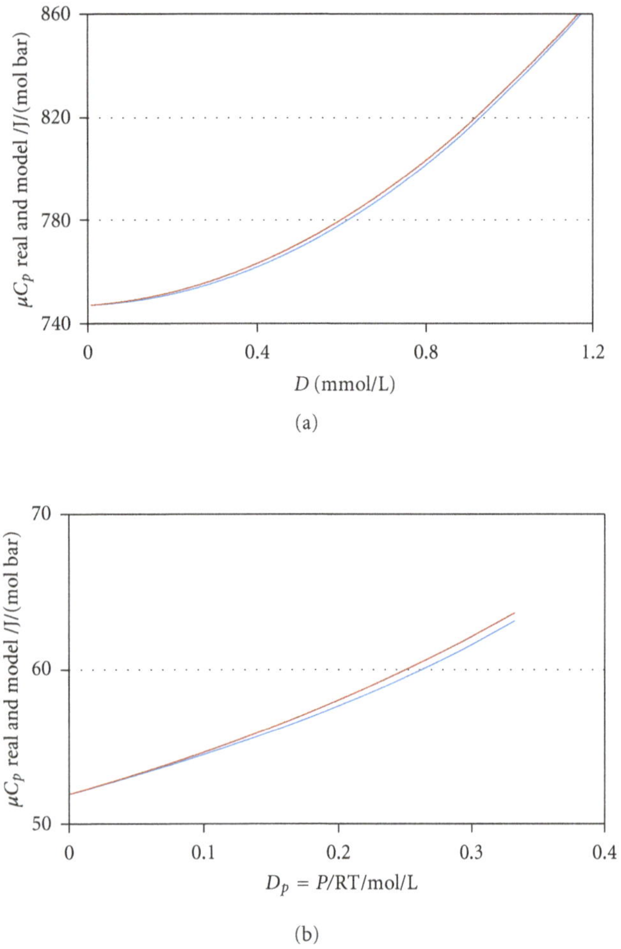

FIGURE 11: The product of the Joule-Thomson coefficient by specific heat at constant pressure versus molar density: (a) in the Water vapor at temperature 300 K, (b) in the Argon gas at temperature 100 K. Blue lines: experimental values; red lines: modeled values by (22) and (24).

We know that $C_2(T)$ differs from the $C_{u2}(T) = V_2 \exp(E_2/T)$ on a constant value C_{v2} that gives zero at differentiation. So we finally have the expression for the μC_p:

$$\mu C_p = C_2 + \frac{K_{u2}}{RT} + 2\left(C_3 + \frac{K_{u3}}{RT}\right) D_m + 3\left(C_4 + \frac{K_{u4}}{RT}\right) D_m^2. \tag{22}$$

The expression (22) predicts a large quadratic density dependence of the μC_p product in the Water vapor. This prediction is confirmed by Figure 11(a). Equation (22) is valid for gases with large E_n/T ratios, like Water or Methanol vapors. But for van der Waals gases, like Argon, the expression for μC_p should take into account the difference between monomer fraction density and total density of the gas.

The quadratic character of the μC_p dependence on density, shown on Figure 11(a), tells about a large share of 4-particle clusters in the Water vapor at room temperatures.

TABLE 2: The values of apparent equilibrium constants C_n and series expansion coefficients K_{un} for potential energy of the Argon gas at $T = 100$ K.

n	C_n	Units	Values	K_{un}	Units	Values
2	C_2	L/mol	0.182778	K_{u2}	JL/mol^2	279.3154
3	C_3	(L/mol)^2	0.066657	K_{u3}	JL^2/mol^3	215.5387
4	C_4	(L/mol)^3	0.036931	K_{u4}	JL^3/mol^4	185.4765

5.3. The Joule-Thomson Coefficient for Van Der Waals Gases. The expansion of the V_{int}, given by (19), with account of (13) and (14), up to the second exponent of the D_m results in

$$V_{\text{int}} = C_2 + \left(2C_3 - 3C_2^2\right) D_m + \left(3C_4 - 10\, C_3 C_2 + 5C_2^3\right) D_m^2. \tag{23}$$

The differentiation of equilibrium constants by T gives $T dC_n/dT = -K_{un}/RT$. The differentiation of D_m gives $T(\partial D_m/\partial T)|_P = -\left(K_{u2}/RT + D_p/D\right) D_p/T$. In the result, we have

$$\begin{aligned}\mu C_p = {} & C_2 + \frac{K_{u2}}{RT} + 2\left(C_3 + \frac{K_{u3}}{RT} - 3C_2^2 - 6C_2\frac{Ku2}{RT}\right) D_m \\ & + \left(3\left(C_4 + \frac{Ku4}{RT}\right) - 10C_3C_2 + 5C_2^3 \right. \\ & \left. -10\left(C_3\frac{K_{u2}}{RT} + C_2\frac{K_{u3}}{RT}\right)\right) D_m^2 + 15\, C_2^2 \frac{K_{u2}}{RT} D_m^2 \\ & + \left(\left(2C_3 - 3C_2^2\right) + 2\left(3C_4 - 10\, C_3C_2 + 5C_2^3\right) D_m\right) \\ & \times \left(\frac{K_{u2}}{RT} + \frac{D_p}{D}\right)\frac{D_p}{T}.\end{aligned} \tag{24}$$

To apply this equation to some real gas, for example, Argon, we should know the values of apparent equilibrium constants and coefficients of the potential energy expansion at some fixed temperatures. For Argon, we have the values for expansion coefficients at $T = 100$ K, shown in Table 2.

The result of these values substitution into (19) is shown on Figure 11(b). The deviations of the predicted values from the experimental values on Figure 11 can be explained by the assumptions made in the derivation of the (22) and (24).

5.4. Molecular Interaction Parameters for Gaseous Alkanes. Applying the described method of equilibrium constants and their parameters determination to the row of Alkanes, we met some deviations of their characteristics from those described in the previous section. First of all, the correlation curve of K_{u2} and C_2 does not look like the straight line. That is clearly seen on Figure 12.

Figure 12 shows that the forms of the correlation curve $K_{u2}(C_2)$ for different representatives of the Alkanes are similar, only the maximal values of the PDT equilibrium constants grow with the number of carbon atoms in the molecule. It is rather understandable.

But if we compare these correlation curves with curves shown on Figure 10 for Argon and Water, we can notice that the sign of the curvature is opposite. In Alkanes, the tangent

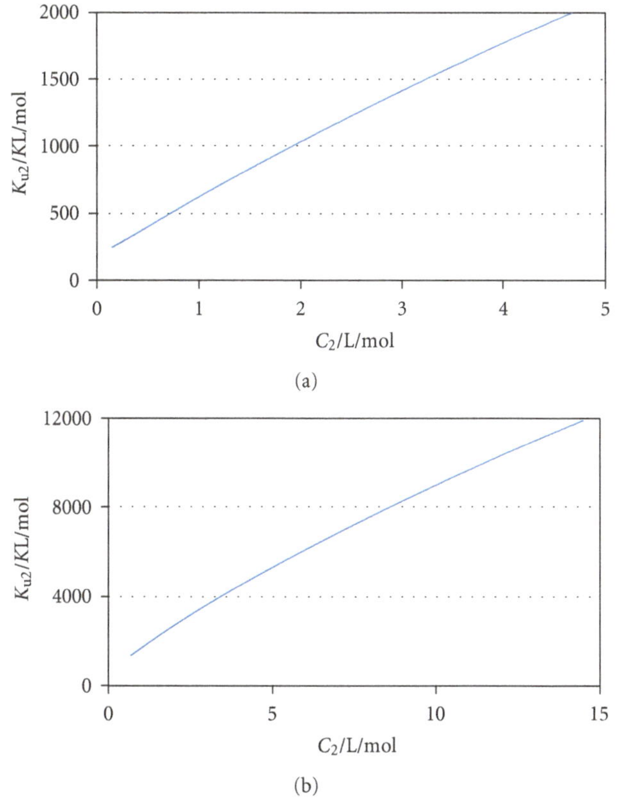

FIGURE 12: Typical correlation curves $K_{u2}(C_2)$ for Alkanes: (a) the correlation curve for Butane; (b) the correlation curve for Decane.

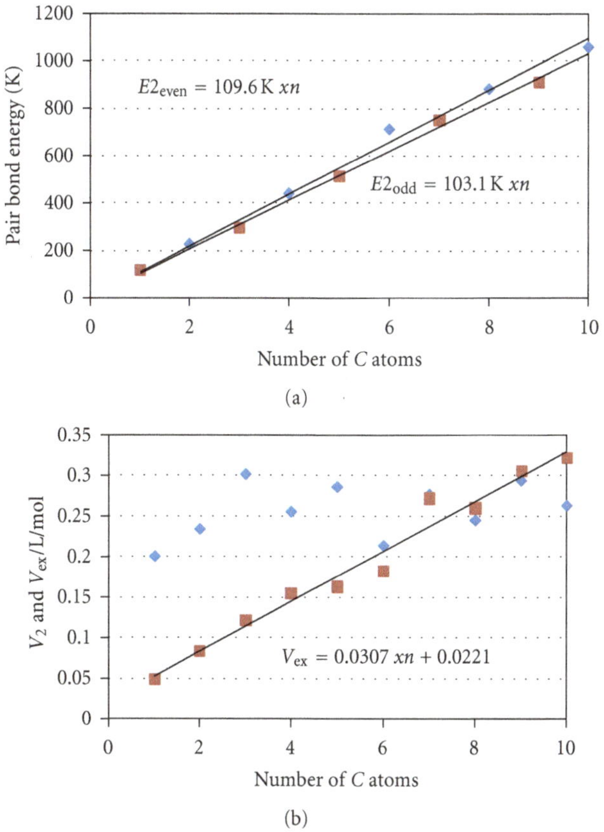

FIGURE 13: Effective pair bond parameters versus the number n of C atoms in Alkanes molecules: (a) effective pair bond energy $E_2(n)$: blue rhombs—even numbers; red squares—odd numbers; (b) effective volume for the attraction zone $V_2(n)$: blue rhombs; Excluded volume $V_{ex}(n)$: red squares.

of the curve's slope at low temperatures becomes lower than at higher temperatures. In many other gases, including Argon and Water, there is a growth of the tangent to lower temperatures. The growing tangent at low temperatures is understandable. It tells about the possible growth of the effective bond energy at low temperatures. But the apparent diminishing of the effective bond energy in Alkanes at low temperatures may signal about some unknown detail of molecular interactions in these gases.

Therefore, the effective interaction parameters have been estimated only for high temperatures, where the correlation curves become straighter. The results are shown on the Figure 13.

Figure 13 shows that, in Alkanes, there is almost linear dependence of the effective bond energy on the number of carbon atoms in the molecule. But this line doubles: for even numbers of C atoms, the bond energies are slightly higher than for odd numbers. The zig-zag dependence of the Alkanes' properties on the number of C atoms n in their molecules is well known. The melting points for the row of Alkanes from Ethane to Decane [1] plotted versus n show also the zig-zag dependence, Figure 14. This paper has a goal to develop new methods of the experimental thermophysical properties analysis. The author hopes that the parallel between the zig-zag dependences for the effective pair bond energy and melting point in Alkanes may stimulate new research works in this direction.

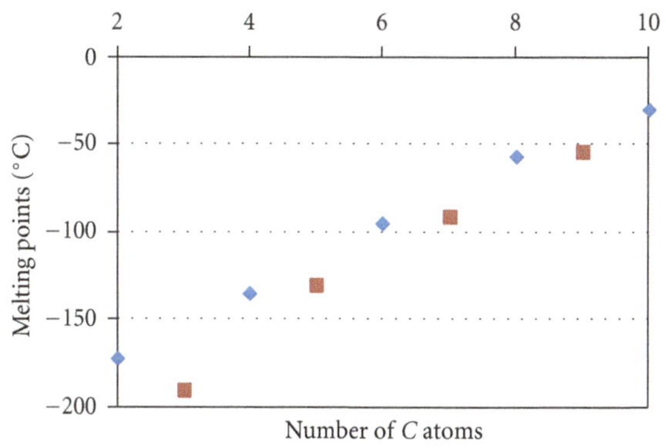

FIGURE 14: Melting points of Alkanes versus the number n of C atoms in molecules.

The values of the attraction zone volume V_2, shown at Figure 13(b), do not exhibit a general dependence on the number of C atoms but have a large dispersion.

The excluded volume grows linearly with growing own volume of molecules that is quite natural.

These facts permit to make conclusions about the predominant dimer configuration in Alkanes. Linear growth of the bond energy with the number of C atoms tells about the parallel position of molecules in the dimer, when all C atoms participate equally in the interaction. The value of the V_2 being almost constant also tells in favor of this configuration: the interacting molecules move along their axes not more than on the C–C distance, independently on the number of C atoms.

6. Conclusions

(i) It was shown that, for real gases, there is an essential difference between the apparent clusters' equilibrium constants responsible for pressure and potential energy.

(ii) The dimer equilibrium constant for potential energy of a real gas looks like the equilibrium constant for molecular compounds and its temperature dependence obeys to the Boltzmann law. Therefore, it corresponds to real physical dimers.

(iii) The dimer equilibrium constant for pressure contains an additional negative term that may be attributed to virtual dimers. Due to this term, it crosses the zero level and becomes negative at temperature over the Boyle point.

(iv) The introduction of the monomer fraction density in a family of thermodynamics variables expands power of the computer-assisted analysis of thermophysical properties of pure gases.

(v) The developed method of the computer-aided analysis of the precise thermophysical values from the NIST databases shows the existence in the Water vapor of tetramers with large bond energy at temperatures lower than the boiling point.

(vi) The investigation of the molecular interactions in Alkanes resulted in the linear dependence of the effective bond energy on the number of C atoms in molecule and led to conclusions about parallel position of molecules in the Alkanes dimers.

References

[1] L. Pauling, *General Chemistry*, W.H. Freeman and Company, San-Francisco, Calif, USA, 1970.

[2] O. Sackur, "Die universelle Bedeutung des sog. elementaren Wirkungsquantums," *Annalen der Physik*, vol. 40, p. 67, 1913.

[3] H. Tetrode, "Die chemische Konstante der Gase und das elementare Wirkungsquantum," *Annalen der Physik*, vol. 343, no. 7, pp. 434–442, 1912.

[4] C. T. Ewing, "Molecular association in sodium, potassium, and cesium vapors at high temperatures," *Journal of Physical Chemistry*, vol. 71, no. 3, pp. 473–477, 1967.

[5] N. B. Vargaftik, Yu. K. Vinogradov, and V. S. Yargin, *Handbook of Physical Properties of Liquids and Gases*, Begell House, 3rd edition, 1996.

[6] Q. Yu and H. Gao, "A simple determination of the NO_2 dimerization equilibrium constant," *Journal of Chemical Education*, vol. 74, no. 2, pp. 233–234, 1997.

[7] NIST Database on Thermophysical Properties of Fluids, http://webbook.nist.gov/chemistry/fluid.

[8] NIST Database of the Thermophysical Properties of Gases Used in the Semiconductor Industry, http://properties.nist.gov/fluidsci/semiprop/.

[9] L. D. Landau and E. M. Lifshitz, *Statistical Physics*, Pergamon Press, London, UK, 1980.

[10] R. Feynman, *Statistical Mechanics; A Set of Lectures*, W.A. Benjamin, Mass, USA, 1972.

[11] J. E. Mayer and G. M. Mayer, *Statistical Mechanics*, John Wiley and Sons, New York, NY, USA, 1977.

[12] I. J. Ford, "Statistical mechanics of nucleation: a review," *Journal of Mechanical Engineering Science C*, vol. 218, no. 8, pp. 883–899, 2004.

[13] P. Schaaf, B. Senger, and H. Reiss, "Defining physical clusters in nucleation theory from the N-particle distribution function," *Journal of Physical Chemistry B*, vol. 101, no. 43, pp. 8740–8747, 1997.

[14] I. Kusaka and D. W. Oxtoby, "Identifying physical clusters in vapor phase nucleation," *Journal of Chemical Physics*, vol. 110, no. 11, pp. 5249–5261, 1999.

[15] D. I. Zhukhovitskii, "Structural transition in hot small clusters," *Journal of Chemical Physics*, vol. 110, no. 16, pp. 7770–7778, 1999.

[16] IUPAC, *Compendium of Chemical Terminology*, The Gold Book, Blackwell Scientific Publications, Oxford, UK, 2nd edition, 1997.

[17] B. Sedunov, "Monomolecular fraction in real gases," in *Proceedings of the Joint European Thermodynamics Conference (JETC IX '07)*, pp. 176–180, 2007.

[18] B. Sedunov, "Monomer fraction in real gases," *International Journal of Thermodynamics*, vol. 11, no. 1, pp. 1–9, 2008.

[19] B. Sedunov, "New approaches to teaching the thermal physics of fluids," in *Proceedings of the 21st IUPAC International Conference on Chemical Thermodynamics (ICCT '10)*, p. 339, Tsukuba, Japan, 2010.

[20] Ch. Kittel, *Thermal Physics*, John Wiley and Sons, New York, NY, USA, 1970.

[21] G. N. Lewis and M. Randall, *Thermodynamics*, McGraw–Hill, New York, NY, USA, 2nd edition, 1961.

[22] L. Belloni, "A Note on fermi's route to fermi-dirac statistics," *Scientia*, vol. 113, pp. 422–430, 1978.

[23] B. Sedunov, "Equilibrium constants for clusters in real gases," in *Proceedings of the 23rd ESAT*, p. 66, Cannes, France, 2008.

[24] R.C. Reid, J. M. Prausnitz, and T. K. Sherwood, *The Properties of Liquids and Gases*, McGraw–Hill, New York, NY, USA, 1977.

[25] J. Wisniak, "The Joule–Thomson coefficient for pure gases and their mixtures," *Journal of Chemical Education*, vol. 4, pp. 51–57, 1999.

22

Separation of Methylene Blue Dye from Aqueous Solution Using Triton X-114 Surfactant

Arunagiri Appusamy,[1] Prabisha Purushothaman,[1] Kalaichelvi Ponnusamy,[1] and Anantharaj Ramalingam[2]

[1]*Department of Chemical Engineering, National Institute of Technology, Tiruchirappalli, Tamilnadu 620015, India*
[2]*Department of Chemical Engineering, Faculty of Engineering, University of Malaya, Kuala Lumpur 50603, Malaysia*

Correspondence should be addressed to Anantharaj Ramalingam; anantharaj@um.edu.my

Academic Editor: Felix Sharipov

In this study, the interaction energy between Triton X-114 surfactant + methylene blue or water and methylene blue + water was investigated using Hartree-Fock (HF) theory with 6-31G* basis set. The results of structures and interaction energies show that these complexes have good physical and chemical interactions at atom and molecular levels. However, the Triton X-114 surfactant + methylene blue complex shows stronger molecular interaction compared to other complexes systems. The order of the interaction energy is 4303.472023 (Triton X-114 surfactant + water) > −1222.962 (methylene blue + water) > −3573.28 (Triton X-114 surfactant + methylene blue) kJ·mole^{-1}. Subsequently, the cloud point extraction was carried out for 15 ppm of methylene blue in a mixture at 313.15 and 323.15 K over the surfactant concentration range from 0.01 M to 0.1 M. From the measured data, the excess molar volume was calculated for both phases. The results show a positive deviation in the dilute phase and a negative deviation in the surfactant rich phase. It is confirmed that the interaction between Triton X-114 and methylene blue is stronger than other complex systems due to the presence of chemical and structural orientation. The concentration of dyes and surfactant in the feed mixture and temperature effect in both phases has been studied. In addition, the thermodynamics feasibility and efficiency of the process have also been investigated.

1. Introduction

The removal of dyes such as methylene blue (MB) from industrial wastewater before discharge is of uttermost importance. Organic dyes colorize other substances in water, making them visible and aesthetically unpleasant [1]. Besides, photosynthetic activities are hampered due to interference with sunlight penetration into water bodies, thus affecting fish and other aquatic organisms. Ghosh and Bhattacharyya [2] reported that MB is not strongly hazardous but it has very harmful effects on living things. Therefore, the removal of MB from waste water is an essential task. Some of the dyes are carcinogenic and mutagenic due to their molecular structure, functional group, and types such as benzidine and metals [3]. There are serious adverse effects with far-reaching consequences that are attributed to MB. The contact with eye can lead to permanent blindness to human and animals, experiencing short period of rapid or difficult breathing when inhaled, burning sensation when ingested through mouth, and causes other adverse conditions such as nausea, vomiting, profuse sweating, and mental confusion.

Due to recent increase in industrial activities such as textile manufacturing, the use of dyes increases rapidly. The most recent statistics [4] indicated rapid use of these dyes with more than 10,000 different types of dyes being produced. They reported a global annual manufacture of these dyestuffs and intermediates estimated to be 7×10^8 kg. Therefore, the removal of dyes from wastewater is gaining research interest. Removal of these dyes from wastewater effluents is problematic as they possess a complex aromatic chemical structure that makes them highly visible and reduced photosynthetic activity in aquatic systems by reducing light

TABLE 1: Equation used in this study.

Name	Formula	
Binding energy (BE)	$BE_{S-D} = TE_{S+D} - (TE_S + TE_D)$	(a)
Excess molar volume (V_E)	$V_E = \sum_{i=1}^{3} \frac{x_i M_i}{\rho_{mix}} - \sum_{i=1}^{3} \frac{x_i M_i}{\rho_i}$	(b)
Phase volume ratio (R_v)	$R_v = \frac{V_s}{V_w}$	(c)
Pre concentration factor (f_c)	$f_c = \frac{V_t}{V_s}$	(d)
Distribution coefficient (K_d)	$K_d = \frac{C_s}{C_w}$	(e)
Cloud point extraction efficiency (η)	$\eta = \frac{C_0 V_t - C_w (V_t - V_s)}{C_0 V_t} \times 100$	(f)
Change in Gibbs free energy (ΔG^0), change in entropy (ΔS^0), and Change in enthalpy (ΔH^0)	$\Delta G^0 = \Delta H^0 - T\Delta S^0$	(g)
	$\log\left(\frac{q_e}{C_e}\right) = \frac{\Delta S^0}{2.303R} - \frac{\Delta H^0}{2.303RT}$ Where: $q_e = \frac{A}{X}$, $A = V_0 C_0 - V_d C_e$, $X = C_s V_0$	(h)

Note: S: surfactant; D: dyes.

penetration. Several methods have been employed. Some of these techniques are chemical coagulation or flocculation, ozonation, adsorption, oxidation processes, nanofiltration, chemical precipitation, ion-exchange, reverse osmosis, and ultrafiltration. All of these methods either are operationally expensive, inefficient in the removal of the dyes, or create disposal problems. In the last decade, increasing interest in the use of aqueous micellar solution has been found in the field of separation science [5]. Cloud point extraction (CPE) can be used for removing dye from aqueous solution. Such a method offers some advantages over conventional liquid-liquid extraction (LLE), including high extraction efficiency, ease of waste disposal, and the use of nontoxic and less dangerous reagents such as surfactant.

Above the cloud point temperature, aqueous solution of a nonionic surfactant separates into two phases, namely, a surfactant rich phase, which has small volume compared to the solution and is called coacervate phase, and the other is dilute bulk aqueous phase containing surfactant concentration slightly above the critical micelle concentration (CMC) [6]. The dye molecules present in aqueous solution of nonionic surfactant are distributed between the two phases above the cloud point temperature [7]. This phenomenon is known as the cloud point extraction (CPE). The dyes are solubilized in the coacervate phase, and the dye free aqueous dilute phase can be disposed of.

The physical properties of binary mixtures have been studied for many reasons. The most important reason is to provide information about the molecular interactions between similar and dissimilar structure of the molecules in the mixture [8]. In addition, the density, viscosity, and refractive index are essential to understand the thermodynamic behavior of solute or solvent in liquid mixtures [9–13]. The study of excess thermodynamic properties is of considerable interest in understanding the intermolecular interactions in liquid mixtures [14]. For example, excess molar volumes, which depend on the composition and/or temperature, are of great importance in understanding the nature of molecular aggregation that exists in the mixtures. V_E is the resultant contributions from several opposing effects [15]. These may be divided arbitrarily into three types, namely, chemical, physical, and structural ones. Physical contributions, which are nonspecific interactions between the real species present in the mixture, contribute a positive term to V_E. The chemical or specific intermolecular interactions result in a volume decrease, and these include charge-transfer type forces and other complex forming interactions. This effect contributes negative values to V_E. The structural contributions are mostly negative and arise from several effects, especially from interstitial accommodation and changes of free volume. In other words, structural contributions arising from the geometrical fitting of one component into the other due to the differences in the free volume and molar volume between components lead to negative contribution to V_E. The increase in excess molar volume with increasing dipole-dipole interactions is stronger in higher hydrogen bonding.

In this work, the binding energy (Table 1) in mixtures of Triton X-114 surfactant + MB, Triton X-114 surfactant + water, and MB + water complex systems were investigated using Hartree-Fock (HF) Theory with 6-31G* basis set. Subsequently, the density, viscosity, and refractive index were measured at different concentrations of surfactant and four different temperatures. Furthermore, the effects of various design parameters such as dye concentration, surfactant concentration, electrolyte concentration, and cloud point temperature on the phase volume ratio, preconcentration

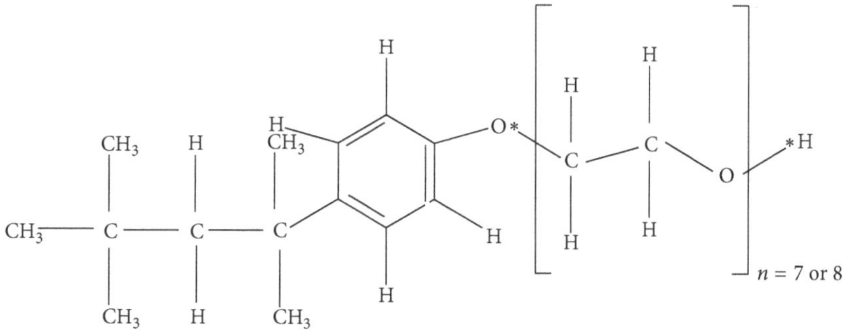

FIGURE 1: Chemical structure of Triton X-114.

factor, distribution coefficient, and extraction efficiency were calculated (Table 1). Lastly, various thermodynamic parameters such as the change in enthalpy, entropy, and Gibbs free energy were also determined (Table 1).

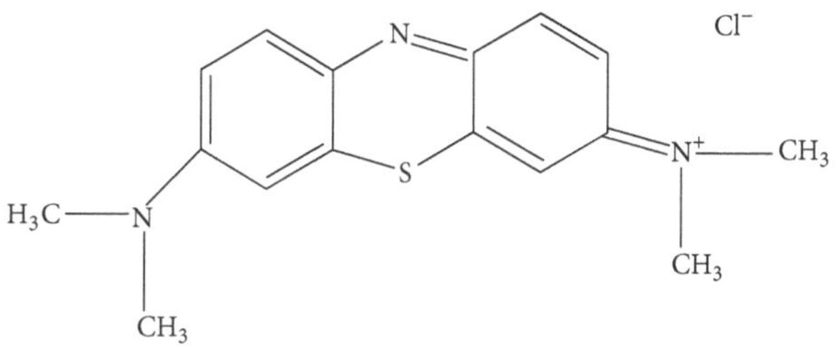

FIGURE 2: Chemical structure of methylene blue (MB) dye.

2. Computational Details

MOLDEN visualization packages [16] have been used to prepare the chemical structure of Triton X-114 surfactant, MB, and water. The complex systems of Triton X-114 surfactant + MB, Triton X-114 surfactant + water, and MB + water were initially drawn by using dummy atoms. After geometry optimization, the complex system does not have dummy atoms, which makes a barrier between these two molecules at the gas phase studied. These similar concepts have been used earlier by Meng et al. [17] and Turner et al. [18] for ab initio calculations. Quantum chemical calculations were performed using Hartree-Fock (HF) Theory with 6-31G* basis set. The geometry optimizations were carried out using HF/6-31G* basis set. For the same basis set, the fundamental vibrational frequency calculations were carried out to ensure the true minimum. The total electronic energies for Triton X-114 surfactant, MB dye, water, and their complex were corrected with the basis set of superposition error using Boys-Bernardi counterpoise techniques [19]. All theoretical calculations were performed using the Gaussian 03 software package [20].

3. Experimental Details

Triton X-114, purchased from Sigma Life Sciences, India, was used as a nonionic surfactant. It is a high purity (>95%) and water soluble liquid. Triton X-114 (octyl phenol poly (ethylene glycol ether)) contains approximately 8-9 ethoxy units per molecule (density at 298.15 K is 1.058 $g{\cdot}mL^{-1}$, Mol. Wt.: 537, λ_{max}: 223 nm) and is abbreviated as Triton X-114. The critical micellar concentration (CMC) of Triton X-114 is 2.1×10^{-4} M at 298.15 K and the cloud point temperature is 297.15 K. MB dye (Mol. Wt.: 319.85, density: 1.757 $g{\cdot}mL^{-1}$, λ_{max}: 665 nm) purchased from Nuchem chemicals, Ahmedabad, India, was used as a solute. The electrolytes used were sodium sulphate and calcium chloride (purchased from Paxmy). JASCO UV spectrophotometer was used for measuring dye concentration and surfactant concentration in dilute phase after phase separation. The density was measured using a 5 mL pycnometer. BROOKFIELD DV-II + PRO viscometer was used to measure the viscosity of the solutions. Refractive index was measured using Abbe Refractometer purchased from Guru Nanak instruments, New Delhi. The chemical structure of Triton X-114 and MB is given in Figures 1 and 2.

50 mL of aqueous micellar solutions was prepared with different concentrations of solute and surfactant. The concentrations of solute (MB) used were 15 ppm, 25 ppm, and 50 ppm. Surfactant concentration was varied from 0.01 M to 0.1 M. 1,000 ppm of dye stock solution was prepared by taking 0.1 g of MB and making it up to 100 mL. Various concentrations of dye solutions were prepared by taking known volumes from stock solution and then diluted appropriately. The calibration of UV spectrophotometer was done using the prepared dye solutions. 50 mL batch solutions of MB and Triton X-114 surfactant were prepared by varying dye concentration and surfactant concentration. The solutions were heated above CPT until phase separation was obtained.

The cloud point of aqueous surfactant solution was determined by heating 50 mL of such molecular solutions in glass tubes. A thermostatic bath made by Technico Laboratory Products Ltd using a Honeywell controller was used to heat the solution. The rate of temperature increase in the water bath was set at 1 K per minute. The cloud point was determined by visual observation of the temperature at which the solution became obvious turbid. While heating, the temperature at which turbidity first appeared was noted down. Heating was continued until the solution became fully turbid. Then, it was cooled down until the turbidity completely disappeared and the solution became transparent.

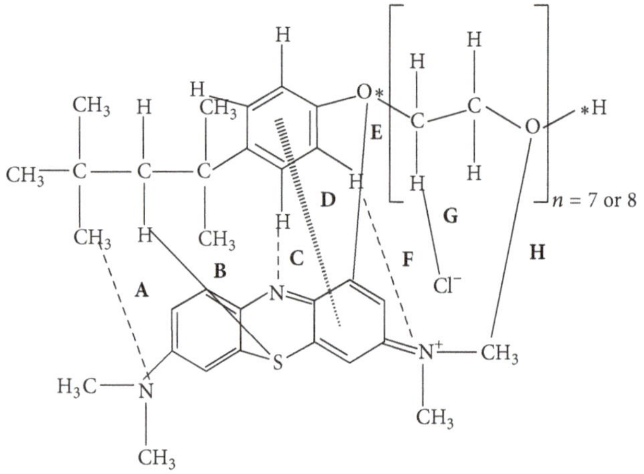

The possible molecular interaction at molecular level:

(A) H (surfactant)-N (methylene blue) interaction
(B) H (surfactant)-S (methylene blue) interaction
(C) H (surfactant)-N (methylene blue) interaction
(D) π (surfactant)- π (methylene blue) interaction
(E) O (surfactant)-H (methylene blue) interaction
(F) H (surfactant)-N (methylene blue) interaction
(G) H (surfactant)-Cl (methylene blue) interaction
(H) O (surfactant)-H (methylene blue) interaction

(a)

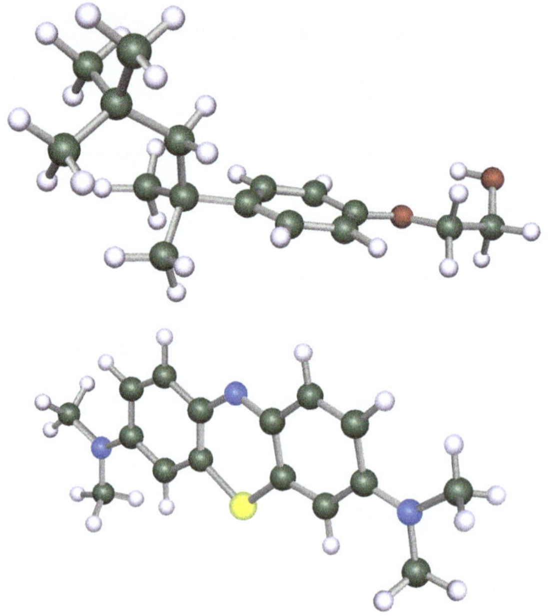

(b)

FIGURE 3: (a) The possible molecular interaction between Triton X-114 surfactant and methylene blue at molecular level. (b) Optimized complex structure of Triton X-114 with methylene blue dye using HF/6-31G* basis set.

The temperature was also noted down. The average of both readings was taken as the CPT of the solution. The experiments were repeated for different combinations of surfactant and solute concentrations.

50 mL of micellar solution containing MB and Triton X-114 was taken. The different concentrations of MB such as 15, 25, and 50 ppm, the concentration of surfactant, were varied from 0.01 M to 0.1 M. Each set of samples was kept in the thermostatic bath and maintained at the operating temperature for 30 min. Since the surfactant density is 1.058 $g{\cdot}cm^{-3}$, therefore, the surfactant rich phase can settle through the aqueous phase. The heated solution was allowed to settle for 1 hr. The volumes of surfactant rich phase and dilute phase were noted down. Then, the concentration of MB in the dilute phase was determined by using UV spectrophotometer. The concentration of surfactant rich phase concentration was obtained from the calculation of material balance. The phase volume ratio, preconcentration factor, distribution coefficient, and extraction efficiency were then determined. The above said parameters were found at different operating temperatures.

4. Results and Discussion

4.1. Binding Energy. The total electronic energy was predicted using HF/6-31G* basis set for Triton X-114 surfactant, Methylene Blue dye, water, Triton X-114 surfactant + Methylene Blue dye (Figures 3(a) and 3(b)), Triton X-114 surfactant + water, Methylene Blue dye + water. This calculation step was performed using Gaussian 03 packages. From this total energy value, the binding energy between Triton X-114 surfactant + MB (Figures 3(a) and 3(b)) and Triton X-114 surfactant + water system was calculated. The binding energy can be defined as the difference between the total electronic energy of the binary system and individual molecules ((a) in Table 1). The binding energy between Triton X-114 surfactant + MB and MB + water is −1222.9620 and −3573.2800 $kJ{\cdot}mole^{-1}$, respectively. This value describes that the Triton X-114 surfactant + MB and MB + water system has reliable chemical and physical interaction. However, the binding energy of Triton X-114 surfactant + MB system is low (−1222.9620 $kJ{\cdot}mole^{-1}$) when compared to MB + water system (−3573.2800 $kJ{\cdot}mole^{-1}$) due to the strength of hydrogen bonds. On the other hand, the presence of MB in water is in ppm level. The binding energy in Triton X-114 surfactant + water is 4303.4720 $kJ{\cdot}mole^{-1}$, which indicates that the distance between Triton X-114 surfactant and water molecules is large due to the absence of chemical and physical interaction. In addition, Triton X-114 surfactant is immiscible with water molecules, which is more favorable for effective separation of MB from water. There are also several chemical and physical interactions occurring between Triton X-114 surfactant and MB such as orbital interaction, charge-charge interaction, and hydrogen bond interaction with neutral molecules, van der Waals interaction between

Table 2: Physical properties of coacervate phase at 313.15 and 323.15 K.

Triton X-114 (M)	Density (ρ) (g·cm^{-3})	Refractive index (nD)	Viscosity (μ) (m·Pas)	Excess molar volume (V_E) (cm^3·mole^{-1})
		313.15 K		
0.01	NA	1.355	NA	NA
0.02	NA	1.358	NA	NA
0.03	1.0110	1.357	NA	−0.10675
0.04	1.0134	1.362	NA	−0.13257
0.05	1.0214	1.363	NA	−0.26556
0.06	1.0258	1.363	NA	−0.35633
0.07	1.0219	1.364	162.2	−0.26946
0.08	1.0233	1.366	164.4	−0.28251
0.09	1.0270	1.377	186.2	−0.35201
0.10	1.0340	1.387	190.0	−0.49725
		323.15 K		
0.01	NA	1.358	NA	NA
0.02	NA	1.360	NA	NA
0.03	NA	1.361	NA	NA
0.04	1.0218	1.364	NA	−0.26332
0.05	1.0240	1.365	NA	−0.29124
0.06	1.0268	1.367	NA	−0.33472
0.07	1.0244	1.370	190.2	−0.27629
0.08	1.0291	1.371	194.4	−0.37310
0.09	1.0322	1.376	193.8	−0.00564
0.10	1.0355	1.379	202.2	−0.49792

alkyl chain, CH-π interaction, π-π interaction, and n-π interaction (Figure 3(a)) [21–24].

4.2. Properties of Dilute Phase and Coacervate Phase. As listed in Table 2, the dilute phase properties were found to be similar to that of water. The dye present in the solution is solubilized in the surfactant micelle present in the coacervate phase. Therefore, the dilute phase only contains small concentration of dye and surfactant, and the major component is water.

Density. The variation of density with surfactant concentration and temperature are given in Table 2. The coacervate phase consists of dye, surfactant, and water molecules. With the increase in surfactant concentration, the surfactant molecules form higher number of micelles, and the size of the micelle also increases [25]. Due to the density difference, the surfactant will settle in the coacervate phase and the surfactant content in the coacervate phase increased. As there are a larger number of micelles, more dye solute will be solubilized in the surfactant micelles. Hence, the density increases with surfactant concentration. With the increase in temperature, the solubility of the dye increased. The maximum amount of solute will be solubilized. The main reason is water molecules escape from the external layers of the micelle. As the water content decreases, the solution becomes very dense.

Viscosity. The variation of viscosity with surfactant concentration and temperature are given in Table 2. The viscosity of Triton X-114 surfactant is 260 cp, which is very high compared to that of water (1 cp). Therefore, as the surfactant concentration increases, the surfactant rich phase becomes more viscous due to molecular attraction between similar structures of the molecules. An increase in the content of water molecules can decrease the viscosity due to the weak hydrogen bond interaction. Thus, at higher temperatures, the viscosity of the solution increases due to the removal of water molecules from the external layers of micelle. It is noted that the viscosity of the mixture strongly depends on the presence of water or moisture.

Refractive Index. It is defined as the ratio of the speed of light in vacuum to the speed of light in the medium. The refractive index increases with surfactant concentration and temperature, as given in Table 2. It can be explained that more dye solubilized at higher temperature and surfactant concentration and there is no mobility of ions in the surfactant rich phase. As the density increases, there is a compact arrangement of the molecules in the phase, and consequently the speed of light through that medium decreases. As a result, the refractive index increases. Thus, the refractive index strongly depends on the amount of dye present in the coacervate phase. A linear regression was sufficient for a reliable extrapolation for all measured quantities.

4.3. Excess Molar Volume of Dilute Phase and Coacervate Phase. The excess molar volumes for both phases were calculated with the help of measured pure component density by using (b) in Table 1. The value of excess molar volume, V_E, was found to be positive in the dilute phase. In Table 2, it is shown that V_E values are positive, which means that there is an expansion after mixing. That is, the actual volume of the mixture is greater than the volume occupied by the individual components. The physical effects of the dye, surfactant, and water molecules present in the dilute phase contribute to the positive value of V_E due to the presence of physical interactions. Hence, the possible physical interactions are (1) hydrophobic interactions between surfactant hydrocarbon chains dye and water curvature-dependent interfacial effects at the micellar core-water interface, (2) steric and electrostatic interactions between surfactant hydrophilic moieties, and (3) disruption of bonds between the molecules, which also makes positive contributions. As the temperature increases, there will be disruption of H-bond between the surfactant molecule and water molecules, which results in volume expansion and hence positive V_E. Excess molar volume in coacervate phase is illustrated in Figure 8. It is seen that V_E values are negative, which means there is contraction after mixing. The actual volume of the mixture is less than the volume occupied by the individual components. Hence, there is a strong intermolecular force of attraction between the molecules in the surfactant rich phase. The chemical and structural effects of the dye, surfactant, and water molecules make a negative contribution. The packing effects or conformational changes of the molecules in the mixtures are more difficult to categorize. The interstitial accommodation gives negative contributions. The calculated excess molar volume values are shown linearly fit which indicated a high degree of experimental data.

4.4. Effect of Surfactant and Solute Concentrations on CPT. Table 3 shows the effect of solute and surfactant concentrations on cloud point temperature. The experiments were carried out by varying the surfactant concentration from 0.01 M to 0.1 M for 15 ppm, 25 ppm, and 50 ppm solute concentrations. The CPT first decreases and then increases with the increase in surfactant concentration. The decrease in CPT with an increase in surfactant concentration is due to the increase in molecular concentration and the phase separation results from the increased molecular interaction. After 0.06 M surfactant concentrations, the CPT starts to increase. This is because of the formation of structured surfactant Triton X-114-MB system present in the micelles at high surfactant concentrations. More energy is required to remove these "free floating" water molecules, and hence CPT increases with surfactant concentration at higher surfactant concentrations. From Table 3, it can be observed that CPT is increased with increase in solute concentration. Because,

TABLE 3: CPT at different solute and surfactant concentration.

Triton X-114 (M)	15 ppm	25 ppm	50 ppm
0.01	296.7	297.2	298.2
0.02	296.45	297.05	297.8
0.03	296.3	297.05	297.55
0.04	296.2	296.7	297.55
0.05	296.2	296.2	296.45
0.06	295.4	296.4	296.6
0.07	295.6	296.45	296.8
0.08	296.65	296.85	297.4
0.09	297.25	296.95	297.6
0.10	297.35	297.25	297.7

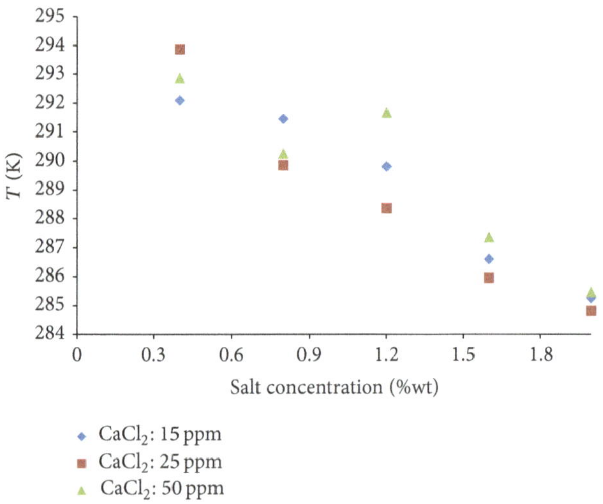

FIGURE 4: Effect of $CaCl_2$ on CPT at 0.1 M Triton X-114 (R^2 values by linear regression: 0.9613 for 15 ppm, 0.9796 for 25 ppm, and 0.8799 for 50 ppm).

polyoxyethlenated alkyl phenol molecules present in Triton X-114. This hydrophobic part depends on a hydrocarbon chain with different number of carbon atoms and may be linear or branched. It may also contain aromatic rings. Therefore, the surfactant concentration increased and the hydrophobicity of the ligand and of the complex formed on the apparent equilibrium constants in the micellar medium [25, 26].

4.5. Effect of Electrolytes on Cloud Point Temperature. Salts can promote or inhibit the dehydration process and therefore either decrease or increase the cloud point. This behavior is said to be the salting-in and salting-out phenomena. In order to study the salting-out effects, $CaCl_2$ and Na_2SO_4 were used. In Figure 4, the salting out effect of $CaCl_2$ is shown at 0.1 M surfactant concentration. The concentration of salt was varied from 2 to 10 wt%. Figure 5 shows the salting out effect of Na_2SO_4 with salt concentration varied from 0.4 to 2 wt% at surfactant concentration 0.1 M and

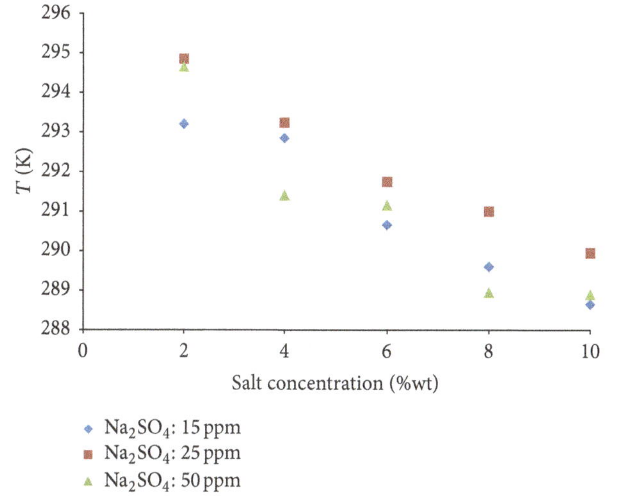

Figure 5: Effect of Na_2SO_4 on CPT at 0.1 M Triton X-114 (R^2 values by linear regression: 0.9557 for 15 ppm, 0.9544 for 25 ppm, and 0.837 for 50 ppm).

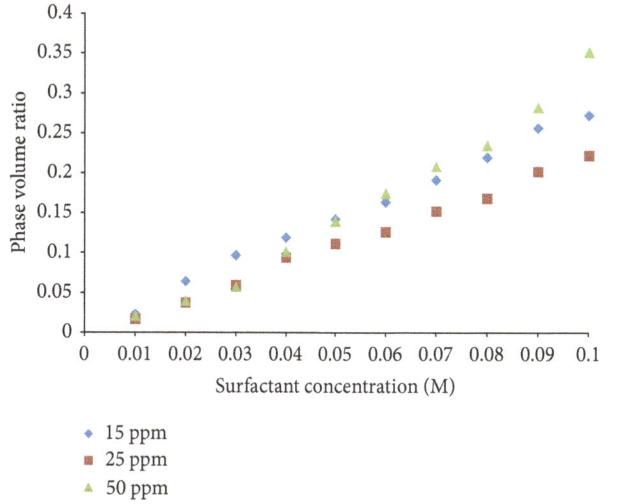

Figure 6: Effect of solute and surfactant concentration on phase volume ratio at 323.15 K (R^2 values by linear regression: 0.9956 for 15 ppm, 0.9824 for 25 ppm, and 0.9941 for 50 ppm).

solute concentrations of 15 ppm, 25 ppm, and 50 ppm. In the figure, it is shown that the cloud point decreases with the increase in Na_2SO_4 and $CaCl_2$ concentrations. The addition of certain salts to the nonionic surfactant solution can depress the cloud point temperature by decreasing the availability of nonassociated water molecules to hydrate the ether oxygen's of the polyethylene chains. They convert free water molecules to aggregate water molecules. Along with EO groups, these salts also have a competition for water molecules. From both figures, one can notice that the Na_2SO_4 salt is more effective than $CaCl_2$. The magnitude of the effects of the anions and cations appears to depend on the radius of the hydrated ions; the smaller the radius, the greater the effects. Here, the anion of the added electrolyte appears to have a much greater effect than the cation in decreasing the solubilizing power of water. Since SO_4^{2-} is a polyvalent ion, it will dehydrate water more quickly from the EO chain. Hence, the effect of Na_2SO_4 is greater than $CaCl_2$ [27]. As can be seen from Figures 4 and 5, a linear regression was sufficient for a reliable extrapolation and gave R^2 range: 0.8799–0.9796 for $CaCl_2$ and 0.8370–0.9557 for Na_2SO_4 at 0.1 M of Triton X-114.

4.6. Effect of Operating Variables on CPE Techniques. Experiments were done by changing the operating variables such as solute and surfactant concentration, as well as operating temperature. The effects of these variations on different design parameters such as phase volume ratio (R_v), preconcentration factor (f_c), efficiency (η%), and distribution coefficient (K_d) were investigated. Based on the CPT, the operating temperatures chosen were 303.15, 313.15, 323.15, and 333.15 K for solute concentrations 15 ppm, 25 ppm, and 50 ppm with surfactant varying from 0.01 M to 0.1 M.

4.6.1. Phase Volume Ratio (R_v). The phase volume ratio, R_v, is defined as the ratio of the volume of the surfactant rich phase to that of the aqueous phase ((c) in Table 1). The volumes of the two phases were measured using graduated cylinders. Experiments were conducted for different solute concentrations (15 ppm, 25 ppm, and 50 ppm) with a surfactant varying from 0.01 M to 0.1 M. Figure 12 shows the effect of surfactant and solute concentration in phase volume ratio. From Figure 6, R_v increases with the increase in surfactant concentration and solute concentration. At constant feed dye concentration and operating temperature (323.15 K), R_v increases with surfactant concentration. At higher surfactant concentrations, all the added surfactant will simply go into the surfactant rich phase and the concentration of surfactant in dilute phase being constant at CMC [26, 28–31]. This increases the coacervate phase volume, thereby increasing the phase volume ratio. Solute concentration only has a negligible effect on R_v. The phase volume ratio shows a slight increase in dye concentration at constant temperature. The plotted data shows a reliable linear regression with R^2 values such as 0.9956 (15 ppm), 0.9824 (25 ppm), and 0.9941 (50 ppm).

Figure 7 shows the effect of temperature on phase volume ratio. R_v decreases with temperature. At elevated temperatures, two opposing phenomena occur. The interaction among the Triton X-114 micelles increases due to dehydration from the external layers of micelles, resulting in a decrease in the volume of surfactant rich phase. Also, the increase of micellar aggregation number and micelle sizes results in increased solubilization of MB in the micelles [25, 26]. Depending on the solute (MB) surfactant (Triton X-114) system and the operating conditions, either of these phenomena will be predominant. At lower surfactant concentrations and high operating temperature (333.15 K), the dehydration of water molecules from external micellar layers may be predominant, thereby increasing the dilute phase volume and consequently decreasing the phase volume ratio. The surfactant concentration increases by keeping the operating temperature fixed at 333.15 K (Table 4). The latter phenomena, that is, the increase in micellar number and size, suppresses

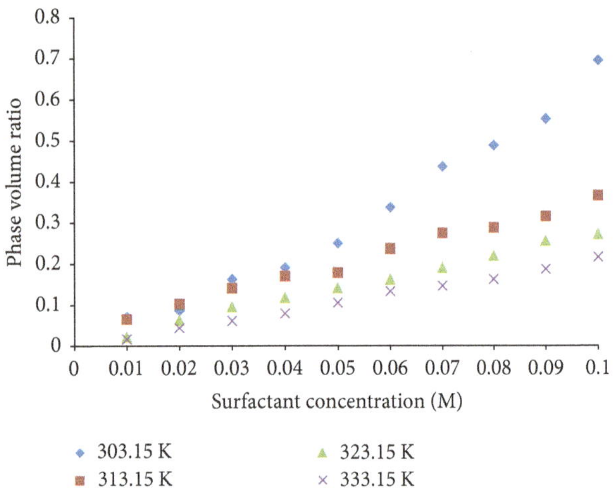

Figure 7: Effect of temperature on the phase volume ratio of solute concentration of 15 ppm (R^2 values by linear regression: 0.9744 for 303.15 K, 0.9896 for 313.15 K, 0.9941 for 323.15 K, and 0.9958 for 333.15 K).

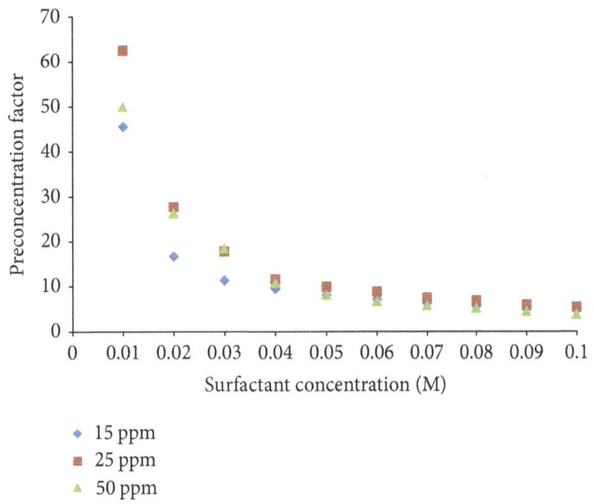

Figure 8: Effect of solute and surfactant concentration on preconcentration factor at 323.15 K (R^2 values by exponential regression: 0.8157 for 15 ppm, 0.8571 for 25 ppm, and 0.9141 for 50 ppm).

the former, thereby increasing R_v. The linear rgression extraplotion was given R^2 value as 0.9744 (303.15 K), 0.9896 (313.15 K), 0.9941 (323.15 K), and 0.9958 (33.15 K).

4.6.2. Preconcentration Factor (f_c). The preconcentration factor, f_c, is defined as the ratio of the volume of bulk solution before phase separation (V_t) to that of surfactant rich phase after phase separation (V_s) ((d) in Table 1). Preconcentration factor is an indication of the ratio of solute concentration in feed to that in the surfactant rich phase. The higher the value of preconcentration factor is, the lesser will be the separation of solute and vice versa. Figure 8 shows the effect of surfactant and solute concentration in preconcentration factor. It is observed that the preconcentration factor decreases with the increase in surfactant concentration. As discussed earlier, surfactant rich phase volume increases with increasing surfactant concentration, thereby decreasing f_c [32, 33] and the similar trend is also observed. As shown in Figure 8, f_c increases only marginally with solute concentration. At 50 ppm MB concentration, the surfactant concentration is not enough to solubilized MB in micelles and therefore the trend is below the 25 ppm of MB line. But, the above 0.9% wt of surfactant concentration increased, concentration of micelles increases, and more MB will be solubilized in micelle. On the other hand, further increase in MB concentration only results in an increase of unsolubilized in micelles as a function of surfactant concentration. Therefore, it can be concluded that the ratios of surfactant and MB play an important role in the removal of the dye.

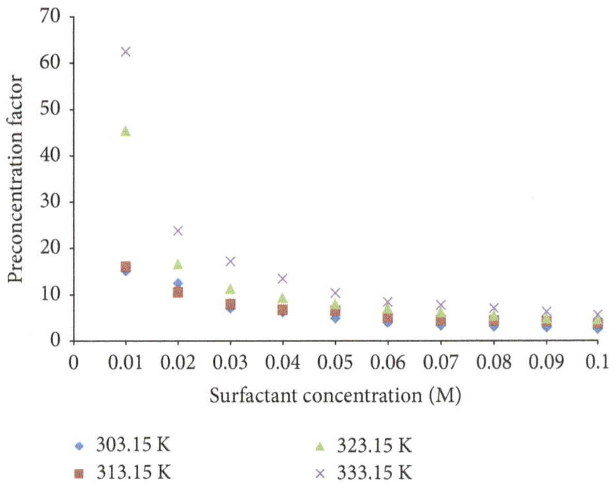

Figure 9: Effect of operating temperature on preconcentration factor at solute concentration of 15 ppm (R^2 values by exponential regression: 0.9155 for 303.15 K, 0.9155 for 313.15 K, 0.8157 for 323.15 K, and 0.8533 for 333.15 K).

Preconcentration factor is only slightly affected by operating temperature as evident from Figure 9. At higher temperature, micellar aggregation number and size increase, thereby increasing surfactant rich phase volume [25, 26]. On the other hand, dehydration from the external layers of micelles at elevated temperatures reduces the surfactant rich phase volume. These two opposing phenomena act against each other so as to keep the volume of surfactant rich phase almost constant. The linear regression R^2 values were obtained greater than 0.9824 for all ppm and greater than 0.8533 for all temperatures.

4.6.3. Distribution Coefficient (K_d). The distribution coefficient or equilibrium partition coefficient, K_d or K_p, is defined as the ratio of the concentration of solute in surfactant rich phase to that of the concentration of solute in dilute phase ((e) in Table 1). The distribution coefficient increases with the increase in the surfactant concentration and the decrease in solute concentration. The distribution of solute depends on the specific solute (MB) surfactant (Triton X-114) interaction. If there is more interaction, then the distribution

TABLE 4: Phase volume ratio, preconcentration factor, distribution coefficient, and efficiency at different temperature.

Triton X-114 (M)	Phase volume ratio (R_v)	Preconcentration factor (f_c)	Distribution coefficient (K_d)	Efficiency (η)
		15 ppm at 303.15 K		
0.01	0.0707	15.1515	0.0747	6.9506
0.02	0.0870	12.5000	0.1170	10.4732
0.03	0.1628	7.1429	0.3227	24.3942
0.04	0.1905	6.2500	0.4399	30.5528
0.05	0.2500	5.0000	0.7335	42.3137
0.06	0.3369	3.9683	0.9449	48.5841
0.07	0.4368	3.2895	1.2283	55.1226
0.08	0.4881	3.0488	1.4877	59.8018
0.09	0.5528	2.8090	1.8535	64.9558
0.1	0.6949	2.4390	2.1537	68.2915
		15 ppm at 313.15 K		
0.01	0.0661	16.1290	0.3026	23.2298
0.02	0.1038	10.6383	0.4615	31.5766
0.03	0.1416	8.0645	0.6656	39.9617
0.04	0.1710	6.8493	0.8322	45.4217
0.05	0.1792	6.5789	1.0131	50.3265
0.06	0.2376	5.2083	1.2835	56.2074
0.07	0.2755	4.6296	1.5056	60.0899
0.08	0.2887	4.4643	1.6316	61.9996
0.09	0.3158	4.1667	1.7639	63.8199
0.1	0.3661	3.7313	2.2204	68.9484
		15 ppm at 323.15 K		
0.01	0.0225	45.4545	0.5087	33.7170
0.02	0.0638	16.6667	0.6784	40.4204
0.03	0.0965	11.3636	0.9359	48.3446
0.04	0.1186	9.4340	1.1001	52.3824
0.05	0.1416	8.0645	1.2567	55.6872
0.06	0.1628	7.1429	1.5427	60.6717
0.07	0.1905	6.2500	1.7026	62.9990
0.08	0.2195	5.5556	1.7842	64.0829
0.09	0.2563	4.9020	2.1293	68.0438
0.1	0.2723	4.6729	2.8259	73.8624
		15 ppm at 333.15 K		
0.01	0.0163	62.5000	0.1154	10.3467
0.02	0.0438	23.8095	0.2317	18.8106
0.03	0.0616	17.2414	0.3434	25.5623
0.04	0.0799	13.5135	0.4562	31.3293
0.05	0.1062	10.4167	0.6249	38.4571
0.06	0.1338	8.4746	0.8351	45.5081
0.07	0.1468	7.8125	0.9366	48.3643
0.08	0.1628	7.1429	1.0632	51.5316
0.09	0.1876	6.3291	1.1840	54.2134
0.1	0.2165	5.6180	1.6252	61.9083
		25 ppm at 303.15 K		
0.01	0.0183	55.5556	0.2773	21.7079
0.02	0.0417	25.0000	0.3484	25.8403
0.03	0.0593	17.8571	0.4019	28.6672
0.04	0.1211	9.2593	0.5676	36.2097
0.05	0.1364	8.3333	0.6415	39.0796
0.06	0.1601	7.2464	0.7831	43.9178
0.07	0.1848	6.4103	0.8082	44.6953
0.08	0.2165	5.6180	1.1515	53.5213
0.09	0.2821	4.5455	1.4388	58.9955
0.1	0.3021	4.3103	1.6148	61.7555

TABLE 4: Continued.

Triton X-114 (M)	Phase volume ratio (R_v)	Preconcentration factor (f_c)	Distribution coefficient (K_d)	Efficiency (η)
		25 ppm at 313.15 K		
0.01	0.0163	62.5000	0.6042	37.6622
0.02	0.0373	27.7778	0.7016	41.2326
0.03	0.0593	17.8571	0.7983	44.3929
0.04	0.0941	11.6279	0.9376	48.3908
0.05	0.1111	10.0000	1.1158	52.7369
0.06	0.1261	8.9286	1.3285	57.0539
0.07	0.1521	7.5758	1.4299	58.8459
0.08	0.1682	6.9444	1.4366	58.9593
0.09	0.2019	5.9524	1.7389	63.4888
0.1	0.2225	5.4945	1.9361	65.9415
		25 ppm at 323.15 K		
0.01	0.0121	83.3333	0.2784	21.7752
0.02	0.0309	33.3333	0.3371	25.2086
0.03	0.0460	22.7273	0.4013	28.6355
0.04	0.0823	13.1579	0.5039	33.5066
0.05	0.1038	10.6383	0.6162	38.1269
0.06	0.1111	10.0000	0.7357	42.3861
0.07	0.1468	7.8125	0.9102	47.6493
0.08	0.1682	6.9444	1.0515	51.2560
0.09	0.1933	6.1728	1.2762	56.0663
0.1	0.2136	5.6818	1.2591	55.7342
		25 ppm at 333.15 K		
0.01	0.0246	41.6667	0.3059	23.4232
0.02	0.0593	17.8571	0.4026	28.7046
0.03	0.0776	13.8889	0.5038	33.5025
0.04	0.1338	8.4746	0.6658	39.9677
0.05	0.1574	7.3529	0.8487	45.9073
0.06	0.1792	6.5789	1.0177	50.4391
0.07	0.2107	5.7471	1.1811	54.1511
0.08	0.2315	5.3191	1.3880	58.1244
0.09	0.2887	4.4643	1.3475	57.4011
0.1	0.3158	4.1667	1.5486	60.7626
		50 ppm at 303.15 K		
0.01	0.0395	26.3158	0.0370	3.5638
0.02	0.0753	14.2857	0.0584	5.5157
0.03	0.1468	7.8125	0.1044	9.4490
0.04	0.2107	5.7471	0.2132	17.5724
0.05	0.2594	4.8544	0.3665	26.8194
0.06	0.3298	4.0323	0.7557	43.0437
0.07	0.4006	3.4965	0.8972	47.2894
0.08	0.4493	3.2258	1.0437	51.0689
0.09	0.6234	2.6042	1.3295	57.0729
0.1	0.7241	2.3810	1.5805	61.2479
		50 ppm at 313.15 K		
0.01	0.0331	31.2500	0.0975	8.8866
0.02	0.0571	18.5185	0.1475	12.8514
0.03	0.0684	15.6250	0.2105	17.3869
0.04	0.1236	9.0909	0.2933	22.6806
0.05	0.1933	6.1728	0.4603	31.5218
0.06	0.2315	5.3191	0.6017	37.5680
0.07	0.2953	4.3860	0.7990	44.4143
0.08	0.3369	3.9683	1.0073	50.1827
0.09	0.3699	3.7037	1.3389	57.2440
0.1	0.4706	3.1250	1.6083	61.6612

TABLE 4: Continued.

Triton X-114 (M)	Phase volume ratio (R_v)	Preconcentration factor (f_c)	Distribution coefficient (K_d)	Efficiency (η)
		50 ppm at 323.15 K		
0.01	0.0204	50.0000	0.1507	13.0993
0.02	0.0395	26.3158	0.2359	19.0858
0.03	0.0571	18.5185	0.3483	25.8306
0.04	0.1013	10.8696	0.5550	35.6927
0.05	0.1390	8.1967	0.6922	40.9070
0.06	0.1737	6.7568	0.9491	48.6952
0.07	0.2077	5.8140	1.0966	52.3033
0.08	0.2346	5.2632	1.1722	53.9629
0.09	0.2821	4.5455	1.4955	59.9275
0.1	0.3514	3.8462	1.7496	63.6310
		50 ppm at 333.15 K		
0.01	0.0183	55.5556	0.0756	7.0287
0.02	0.0309	33.3333	0.0982	8.9438
0.03	0.0684	15.6250	0.1437	12.5679
0.04	0.0917	11.9048	0.2241	18.3057
0.05	0.1211	9.2593	0.3609	26.5204
0.06	0.1416	8.0645	0.4648	31.7317
0.07	0.1574	7.3529	0.6028	37.6095
0.08	0.1962	6.0976	0.7631	43.2826
0.09	0.2077	5.8140	0.8053	44.6087
0.1	0.2255	5.4348	1.0177	50.4385

coefficient will be high. Figure 10 shows the effect of surfactant and solute concentration on distribution coefficient [32, 33]. The experiments were carried out at 303.15 K, 313.15 K, 323.15 K, and 333.15 K (Table 4). At fixed surfactant and solute concentrations, K_d increases with operating temperature. As the temperature increases, micellar interaction, which is repulsive at lower temperatures, becomes attractive and hence micellar aggregation number increases. This results in increased solubilization of dye into the surfactant rich phase, thus increasing the dye concentration of the phase. Hence, distribution coefficient increases with operating temperature as evident in Figure 11. The experimental data was fitted by linear regression extrapolation and R^2 values are greater than 0.951 for all surfactant concentration and temperatures.

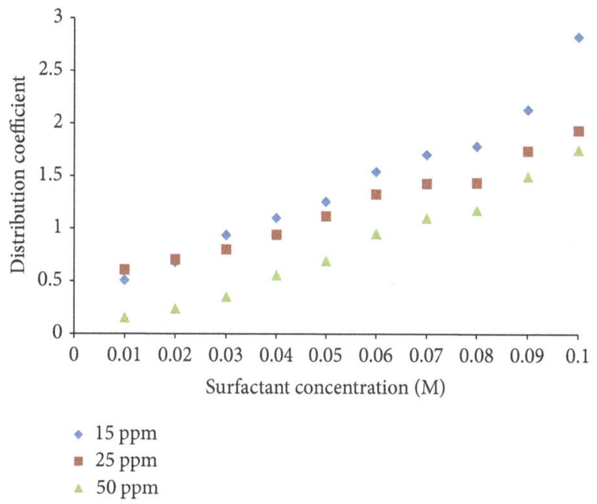

FIGURE 10: Effect of surfactant and solute concentration on distribution coefficient at 323.15 K (R^2 values by linear regression: 0.951 for 15 ppm, 0.9811 for 25 ppm, and 0.983 for 50 ppm).

4.6.4. Process Efficiency (η). The recovery efficiency of solute, η, can be characterized as the percentage of solute extracted from the bulk solution into the surfactant rich phase ((f) in Table 1). Figures 12 and 13 show the effect of surfactant and solute concentration on extraction efficiency. It increases with the increase in surfactant concentration and decreases with solute concentration. As feed surfactant concentration increases, more dye will be solubilized into the micelles, thereby increasing the efficiency. When solute concentration increases at fixed surfactant concentration and operating temperature, all the added solute will remain in the dilute phase, thereby decreasing the extraction efficiency [26, 28–31].

Experiments were done for various operating temperatures such as 303.15 K, 313.15 K, 323.15 K, and 333.15 K by varying the surfactant concentration and keeping the dye concentration fixed at 15 ppm (Table 4 and Figure 12). As shown in Figure 13, the efficiency of extraction increases with operating temperature. This is due to the increased solubilization of dye resulting from increased micellar size and aggregation number [25, 26]. This similar trend has been observed in the study of Triton X-114-Chrysoidine system [30]. However, after 323.15 K, the effect of temperature is insignificant. It is because dehydration of water occurs from the micelle interface at higher temperatures and micelle interaction increases. As a result, the solubilization of dye cannot occur at the micelle water interface, which decreases

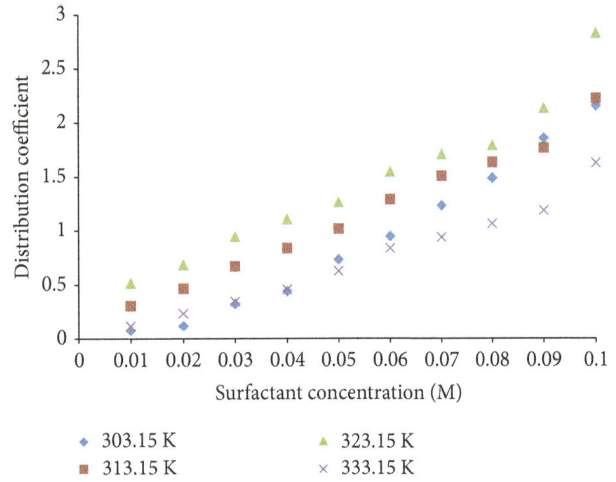

FIGURE 11: Effect of operating temperature on distribution coefficient at solute concentration of 15 ppm (R^2 values by linear regression: 0.9754 for 303.15 K, 0.951 for 313.15 K, 0.9878 for 323.15 K, and 0.9699 for 333.15 K).

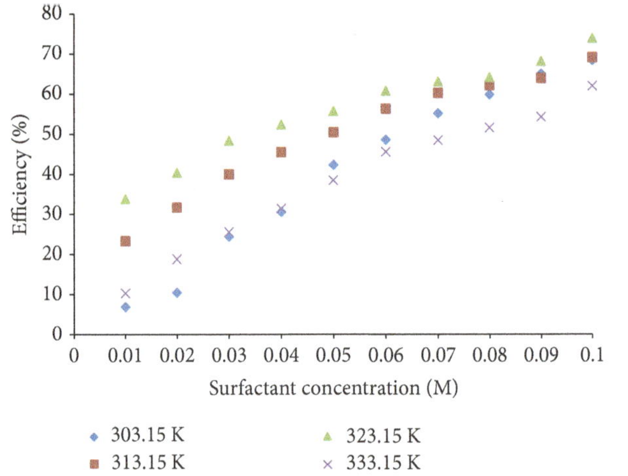

FIGURE 13: Effect of operating temperature of efficiency at solute concentration of 15 ppm (R^2 values by exponential regression: 0.9745 for 303.15 K, 0.9573 for 313.15 K, 0.9644 K for 323.15 K, and 0.9776 for 333.15 K).

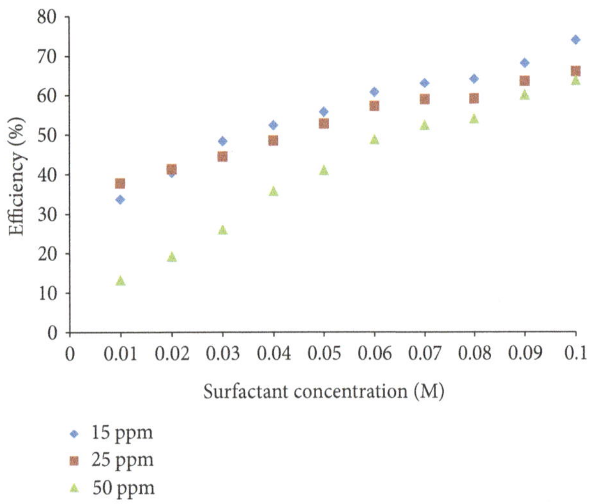

FIGURE 12: Effect of surfactant and solute concentration on efficiency at 323.15 K (R^2 values by linear regression: 0.9649 for 15 ppm, 0.9797 for 25 ppm, and 0.9746 for 50 ppm).

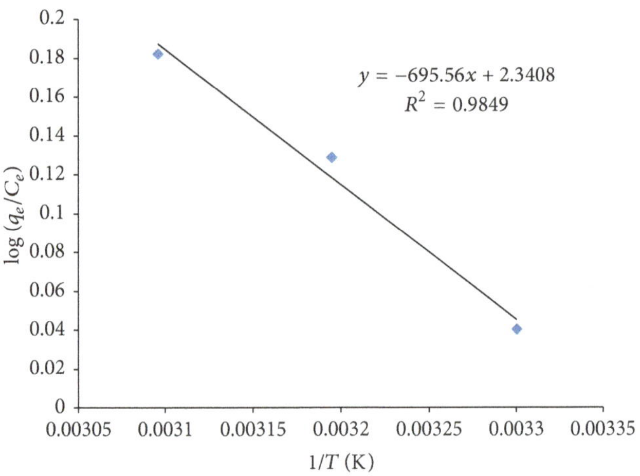

FIGURE 14: Plot of log (q_e/C_e) versus $1/T$ at surfactant concentration of 0.07 M.

the efficiency. The linear regression extrapolation shows R^2 values such as the following: for solute concentration on efficiency, 0.9649 (15 ppm), 0.9792 (25 ppm), and 0.9746 (50 ppm) and for operating temperature on efficiency, 0.9745 (303.15 K), 0.9573 (313.15 K), 0.9644 (323.15 K), and 0.9776 (333.15 K), respectively.

4.7. Determination of Thermodynamic Parameters. In any process, the study of thermodynamics is important since it is an indication of the feasibility of the process. It serves to identify the extent to which a process can be preceded before attaining equilibrium. Thermodynamic data may also be useful to establish the possible mechanism for CPE of various solutes. The thermodynamic parameters ΔH^0, ΔS^0, and ΔG^0 for the CPE of MB using Triton X-114 were calculated using (g) and (h) in Table 1. ΔS^0 and ΔH^0 were obtained from a plot of $\log(q_e/c_e)$ versus $(1/T)$ from (h) in Table 1. Once these two parameters were obtained, ΔG^0 was determined from (g) in Table 1. $\log(q_e/c_e)$ versus $1/T$ graphs were plotted for various surfactant concentrations as shown in Figure 14, where the slopes and intercepts were obtained to find ΔH^0 and ΔS^0.

The Change in Enthalpy (ΔH^0). The variations of enthalpy change (ΔH^0) during CPE of MB at different concentrations of Triton X-114 are as shown in Figure 15. From the figure, it may be seen that the value of ΔH^0 increases with the Triton X-114 concentration. The positive values of ΔH^0 indicate that the solubilization of dye is endothermic in nature (Table 5). The endothermic nature is also indicated by the increase in the amount of solubilization with temperature. The increase in ΔH^0 value in initial surfactant concentration can be accounted for an increase in solubilization capacity [30]. The

TABLE 5: Change in entropy, change in enthalpy, and change in Gibbs free energy at different surfactant concentration and different temperature.

Triton X-114 (M)	T (K)	$\log(q_e/C_e)$	ΔS (J/mole K)	ΔH (J/mole)	$-\Delta G$ (J/mole)
0.01	303.15	−0.3306	253.5656	78415.2053	1919.0120
	313.15	0.2769	253.5656	78415.2053	950.8279
	323.15	0.5025	253.5656	78415.2053	3486.4839
	333.15	−0.1417	253.5656	78415.2053	6022.1400
0.02	303.15	−0.4368	230.2635	71928.1536	−2158.3041
	313.15	0.1592	230.2635	71928.1536	144.3312
	323.15	0.3266	230.2635	71928.1536	2446.9665
	333.15	−0.1401	230.2635	71928.1536	4749.6018
0.03	303.15	−0.1723	140.6855	43483.1595	−855.4407
	313.15	0.1422	140.6855	43483.1595	551.4147
	323.15	0.2902	140.6855	43483.1595	1958.2701
	333.15	−0.1452	140.6855	43483.1595	3365.1255
0.04	303.15	−0.1626	120.8778	37434.5773	−808.5972
	313.15	0.1143	120.8778	37434.5773	400.1810
	323.15	0.2354	120.8778	37434.5773	1608.9592
	333.15	−0.1468	120.8778	37434.5773	2817.7375
0.05	303.15	−0.0375	71.8343	21946.4542	−180.6514
	313.15	0.1028	71.8343	21946.4542	537.6920
	323.15	0.1963	71.8343	21946.4542	1256.0353
	333.15	−0.1071	71.8343	21946.4542	1974.3786
0.06	303.15	−0.0067	65.9964	19993.4457	3.4541
	313.15	0.1263	65.9964	19993.4457	663.4178
	323.15	0.2062	65.9964	19993.4457	1323.3815
	333.15	−0.0603	65.9964	19993.4457	1983.3452
0.07	303.15	0.0403	44.8196	13317.9861	262.3618
	313.15	0.1287	44.8196	13317.9861	710.5581
	323.15	0.1821	44.8196	13317.9861	1158.7544
	333.15	−0.0774	44.8196	13317.9861	1606.9507
0.08	303.15	0.0655	25.6553	7394.6262	378.9162
	313.15	0.1056	25.6553	7394.6262	635.4688
	323.15	0.1444	25.6553	7394.6262	892.0213
	333.15	−0.0804	25.6553	7394.6262	1148.5739
0.09	303.15	0.1098	20.0911	5548.8418	538.7604
	313.15	0.0883	20.0911	5548.8418	739.6713
	323.15	0.1702	20.0911	5548.8418	940.5823
	333.15	−0.0848	20.0911	5548.8418	1141.4932
0.1	303.15	0.1293	38.3460	10958.1008	660.7315
	313.15	0.1425	38.3460	10958.1008	1044.1913
	323.15	0.2472	38.3460	10958.1008	1427.6511
	333.15	0.0070	38.3460	10958.1008	1811.1109

experimental data was fit with linear regression and 0.8383 R^2 value.

The Change in Entropy (ΔS^0). The variations of entropy change (ΔS^0) at different concentrations of Triton X-114 are shown in Figure 16. The entropy changes are positive (Table 5), which reflects a good affinity of dye molecules towards surfactant micelles. For all the systems, the change in entropy (ΔS^0) increases with surfactant concentration. Entropy depends on insolubilized dye molecules and free surfactant molecules in the CPE system. The increase in ΔS^0 value in surfactant concentration is due to the increase of free surfactant molecule in the dilute phase. On the other hand, CMC of the surfactant molecule decreases with the increase in dye concentration at a fixed surfactant concentration. This similar trend has been observed by Purkait et al. [31]. The linear regression R^2 value is 0.8036.

The Change in Gibbs Free Energy (ΔG^0). The values of Gibbs free energy (ΔG^0) with temperature have been calculated by knowing the enthalpy of solubilization (ΔH^0) and the entropy of solubilization (ΔS^0) at different surfactant concentrations. From Figure 17, ΔG^0 increases with temperature and decreases with surfactant concentration. In all the cases calculated, the values of free energy changes ΔG^0 were found to be negative as shown in Figure 17 and it is given in Table 5.

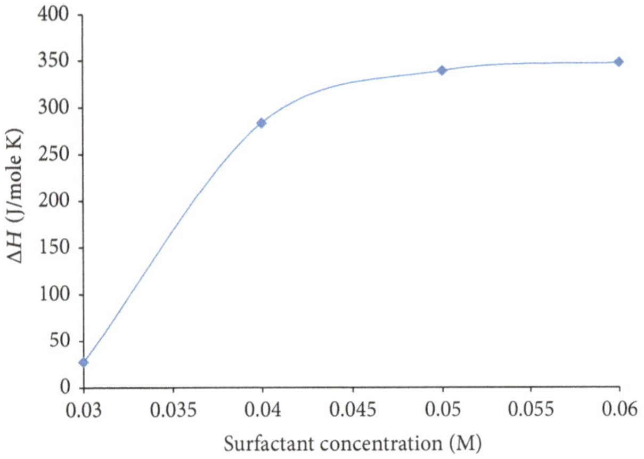

FIGURE 15: Change in enthalpy versus surfactant concentration in 15 ppm of dye (R^2 value is 0.8383).

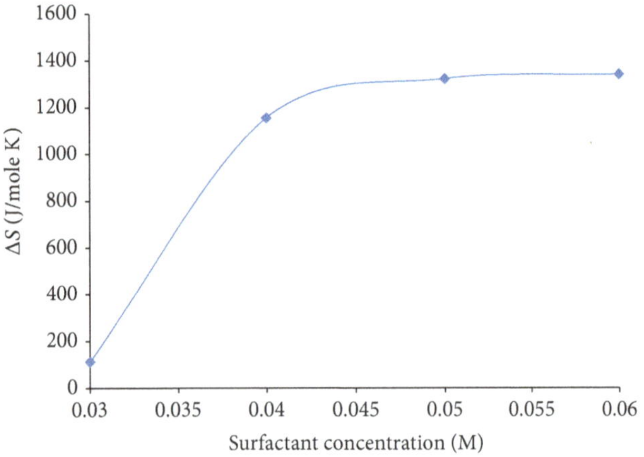

FIGURE 16: Change in entropy versus surfactant concentration in 15 ppm of MB dye (R^2 value is 0.8036).

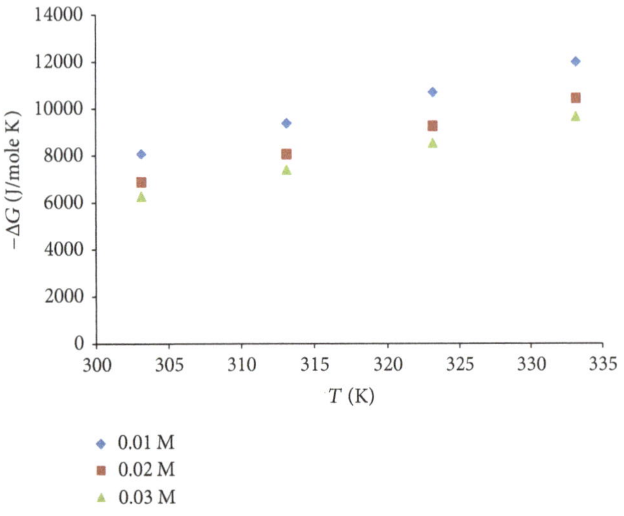

FIGURE 17: Change in Gibbs free energy versus temperature at 0.01 M, 0.02 M, and 0.03 M of MB dye (R^2 value is 1 for all).

The negative values of ΔG^0 indicate that the dye solubilization process is spontaneous and thermodynamically favorable. The increase in negative values of ΔG^0 with temperature implies the greater driving force of solubilization, which is confirmed by the greater extent of dye extraction with the increase in temperature [31]. The linear relationship between negative ΔG^0 as function of temperature was obtained R^2 value as unity.

5. Conclusions

A quantum chemical calculation in this study of the binding energy between Triton X-114 surfactant and MB/Triton X-114 surfactant + water/MB + water system has been investigated. The results show that the interactions of Triton X-114 surfactant + MB and MB + water depend mainly on hydrogen bond interaction, CH-π interaction, and π-π interaction. The absorption capacity of MB dye in the surfactant depends strongly on the structure, strength of hydrogen bonding, and a hetero atom in the surfactant. However, the calculated binding energy is directly proportional to the distance, which shows that the hydrogen bond interaction is the dominant interaction in the attractions. In addition, the cloud point extraction was carried out at a cloud point temperature, where the mixture of surfactant, water, and MB exhibited two phases. The density, viscosity, and refractive index of pure and two phases of the binary mixture were measured. From this measured density, the excess molar volume has been calculated, which confirmed the type of interaction occurred between similar and dissimilar compounds in two different phases. Furthermore, the design parameters and thermodynamics feasibility factors such as phase volume ratio, preconcentration factor, distribution coefficient, the efficiency of the process, change in enthalpy (ΔH^0), change in entropy (ΔS^0), and change in Gibbs free energy (ΔG^0) have been investigated. It is concluded that the negative value of ΔG^0 indicates that extraction is spontaneous and thermodynamically favorable.

Nomenclature

CPE: Cloud point extraction
CPT: Cloud point temperature
MB: Methylene blue
TX-114: Triton X-114
BE: Binding energy
V_E: Excess molar volume
x_i: Moles of component i
M_i: Molecular weight of component i
ρ_i: Density of component in $g{\cdot}cm^{-3}$
ρ_{mix}: Density of mixture in $g{\cdot}cm^{-3}$
R_v: Phase volume ratio
V_s: Volume of surfactant rich phase
V_w: Volume of aqueous phase
f_c: Preconcentration factor
V_t: Volume of bulk solution before phase separation

V_s: Surfactant rich phase after phase separation
K_d: Distribution coefficient
C_s: Concentration of solute in surfactant rich phase
C_w: Concentration of solute in dilute phase
η: Efficiency of solute
C_0: Initial concentration of solute in micellar solution
C_W: Concentration of solute in dilute phase
V_t: Total feed volume
V_s: Volume of surfactant rich phase
ΔG^0: Change in Gibbs free energy
ΔS^0: Change in entropy
ΔH^0: Change in enthalpy
T: Temperature in K
q_e: Moles of dye solubilized per mole of non-ionic surfactant
C_e: Equilibrium concentration of dye (moles/L) before the completion of two phases
A: Moles of dye solubilized in the micelles
V_0: Volume of feed solution
V_d: Dilute phase after CPE
C_s: Concentration of surfactant in feed.

Conflict of Interests

The authors declare that there is no conflict of interests regarding the publication of this paper.

Acknowledgments

The work is partially supported by the Tamil Nadu State Council for Science and Technology (TNSCST/S&T project/VR/ES/2009-2010). The authors wish to thank the Tamil Nadu State Council Science and Technology for their financial support.

References

[1] M. I. El-Khaiary, "Kinetics and mechanism of adsorption of methylene blue from aqueous solution by nitric-acid treated water-hyacinth," *Journal of Hazardous Materials*, vol. 147, no. 1-2, pp. 28–36, 2007.

[2] D. Ghosh and K. G. Bhattacharyya, "Adsorption of methylene blue on kaolinite," *Applied Clay Science*, vol. 20, no. 6, pp. 295–300, 2002.

[3] E. Lorenc-Grabowska and G. Gryglewicz, "Adsorption characteristics of Congo Red on coal-based mesoporous activated carbon," *Dyes and Pigments*, vol. 74, no. 1, pp. 34–40, 2007.

[4] M. Doğan, H. Abak, and M. Alkan, "Adsorption of methylene blue onto hazelnut shell: kinetics, mechanism and activation parameters," *Journal of Hazardous Materials*, vol. 164, no. 1, pp. 172–181, 2009.

[5] N. D. Gullickson, J. F. Scamehornand, and J. H. Harwell, *Liquid-Coacervate Extraction-Surfactant Based Separation Processes*, Marcel Dekker, New York, NY, USA, 1989.

[6] W. Kimchuwanit, S. Osuwan, J. F. Scamehorn, J. H. Harwell, and K. J. Haller, "Use of a micellar-rich coacervate phase to extract trichloroethylene from water," *Separation Science and Technology*, vol. 35, no. 13, pp. 1991–2002, 2000.

[7] M. J. Rosen, *Surfactant and Interfacial Phenomena*, John Wiley & Sons, New York, NY, USA, 2nd edition, 2000.

[8] U. S. Vural, V. Muradoglu, and S. Vural, "Excess molar volumes, and refractive index of binary mixtures of glycerol + methanol and glycerol + water at 298.15 k and 303.15 K," *Bulletin of the Chemical Society of Ethiopia*, vol. 25, no. 1, pp. 111–118, 2011.

[9] R. Francesconi, F. Comelli, and C. Castellari, "Excess molar enthalpies and excess molar volumes of binary mixtures containing dialkyl carbonates + anisole or phenetole at (288.15 and 313.15) K," *Journal of Chemical and Engineering Data*, vol. 45, no. 4, pp. 544–548, 2000.

[10] U. S. Vural, H. Yuruk, and V. Muradoglu, "Excess volumetric properties of mixtures epichlorohydrine + acetone or chloroform at 291.15 and 296.15 K," *Russian Journal of Physical Chemistry A*, vol. 77, no. 7, pp. 1091–1094, 2003.

[11] U. S. Vural, V. Muradoglu, and H. Yuruk, "Excess volumetric properties of mixtures (epichlorohydrine + toluene and epichlorohydrine + xylene) at 298.15, 308.15, 323.15, and 338.15 K," *Russian Journal of Physical Chemistry*, vol. 78, p. 33, 2004.

[12] U. S. Vural, "Excess molar volumes and viscosities of binary mixtures of epichlorohydrine and alcohols," *Russian Journal of Physical Chemistry A*, vol. 79, no. 7, pp. 1096–1101, 2005.

[13] R. K. Hind, E. McLaughlin, and A. R. Ubbelohde, "Structure and viscosity of liquids. Viscosity-temperature relationships of pyrrole and pyrrolidine," *Transactions of the Faraday Society*, vol. 56, pp. 331–334, 1960.

[14] M. Domínguez, J. Pardo, M. C. López, F. M. Royo, and J. S. Urieta, "Viscosities of the ternary mixture (1-butanol + n-hexane + 1-butylamine) at the temperatures 298.15 and 313.15 K," *Fluid Phase Equilibria*, vol. 124, no. 1-2, pp. 147–159, 1996.

[15] I. Frenkel, *Kinetic Theory of Liquids*, Oxford University Press, London, UK, 1946.

[16] G. Schaftenaar and J. H. Noordik, "Molden: a pre- and post-processing program for molecular and electronic structures," *Journal of Computer-Aided Molecular Design*, vol. 14, no. 2, pp. 123–134, 2000.

[17] Z. Meng, A. Dölle, and W. R. Carper, "Gas phase model of an ionic liquid: semi-empirical and ab initio bonding and molecular structure," *Journal of Molecular Structure*, vol. 585, pp. 119–128, 2002.

[18] E. A. Turner, C. C. Pye, and R. D. Singer, "Use of ab initio calculations toward the rational design of room temperature ionic liquids," *Journal of Physical Chemistry A*, vol. 107, no. 13, pp. 2277–2288, 2003.

[19] S. Boys and F. Bernardi, "The calculation of smaller molecular intercations by the differences of separate total energies, some procedure with reduced errors," *Molecular Physics*, vol. 19, pp. 553–566, 1970.

[20] M. J. Frisch, G. W. Trucks, H. B. Schlegel et al., *Gaussian 03*, Revision C.02, Gaussian, Pittsburgh, Pa, USA, 2004.

[21] A. M. Fernandes, M. A. A. Rocha, M. G. Freire, I. M. Marrucho, J. A. P. Coutinho, and L. M. N. B. F. Santos, "Evaluation of cation-anion interaction strength in ionic liquids," *The Journal of Physical Chemistry B*, vol. 115, no. 14, pp. 4033–4041, 2011.

[22] R. Lü, J. Lin, and Z. Qu, "Theoretical study on interactions between thiophene/dibenzothiophene/cyclohexane/toluene and 1-methyl-3-octylimidazolium tetrafluoroborate," *Structural Chemistry*, vol. 24, no. 2, pp. 507–515, 2013.

[23] O. Castellano, R. Gimon, and H. Soscun, "Theoretical study of the σ-π and π-π interactions in heteroaromatic monocyclic molecular complexes of benzene, pyridine, and thiophene dimers: implications on the resin-asphaltene stability in crude oil," *Energy and Fuels*, vol. 25, no. 6, pp. 2526–2541, 2011.

[24] R. Anantharaj and T. Banerjee, "Quantum chemical studies on the simultaneous interaction of thiophene and pyridine with ionic liquid," *AIChE Journal*, vol. 57, no. 3, pp. 749–764, 2011.

[25] R. Carabias-Martínez, E. Rodríguez-Gonzalo, B. Moreno-Cordero, J. L. Pérez-Pavón, C. García-Pinto, and E. Fernández Laespada, "Surfactant cloud point extraction and preconcentration of organic compounds prior to chromatography and capillary electrophoresis," *Journal of Chromatography A*, vol. 902, no. 1, pp. 251–265, 2000.

[26] A. Kitahara, K. Watanabe, K. Kon-no, and T. Ishikawa, "Mechanism of solubilization of water by oil-soluble surfactant solutions I. Anionic surfactants. Part II. Cationic Surfactants, appears on pp. 1–5 of this issue," *Journal of Colloid And Interface Science*, vol. 29, no. 1, pp. 48–54, 1969.

[27] M. K. Purkait, S. S. Vijay, S. DasGupta, and S. De, "Separation of congo red by surfactant mediated cloud point extraction," *Dyes and Pigments*, vol. 63, no. 2, pp. 151–159, 2004.

[28] M. K. Purkait, S. Das Gupta, and S. De, "Determination of design parameters for the cloud point extraction of congo red and eosin dyes using TX-100," *Separation and Purification Technology*, vol. 51, no. 2, pp. 137–142, 2006.

[29] M. K. Purkait, S. Das Gupta, and S. De, "Determination of thermodynamic parameters for the cloud point extraction of different dyes using TX-100 and TX-114," *Separation and Purification Technology*, vol. 244, pp. 130–138, 2009.

[30] M. K. Purkait, S. DasGupta, and S. De, "Performance of TX-100 and TX-114 for the separation of chrysoidine dye using cloud point extraction," *Journal of Hazardous Materials*, vol. 137, no. 2, pp. 827–835, 2006.

[31] M. K. Purkait, S. Das Gupta, and S. De, "Cloud point extraction of toxic eosin dye using Triton X-100," *Water Research*, vol. 39, pp. 3885–3890, 2005.

[32] B. Yao and L. Yang, "Equilibrium partition of polycyclic aromatic hydrocarbons in cloud point extraction with a silicone surfactant," *Journal of Colloid and Interface Science*, vol. 319, no. 1, pp. 316–321, 2008.

[33] D. Sicilia, S. Rubio, and D. Pérez-Bendito, "Evaluation of the factors affecting extraction of organic compounds based on the acid-induced phase cloud point approach," *Analytica Chimica Acta*, vol. 460, no. 1, pp. 13–22, 2002.

23

Thermal Performance and Economic Analysis of 210 MWe Coal-Fired Power Plant

Ravinder Kumar,[1] Avdhesh Kr. Sharma,[1] and P. C. Tewari[2]

[1] *Department of Mechanical Engineering, D.C.R University of Science and Technology, Murthal, Sonepat, India*
[2] *Department of Mechanical Engineering, N.I.T Kurukshetra, Haryana, India*

Correspondence should be addressed to Ravinder Kumar; rav.chauhan@yahoo.co.in

Academic Editor: Mohammad Al-Nimr

This paper presents the thermal and economic performance of a 210 MWe coal-fired power plant situated in North India. Analysis is used to predict coal consumption rate, overall thermal efficiency, mass flow rate of steam through boiler, and Net present value (NPV) of plant for given load. Thermodynamic analysis was carried out using mass and energy equations followed by empirical correlations. Predicted mass flow rate of steam, coal consumption rate, and thermal efficiency give fair agreement with plant operating data. The economic analysis includes operational activities such as equipment cost, fuel cost, operations and maintenance cost, revenue, and plant net present value. From economic point of view, the effect of condensate extraction pump redundancy on net present value is observed to be sensitive than boiler feed pump redundancy.

1. Introduction

Thermal power plant working is based on Rankine cycle, where the thermal efficiency of cycle can be thermodynamically improved by increasing the mean temperature of heat addition, that is, by introducing feedwater heating systems. Numerous researchers [1–3] have reported enhancement in thermal efficiency by dividing overall enthalpy equally via feedwater heaters. It was proposed by [4] that the thermodynamic performance of Rankine cycle power plant can be improved by reducing volumetric flow rate of steam. Later, researchers tried to improve the efficiency of the plant by increasing steam pressure, which resulted in degradation of the steam quality at the turbine exhaust. To overcome such problems, steam reheating was introduced after high pressure turbine exhaust, which leads to decrease of the moisture content at low-pressure turbine exhaust. The effect of reheat pressure on cycle efficiency was investigated [5]. They reported best performance of steam power plant at optimal reheat pressure. All the modifications/improvements in Rankine cycle (like feed water heating and reheating) lead to a substantial improvement in cycle efficiency. The study [6] on energy analysis of 250 MWe Hamedan steam power plant suggested that energy loss mainly occurs in the condenser.

The economic analysis of the plant has been carried out in past on the basis of initial capital investment, operating costs, annual revenue, and profit obtained. The Net present value of plant has been evaluated in the literature [7]. The equipment include steam turbine, boiler, generators, and other auxiliary components such as pumps, condensers, and so forth. Researchers reported considerable work on economic analysis using net present worth method in the various process industries. Sensitivity analysis for the capacity improvement of a combined cycle power plant (100–600 MW) concerning economic performance has been studied [8]. Economic feasibility and financial risk of refuse derived fuel (RDF) production plants have been instead evaluated on the basis of the Net present value index over a capacity range of 25–200 t/h comparing either single RDF production plants or facilities integrating also compost production and/or electricity generation [9]. The Net present value (NPV) approach was implemented to determine the economic manufacturing quantities for an unreliable production system over an infinite planning horizon [10]. The feasibility of using biomass to

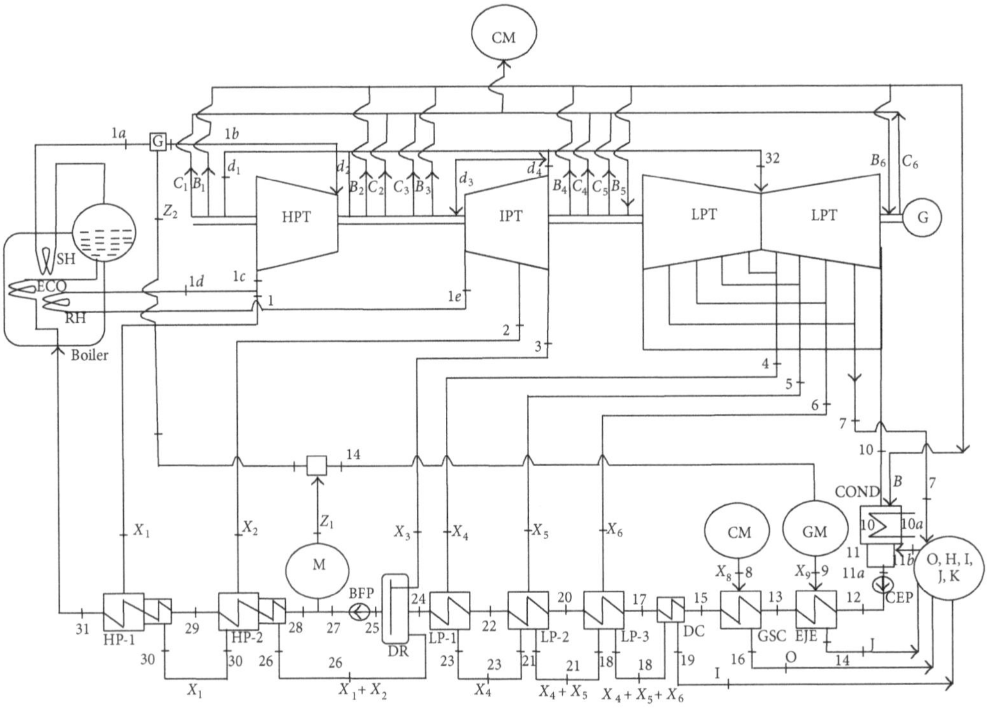

FIGURE 1: Steam turbine cycle based coal-fired power plant.

provide electricity in combustion and gasification plants was investigated and evaluated [11]. The study of economic feasibility of constructing a 560 MW coal-fired power plant in Turkey, using real options theory, was discussed [12]. A parametric study concerning the use of combined cycle technologies for power generation and cost-benefit analysis was carried out using the independent power producers optimization algorithm in which the electricity unit cost was calculated by independent power producers in Cyprus [13]. Performance, cost, and emissions data for coal and natural gas-fired power plants were presented, based on information from studies carried out recently for the IEA Greenhouse Gas R&D Program by major engineering contractors and process licensors [14]. A new methodology is presented for new design of power plants, which combines the benefits of thermodynamics, economics, and mathematical optimization [15]. In order to account for the cost of the investment required, the total capital cost must be placed on an annual basis. The annual cost consists due to interest accumulated on the investment, depreciation, maintenance, insurance, and taxes. The equipment life deteriorates with time and its depreciation cost is associated with it and thus, loses value [16].

This work presents the thermal and economic analysis of thermal power plant using thermodynamic analysis, and economic analysis based on Net present value approach. The predicted results agree with plant operating data.

2. Coal-Fired Power Plant

A schematic diagram of 210 MWe unit of a coal-fired power plant is shown in Figure 1. In power plants, several physical, chemical, and mechanical processes are conducted to transfer the energy, stored in fossil fuel, into electrical energy. This energy conversion is divided into several stages. Thermal power plant (Figure 5) uses coal as feedstock to convert it into mechanical energy through the expansion of steam from a high pressure in a suitable prime mover called steam turbine. Generator coupled with turbine produces electrical energy. Coal received from collieries in the rail wagons is mechanically unloaded by wagon tippler and transported by belt conveyor system to the boiler raw coal bunkers. The crushed coal, when not required for raw coal bunkers, is carried to coal storage area through belt conveyor and telescopic chute. The quantity of coal from coal bunkers to coal mill is regulated through raw coal chain feeder, where coal is pulverized into the fine powder form. The pulverized coal is then sucked by vapor fan and finally stored in the pulverized coal bunkers. The pulverized coal is then pushed to the boiler furnace, which is comprised of water tube walls all around through which water circulates. This chemically treated water running through the walls of boiler furnace gets evaporated at high temperature into steam by getting furnace heat. This steam is further heated in the superheater (SH). The superheated steam produced in the superheater enters into the high pressure Turbine (HPT). After expansion in HPT, cold reheat steam is divided into two streams, one is sent for reheating in reheater (RH), and another is sent towards high pressure feedwater heater (HP-1). Steam then passes through intermediate pressure turbine (IPT) and low pressure turbine (LPT) respectively. The incoming stream of steam towards IPT from HPT after RH is divided into three streams. One is sent towards intermediate pressure feedwater heater (HP-2), and the other two are sent towards Deaerator (DR) and LPT respectively. Similarly the incoming stream of steam towards

LPT from IPT is divided into four streams, and out of these four streams, three streams are sent towards low pressure feedwater heater (LP-1, LP-2 and LP-3) and one stream is sent towards condenser, respectively. The steam after doing useful work in the turbine is condensed in condenser. The condensate is sent by condensate extraction pumps (CEP) towards gland steam cooler (GSC), drain cooler (DC) and remaining is sent towards low pressure feedwater heaters. Since the extracted steam upon condensation gets subcooled so the drain cooler (DC) is used. From the last low pressure feedwater heater (LP-1) outlet, the condensate enters in deaerator shell. Boiler feed pump (BFP) supplies this condensate quantity from deaerator (DR) to Low Pressure Feedwater Heaters and High Pressure Feed Water Heaters respectively. Boiler feed pump (BFP) is a multistage pump provided for pumping feedwater (FW) to economizer. Three pumps each of 50% of total capacity are provided out of which two pumps work in parallel and third will be reserve. After HP-1 the condensate passes through economizer (ECO) and finally it enters into the boiler drum. Hence the cycle is completed.

3. Thermodynamic Modeling

Energy and mass balance equations have been used for the mathematical modeling of each component. The empirical relations have been also derived based on thermodynamic relations.

3.1. High Pressure Feedwater Heater (HP-1). High pressure feedwater heater receives superheated steam bled from the turbine at state 1, the steam is first desuperheated then condensed and finally subcooled, whereas the feedwater gets heated as shown in Figure 2. The schematic diagram of high pressure feedwater heater "HP-1" (for $i = 1$) is shown in Figure 3.

Mass balance:

$$\sum x_{\text{in}} = \sum x_{\text{out}}. \tag{1}$$

Energy balance:

$$x_i \times h_{s(i)} + 1 \times h_{f(i+1)} = 1 \times h_{f(i)} + x_i \times h_{c(i)},$$

$$x_i = \frac{\left(h_{f(i)} - h_{f(i+1)}\right)}{\left(h_{s(i)} - h_{c(i)}\right)}, \tag{2}$$

where $h_{f(i)} = h_{(i)l} - \text{TTD} \times \text{Cpw}$ and $h_{c(i)} = h_{f(i+1)} + \text{ETD} \times \text{Cpw}$. "$x$" is the fractional mass flow with total steam flow from boiler. TTD and ETD are terminal and entry temperature difference of feedwater heaters. The value of TTD and ETD is taken to be 5°C and 2°C, respectively. The pressure for each turbine extraction is supplied empirically using design data of 210 MWe thermal power plants (see Table 1) as given by (3) as

$$p_i = 0.189\,\text{MWe} + 1.476. \tag{3}$$

The formulations of high pressure feedwater heater (HP-2) (Figure 13), deaerator (Figure 14), low pressure feedwater heaters (LP-1, LP-2, LP3) (Figures 15, 16, and 17), drain cooler (Figure 18), gland steam condenser (Figure 19), and ejector (Figure 20) are detailed in Appendix A.

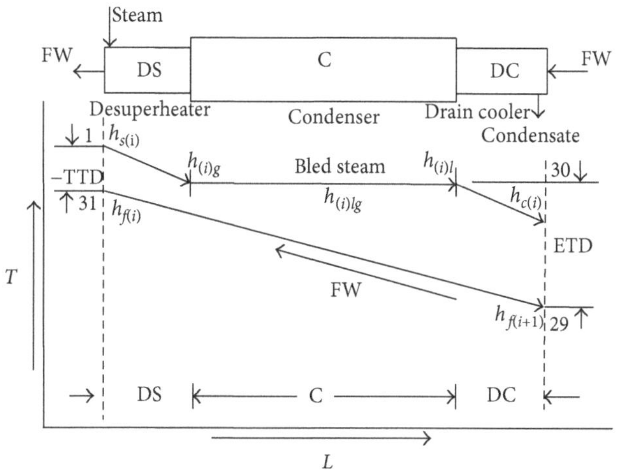

FIGURE 2: Temperature-length diagram for HP-1 (DS: desuperheater, C: condenser; DC: drain cooler).

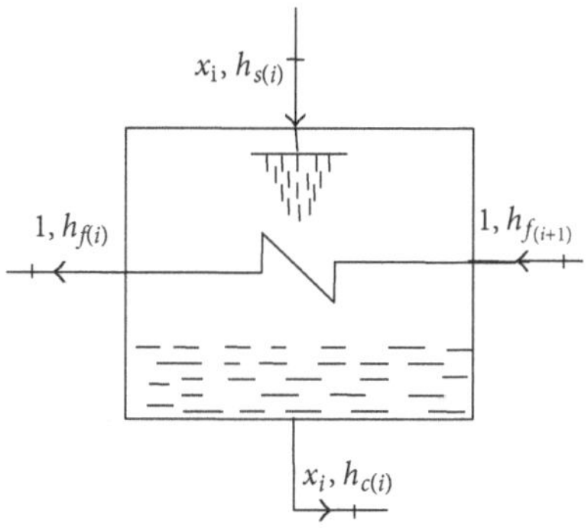

FIGURE 3: Schematic diagram of high pressure feedwater heater (HP-1).

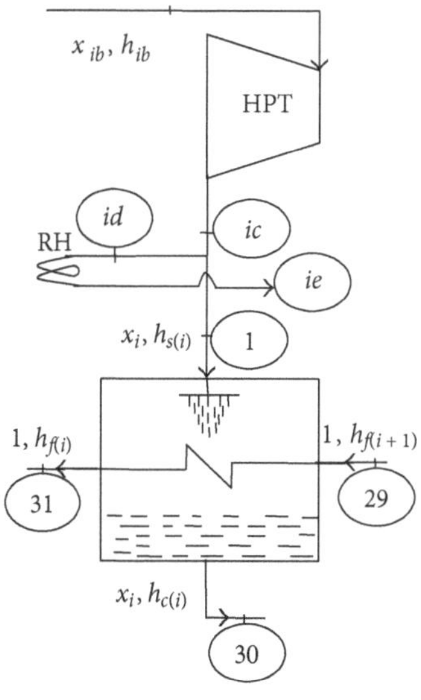

FIGURE 4: Schematic diagram of turbine-cum-high pressure feedwater heater.

Table 1: Design data of pressure (bar) of bled stream at various locations from turbine.

Sr. no.	MWe	Pressure (bar) of bled stream from the turbine					
		P_1	P_2	P_3	P_4	P_5	P_6
1.	126	25.36	10.60	4.37	1.589	0.558	0.288
2.	168	33.26	13.84	5.28	1.961	0.695	0.356
3.	210	41.27	16.25	6.43	2.396	0.853	0.431

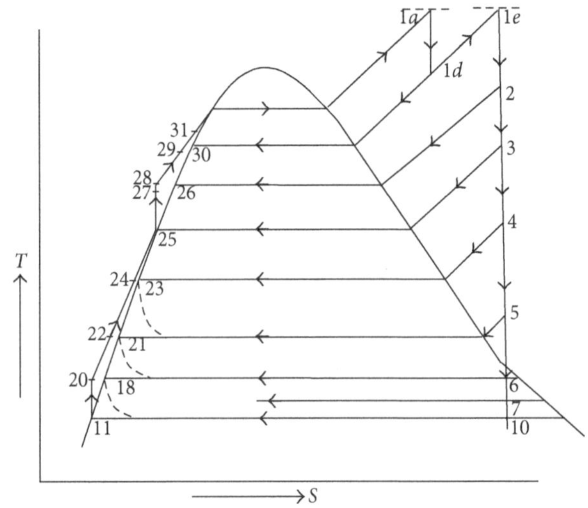

Figure 5: T-S diagram of thermal power plant.

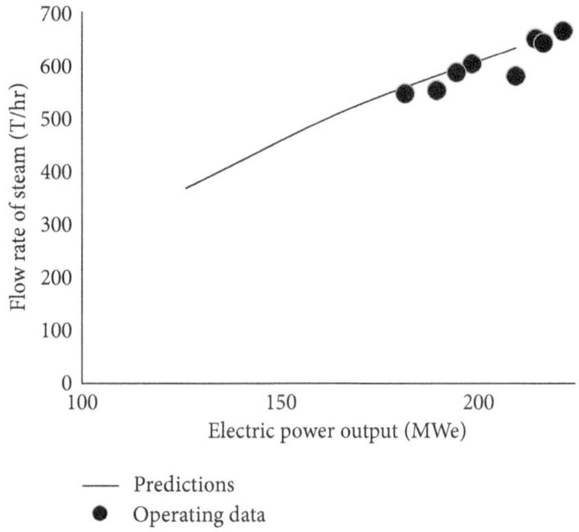

Figure 7: Mass flow rate of steam versus plant output.

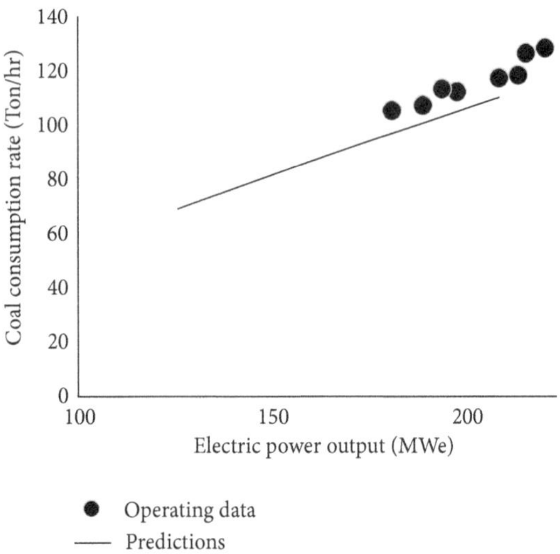

Figure 6: Coal consumption (T/hr) versus plant output.

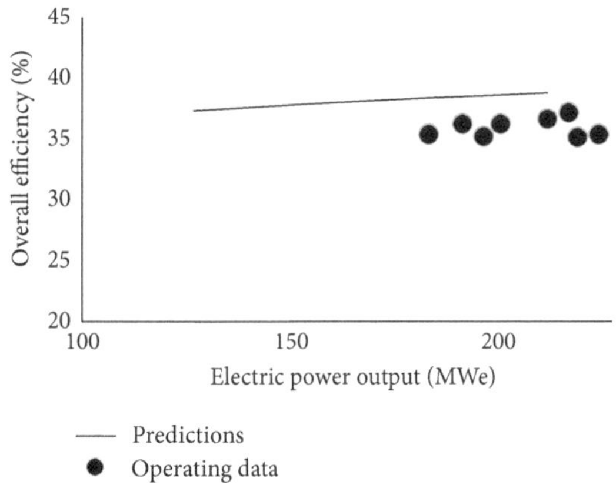

Figure 8: Overall efficiency versus plant output.

3.2. Steam Turbine. A steam turbine is one module that extracts thermal energy from pressurized steam and converts it into useful mechanical work. Steam turbine is condensing, tandem compounded, horizontal, reheat type, and single shaft machine. It has got separate high pressure, and intermediate and low-pressure parts. The HP part is a single cylinder and IP & LP parts are double flow cylinders. The condensate which is leaving the first HP-1 feedwater heater is mixed with the subsequent feedwater heater and then the total condensate is mixed with the next closed feedwater heater and then the resultant is normally dumped into the deaerator. In the modern thermal power plants, the modeling of steam turbine is carried out along with feedwater heaters. Thus, steam turbine-feedwater heater can be treated as mathematical element to describe the thermal power plant. The schematic diagram of steam turbine-cum-high pressure feedwater heater is shown in Figure 4.

Thus, the power output of the turbine can be written in terms of mass flow rate of working substance and enthalpy drop as

$$P_{op} = m_w \Delta h, \tag{4}$$

where $\Delta h = h_{\text{in}} - h_{\text{out}}$.

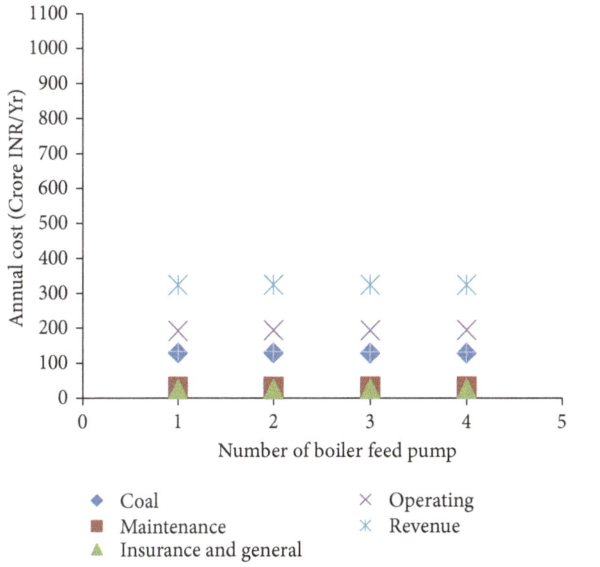

FIGURE 9: Annual cost (Crore INR/Yr) versus number of boiler feed pump.

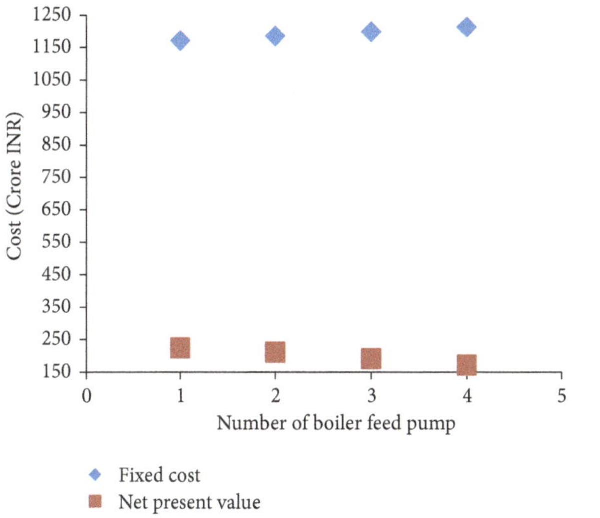

FIGURE 10: Cost (Crore INR) versus number of boiler feed pump.

Mass balance:

$$x_{1b} = \sum_{k=1}^{2} x_{B_k} + \sum_{k=1}^{2} x_{C_k} + \sum_{k=1}^{2} x_{d_k} + x_{1c}, \quad (5a)$$

$$x_{1c} = x_{1d} + x_{1e}. \quad (5b)$$

The notation of subscripts B_k, C_k, d_k $1c$, $1d$, and $1e$ represents the various states in Figure 1.

Work done:

$$W_{t1} = x_{1b}\left(h_{1b} - h_{1c}\right). \quad (6)$$

Accordingly, formulation of intermediate and low pressure turbine can be done and final work can be written as

$$W_t = W_{t1} + W_{t2} + W_{t3}. \quad (7)$$

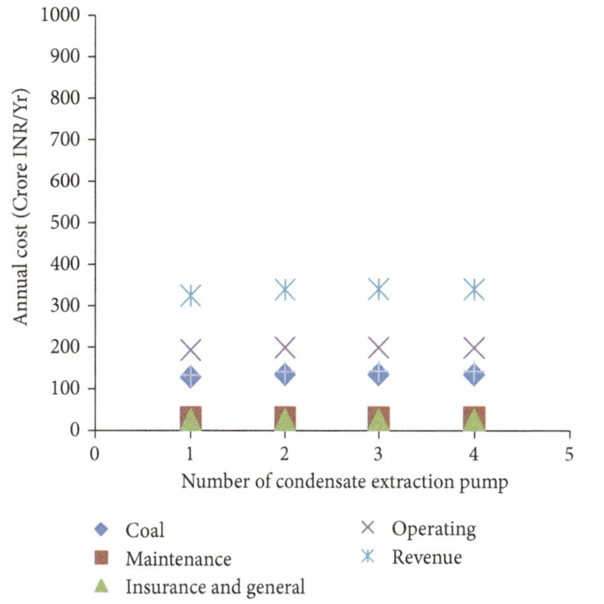

FIGURE 11: Annual cost (Crore INR/Yr) versus number of condensate extraction pump.

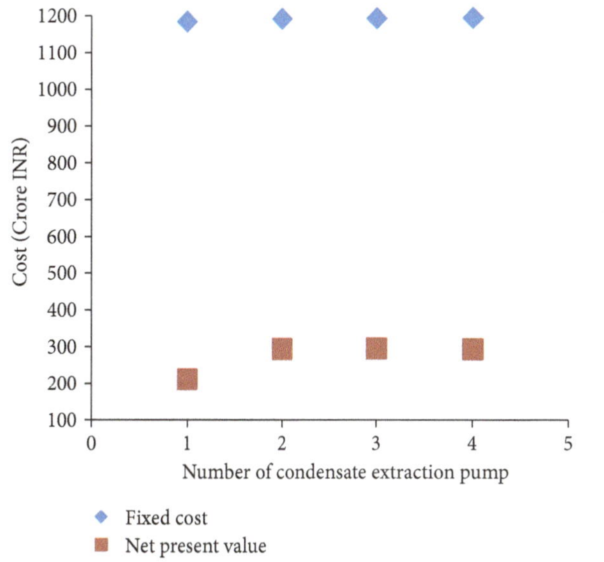

FIGURE 12: Cost (Crore INR) versus number of condensate extraction pump.

3.3. Condenser. In the analysis, it was assumed that the balance steam from turbine after subtraction of bleeds is condensed all in condenser. Thus mass balance can be written as

$$x_{10a} = x_{10} + \sum_{k=1}^{6} x_{B_k}, \quad (8a)$$

$$x_{11} = x_{10a} + x_{11b}. \quad (8b)$$

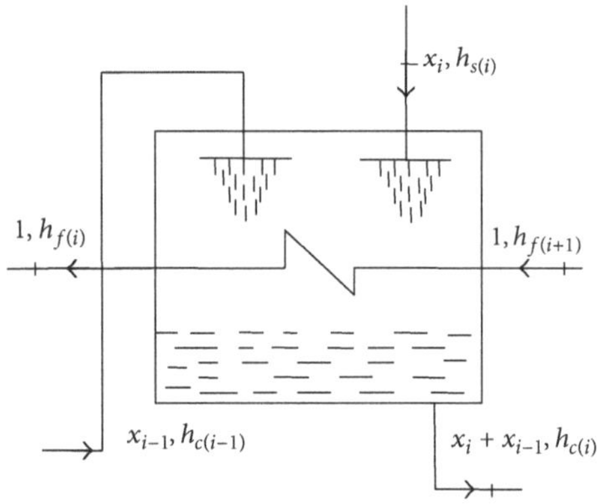

FIGURE 13: Mass and energy balance for high pressure feedwater heater (HP-2).

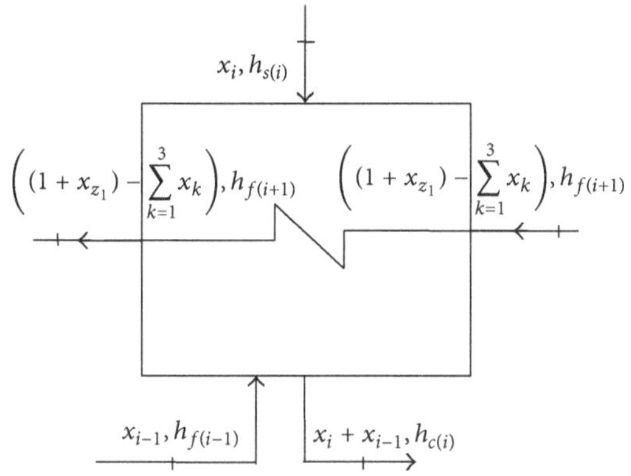

FIGURE 16: Mass and energy balance for low pressure feedwater heater (LP-2).

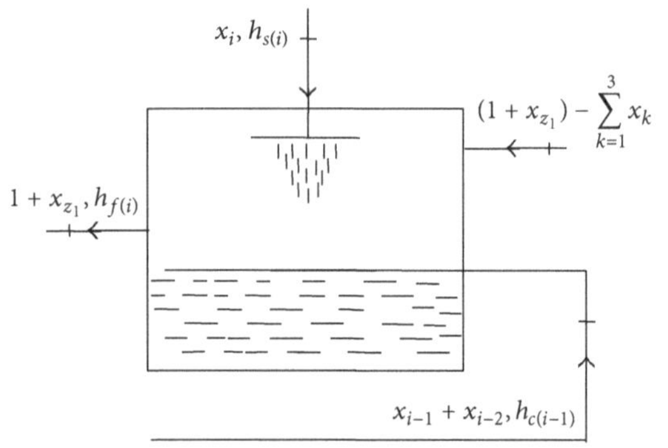

FIGURE 14: Mass and energy balance for deaerator (DR).

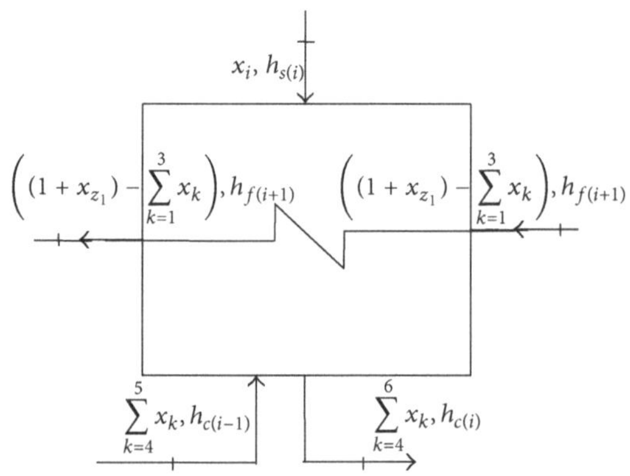

FIGURE 17: Mass and energy balance for low pressure feedwater heater (LP-3).

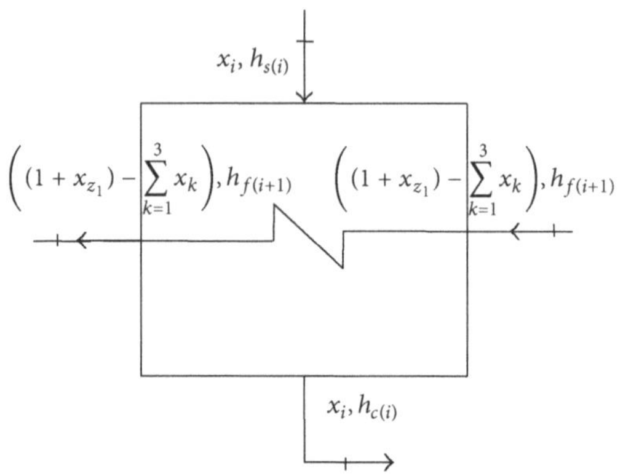

FIGURE 15: Mass and energy balance for low pressure feedwater heater (LP-1).

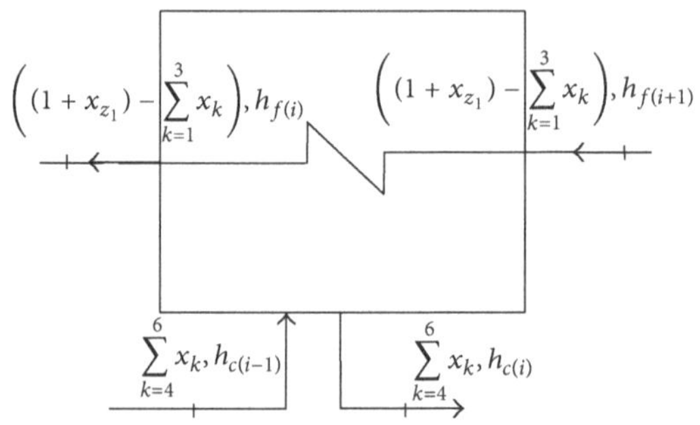

FIGURE 18: Mass and energy balance for drain cooler (DC).

Mass consumption rate of coal can be written in terms of unit mass flow rate of water (m_{uw}) as

$$m_{\mathrm{coal}} = \frac{m_{\mathrm{uw}}\left(x_{1a}\left(h_{1a}-h_{31}\right)+x_{1d}\left(h_{1e}-h_{1d}\right)\right)}{\left(\eta_{\mathrm{boiler}}\mathrm{CV}_{\mathrm{coal}}\right)}. \tag{8c}$$

η_{boiler} is the boiler efficiency; it is fixed at 0.86 in the present calculations [17].

The overall plant efficiency can be defined in terms of plant capacity (MWe) as

$$\eta_{\mathrm{plant}} = 1000 \times \frac{\mathrm{MWe}}{\left(m_{\mathrm{coal}} \times \mathrm{CV}_{\mathrm{coal}}\right)}. \tag{9}$$

4. Economic Analysis

The economic analysis of the plant has been carried out on the basis of initial capital investment, operating costs, and annual revenue. The equipment cost include boiler, steam turbine, condenser, generator, and auxiliary equipment such

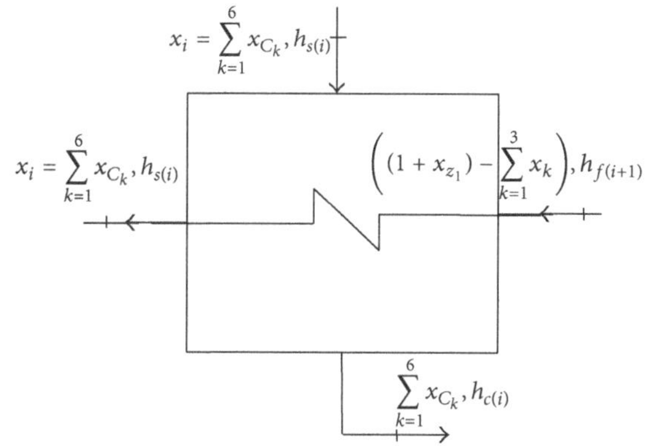

Figure 19: Mass and energy balance for gland steam condenser (GSC).

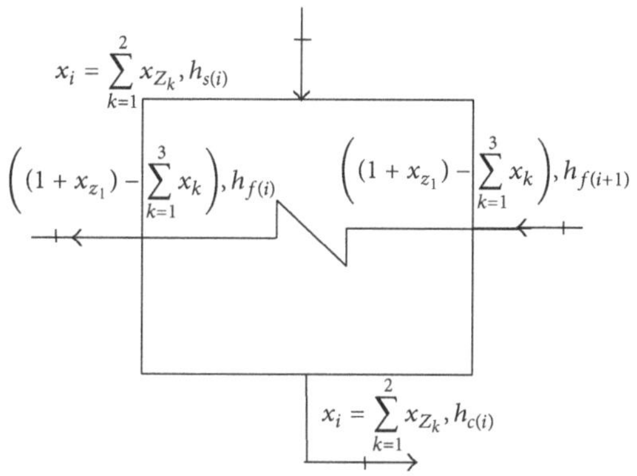

Figure 20: Mass and energy balance for ejector (EJE).

as condensate extraction pump, feed water pump, and so forth. Thus, fixed (equipment) cost, C_F, can be written in terms of redundancies of respective components (if any) as

$$C_F = f\left(C_{\text{boiler}} + C_{\text{turbine}} + C_{\text{condenser}} + n_{\text{ce}}\left(C_{\text{condpump}}\right) + n_{\text{fb}}\left(C_{\text{boilerfeedpump}}\right) + C_{\text{generator}}\right). \quad (10a)$$

C_F represents fixed cost. The coal storage, ash handling, electrical works, civil works, and fumes treatment costs are neglected in the present work. The initial cost for each equipment/component C_{Fi} has been obtained from [11] empirically in following form

$$C_{Fi} = a(\text{MWe})^b, \quad (10b)$$

where the values of constants a and b are taken from Table 2.

The total operating costs include maintenance, insurance, and general costs, total operating labor, and purchase of coal feedstock as [11]

$$C_{\text{OP}} = \left(C_{\text{maint}} + C_{\text{ins}} + n_L C_{\text{lab}} + m_{\text{coal}} C_{\text{coal}}\right), \quad (11)$$

where C is cost in (INR), while subscripts OP, maint, ins, coal, ep, and lab correspond to operating, maintenance, insurance, coal feedstock, electricity/MWe, and individual labour cost, respectively. Here maintenance and insurance costs are taken to be 3% and 2.5% of the total fixed cost [11]. Average personnel salary on annual basis is deduced from plant data as given in Table 3.

Table 2: Constants for purchased equipment cost estimation [11].

Equipments	a	b
Generator	138300	0.6107
Steam turbine	633000	0.398
Condensate extraction pumps	9000	0.4425
Steam boiler	1340000	0.694
Boiler feed pumps	35000	0.6107
Condenser	398000	0.333

Table 3: Operating data for calculating average salary of plant personnel in a typical 210 MWe plant unit.

Total personnel deployed in 210 MWe unit	Total monthly salary of personnel (Rs.)	Average monthly salary (Rs.)
233	6025000	25858.36

Annual revenue obtained from the electricity can be evaluated as

$$R_{\text{an}} = f_{\text{MWe}} \times \text{hrs} \times \text{MWe} \times C_{\text{ep}}. \quad (12)$$

Thus, Net present value (NPV) can be written as

$$\text{NPV} = \sum_{l=1}^{pl} \frac{\left(R_{\text{an}} - C_{\text{OP}}\right)_l}{(1 + D)^l} - C_F. \quad (13)$$

All reference values of data collected for the analysis are tabulated in Table 4. The taxes and financial charges have been neglected in this work.

5. Results and Discussion

The above analysis was used to predict the coal consumption rate, overall thermal efficiency, mass flow rate of steam through boiler, and Net present value of thermal power plant for given plant capacity and redundancy on pumps. The predictions of mass flow rate, overall plant efficiency, and coal consumption rate were compared with operating plant data from plant for validation as shown in Figures 6, 7, and 8. The coal consumption rate and mass flow rate of steam increase with electric power output of plant. However, overall efficiency is not showing any significant increase with electric power output. A fair agreement between predictions and operating data has been shown graphically at wide range of load conditions of plant.

The effect of employing redundancy (excessive) units to boiler feed pumps and condensate extraction pump on annual cost of plant and Net present value is shown in Figures 9, 10, 11, and 12, which reflects that equipment, maintenance,

TABLE 4: Input data collected from literature and plant records.

Parameter	n_L	Lab	C_{coal}	CV_{coal}	f	pl	C_{ep}	f_{Mwe}	D	hrs/yr
Value	233	25858.36 (INR)	2696.81 (INR/Tone)	17656 (kJ/kg)	1.87	20 (year)	4000 (Rs/MW-hr)	0.9	9%	4555

insurance, operating, coal, and revenue increase with any increase in pump redundancy in various subsystems. The Net present value decreases with any addition of boiler feed pump redundancy in water circulation subsystem (Figure 10), while it shows marginally increasing trend with condensate extraction pump redundancy in condenser unit (Figure 12).

The effect of condensate extraction pump redundancy on Net present value is comparatively higher as compared with case of boiler feed pump. It is expected due to lower initial cost of condensate extraction pump as compared with the cost of boiler feed pump.

6. Conclusions

The thermal and economic analysis of a 210 MWe coal-fired power plant was carried out to predict the coal consumption rate, overall thermal efficiency, mass flow rate of steam through boiler, and Net present value of thermal power plant for given plant load and redundancy of boiler feed pump and condensate extraction pump. Thermodynamic modeling was carried out for evaluating the thermal performance. Predictions of mass flow rate of steam, coal consumption rate, and thermal efficiency were compared against plant operating data for validation. Economic analysis includes equipment cost, fuel cost, operations and maintenance cost, revenue, and plant Net present value. The redundancy due to condensate extraction pump is sensitive to Net present value.

Appendix

A. High Pressure Feedwater Heater (HP-2)

In the same manner for $i = 2$,

Energy balance:

$$x_i = \frac{\left(h_{f(i)} - h_{f(i+1)}\right)}{\left(h_{s(i)} - h_{c(i)}\right)} + x_{i-1}\left(\frac{\left(h_{c(i)} - h_{c(i-1)}\right)}{\left(h_{s(i)} - h_{c(i)}\right)}\right). \quad \text{(A.1)}$$

Empirical relations:

$$p_i = 0.077 \times \text{MWe} + 0.83, \quad \text{(A.2)}$$

$$h_{f(i)} = h_{(i)l} - \text{TTD} \times \text{Cpw}, \quad \text{(A.3)}$$

$$h_{c(i)} = h_{f(i+1)} + \text{ETD} \times \text{Cpw}. \quad \text{(A.4)}$$

B. Deaerator (DR)

Energy balance, for $i = 3$:

$$x_i = \left(\left(1 + x_{z_1}\right) \times \frac{\left(h_{f(i)} - h_{f(i+1)}\right)}{\left(h_{s(i)} - h_{f(i+1)}\right)}\right) - \left(\left(x_{i-1} + x_{i-2}\right) \times \frac{\left(h_{c(i-1)} - h_{f(i+1)}\right)}{\left(h_{s(i)} - h_{f(i+1)}\right)}\right) \quad \text{(B.1)}$$

$$p_i = 0.0312 \times \text{MWe} + 0.4367. \quad \text{(B.2)}$$

C. Low Pressure Feedwater Heater (LP-1)

Energy balance, for $i = 4$:

$$x_i = \frac{\left(\left(\left(1 + x_{z_1}\right) - \sum_{k=1}^{3} x_k\right) \times \left(h_{f(i)} - h_{f(i+1)}\right)\right)}{\left(h_{s(i)} - h_{c(i)}\right)}, \quad \text{(C.1)}$$

$$p_i = 0.0111 \times \text{MWe} + 0.189, \quad \text{(C.2)}$$

$$h_{f(i)} = h_{(i)l} - \text{TTD} \times \text{Cpw}, \quad \text{(C.3a)}$$

$$h_{c(i)} = h_{(i)l} - \text{ETD} \times \text{Cpw}. \quad \text{(C.3b)}$$

D. Low Pressure Feedwater Heater (LP-2)

Energy balance, for $i = 5$:

$$x_i = \left(\frac{\left(\left(1 + x_{z_1}\right) - \sum_{k=1}^{3} x_k\right) \times \left(h_{f(i)} - h_{f(i+1)}\right)}{\left(h_{s(i)} - h_{c(i)}\right)}\right) + \left(x_{i-1} \times \frac{\left(h_{c(i)} - h_{f(i-1)}\right)}{\left(h_{s(i)} - h_{c(i)}\right)}\right), \quad \text{(D.1)}$$

$$p_i = 0.00405 \times \text{MWe} + 0.048, \quad \text{(D.2)}$$

$$h_{f(i)} = h_{(i)l} - \text{TTD} \times \text{Cpw}, \quad \text{(D.3a)}$$

$$h_{c(i)} = h_{(i)l} - \text{ETD} \times \text{Cpw}. \quad \text{(D.3b)}$$

E. Low Pressure Feedwater Heater (LP-3)

Energy balance, for $i = 6$:

$$x_i = \left(\frac{\left(\left(1 + x_{z_1}\right) - \sum_{k=1}^{3} x_k\right) \times \left(h_{f(i)} - h_{f(i+1)}\right)}{\left(h_{s(i)} - h_{c(i)}\right)}\right) + \left(\sum_{k=4}^{5} x_k \times \frac{\left(h_{c(i)} - h_{c(i-1)}\right)}{\left(h_{s(i)} - h_{c(i)}\right)}\right), \quad \text{(E.1)}$$

$$p_i = 0.00198 \times \text{MWe} + 0.039, \quad \text{(E.2)}$$

$$h_{f(i)} = h_{(i)l} - \text{TTD} \times \text{Cpw}, \quad \text{(E.3a)}$$

$$h_{c(i)} = h_{(i)l} - \text{ETD} \times \text{Cpw}. \quad \text{(E.3b)}$$

F. Drain Cooler (DC)

Energy balance, for $i = 7$:

$$h_{f(i)} = \left(\sum_{k=4}^{6} x_k \times \frac{(h_{c(i-1)} - h_{c(i)})}{((1 + x_{z_1}) - \sum_{k=1}^{3} x_k)} \right) + h_{f(i+1)}. \quad \text{(F.1)}$$

G. Gland Steam Condenser (GSC)

Energy balance, for $i = 8$:

$$h_{f(i)} = \left(\sum_{k=1}^{6} x_{C_k} \times \frac{(h_{s(i)} - h_{c(i)})}{((1 + x_{z_1}) - \sum_{k=1}^{3} x_k)} \right) + h_{f(i+1)}. \quad \text{(G.1)}$$

H. Ejector (EJE)

Energy balance, for $i = 9$:

$$h_{f(i)} = \left(\sum_{k=1}^{2} x_{z_k} \times \frac{(h_{s(i)} - h_{c(i)})}{((1 + x_{z_1}) - \sum_{k=1}^{3} x_k)} \right) + h_{f(i+1)}. \quad \text{(H.1)}$$

Nomenclature

C_{coal}: Coal cost (INR/tone)
c_e: Condensate extraction pump
CV_{coal}: Calorific value of coal (kJ/kg)
D: Discount rate (%/yr)
f: Factor for direct installation, auxiliary, instrumentation and control, engineering, and plant start-up
f_{MWe}: % of net electric energy power plant output assumed as 90% of MWe available for sale
x_i: Fractional mass flow rate of steam at "ith" state (tone/hr)
$h_{s(i)}$: Enthalpy of superheated steam (kJ/kg)
$h_{f(i)}$: Enthalpy of feedwater at exit (kJ/kg)
$h_{f(i+1)}$: Enthalpy of feedwater at entry (kJ/kg)
$h_{c(i)}$: Enthalpy of condensate (kJ/kg)
m_{coal}: Annual coal consumption (tone/hr)
hrs: Annual operating hours of the plant (hrs/yr)
n_{ce}: Number of sub-systems (redundancy) of condensate extraction pumps
n_{fb}: Number of sub-systems (redundancy) of boiler feed pumps
n_L: Total number of employed personnel
P_i: Pressure of bled steam from turbine at "ith" state (bar)
pl: Plant life (years)
P_{npv}: (INR)
R_{an}: Annual revenue earned (INR/yr)
f_b: Boiler feed pump
η_{plant}: Overall thermal efficiency of the plant (%).

Conflict of Interests

The authors declare that they have no conflict of interests regarding the publication of this paper.

References

[1] R. W. Haywood, "A generalized analysis of the regenerative steam cycle for a finite number of heaters," *Proceedings of the Institution of Mechanical Engineers A*, vol. 161, pp. 157–164, 1949.

[2] C. D. Weir, "Optimization of heater enthalpy rises in feed-heating trains," *Proceedings of the Institution of Mechanical Engineers A*, vol. 174, pp. 769–796, 1960.

[3] J. H. Horlock, "Simplified analyses of some vapour power cycles," *Proceedings of the Institution of Mechanical Engineers A*, vol. 210, no. 3, pp. 191–202, 1996.

[4] G. Angelino, C. Invernizzi, and G. Molteni, "The potential role of organic bottoming rankine cycles in steam power stations," *Proceedings of the Institution of Mechanical Engineers*, vol. 213, no. 2, pp. 75–81, 1999.

[5] A. Kostyuk and V. Frolov, *Steam and Gas Turbines*, Mir Publishers, Moscow, Russia, 1985.

[6] M. Ameri, P. Ahmadi, and A. Hamidi, "Energy, exergy and exergoeconomic analysis of a steam power plant: a case study," *International Journal of Energy Research*, vol. 33, no. 5, pp. 499–512, 2009.

[7] D. S. Remer and A. P. Nieto, "A compendium and comparison of 25 project evaluation techniques. Part 1: net present value and rate of return methods," *International Journal of Production Economics*, vol. 42, no. 1, pp. 79–96, 1995.

[8] S. Boonnasa and P. Namprakai, "Sensitivity analysis for the capacity improvement of a combined cycle power plant (100–600 MW)," *Applied Thermal Engineering*, vol. 28, no. 14-15, pp. 1865–1874, 2008.

[9] A. C. Caputo and P. M. Pelagagge, "RDF production plants: II. Economics and profitability," *Applied Thermal Engineering*, vol. 22, no. 4, pp. 439–448, 2002.

[10] B. C. Giri and T. Dohi, "Optimal lot sizing for an unreliable production system based on net present value approach," *International Journal of Production Economics*, vol. 92, no. 2, pp. 157–167, 2004.

[11] A. C. Caputo, M. Palumbo, P. M. Pelagagge, and F. Scacchia, "Economics of biomass energy utilization in combustion and gasification plants: effects of logistic variables," *Biomass and Bioenergy*, vol. 28, no. 1, pp. 35–51, 2005.

[12] R. Madlener and S. Stoverink, "Power plant investments in the Turkish electricity sector: a real options approach taking into account market liberalization," *Applied Energy*, vol. 97, pp. 124–134, 2012.

[13] A. Poullikkas, "Parametric study for the penetration of combined cycle technologies into Cyprus power system," *Applied Thermal Engineering*, vol. 24, no. 11-12, pp. 1697–1707, 2004.

[14] J. Davison, "Performance and costs of power plants with capture and storage of CO_2," *Energy*, vol. 32, no. 7, pp. 1163–1176, 2007.

[15] J. Manninen and X. X. Zhu, "Thermodynamic analysis and mathematical optimisation of power plants," *Computers and Chemical Engineering*, vol. 22, supplement 1, no. 1, pp. S537–S544, 1998.

[16] M. Peters, K. Timmerhaus, and R. West, *Plant Design and Economics for Chemical Engineers*, McGraw-Hill, New York, NY, USA, 1991.

[17] *Thermal Power Plant Efficiency and Performance Monitoring*, National Power Training Institute (NPTI), Faridabad, India, 2000.

24

Thermoacoustic, Volumetric, and Viscometric Investigations in Binary Liquid System of Cyclohexanone with Benzyl Benzoate at $T = 308.15, 313.15$, and 318.15 K

Sk. Md Nayeem,[1] M. Kondaiah,[2] K. Sreekanth,[3] and D. Krishna Rao[4]

[1] *Department of Physics, KRK Govt. Degree College, Addanki 523 201, India*
[2] *Department of Physics, NM Govt. Degree College, Jogipet, Medak District, Telangana 502270, India*
[3] *Department of Physics, PBN College, Guntur District, Nidubrolu 522124, India*
[4] *Department of Physics, Acharya Nagarjuna University, Nagarjuna Nagar 522 510, India*

Correspondence should be addressed to D. Krishna Rao; krdhanekula@yahoo.co.in

Academic Editor: Ahmet Z. Sahin

Ultrasonic velocities (u), densities (ρ), and viscosities (η) of binary liquid mixtures of cyclohexanone with benzyl benzoate, including pure liquids, over the entire composition range have been measured at 308.15 K, 313.15 K, and 318.15 K. Using the experimental results, parameters such as molar volume (V_m), isentropic compressibility (k_s), intermolecular free length (L_f), acoustic impedance (Z), internal pressure (π_i), enthalpy (H), Gibbs free energy of activation of viscous flow (G^{*E}), and excess/deviation properties of these including partial molar volumes ($\overline{V}_{m,1}$ and $\overline{V}_{m,2}$), excess partial molar volumes ($\overline{V}^E_{m,1}$ and $\overline{V}^E_{m,2}$), partial molar volume of the components at infinite dilution ($\overline{V}^{\infty}_{m,1}$, $\overline{V}^{\infty}_{m,2}$), and excess partial molar volume at infinite dilution ($\overline{V}^{E,\infty}_{m,1}$ and $\overline{V}^{E,\infty}_{m,2}$) have been computed. The observed negative values of V^E_m, Δk_s, L^E_f, and π^E_i and positive values of z^E, H^E, ΔG^{*E}, $\Delta\eta$, and Δu for all the liquid mixtures studied clearly indicate the presence of strong dipole-dipole-type interactions, fitting of smaller molecules into bigger molecules. Further theoretical values of sound velocity and viscosity in the mixtures have been evaluated using various theories and have been compared with experimental values to verify the applicability of such theories to the systems studied.

1. Introduction

Volumetric, viscometric, and ultrasonic investigations of liquid mixtures are of considerable importance in understanding the intermolecular interactions occurring among component molecules and they find application in several industrial and technological processes [1, 2]. The work on medicinally used chemical compounds requires the attention of the society in all aspects including ultrasonic behaviour. Benzyl benzoate is a carboxylate ester which is used in oily injections and as an insect repellent and as acaricide and pediculicide in veterinary hospitals. It is an effective and inexpensive topical treatment for human scabies. It is a polar (μ = 2.06 D) molecule ($C^+{=}O^-$) with the structure shown in Figure 12. Behaviour of benzyl benzoate in many liquids such as aliphatic alkanes, aromatic alkanes, aliphatic alcohols, substituted benzenes, acetates, ketones, and DMSO (super solvent) has been thoroughly studied ultrasonically [3–6].

Ketone is an organic compound that contains a carbonyl group and two aliphatic or aromatic substituents containing the chemical formula $RCOR^1$. Here, R and R^1 may be same or different incorporated into a ring (alkyl, aryl, and heterocyclic radicals). Cyclohexanone is a ketone liquid. The chemical reactivity of the carbonyl group ($C^+{=}O^-$) plays important role in chemical reactions and is influenced considerably by steric effects. The greater electronegativity of O^- and high dipole moment make ketones polar (μ = 3.25 D). The resonance structure shown in Figure 12 illustrates this polarity. Thus a study on thermophysical properties data of binary liquid mixtures has attracted considerable interest in the literature

[7–10]. Cyclohexanone and its derivatives are used for the synthesis of pharmaceuticals, dyes, herbicides, pesticides, plasticizers, and rubber chemicals.

The liquids under investigation have been chosen on the basis of their medicinal and industrial applications. These applications have greatly stirred the need for extensive information on the thermodynamic, acoustic, and transport properties of these solvents and their mixtures [11–13]. Ultrasonic and density data for binary mixtures of benzyl benzoate with Isomers of butanol have been studied previously in our laboratory [14]. The literature survey reveals that Madhuri et al. [15] have reported ultrasonic, volumetric, and viscometric studies of benzyl benzoate with acetonitrile and benzonitrile. Recently, Sri et al. [16] reported ultrasonic velocity and density in binary liquids of certain aldehydes and esters. In the present investigation, a detailed study of the binary mixture of benzyl benzoate (BB) with cyclohexanone (CH) at three temperatures 308.15 K, 313.15 K, and 318.15 K is aimed. From the experimentally measured data of ultrasonic velocity (u), densities (ρ), and viscosities (η), thermodynamic and other related parameters like molar volume, isentropic compressibility, acoustic impedance, free length, relaxation time, internal pressure, enthalpy, and Gibbs free energy of activation of viscous flow are computed and in terms of some of these excess/deviation parameters, the nature of molecular interactions is predicted in the binary mixtures. An evaluation of velocities and viscosities using three different empirical theories is also attempted at all the three temperatures.

2. Experimental Section

Benzyl benzoate (BB) and cyclohexanone (CH) used in the present study were the AR grade products from LOBA Chemicals, India, and were purified by standard methods described in the literature [17–19]. The mass fraction purity of liquids obtained is >0.995. Before use, the chemicals were stored over 0.4 nm molecular sieves approximately for 72 h to remove water content and degassed and latter kept in air tight bottles. The mass measurements are performed with a METTLER TOLEDO (Switzerland make) AB135-S/FACT digital balance with an accuracy of ±0.01 mg. The uncertainty in the mole fraction is 10^{-4}. The ultrasonic velocity of pure liquids and their binary mixtures has been measured by using a multifrequency ultrasonic interferometer (M-82 model) supplied by Mittal Enterprises, New Delhi, at a fixed frequency of 2 MHz with an accuracy of ±0.2%. The temperature of liquid sample in the interferometer cell is maintained constant by circulating water pumped from constant temperature water bath. In the present study, the constant temperature water bath (digital electronic) supplied by Concord Instruments Co., Ltd., Chennai (RAAGA type), has been used. The instrument can maintain temperature to ±0.01 K as per its specifications.

Densities and viscosities of pure liquids and liquid solutions are determined using 5 cm^3 two stem double walled Parker & Parker-type pycnometer [20] and Ostwald viscometer which is standardized as described by Naidu and Prasad [21] using triply distilled water, respectively. The estimated accuracy in measuring the density is 3 in 10^5 parts. The uncertainty in the viscosity measurement is ±0.2%. The detailed description of measurement of density and viscosity is discussed in our previous papers [22–25]. The densities, velocities, and viscosities of pure liquids in this investigation at temperature of 308.15 K, 313.15 K, and 318.15 K are compiled in Table 1 together with the literature data [15, 26] available. These results are found to be in good agreement with reported data.

3. Results and Discussion

The measured values of density, velocity, and viscosity and the calculated values of molar volume, isentropic compressibility, free length, acoustic impedance, internal pressure, relaxation time, enthalpy, and Gibbs free energy of activation of viscous flow are calculated using standard relations and are presented in Table 2 with mole fraction of cyclohexanone in benzyl benzoate. In the present investigation, the values of u and z increase and those of k_s and L_f decrease as the mole fraction increases. The variation of ultrasonic speed in a solution depends upon the increase or decrease of L_f after mixing the components. This is due to the fact that ultrasonic speed increases if the intermolecular free length decreases and vice versa. The decrease in the values of k_s and L_f with mole fraction in the present study indicates significant interactions between CH and BB molecules.

Study of deviation/excess properties plays vital role in the study of molecular interactions. The nonlinear variation of excess/deviation properties with mole fraction of BB is responsible for nonideality in the systems of binary liquid mixture [15]. The factors responsible for such departure from ideality may either be due to the presence of intermolecular forces between the constituents in the mixture or due to compound formation between solute and solvent, or as a result of association of either to form complex molecules [27]. These excess parameters throw light upon the strength of interaction and their variation with mole fraction finds application in typifying the physicochemical behaviour of liquid mixtures [28].

The deviation/excess properties have been calculated using the relation

$$Y^E = Y_{\text{real}} - Y_{\text{ideal}}, \tag{1}$$

where Y^E represent deviation/excess value of the parameter, Y_{real} is the experimental value of the parameter of liquid mixture, and Y_{ideal} is the ideal value of the parameter computed theoretically. The $Y^E = V_m^E$ is excess molar volume, Z^E is excess specific acoustic impedance, Δk_s is deviation in isentropic compressibility, L_f^E is excess intermolecular free length, π_i^E is excess internal pressure, H^E is excess enthalpy, ΔG^{*E} is excess Gibbs free energy of activation of viscous flow, $\Delta\eta$ is deviation in viscosity, and Δu is deviation in velocity.

The deviation in isentropic compressibility (Δk_s) has been calculated from

$$\Delta k_s = k_s - (\Phi_1 k_{s1} + \Phi_2 k_{s2}), \tag{2}$$

TABLE 1: Comparison of experimental values of ultrasonic velocity (u), density (ρ), and viscosity, (η) of pure liquids with the literature values concerned at 308.15, 313.15, and 318.15 K.

Liquid	Temp/K	$\rho/(\text{kg}\cdot\text{m}^{-3})$		$\eta/(10^{-3}\ \text{N}\cdot\text{s}\cdot\text{m}^{-2})$		$u/(\text{m}\cdot\text{s}^{-1})$	
		Exptl.	Lit.	Exptl.	Lit.	Exptl.	Lit.
Cyclohexanone	308.15	939.60	939.6 (26)	1.594	1.6013 (26)	1362.00	1362.0 (26)
	313.15	934.70	934.7 (26)	1.472	1.4561 (26)	1348.20	1348.2 (26)
	318.15	922.30	922.5 (26)	1.370	1.3101 (26)	1275.00	1272.8 (26)
Benzyl benzoate	308.15	1114.54	1114.5 (15)	7.118	—	1493.30	1493.3 (15)
	313.15	1109.70	1109.7 (15)	6.261	—	1471.40	1471.4 (15)
	318.15	1104.80	—	5.550	—	1440.00	—

where Φ_1, Φ_2 and k_{s1}, k_{s2} are the volume fractions and isentropic compressibilities of components 1 and 2, respectively. Since k_s is not additive on mole fraction but is additive on volume fraction, hence such values are calculated using volume fraction Φ_i:

$$\Phi_i = \frac{x_i V_i}{\sum x_i V_i}. \tag{3}$$

The excess/deviation properties have been fitted to a Redlich-Kister-type polynomial equation [29]:

$$Y^E = x(1-x)\sum_{i=0}^{j} A_i (1-2x)^i, \tag{4}$$

where x is the mole fraction of CH and for evaluating Δk_s, x is replaced by volume fraction (Φ_i) and A_i are the adjustable parameters of the function and are determined using the least square method. In the present investigation "i" values taken are from 0 to 4. The corresponding standard deviations $\sigma(Y^E)$ were calculated using the expression

$$\sigma\left(Y^E\right) = \left[\frac{\sum\left(Y^E_{\text{exp}} - Y^E_{\text{cal}}\right)^2}{(m-n)}\right]^{1/2}, \tag{5}$$

where "m" is the total number of experimental points and "n" is the number of coefficients in (4). The calculated values of the coefficients A_i along with the standard deviations (σ) are given in Table 3.

Figure 1 represents the variation of excess molar volume (V_m^E) with mole fraction of CH. The sign of V_m^E depends upon the contraction and expansion of volume of the liquids due to mixing. The excess molar volume is the resultant contribution from several opposing effects, namely, chemical, physical, and structural [30]. The chemical or specific interactions result in volume contractions, leading to negative excess molar volume, and these include charge-transfer complexes, dipole-dipole, dipole-induced dipole interactions and formation of H-bonding between component molecules. The physical interactions or nonspecific interactions are weak and these include breaking of the structure of one or both of the components in a solution, that is, the loss of dipolar association between the molecules (dispersion forces), steric hindrance of the molecules and H-bond rupture, and stretching of self-associated molecules (like alcohols). The structural contributions are mostly negative and arise from several effects such

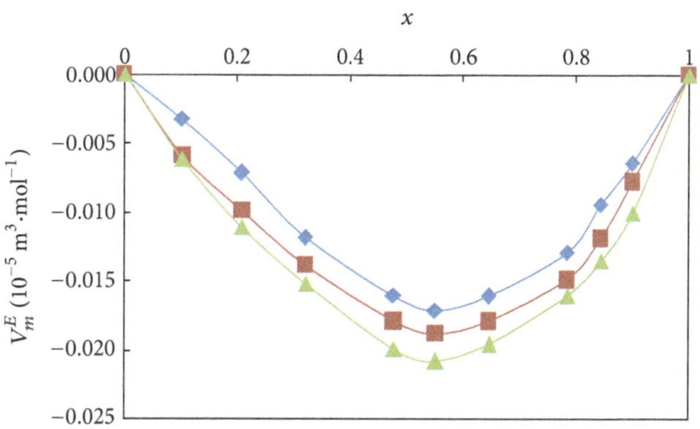

FIGURE 1: Variation of excess molar volume, V_m^E, with mole fraction, x, of CH in BB at (◆) 308.15 K, (■) 313.15 K, and (▲) 318.15 K.

as interstitial accommodation and geometrical fitting of one component into another due to the differences in the molar volume and free volume between components. In the present investigation, the sign of V_m^E is found to be negative over the entire composition range.

The nature of existing interaction in a binary liquid can also be analysed by knowing their individual chemical and physical properties. It is well known that BB and CH are polar and their dipole moment values follow: $\mu_{\text{CH}} = 3.25\,\text{D} > \mu_{\text{BB}} = 2.06\,\text{D}$. In the present study, possibility of Keesom dipole-dipole van der Waals forces which arise due to the dipole moment of the components gives strong interactions leading to negative values of V_m^E [31]. Moreover, the liquids BB and CH lack hydroxyl groups; hence the possibility for formation of intermolecular hydrogen bonds is trifling. Further, the molar volumes of CH and BB are 10.4459 and 19.0428 ($\times 10^{-5}$ m^3·mol^{-1}), respectively, at 308.15 K. This implies that geometrical fitting of smaller molecules into the voids created by the bigger molecules is most favorable. The V_m^E values in binary mixture are thereby decreased with increasing temperature. Negative V_m^E values indicate strong interactions following the order 318.15 K > 313.15 K > 308.15 K. According to Rastogi et al. [32], such behaviour may arise due to the fact that, as temperature increases, thermal energy activates the molecule; this would increase the rate of association of unlike molecules. Similar type of trend was observed by Baragi et al. [33] in methyl cyclohexane with alkanes.

Table 2: Experimental values of densities (ρ), velocities (u), and viscosities (η) with calculated properties of molar volume, (V_m), acoustic impedance, (Z), isentropic compressibility, (k_s), free length, (L_f), internal pressure, (π_i), enthalpy, (H), and Gibbs free energy of activation of viscous flow, (G^{*E}), with mole fraction, x, of cyclohexanone (x) in benzyl benzoate at T = 308.15 K, 313.15 K, and 328.15 K.

x	ρ/(kg·m^{-3})	u/(m·s^{-1})	η/(10^{-3} N·s·m^{-2})	V_m/(10^{-5} m^3·mol^{-1})	Z^E/(10^6 kg·m^{-2}·s^{-1})	k_s/(10^{-10} Pa^{-1})	L_f/(10^{-10} m)	π_i^E/(10^6 Pa)	H/(J·mol^{-1})	G^{*E}/(kJ·mol^{-1})
308.15 K										
1.0000	939.60	1362.00	1.594	10.4459	1.2797	5.7372	0.5016	521.97	54525.20	−39.9867
0.9021	969.35	1385.94	1.878	11.2775	1.3435	5.3706	0.4853	505.62	57021.57	−39.3711
0.8443	984.74	1397.80	2.071	11.7709	1.3765	5.1973	0.4775	499.07	58745.42	−39.0100
0.7852	999.11	1409.21	2.289	12.2765	1.4080	5.0400	0.4702	493.94	60639.08	−38.6459
0.6470	1028.33	1433.09	2.879	13.4610	1.4737	4.7349	0.4557	486.22	65450.91	−37.8231
0.5522	1045.47	1447.20	3.347	14.2748	1.5130	4.5670	0.4476	483.19	68975.81	−37.2863
0.4789	1057.32	1456.83	3.746	14.9058	1.5403	4.4563	0.4421	481.64	71793.50	−36.8876
0.3232	1079.23	1473.61	4.696	16.2491	1.5904	4.2669	0.4326	479.88	77976.77	−36.0876
0.2112	1092.84	1482.60	5.466	17.2160	1.6203	4.1629	0.4273	479.51	82553.29	−35.5502
0.1035	1104.50	1488.79	6.276	18.1468	1.6444	4.0847	0.4233	479.63	87038.88	−35.0614
0.0000	1114.54	1493.30	7.118	19.0428	1.6643	4.0236	0.4201	479.96	91398.24	−34.6154
313.15 K										
1.0000	934.70	1348.89	1.472	10.5007	1.2608	5.8800	0.5078	502.22	52736.76	−40.1779
0.9021	964.24	1377.00	1.707	11.3373	1.3278	5.4694	0.4898	481.96	54641.93	−39.6016
0.8443	979.69	1390.87	1.870	11.8317	1.3626	5.2763	0.4811	473.72	56050.30	−39.2590
0.7852	994.09	1401.82	2.054	12.3385	1.3935	5.1190	0.4738	467.57	57692.00	−38.9103
0.6470	1023.27	1422.33	2.558	13.5275	1.4554	4.8306	0.4603	458.59	62036.14	−38.1127
0.5522	1040.41	1434.12	2.962	14.3443	1.4921	4.6733	0.4527	455.16	65289.63	−37.5869
0.4789	1052.26	1442.36	3.307	14.9774	1.5178	4.5680	0.4476	453.41	67909.90	−37.1940
0.3232	1074.26	1457.01	4.135	16.3243	1.5652	4.3849	0.4386	451.48	73702.04	−36.4016
0.2112	1087.89	1464.42	4.809	17.2944	1.5931	4.2863	0.4336	451.19	78030.81	−35.8666
0.1035	1099.62	1468.85	5.520	18.2273	1.6152	4.2150	0.4300	451.53	82303.37	−35.3788
0.0000	1109.70	1471.40	6.261	19.1259	1.6328	4.1623	0.4273	452.17	86483.25	−34.9328
318.15 K										
1.0000	933.20	1275.00	1.370	10.5176	1.1898	6.5918	0.5377	497.80	52356.76	−40.3579
0.9021	962.12	1309.97	1.567	11.3623	1.2604	6.0568	0.5154	472.78	53719.01	−39.8149
0.8443	977.21	1326.76	1.704	11.8617	1.2965	5.8133	0.5050	462.25	54831.30	−39.4904
0.7852	991.40	1342.23	1.860	12.3720	1.3307	5.5988	0.4956	453.82	56147.48	−39.1584
0.6470	1020.06	1372.34	2.290	13.5702	1.3999	5.2053	0.4778	440.75	59810.82	−38.3889
0.5522	1036.87	1388.73	2.639	14.3933	1.4399	5.0008	0.4683	435.57	62693.06	−37.8743
0.4789	1048.47	1399.38	2.940	15.0316	1.4672	4.8705	0.4622	432.95	65079.84	−37.4865
0.3232	1070.00	1417.30	3.668	16.3892	1.5165	4.6525	0.4517	430.03	70478.74	−36.6981
0.2112	1083.31	1427.23	4.265	17.3675	1.5461	4.5317	0.4458	429.22	74544.92	−36.1631
0.1035	1094.79	1435.30	4.895	18.3077	1.5714	4.4338	0.4410	428.90	78523.17	−35.6752
0.0000	1104.80	1440.00	5.550	19.2107	1.5909	4.3651	0.4376	429.05	82424.53	−35.2306

TABLE 3: Coefficients A_i of Redlich-Kister-type polynomial equation and the corresponding standard deviations, σ, of the system CH + BB at T = 308.15 K, 313.15 K, and 318.15 K.

Property	A_0	A_1	A_2	A_3	A_4	σ
At 308.15 K						
$V_m^E/(10^{-5}\ m^3 \cdot mol^{-1})$	−0.1852	−1.2716	−0.6378	5.1475	1.5786	0.0442
$Z^E/(10^6\ kg \cdot m^{-2} \cdot s^{-1})$	0.2425	−0.0386	0.0089	0.0046	0.1483	0.0001
$\Delta k_s/(10^{-10}\ Pa^{-1})$	−1.5704	0.4932	−0.1605	0.0213	−0.1559	0.0003
$L_f^E/(10^{-10}\ m)$	−0.0727	0.0009	0.1215	0.2317	−0.3916	0.0060
$\pi_i^E/(10^6\ Pa)$	−81.96	72.84	222.75	−575.16	−400.46	6.1774
$H^E/(J \cdot mol^{-1})$	−7945.70	3556.51	−1661.40	519.50	−462.21	0.4439
$\Delta G^{*E}/(kJ \cdot mol^{-1})$	1.2005	0.0541	−0.1314	0.0720	−0.0263	0.0001
$\Delta\eta/(mN \cdot s \cdot m^{-2})$	−2.912	−0.001	0.001	0.003	−0.002	0.0002
$\Delta u/(m \cdot s^{-1})$	106.13	−6.91	0.50	3.72	11.64	0.03
At 313.15 K						
$V_m^E/(10^{-5} m^3 \cdot mol^{-1})$	−0.0473	−0.0514	−0.1583	0.4146	0.1953	0.0046
$Z^E/(10^6\ kg \cdot m^{-2} \cdot s^{-1})$	0.255	−0.0511	0.0606	−0.0826	−0.0177	0.0002
$\Delta k_s/(10^{-10}\ Pa^{-1})$	−1.6905	0.6082	−0.5890	0.6564	0.0583	0.0012
$L_f^E/(10^{-10}\ m)$	−0.0738	0.0240	−0.0262	0.0258	0.0073	0.0001
$\pi_i^E/(10^6\ Pa)$	−93.42	55.97	−36.68	24.88	−2.77	0.0259
$H^E/(J \cdot mol^{-1})$	−9845.27	4398.17	−3040.88	2073.36	435.13	3.1835
$\Delta G^{*E}/(kJ \cdot mol^{-1})$	0.9980	0.1547	−0.1695	0.0906	−0.0263	0.0001
$\Delta\eta/(mN \cdot s \cdot m^{-2})$	−2.644	−0.001[	0.001	0.001	−0.001	0.0003
$\Delta u/(m \cdot s^{-1})$	119.47	−21.79	58.41	−74.26	−24.87	0.15
At 318.15 K						
$V_m^E/(10^{-5} m^3 \cdot mol^{-1})$	−0.0656	0.0232	0.0195	0.0245	0.0006	0.0002
$Z^E/(10^6\ kg \cdot m^{-2} \cdot s^{-1})$	0.2772	−0.0793	−0.0024	0.0463	0.0444	0.0003
$\Delta k_s/(10^{-10}\ Pa^{-1})$	−2.2868	1.0124	−0.3349	0.1037	−0.3589	0.0006
$L_f^E/(10^{-10}\ m)$	−0.0952	0.0378	−0.0091	−0.0011	−0.0163	0.0001
$\pi_i^E/(10^6\ Pa)$	−119.32	72.47	−29.02	9.38	−10.37	0.0120
$H^E/(J \cdot mol^{-1})$	−10268.50	8865.00	−13730.10	−17705.21	15648.81	2.5820
$\Delta G^{*E}/(kJ \cdot mol^{-1})$	0.7885	0.2677	−0.1434	0.0399	0.0029	0.0001
$\Delta\eta/(mN \cdot s \cdot m^{-2})$	−2.436	0.007	0.068	−0.001	0.003	0.0005
$\Delta u/(m \cdot s^{-1})$	156.09	−49.22	2.17	4.21	37.00	0.06

The strong molecular interactions in the presented binary system are well reflected in the properties of partial molar volumes. The partial molar volumes $\overline{V}_{m,1}$ of component 1 (CH) and $\overline{V}_{m,2}$ of component 2 (BB) in the mixtures have been calculated by using the following equations:

$$\overline{V}_{m,1} = V_m^E + V_1^* + x_2 \left(\frac{\partial V_m^E}{\partial x}\right)_{T,P},$$
$$\overline{V}_{m,2} = V_m^E + V_2^* + x_1 \left(\frac{\partial V_m^E}{\partial x}\right)_{T,P}, \quad (6)$$

where V_1^* and V_2^* are the molar volumes of the pure components of CH and BB, respectively. By differentiating (4) and then rearranging one can get the following equations for $\overline{V}_{m,1}$ and $\overline{V}_{m,2}$.

The derivates in the above equations are obtained by differentiating Redlich-Kister equation (8) which leads to the following equations for $\overline{V}_{m,1}$ and $\overline{V}_{m,2}$:

$$\overline{V}_{m,1} = V_1^* + x_2^2 \sum_{i=0}^{j} A_i (x_2 - x_1)^i - 2x_1 x_2^2 \sum_{i=1}^{j} A_i (x_2 - x_1)^{i-1}, \quad (7)$$

$$\overline{V}_{m,2} = V_2^* + x_1^2 \sum_{i=0}^{j} A_i (x_2 - x_1)^i - 2x_2 x_1^2 \sum_{i=1}^{j} A_i (x_2 - x_1)^{i-1}. \quad (8)$$

The excess partial molar volumes are then given by

$$\overline{V}_{m,1}^E = \overline{V}_{m,1} - V_1^*,$$
$$\overline{V}_{m,2}^E = \overline{V}_{m,2} - V_2^*. \quad (9)$$

The values of $\overline{V}_{m,1}$ and $\overline{V}_{m,2}$ are tabulated in Table 4. This table reproduces the values of $\overline{V}_{m,1}$ and $\overline{V}_{m,2}$ for both the components in the mixtures are smaller than their individual molar volumes in the pure state; this indicates that shrinkage of volume takes place on mixing CH with BB. Figures 2 and 3 represent the disparity of excess partial molar volumes of $\overline{V}_{m,1}^{E}$ and $\overline{V}_{m,2}^{E}$, respectively, in the binary mixture. Examination of these figures not only reveals the existence of strong forces between the unlike molecules but also supports the deductions drawn from excess molar volume.

The partial molar volumes and excess partial molar volumes of the components at infinite dilution, $\overline{V}_{m,1}^{\infty}$, $\overline{V}_{m,2}^{\infty}$, $\overline{V}_{m,1}^{E,\infty}$, and $\overline{V}_{m,2}^{E,\infty}$, respectively, were obtained by putting $x = 0$ in (7) and $x = 1$ in (8):

$$\begin{aligned}\overline{V}_{m,1}^{E,\infty} &= A_0 + A_1 + A_2 + A_3 + \cdots = \overline{V}_{m,1}^{\infty} - V_1^{*},\\ \overline{V}_{m,2}^{E,\infty} &= A_0 - A_1 + A_2 - A_3 + \cdots = \overline{V}_{m,2}^{\infty} - V_2^{*}.\end{aligned} \quad (10)$$

The pertinent values of $\overline{V}_{m,1}^{\infty}$, $\overline{V}_{m,2}^{\infty}$, $\overline{V}_{m,1}^{E,\infty}$, and $\overline{V}_{m,2}^{E,\infty}$ are reported in Table 5. This table shows that these values are negative, from which we conclude that strong interactions exist among the unlike molecules of the mixtures, which support the trends of V_m^E values observed in these mixtures.

Figures 4 and 5 represent the deviation in isentropic compressibility (Δk_s) and excess intermolecular free length (L_f^E), respectively. These values are found to be negative for the system under investigation at all temperatures. The nature of Δk_s, L_f^E plays vivacious role in assessing the compactness due to molecular rearrangement. The extent of molecular interactions in liquid mixtures may be due to charge transfer, dipole-induced-dipole, dipole-dipole interactions, and interstitial accommodation, leading to more compact structure making Δk_s and L_f^E negative. The values of the excess functions Δk, L_f^E depend upon several physical and/or chemical contributions [34–37]. The physical contribution consists of dispersion forces or weak dipole-dipole interaction that leads to positive values of Δk_s and L_f^E. Another factor, which involves a physical contribution, is the geometrical effect allowing the fitting of molecules of two different sizes into each other's structure resulting in negative Δk_s, L_f^E values. The strength of the interactions increases with temperatures as observed from more negative values of the excess parameters. They are negative throughout and become more negative at all concentrations as the temperature is increased. In hetero-molecular interaction between the component molecules of the mixtures, Fort and Moore [38] found that the negative value of excess compressibility indicates greater interaction between the components of the mixtures due to the formation of hydrogen bond. Thus the negative Δk_s values for binary mixtures indicate strong interactions between BB and CH.

According to Singh et al. [39], negative values of excess intermolecular free length L_f^E indicate that sound waves cover longer distances due to decrease in intermolecular free length ascribing the dominant nature of hydrogen bond interaction between unlike molecules. Fort and Moore indicated that the positive values of excess free length should be attributed to the dispersive forces, and negative excess values should be due

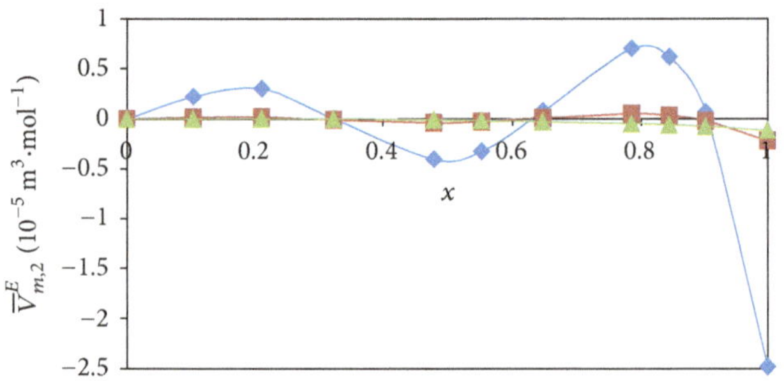

FIGURE 2: Variation of excess partial molar volume, $\overline{V}_{m,2}^{E}$, with mole fraction, x, of CH in BB (◆) 308.15 K, (■) 313.15 K, and (▲) 318.15 K.

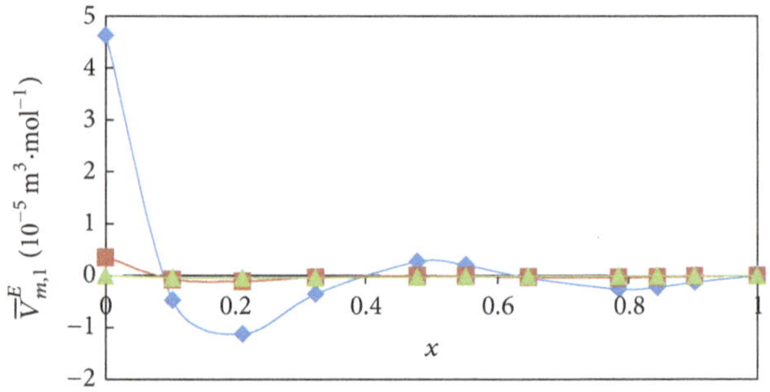

FIGURE 3: Variation of excess partial molar volume, $\overline{V}_{m,1}^{E}$, with mole fraction, x, of CH in BB (◆) 308.15 K, (■) 313.15 K, and (▲) 318.15 K.

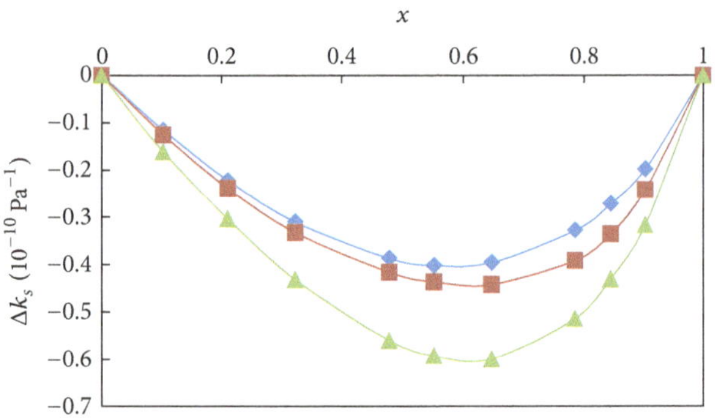

FIGURE 4: Variation of deviation in isentropic compressibility, Δk_s, with mole fraction, x, of CH in BB (◆) 308.15 K, (■) 313.15 K, and (▲) 318.15 K.

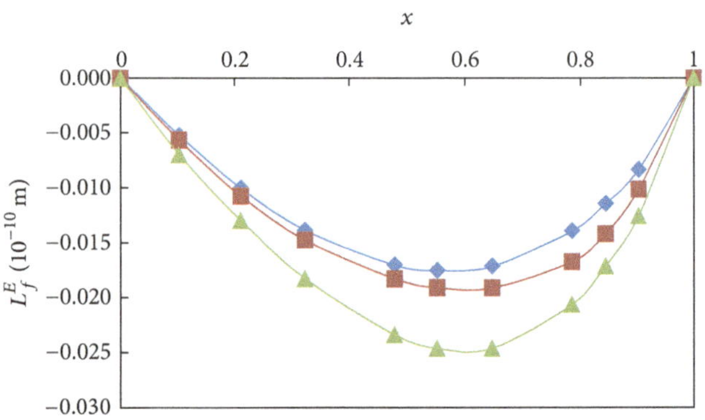

FIGURE 5: Variation of excess free length, L_f^E, with mole fraction, x, of CH in BB (◆) 308.15 K, (■) 313.15 K, and (▲) 318.15 K.

TABLE 4: Mole fraction (x) and values of partial molar volume $\overline{V}_{m,2}/(10^{-5}\ m^3\cdot mol^{-1})$, $\overline{V}_{m,1}/(10^{-5}\ m^3\cdot mol^{-1})$ of the binary system at 308.15, 313.15, and 318.15 K.

x	T = 308.15 K		T = 313.15 K		T = 318.15 K	
	$\overline{V}_{m,2}$	$\overline{V}_{m,1}$	$\overline{V}_{m,2}$	$\overline{V}_{m,1}$	$\overline{V}_{m,2}$	$\overline{V}_{m,1}$
1.0000	16.5603	10.4459	18.9107	10.5007	19.0979	10.5176
0.9021	19.1068	10.3190	19.1066	10.4880	19.1381	10.5161
0.8443	19.6593	10.2239	19.1608	10.4759	19.1540	10.5144
0.7852	19.7480	10.1852	19.1786	10.4667	19.1663	10.5123
0.6470	19.1208	10.4029	19.1386	10.4730	19.1864	10.5058
0.5522	18.7192	10.6493	19.1017	10.4931	19.1966	10.4996
0.4789	18.6383	10.7183	19.0898	10.5035	19.2030	10.4935
0.3232	19.0412	10.0898	19.1175	10.4635	19.2123	10.4792
0.2112	19.3430	9.3218	19.1459	10.3961	19.2141	10.4735
0.1035	19.2619	9.9694	19.1433	10.4335	19.2124	10.4826
0.0000	19.0428	15.0774	19.1259	10.8536	19.2107	10.5198

TABLE 5: Values of partial molar volume of the components at infinite dilution ($\overline{V}^{\infty}_{m,1}$, $\overline{V}^{\infty}_{m,2}$) and excess partial molar volume at infinite dilution ($\overline{V}^{E,\infty}_{m,1}$, $\overline{V}^{E,\infty}_{m,2}$) for the system CH + BB at 308.15, 313.15, and 318.15 K.

Temp. (K)	$\overline{V}^{E,\infty}_{m,1}$	$\overline{V}^{E,\infty}_{m,2}$ /$(10^{-5}\ m^3\cdot mol^{-1})$	$\overline{V}^{\infty}_{m,1}$	$\overline{V}^{\infty}_{m,2}$
308.15	4.6315	−3.1203	15.0774	15.9225
313.15	0.3529	−0.3735	10.8536	18.7524
318.15	0.0022	−0.0933	10.5197	19.1173

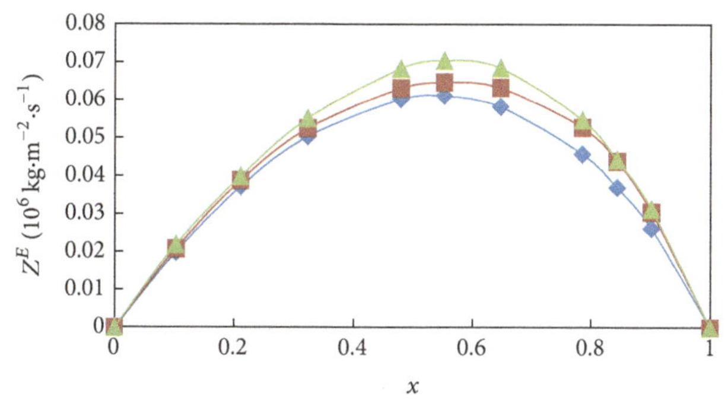

FIGURE 6: Variation of excess acoustic impedance, Z^E, with mole fraction, x, of CH in BB (◆) 308.15 K, (■) 313.15 K, and (▲) 318.15 K.

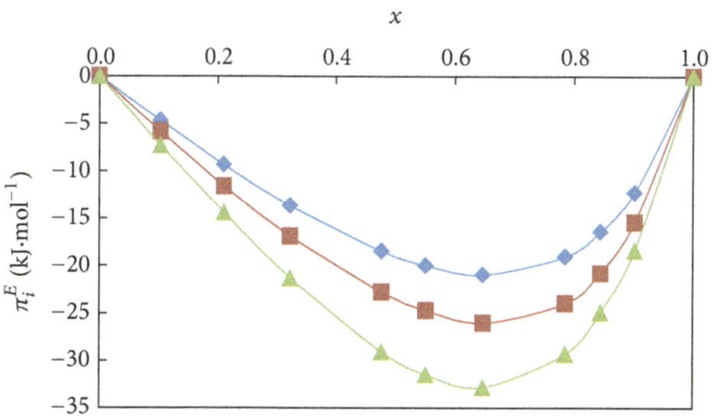

FIGURE 7: Variation of excess internal pressure, π_i^E, with mole fraction, x, of CH in BB (◆) 308.15 K, (■) 313.15 K, and (▲) 318.15 K.

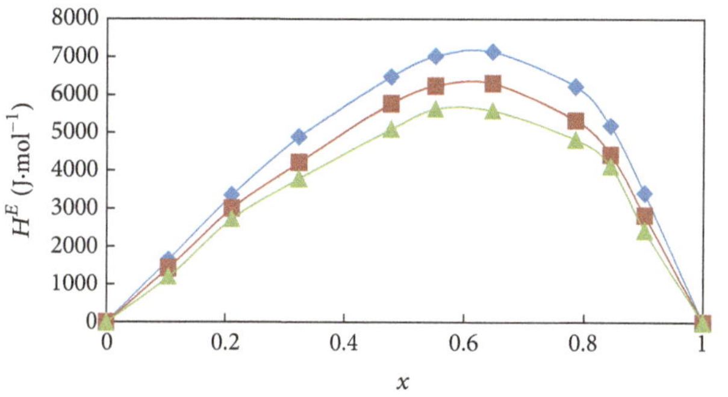

FIGURE 8: Variation of excess enthalpy, H^E, with mole fraction, x, of CH in BB (◆) 308.15 K, (■) 313.15 K, and (▲) 318.15 K.

to hydrogen bond formation or dipole-dipole interactions between solute and solvent. In the present study the negative contribution in all the systems prevails in the existence of greater interactions. Nain et al. [40] also concluded a similar observation. From Figure 6, the excess values of Z^E are found to be positive for the system at all temperatures under investigation. According to Kondaiah and Rao [41], the nature of the positive values of Z^E is attributed to specific interaction between the heteromolecules. This further supports our earlier finding.

The variations of excess internal pressure (π_i^E) and excess enthalpy (H^E) are presented in Figures 7 and 8 for the present system at all the three temperatures, respectively. From these figures it is observed that π_i^E is negative and H^E is positive over the entire composition range of CH. In accordance with Kondaiah et al. [42], the negative values of π_i^E indicate the existence of strong specific interactions and according to Kumar and Rao [43], positive value of ΔH suggests strong bonding and exothermic reactions. Thus, it can be concluded that strong specific interactions are operative in the systems

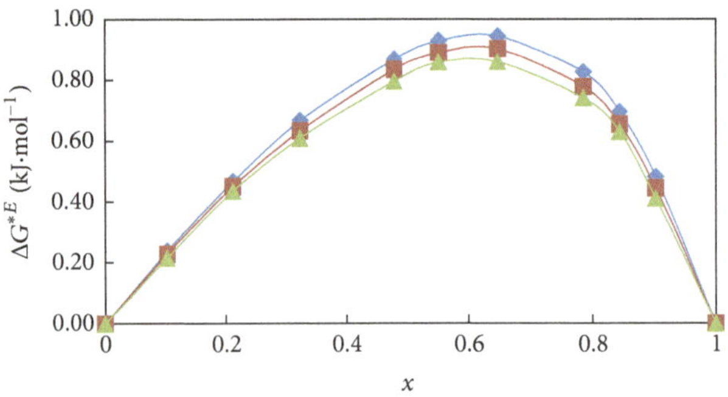

FIGURE 9: Variation of excess Gibbs free energy of activation of viscous flow, ΔG^{*E}, with mole fraction, x, of CH in BB (◆) 308.15 K, (■) 313.15 K, and (▲) 318.15 K.

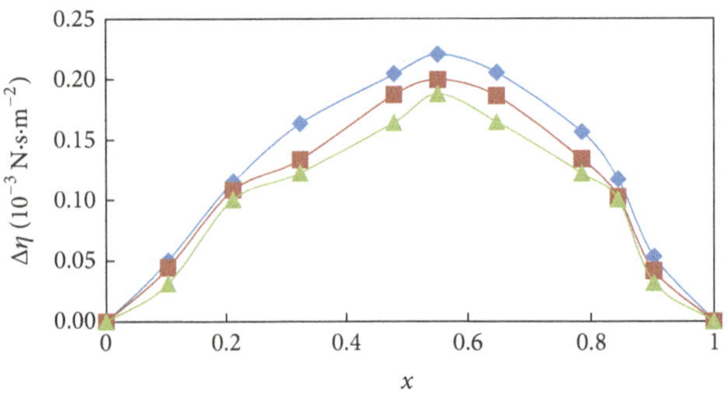

FIGURE 10: Variation of deviation in viscosity, $\Delta\eta$, with mole fraction, x, of CH in BB (◆) 308.15 K, (■) 313.15 K, and (▲) 318.15 K.

investigated. These conclusions support our earlier findings of strong interactions. The variation of excess Gibbs free energy of activation of viscous flow ΔG^{*E} with mole fraction of CH has been presented in Figure 9 at all temperatures. These values are positive over the entire range of composition of CH. According to Iloukhani et al. [44, 45], negative values of ΔG^{*E} indicate the presence of weak physical forces such as dispersive forces in the system. On the other hand positive values of it suggest strong specific interactions. This also supports the conclusions drawn from our earlier findings.

According to Fort and Moore [46], deviation in viscosity tends to become more positive as the strength of the interaction increases. The deviation in viscosity variation gives a qualitative estimation of the strength of the intermolecular interactions. The deviation in viscosities may be generally explained by considering the following factors. (i) The difference in size and shape of the component molecules and the loss of dipolar association in pure component may contribute to a decrease in viscosity and (ii) specific interactions between unlike components such as hydrogen bond formation or dipole-dipole interactions or charge-transfer complexes may cause increase in viscosity in mixtures compared to in pure components [47]. The former effect produces negative deviation in viscosity and latter effect produces positive deviation in viscosity. A perusal of Figure 10 shows that the values of $\Delta\eta$ are positive for the present system at three temperatures. The positive $\Delta\eta$ values decrease with increasing temperature. The positive $\Delta\eta$ values indicate specific interactions while the negative $\Delta\eta$ values indicate dispersion forces [48–50]. In the present investigation, the positive values of all binary systems may be attributed to the dipole-dipole interactions. The sign and magnitude of Δu play important roles in describing molecular rearrangements as a result of molecular interactions occurring among the component molecules in the mixtures. The excess ultrasonic speed with mole fraction of CH is shown in Figure 11. Here we observed that the Δu values are positive for all binary systems over the entire range of composition at all the studied temperatures. Positive deviations indicate the increasing strength of interaction between component molecules of binary liquid mixtures. According to Reddy et al. [51], strong interactions among the components of a mixture lead to the formation of molecular aggregates and more compact structures; then sound will travel faster through the mixture by means of longitudinal waves and hence the ultrasonic speed deviations with respect to a linear behaviour will be positive while if the structure-breaking factor in the mixture predominates resulting expansion, then the speed of sound through the mixture will be slower resulting in negative deviation in the speed of sound. According to Ali et al. [52], more positive values mean much more strong interactions between the molecules. Therefore, based on the above references for the present scenario, interactions are found to be in the order 318.15 K > 313.15 K > 308.15 K; this again emphasizes our view regarding the interaction between CH and BB.

In the present study, theoretical sound velocities have been evaluated by considering CH as one component and BB as the other component in the binary mixture at all

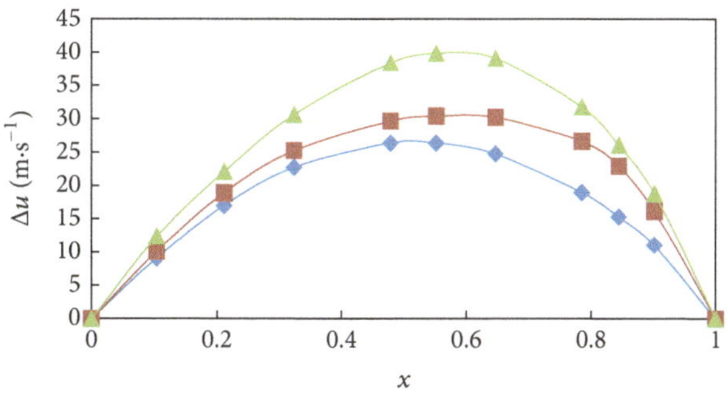

FIGURE 11: Variation of deviation in ultrasonic velocity, Δu, with mole fraction, x, of CH in BB (◆) 308.15 K, (■) 313.15 K, and (▲) 318.15 K.

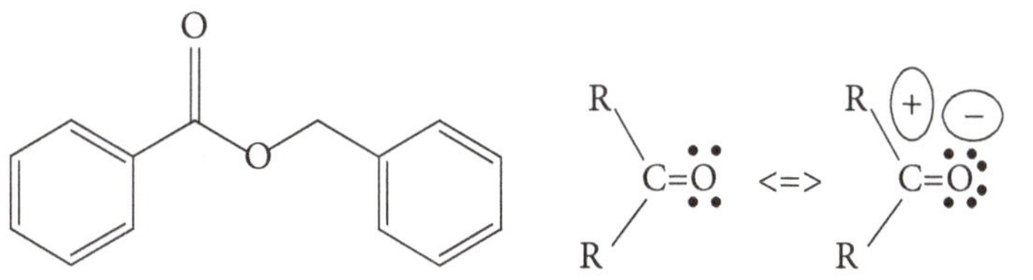

FIGURE 12

the three temperatures. Such an evaluation of theoretical sound velocity is useful to verify the applicability of various assumes of the theories of liquid mixtures and to arrive at some useful implications regarding the (strength of) molecular interactions between component liquids in some cases. The theoretical values of ultrasonic velocity obtained using various theories along with the experimental velocity and percentage deviation are summarized in Table 6. Nomoto [53] established the following relation for sound velocity based on the supposition of the linearity of the molecular sound velocity and the additives of molar volume:

$$U_N = \left\{ \frac{(\sum x_i R_i)}{(\sum x_i V_i)} \right\}^3. \tag{11}$$

Van Dael [54] obtained the ideal mixture relation

$$\sum \left(\frac{x_i M_i}{u_i^2} \right) = \left\{ \frac{1}{\sum x_i M_i} \right\} \left\{ \frac{1}{U_V} \right\}^2. \tag{12}$$

The impedance dependence relation used by Baluja and Parsania [55] is given below:

$$U_{\text{imp}} = \frac{\sum x_i Z_i}{\sum x_i \rho_i}, \tag{13}$$

where x_i is mole fractions, M_i is molecular weight, R is the molar sound speed, and Z_i is acoustic impedance of CH and BB, respectively, at three temperatures.

Percentage deviation in ultrasonic speed is given by

$$\% \Delta u = 100 * \left[1 - \frac{U_{\text{cal}}}{u_{\text{expt}}} \right]. \tag{14}$$

On observing Table 6, among all the empirical theories, Nomoto's relation gives the best estimate of experimental values of sound velocity in the system followed by impedance's relation. In the present binary systems, the difference between experimental and theoretical velocities is greater where the mole fraction of CH varies in the region 0.4 to 0.7. Hence it can be qualitatively inferred that the strength of interaction in the binary mixtures is more in this range of composition of binary mixtures.

The dynamic viscosities of the liquid mixtures have been calculated using the several empirical relations.

Grunberg and Nissan [56] proposed the following equation for the measurement of viscosity of liquid mixtures:

$$\ln \eta = x_1 \ln \eta_1 + x_2 \ln \eta_2 + x_1 x_2 G_{12}, \tag{15}$$

where G_{12} is an interaction parameter and it is the function of components 1, 2 and temperature.

Hind et al. [57] suggested an equation for the viscosity of binary liquid mixtures as

$$\eta = x_1^2 \eta_1 + x_2^2 \eta_2 + 2 x_1 x_2 H_{12}, \tag{16}$$

where H_{12} is an interaction parameter and is attributed to unlike pair interaction.

Katti and Chaudhri [58] proposed the following equation:

$$\ln(\eta V) = x_1 \ln(\eta_1 V_1) + x_2 \ln(\eta_2 V_2) + \frac{x_1 x_2 W_{\text{vis}}}{\text{RT}}, \tag{17}$$

where $(W_{\text{vis}}/\text{RT})$ is an interaction parameter.

The theoretical viscosity values using (15) to (17) along with the percentage error are compiled in Table 7. The evaluated values of parameters G_{12}, H_{12}, and W_{vis}/RT and standard deviations σ ($10^{-3}\,\text{kg}^{-1}\cdot\text{s}^{-1}$) are presented in Table 8.

Table 6: Experimental and theoretical velocities from various empirical relations and percentage deviation in velocities with molefraction, x, of CH with BB at $T = 308.15$, 313.15, and 318.15 K.

x	u_{expt}	U_N	U_V	U_I	$\%U_N$	$\%U_V$	$\%U_I$
308.15 K							
1.0000	1362.00	1362.00	1362.00	1361.96	0.00	0.00	0.00
0.9021	1385.94	1383.13	1331.26	1376.93	−0.20	−3.95	−0.65
0.8443	1397.80	1394.27	1318.02	1385.52	−0.25	−5.71	−0.88
0.7852	1409.21	1404.79	1307.85	1394.13	−0.31	−7.19	−1.07
0.6470	1433.09	1426.46	1296.25	1413.55	−0.46	−9.55	−1.36
0.5522	1447.20	1439.35	1297.68	1426.33	−0.54	−10.33	−1.44
0.4789	1456.83	1448.39	1304.03	1435.94	−0.58	−10.49	−1.43
0.3232	1473.61	1465.38	1333.84	1455.57	−0.56	−9.49	−1.22
0.2112	1482.60	1476.03	1371.26	1469.08	−0.44	−7.51	−0.91
0.1035	1488.79	1485.25	1423.35	1481.61	−0.24	−4.40	−0.48
0.0000	1493.30	1493.30	1493.30	1493.26	0.00	0.00	0.00
313.15 K							
1.0000	1348.89	1348.89	1348.89	1348.88	0.00	0.00	0.00
0.9021	1377.00	1368.65	1318.18	1362.86	−0.61	−4.27	−1.03
0.8443	1390.87	1379.07	1304.90	1370.88	−0.85	−6.18	−1.44
0.7852	1401.82	1388.91	1294.64	1378.91	−0.92	−7.65	−1.63
0.6470	1422.33	1409.16	1282.67	1397.04	−0.93	−9.82	−1.78
0.5522	1434.12	1421.20	1283.67	1408.97	−0.90	−10.49	−1.75
0.4789	1442.36	1429.65	1289.58	1417.93	−0.88	−10.59	−1.69
0.3232	1457.01	1445.52	1318.07	1436.24	−0.79	−9.54	−1.43
0.2112	1464.42	1455.48	1354.08	1448.84	−0.61	−7.53	−1.06
0.1035	1468.85	1464.09	1404.28	1460.53	−0.32	−4.40	−0.57
0.0000	1471.40	1471.60	1471.61	1471.38	0.00	0.00	0.00
318.15 K							
1.0000	1275.00	1275.00	1275.00	1275.22	0.00	0.00	0.00
0.9021	1309.97	1301.37	1247.66	1293.99	−0.66	−4.76	−1.22
0.8443	1326.76	1315.31	1236.19	1304.75	−0.86	−6.83	−1.66
0.7852	1342.23	1328.49	1227.68	1315.54	−1.02	−8.53	−1.99
0.6470	1372.34	1355.68	1219.56	1339.89	−1.21	−11.13	−2.36
0.5522	1388.73	1371.88	1223.20	1355.93	−1.21	−11.92	−2.36
0.4789	1399.38	1383.27	1231.24	1367.99	−1.15	−12.02	−2.24
0.3232	1417.30	1404.69	1265.00	1392.63	−0.89	−10.75	−1.74
0.2112	1427.23	1418.15	1306.03	1409.60	−0.64	−8.49	−1.23
0.1035	1435.30	1429.81	1362.97	1425.35	−0.38	−5.04	−0.69
0.0000	1440.00	1440.00	1440.00	1439.98	0.00	0.00	0.00

The estimated values of σ are indicating that the viscosities are well correlated by all the three viscosity models. Prolongo et al. [59] reported positive values of interaction parameters corresponding to systems with negative excess molar volumes. This is consistent with our results. Among all of these models, the viscosity representations obtained from the Hind model are in best agreement with the experimental viscosity data.

4. Conclusions

(i) Ultrasonic velocities, densities, and viscosities for binary liquids of CH with BB are determined experimentally at $T = 308.15$ K, 313.15 K, and 318.15 K over the entire composition range.

(ii) The values of V_m^E, Δk_s, L_f^E, Z^E, π_i^E, H^E, ΔG^{*E}, and Δu are calculated from experimental results at all three temperatures. The excess/deviation properties are fitted to Redlich-Kister-type polynomial and corresponding standard deviations are evaluated. The observed negative values of V_m^E, Δk_s, L_f^E, and π_i^E and positive values of Z^E, H^E, ΔG^{*E}, $\Delta\eta$, and Δu for all the liquid mixtures studied clearly indicate dipole-dipole-type interactions and the geometrical fitting

TABLE 7: Experimental and theoretical values of viscosity from various empirical relations and percentage deviation in viscosity in CH + BB at T = 308.15, 313.15, and 318.15 K.

T (K)	Exptl. viscosity	Gruenberg	Hind	Katti	%G	%H	%K
	1.594	1.594	1.594	1.595	0.000	0.000	0.000
	1.878	1.891	1.878	1.891	0.698	0.005	0.692
	2.071	2.086	2.071	2.086	0.685	−0.014	0.676
	2.289	2.302	2.289	2.302	0.537	−0.009	0.524
	2.879	2.877	2.879	2.877	−0.052	0.007	−0.069
308.15 K	3.347	3.333	3.347	3.333	−0.445	−0.006	−0.448
	3.746	3.721	3.746	3.721	−0.649	0.003	−0.667
	4.696	4.658	4.695	4.658	−0.794	−0.002	−0.809
	5.460	5.430	5.466	5.431	−0.662	−0.009	−0.658
	6.276	6.253	6.276	6.253	−0.362	0.006	−0.366
	7.118	7.118	7.118	7.118	0.000	0.000	0.000
	1.472	1.472	1.472	1.472	0.000	0.000	0.000
	1.707	1.724	1.707	1.724	0.966	−0.023	0.937
	1.870	1.890	1.870	1.890	1.075	0.021	1.070
	2.054	2.073	2.055	2.073	0.886	0.005	0.876
	2.558	2.560	2.559	2.560	0.070	0.004	0.039
313.15 K	2.962	2.947	2.963	2.947	−0.513	0.007	−0.540
	3.307	3.279	3.308	3.278	−0.871	0.006	−0.907
	4.135	4.084	4.135	4.084	−1.214	0.010	−1.233
	4.809	4.757	4.809	4.756	−1.094	−0.006	−1.123
	5.520	5.484	5.520	5.484	−0.665	−0.005	−0.670
	6.261	6.261	6.261	6.262	0.000	0.000	0.000
	1.370	1.370	1.370	1.370	0.000	0.000	0.000
	1.567	1.585	1.566	1.586	1.122	−0.128	1.148
	1.704	1.726	1.703	1.728	1.344	−0.059	1.408
	1.860	1.883	1.860	1.884	1.237	0.000	1.290
	2.290	2.298	2.294	2.301	0.380	0.175	0.480
318.15 K	2.639	2.629	2.644	2.632	−0.345	0.189	−0.265
	2.940	2.914	2.945	2.917	−0.867	0.170	−0.782
	3.668	3.612	3.671	3.615	−1.548	0.055	−1.472
	4.265	4.202	4.265	4.205	−1.496	−0.023	−1.430
	4.895	4.848	4.893	4.850	−0.974	−0.061	−0.940
	5.550	5.550	5.550	5.550	0.000	0.000	0.000

TABLE 8: Various interaction parameters calculated from various empirical relations and the corresponding standard deviations ($\sigma/10^{-3}$ N·s·m^{-2}) of the present system CH + BB at T = 308.15, 313.15, and 318.15 K.

Temp. (K)	G_{12}	σ	H_{12}	σ	W_{vis}/RT	σ
308.15	0.272	0.020	0.003	0.003	0.004	0.020
313.15	0.186	0.028	0.003	0.003	0.003	0.028
318.15	0.103	0.032	0.002	0.030	0.003	0.031

of molecules leading to strong molecular interactions between CH and BB. These strong interactions increase with temperature.

(iii) The values of $\overline{V}_{m,1}$, $\overline{V}_{m,2}$ and $\overline{V}^E_{m,1}$, $\overline{V}^E_{m,2}$ are calculated from experimental data. The observed lower partial molar volumes in the liquid mixture when compared to the respective molar volumes of pure components indicate that strong interactions present in the system support the trends of V^E_m values very well.

(iv) The ultrasonic velocities/viscosities computed from three different velocity/viscosity theories were correlated with the experimentally measured ultrasonic velocities and their percentage deviations were evaluated. Among all the empirical theories Nomoto's velocity/Hind viscosity relation gives the best estimate of experimental values of sound velocity/viscosity in the system at all temperatures.

Conflict of Interests

The authors declare that there is no conflict of interests regarding the publication of this paper.

Acknowledgments

Sk. Md Nayeem is highly thankful to U.G.C., New Delhi, and Government of India for sanction of financial grant under XII Plan towards MRP (MRP-4671/14(SERO/UGC)).

References

[1] J. Rajasekhar and P. R. Naidu, "Speed of sound of 1, 3-dichloro benzene + methyl ethyl ketone + 1-alkanols at 303.15 K," *Journal of Chemical Engineering Data*, vol. 48, pp. 373–375, 1996.

[2] S. L. Oswal, P. Oswal, and R. P. Phalak, "Speed of sound, isentropic compressibilities, and excess molar volumes of binary mixtures containing p-dioxane," *Journal of Solution Chemistry*, vol. 27, no. 6, pp. 507–520, 1998.

[3] M. Geetalakshmi, *Molecular interaction in the binary liquid mixtures-an ultrasonic study [Ph.D. thesis]*, Sri Venkateswara Univeristy, Tirupati, India, 2006.

[4] M. Geetalakshmi, P. S. Naidu, and K. R. Prasa, "Molecular interactions in binary mixtures of benzyl benzoate with alcohols," *Journal of Pure & Applied Ultrasonics*, vol. 30, pp. 18–23, 2008.

[5] J. Sivasankar, M. Geetalakshmi, P. S. Naidu, and K. R. Prasad, "Excess parameters of benzyl benzoate in dichloromethane isobutanol," *Journal of Pure & Applied Ultrasonics*, vol. 29, no. 2, pp. 82–88, 2007.

[6] P. Ramesh, *Ultrasonic study of the binary liquid mixtures (Benzyl Benzoate + Substituted Benzenes) [M.S. thesis]*, Sri Venkateswara Univeristy, Tirupati, India, 2008.

[7] B. Gonzalez, D. Dominguez, and J. Tojo, "Physical properties of the binary systems methyl cuclopentane with ketones (acetone, butone and 2-pentanone) at $T = (293.15, 298.15, 303.15)$ K," *Journal of Chemical Thermodynics*, vol. 38, no. 6, pp. 707–716, 2006.

[8] P. M. Krishna, B. R. Kumar, B. Sathyanarayana, K. S. Kumar, and N. Satyanarayana, "Densities and speeds of sound for binary liquid mixtures of thiolane-1,1-dioxide with butanone, pentan-2-one, pentan-3-one, and 4-methyl-pentan-2-one at T = (303.15 or 308.15 or 313.15) K," *Journal of Chemical and Engineering Data*, vol. 54, no. 6, pp. 1947–1950, 2009.

[9] P. G. Kumari, P. Venkatesu, M. V. P. Rao, M.-J. Lee, and H.-M. Lin, "Excess molar volumes and ultrasonic studies of *N*-methyl-2-pyrrolidone with ketones at $T = 303.15$ K," *The Journal of Chemical Thermodynamics*, vol. 41, no. 5, pp. 586–590, 2009.

[10] M. Sreenivasulu and P. R. Naidu, "Excess volumes and isentropic compressibility of binary mixtures of butyl amine and cyclic ketones," *Australian Journal of Chemistry*, vol. 32, pp. 1649–1652, 1979.

[11] S. L. Oswal and H. S. Desai, "Studies of viscosity and excess molar volume of binary mixtures. 3. 1-Alkanol+di-n-propylamine, and+di-n-butylamine mixtures at 303.15 and 313.15 K," *Fluid Phase Equilibria*, vol. 186, no. 1-2, pp. 81–102, 2001.

[12] S. Thirumaran and N. Karthikeyan, "Thermo-acoustical and excess thermodynamic studies of ternary liquid mixtures of substituted benzenes in aqueous mixed solvent systems at 303.15, 308.15 and 313.15 K," *International Journal of Chemical Research*, vol. 3, no. 3, pp. 83–98, 2011.

[13] M. V. Rathnam, R. T. Sayed, K. R. Bhanushali, and M. S. S. Kumar, "Molecular interaction study of binary mixtures of methyl benzoate: viscometric and ultrasonic study," *Journal of Molecular Liquids*, vol. 166, pp. 9–16, 2012.

[14] S. M. Nayeem and D. K. Rao, "Ultrasonic investigations of molecular interaction in binary mixtures of benzyl benzoate with isomers of butanol," *International Journal of Pharma Research & Review*, vol. 3, no. 2, pp. 65–78, 2014.

[15] N. J. Madhuri, P. S. Naidu, J. Glory, and K. R. Prasad, "Ultrasonic investigations of molecular interaction in binary mixtures of benzyl benzoate with acetonitrile and benzonitrile," *E-Journal of Chemistry*, vol. 8, no. 1, pp. 457–469, 2011.

[16] S. Sri, Z. Begum, and C. Rambabu, "Theoretical evaluation of ultrasonic velocities in binary liquid mixtures of anisic aldehyde with some esters at different temperatures using different theories and models," *Journal of Thermodynamic Catalysis*, vol. 4, pp. 3–8, 2013.

[17] A. I. Vogel, *Text Book of Organic Chemistry*, John Wiley & Sons, New York, NY, USA, 5th edition, 1989.

[18] J. A. Riddick, W. B. Burger, and T. K. Sankano, *Techniques in Chemistry*, vol. 2, John Wiley & Sons, New York, NY, USA, 4th edition, 1986.

[19] B. S. Furniss, A. J. Hannaford, V. S. Rogers, P. W. G. Smith, and A. R. Tachell, *Vogel's Textbook of Practical Organic Chemistry*, Longman, 4th edition, 1980.

[20] H. C. Parker and E. W. Parker, "Densities of certain aqueous potassium chloride solutions as determined with a new pyknometer," *Physical Chemistry*, vol. 29, pp. 130–137, 1925.

[21] P. S. Naidu and K. R. Prasad, "Ultrasonic velocity and allied parameters in solutions of cypermethrin with xylene and ethanol," *Indian Journal of Pure and Applied Physics*, vol. 42, no. 7, pp. 512–517, 2004.

[22] M. Kondaiah, K. Sreekanth, D. Sravana Kumar, and D. Krishna Rao, "Volumetric and viscometric properties of propanoic acid in equimolar mixtures of N,N-dimethyl formamide + alkanols at T/K = 303.15, 313.15, and 323.15," *Journal of Solution Chemistry*, vol. 42, no. 3, pp. 494–515, 2013.

[23] M. Kondaiah, D. S. Kumar, K. Sreekanth, and D. K. Rao, "Densities and viscosities of binary mixtures of propanoic acid with *N*, *N*-dimethylaniline and *N*, *N*-diethylaniline at $T = (303.15, 313.15, \text{and } 323.15)$ K," *Journal of Chemical and Engineering Data*, vol. 57, no. 2, pp. 352–357, 2012.

[24] K. Sreekanth, M. Kondaiah, D. Sravana Kumar, and D. K. Rao, "Influence of temperature on thermodynamic properties of acid-base liquid mixtures," *Journal of Thermal Analysis and Calorimetry*, vol. 110, no. 3, pp. 1341–1352, 2012.

[25] M. Kondaiah, K. Sreekanth, S. M. Nayeem, and D. K. Rao, "Ultrasonic, volumetric and viscometric study of aqueous electrolyte solutions," *Journal of Chemical, Biological & Physical Sciences*, vol. 4, no. 3, pp. 2402–2415, 2014.

[26] M. Sri Lakshmi, R. Ramesh Raju, C. Rambabu, G. V. Rama Rao, and K. Narendra, "Study of molecular interactions in binary liquid mixtures containing higher alcohols at different temperatures," *Research & Review: Journal of Chemistry*, vol. 2, no. 1, pp. 12–24, 2013.

[27] B. Jacobson, "Intermolecular free lengths in the liquid state. I. Isentropic and isothermal compressibilities," *Acta Chemica Scandinavia*, vol. 6, pp. 1485–1498, 1952.

[28] S. Glasstone, *Text Book of Physical Chemistry*, Macmillan, 2nd edition, 1995.

[29] O. Redlich and A. T. Kister, "Algebraic representation of thermodynamic properties and the classification of solutions," *Industrial & Engineering Chemistry*, vol. 40, no. 2, pp. 345–348, 1948.

[30] M. G. Sankar, V. Ponneri, K. S. Kumar, and S. Sakamuri, "Molecular interactions between amine and cyclic ketones at different temperatures," *Journal of Thermal Analysis and Calorimetry*, vol. 115, no. 2, pp. 1821–1827, 2014.

[31] S. Gadžurić, A. Nikolić, M. Vraneš, B. Jović, M. Damjanović, and S. Dožić, "Volumetric properties of binary mixtures of N-ethylformamide with tetrahydrofuran, 2-butanone, and ethylacetate from T = (293.15 to 313.15) K," *Journal of Chemical Thermodynamics*, vol. 51, pp. 37–44, 2012.

[32] M. Rastogi, A. Awasthi, M. Gupta, and J. P. Shukla, "Ultrasonic investigations of X...HO bond complexes," *Indian Journal of Pure and Applied Physics*, vol. 40, no. 4, pp. 256–263, 2002.

[33] J. G. Baragi, V. K. Mutalik, and S. B. Mekali, "Molecular interaction studies in mixtures of methyl cyclohexane with alkanes: a theoretical approach," *International Journal of Pharmacy and Biological Sciences*, vol. 3, no. 2, pp. 185–197, 2013.

[34] R. Mehra, A. Gupta, and R. Israni, "Excess volume, excess isentropic compressibility, partial molar volume and specific heat ratio of ternary mixtures of cyclohexane + decane + alkanols at 298.15 and 308.15 K," *Indian Journal of Chemistry*, vol. 40, no. 5, pp. 505–508, 2001.

[35] K. K. Ohomuru and S. J. Murakami, "Speeds of sound, excess molar volumes, and isentropic compressibilities of "$x\mathrm{CH_3COC_2H_5} + (1-x)\mathrm{C_6H_6}$", "$x\mathrm{CH_3COC_2H_5} + (1-x)\mathrm{C_6H_5CH_3}$", "$x\mathrm{CH_3COC_2H_5} + (1-x)\mathrm{C_6H_5Cl}$", and "$x\mathrm{C_2H_5COC_2H_5} + (1-x)\mathrm{C_6H_6}$" at 298.15 K," *Chemical Thermodynamics*, vol. 19, pp. 171–176, 1987.

[36] P. Venkatesu and M. V. P. Rao, "Excess volumes of ternary mixtures of *N,N*-dimethylformamide + methyl isobutyl ketone + 1-alkanols at 303.15 K," *Fluid Phase Equilibria*, vol. 98, pp. 173–178, 1994.

[37] M. V. Rathnam, M. S. S. Kumar, R. T. Sayed, and D. R. Ambavadekar, "Study on Thermo Physical properties of binary mixtures of Butyl Propionate and ketones at 303.15 and 313.15 K," *European Chemical Bulletin*, vol. 3, no. 2, pp. 166–172, 2014.

[38] R. J. Fort and W. R. Moore, "Adiabatic compressibilities of binary liquid mixtures," *Transactions of the Faraday Society*, vol. 61, pp. 2102–2111, 1965.

[39] S. Singh, I. Vibhu, M. Gupta, and J. P. Shukla, "Excess acoustical and volumetric properties and the theoretical estimation of the excess thermodynamic functions of binary liquid mixtures," *Chinese Journal of Physics*, vol. 45, no. 4, pp. 412–424, 2007.

[40] A. K. Nain, R. Sharma, A. Ali, and S. Gopal, "Densities and volumetric properties of ethyl acrylate + 1-butanol, or 2-butanol, or 2-methyl-1-propanol, or 2-methyl-2-propanol binary mixtures at temperatures from 288.15 to 318.15 K," *Journal of Molecular Liquids*, vol. 144, no. 3, pp. 138–144, 2009.

[41] M. Kondaiah and D. K. Rao, "Experimental study of the excess properties for (1, 2-dichloroethane or dichloromethane with N, N-dimethyl formamide or dimethyl sulfoxide)," *Journal of Molecular Liquids*, vol. 195, pp. 110–115, 2014.

[42] M. Kondaiah, K. Sreekanth, S. M. Nayeem, and D. K. Rao, "Ultrasonic, volumetric and viscometric study of aqueous-electrolyte solutions," *Journal Chemical, Biological & Physical Sciences*, vol. 4, no. 3, pp. 2401–2415, 2014.

[43] D. S. Kumar and D. K. Rao, "Study of molecular interactions and ultrasonic velocity in mixtures of some alkanols with aqueous propylene glycol," *Indian Journal of Pure and Applied Physics*, vol. 45, no. 3, pp. 210–220, 2007.

[44] H. Iloukhani, H. A. Zarei, and M. Behroozi, "Thermodynamic studies of ternary mixtures of diethylcarbonate (1) + dimethylcarbonate (2) + N,N-dimethylacetamide (3) at 298.15 K," *Journal of Molecular Liquids*, vol. 135, no. 1–3, pp. 141–145, 2007.

[45] H. Iloukhani and Z. Rostami, "Measurement of some thermodynamic and acoustic properties of binary solutions of *N,N*-dimethylformamide with 1-alkanols at 30°C and comparison theories," *Journal of Solution Chemistry*, vol. 32, no. 5, pp. 451–462, 2003.

[46] R. J. Fort and W. R. Moore, "Viscosities of binary liquid mixtures," *Transactions of the Faraday Society*, vol. 62, pp. 1112–1119, 1966.

[47] L. Pikkarainen, "Densities and viscosities of binary mixtures of N,N-dimethylacetamide with aliphatic alcohols," *Journal of Chemical & Engineering Data*, vol. 28, no. 3, pp. 344–347, 1983.

[48] R. Palepu, J. Oliver, and D. Campbell, "Thermodynamic and transport properties of o-chlorophenol with aniline and N-alkylanilines," *Journal of Chemical and Engineering Data*, vol. 30, no. 3, pp. 355–360, 1985.

[49] R. Palepu, J. Oliver, and B. Mackinnon, "Viscosities and densities of binary liquid mixtures of m-cresol with substituted anilin," *Canadian Journal of Chemistry*, vol. 63, pp. 1024–1030, 1985.

[50] T. M. Reed and T. E. Taylor, "Viscosities of liquid mixtures," *Journal of Physical Chemistry*, vol. 63, no. 1, pp. 58–67, 1959.

[51] Y. Reddy, P. S. Naidu, and K. R. Prasad, "Ultrasonic study of acetophenones in the binary-mixtures containing isopropanol as common component," *Indian Journal of Pure & Applied Physics*, vol. 32, no. 12, pp. 958–963, 1994.

[52] A. Ali, F. Nabi, and M. Tariq, "Volumetric, viscometric, ultrasonic, and refractive index properties of liquid mixtures of benzene with industrially important monomers at different temperatures," *International Journal of Thermophysics*, vol. 30, no. 2, pp. 464–474, 2009.

[53] O. Nomoto, "Empirical formula for sound velocity in liquid mixtures," *Journal of the Physical Society of Japan*, vol. 13, no. 12, pp. 1528–1532, 1958.

[54] W. van Dael, *Thermodynamic Properties and Velocity of Sound*, chapter 5, Butterworth, London, UK, 1975.

[55] S. Baluja and P. H. Parsania, "Acoustical properties of 3-α- furyl acrylic acid in protic and aprotic solvents," *Asian Journal of Chemistry*, vol. 7, pp. 417–423, 1995.

[56] L. Grunberg and A. H. Nissan, "Mixture law for viscosity," *Nature*, vol. 164, no. 4175, pp. 799–800, 1949.

[57] R. K. Hind, E. McLaughlin, and A. R. Ubbelohde, "Structure and viscosity of liquids. Camphor + pyrene mixtures," *Transactions of the Faraday Society*, vol. 56, pp. 328–330, 1960.

[58] P. K. Katti and M. M. Chaudhri, "Viscosities of binary mixtures of benzyl acetate with dioxane, aniline, and m-cresol," *Journal of Chemical and Engineering Data*, vol. 9, no. 3, pp. 442–443, 1964.

[59] M. G. Prolongo, R. M. Masegosa, I. Hernández-Fuentes, and A. Horta, "Viscosities and excess volumes of binary mixtures formed by the liquids acetonitrile, pentyl acetate, 1-chlorobutane, and carbon tetrachloride at 25°C," *The Journal of Physical Chemistry*, vol. 88, no. 10, pp. 2163–2167, 1984.

Temperature Dependence of the Raman Frequency of an Internal Mode for SiO_2-Moganite Close to the α-β Transition

Mustafa Cem Lider and Hamit Yurtseven

Department of Physics, Middle East Technical University, 06531 Ankara, Turkey

Correspondence should be addressed to Hamit Yurtseven, hamit@metu.edu.tr

Academic Editor: Mario Amzel

The temperature dependence of the 501 cm^{-1} frequency of the vibrational mode is analyzed for SiO_2-moganite. The experimental data for the heating and cooling cycles of moganite from the literature is used for our analysis. The coexistence of α-β moganite is obtained over a finite temperature interval, and the α-β moganite transition at around 570 K is studied, as observed experimentally.

1. Introduction

SiO_2-moganite has been studied using various techniques over the years since it was discovered [1, 2] and approved as a mineral species in 1999. Its structure has been determined in some previous studies [3, 4]. It exhibits a displacive phase transition from the low temperature α-moganite (monoclinic) to the high temperature β-moganite at around 573 K. Its space group in the α phase is I2/a whereas in the β phase it is Imab. During the phase transformation, hysteresis occurs for the cooling and heating cycles, which indicates a first-order type as opposed to the α-β transition in quartz (T_C = 846 K). This transition in quartz is considered to be of a second order in the presence of the soft modes [5]. It has been observed that the 207 cm^{-1} mode disappears completely above the α-β quartz [6]. More recently, the experimental analyses have identified a first-order character to the phase transition between α-quartz and the intermediate incommensurate phase [7–9]. In our recent study, we have calculated the temperature dependence of the Raman frequency shifts and the linewidths for the optical lattice vibrations of the 128 cm^{-1} and 466 cm^{-1} in the α-phase of quartz from the anharmonic self-energy [10]; we have also calculated [11] the temperature dependence of the damping constant (Raman bandwidths) for the lattice modes of 147 cm^{-1} and 207 cm^{-1} close to the α-β transition in quartz by considering the soft mode behavior in this crystal.

In regard to the SiO_2-moganite system, there is no indication of soft mode behavior in this mineral. Using hard mode spectroscopy [12], it has been observed that the 501 cm^{-1} Raman mode is the highest intensity mode in moganite [13] whereas in quartz the highest intensity mode occurs at 465 cm^{-1}. It has also been pointed out that a relatively intense mode at 463 cm^{-1} in moganite overlaps the 465 cm^{-1} mode in quartz [13]. As observed experimentally [13], the 501 cm^{-1} mode can be associated with the structural phase transition in moganite. Its spontaneous vibrational displacement is expected to vary linearly with the square of the order parameter, which is usually the case in the behavior of hard modes in the Raman spectra for a second-order transition [12].

In the previous study [13], the temperature dependence of the Raman frequency and bandwidth for this symmetric stretching bending vibration (501 cm^{-1} B_{3g} mode) have been measured for the α-β transition in moganite experimentally. The temperature dependence of the unit cell volume of moganite has also been measured for the α-β transition in this mineral previously [13].

In this study, we calculate the Raman frequency of this internal mode using the volume data [13] through the mode

Table 1: Values of the coefficients for the experimental unit-cell volume [13] according to (3) for the α-β phases of moganite. The V_1 value which was obtained by extrapolating the $V(T)$ data [13] at around $T = 303$ K is also given here.

V (Å^3)	a (Å^3)	$b \times 10^{-2}$ (Å^3/K)	$-c \times 10^{-6}$ (Å^3/K^2)	V_1 (Å^3)	Temperature interval (K)
α-β phase	111.71	1.09	6.61	114.41	$297.5 < T < 868.3$

Grüneisen parameter for the α-β transition in moganite. By determining the mode Grüneisen parameter using the vibrational frequency [13] and the unit cell volume [13] data, the Raman frequencies of the 501 cm^{-1} mode are predicted for the α-β transition in moganite.

In Section 2 we give our calculations and results. Discussion and conclusions are given in Sections 3 and 4, respectively.

2. Calculations and Results

The Raman frequency ν can be calculated from the crystal volume V through the isobaric Grüneisen parameter γ_p defined as

$$\gamma_p = -\left(\frac{\partial \ln \nu/\partial T}{\partial \ln V/\partial T}\right)_p, \quad (1)$$

where T and p are the temperature and pressure, respectively.

Thus, using the temperature dependence of the volume the Raman frequency can be predicted by solving (1) as

$$\gamma_p(T) = \nu_0 + \nu_1 \exp\left[-\gamma_p \ln\left(\frac{V(T)}{V_1}\right)\right], \quad (2)$$

where ν_1 and V_1 are the values of the Raman frequency and crystal volume, respectively, at a constant temperature. ν_0 represents the background frequency of the Raman mode considered.

We calculated here the temperature dependence of the 501 cm^{-1} vibrational frequency for heating and cooling cycles of moganite using the unit-cell volume data of this crystal [13]. For this calculation, we first analyzed the unit-cell volume data [13] at various temperatures for the α-β phases of moganite according to a quadratic equation

$$V = a + bT + cT^2, \quad (3)$$

where a, b, and c are constants. Table 1 gives the values of those coefficients from our fit. We plot the observed unit-cell volume [13] as a function of temperature for the α-β phases of moganite according to (3) with the coefficients given in Figure 1.

In order to predict the temperature dependence of the 501 cm^{-1} vibrational mode of moganite in the α-β phases, from the unit-cell volume data [13], we needed to determine the isothermal mode Grüneisen parameter γ_p according to (2). For this determination, we used the experimental data [13] for the Raman mode of 501 cm^{-1} in the α-β phases of moganite. For the heating and cooling cycles of the 501 cm^{-1} vibrational frequency for the temperature interval of 300 to 900 K in the α-β phases of moganite, a quadratic equation

$$\nu(T) = a_0 + a_1 T + a_2 T^2 \quad (4)$$

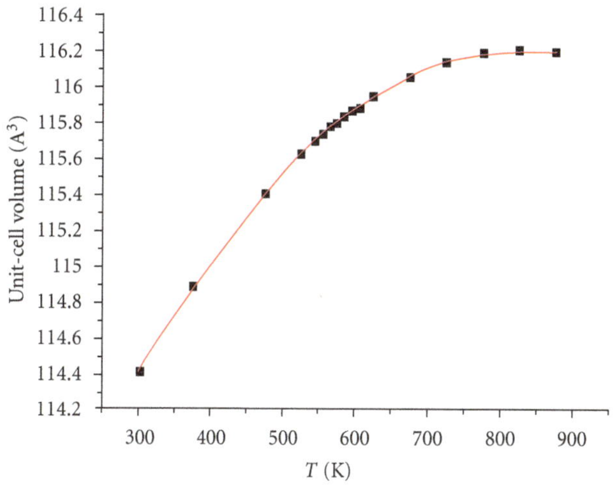

Figure 1: Experimental unit-cell volume [13] as a function of temperature for the the α-β phases of moganite. Solid curve represents (3) fitted to the experimental data.

was used with constants a_0, a_1, and a_2. By fitting (4) to the experimental frequencies [13], the values of the coefficients were determined for both heating and cooling cycles, as given in Table 2. Thus, from this fit of (4) to the experimental vibrational data [13] and the experimental data for the unit-cell volume [13], we were able to determine the value of the isobaric mode Grüneisen parameter as $\gamma_p = 0.75$ for this Raman mode. By assuming that the Grüneisen parameter (γ_p) remains constant throughout the α-β phases of moganite, the Raman frequencies of the 501 cm^{-1} mode were then calculated at various temperatures using the unit-cell volume data [13] according to (2). From our fitting, the background frequency was obtained as $\nu_0 = 0.72$ cm^{-1} for both heating and cooling cycles in moganite. Values of V_1 and ν_1 which were obtained from the extrapolations of the $V(T)$ and $\nu(T)$ data at $T = 303$ K are given in Tables 1 and 3, respectively. In Figures 2 and 3, we plot our calculated vibrational frequencies of 501 cm^{-1} mode as a function of temperature for the cooling and heating cycles, respectively, in the α-β phases of moganite. The observed frequency data [13] are also given in those plots.

3. Discussion

We calculated here the Raman frequencies of the symmetric stretching-bending mode (B_{3g}) as a function of temperature from the experimental unit-cell volume data [13] for the α-β transition in moganite, as shown in Figure 2 (cooling) and Figure 3 (heating). These figures show that our predicted frequencies of this particular internal mode agree with

TABLE 2: Values of the coefficients from the fit of (4) to the experimental data [13] for heating and cooling cycles for the vibrational frequency of 501 cm^{-1} in the α-β phases of moganite.

Vibrational frequency ω (501 cm^{-1})	a_0 (cm^{-1})	$-a_1 \times 10^{-2}$ (cm^{-1}/K)	$a_2 \times 10^{-5}$ (cm^{-1}/K^2)	Temperature interval (K)
Heating	507.0	2.57	1.29	$301.6 < T < 873.8$
Cooling	506.84	2.63	1.39	$303.5 < T < 875.3$

TABLE 3: Values of the isobaric mode Grüneisen parameter γ_P, extrapolated value ν_1 at around $T = 303$ K and background frequency ν_0 for the vibrational frequency indicated in the α-β phases of moganite.

Vibrational frequency	γ_P	ν_1 (cm^{-1})	ν_0 (cm^{-1})
501 cm^{-1}	0.75	500.1	0.72

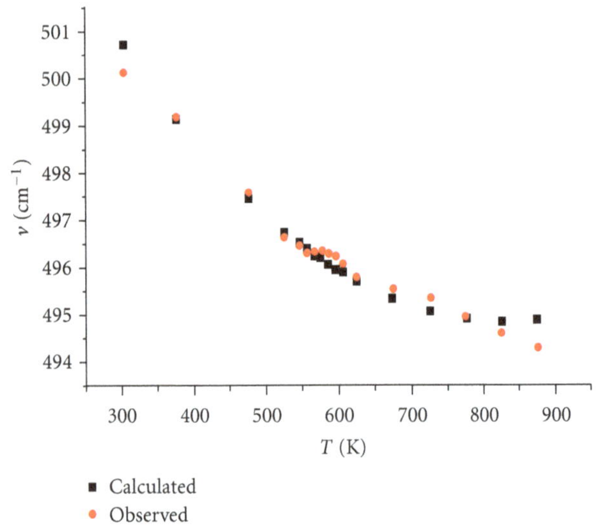

FIGURE 2: Vibrational frequency calculated as a function of temperature according to (2) for the cooling cycle in the α-β phases of moganite. Observed vibrational frequencies are also shown here.

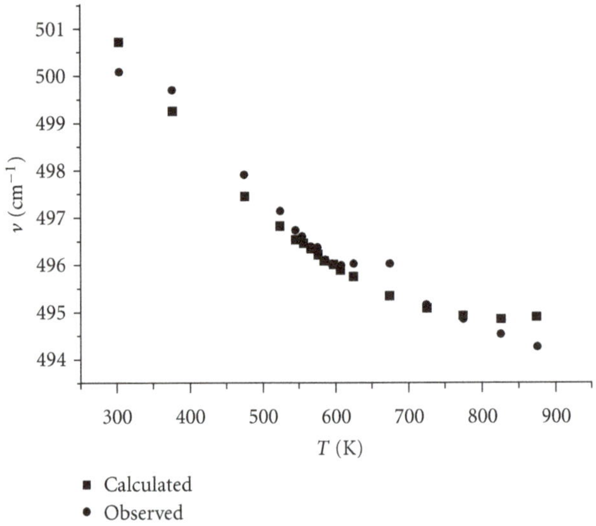

FIGURE 3: Vibrational frequency calculated as a function of temperature according to (2) for the heating cycle in the α-β phases of moganite. Observed vibrational frequencies are also shown here.

the observed frequencies [13] except in the hysteresis region where the high- and low-temperature modifications coexist over a finite transition interval. It has been suggested that there occurs an intermediate phase over a 1.3 K interval between the α and β phases in quartz [14]. The intermediate phase may also occur for the α-β transition in moganite. However, since the experimental unit-cell volume [13] does not give any indication of an intermediate phase in moganite as shown in Figure 1, the Raman frequencies of the 501 cm^{-1} mode which were calculated through (2) do not also show such an intermediate phase over the α-β transition in this material. This intermediate phase which can exist in moganite has been considered in quartz [14] as consisting of ordered arrays of Dauphine microtwins. However, the moganite twins result from the loss of mirror symmetry, whereas the Dauphine twins of quartz result from a loss of 2-fold symmetry along the c-axis.

As studied previously, the α-β transition in moganite is the displacive transition, and it can be associated with the internal mode of 501 cm^{-1}, as observed experimentally [13]. It has been pointed out that the variations observed for the 501 cm^{-1} peak strongly support the structural transition in moganite [14]. Thus, as a hard mode the square of the Raman frequency and also of the linewidth should be associated with the order parameter linearly, as suggested previously [13]. In fact, the observed behavior of the Raman frequency for the 501 cm^{-1} mode, which increases with decreasing temperature (Figures 2 and 3), indicates the temperature dependence of the order parameter for the α-β transition in moganite. Also, the linewidths of this Raman mode decrease with decreasing temperature, as observed experimentally [13], which can be associated with the order parameter in moganite. As stated previously, a linear variation of the order parameter with square of the frequency (spontaneous vibrational displacement) and of the linewidth (spontaneous vibrational broadening) is considered for a second-order transition in the hard mode Raman spectroscopy [15]. In the presence of the coexistence of the α and β phases, and also a possible intermediate phase in moganite, which indicates a first-order displacive transition the temperature dependence of the Raman frequency and of the linewidth for the 501 cm^{-1} mode can be examined within the framework of a second-order transition in this mineral. In fact, we examined the temperature dependence of the Raman bandwidths of this mode by calculating the damping constant in terms of the order parameter below the transition temperature (T_C = 573 K) for the α-β transition in moganite. By calculating the temperature dependence of the order parameter ($T < T_C$) from the molecular field theory [16], the Raman bandwidths of the 501 cm^{-1} mode were predicted using the soft mode-hard mode coupling model [17, 18] and the energy fluctuation model [19] for the α-β transition in moganite. Our calculation of the damping constant of the 501 cm^{-1} mode failed using both models studied since it diverges as the α-β transition temperature was approached, which was not in agreement with the observed [13] Raman bandwidths. This is due to the fact that the vibrational frequency decreases anomalously as the T_C is approached from the low-temperature phase which is essentially the soft mode behavior according to a power-law formula:

$$\omega \sim (T_C - T)^{1/2}. \quad (5)$$

Our calculation indicated that the 501 cm^{-1} mode does not exhibit a soft mode behavior for the α-β transition in moganite. For the α-β transition in quartz, we were able to predict the divergence of the 147 cm^{-1} and 207 cm^{-1} Raman modes in the vicinity of the transition temperature (T_C = 846 K) using the models considered above, as studied in our recent work [11]. We also indicated in that work [11] that the 147 cm^{-1} Raman mode exhibits a soft mode behavior for the α-β transition in quartz. In the case of moganite, the temperature dependence of the vibrational frequency and of the linewidth of the 501 cm^{-1} mode can be further investigated to explain its α-β transition.

4. Conclusions

The temperature dependence of the Raman frequency for the internal mode (501 cm^{-1}) was calculated using the unit-cell volume data through the mode Grüneisen parameters for the α-β transition in moganite. Our calculated Raman frequencies of this mode agree with the observed data in a large temperature interval for this mineral. This shows that the observed behavior of the internal mode associated with the displacive α-β transition in moganite can be predicted adequately by the method of calculation given in this study.

References

[1] O. W. Flörke, J. B. Jones, and H. U. Schmincke, "A new microcrystalline silica from Gran Canaria," *Zeitschrift für Kristallographie*, vol. 143, pp. 156–165, 1976.

[2] O. W. Flörke, U. Flörke, and U. Giese, "Moganite, a new microcrystalline silica-mineral," *Neues Jahrbuch Für Mineralogie Abhandlungen*, vol. 149, pp. 325–336, 1984.

[3] G. Miehe and H. Graetsch, "Crystal structure of moganite: a new structure type for silica," *European Journal of Mineralogy*, vol. 4, no. 4, pp. 693–706, 1992.

[4] P. J. Heaney and J. E. Post, "The widespread distribution of a novel silica polymorph in microcrystalline quartz varieties," *Science*, vol. 255, no. 5043, pp. 441–443, 1992.

[5] C. V. Raman and T. M. K. Nedungadi, "The α-β transformation of quartz," *Nature*, vol. 145, no. 3665, p. 147, 1940.

[6] S. M. Shapiro, D. C. O'Shea, and H. Z. Cummins, "Raman scattering study of the alpha-beta phase transition in quartz," *Physical Review Letters*, vol. 19, no. 7, pp. 361–364, 1967.

[7] P. J. Heaney and D. R. Veblen, "An examination of spherulitic dubiomicrofossils in Precambrian banded iron formations using the transmission electron microscope," *Precambrian Research*, vol. 49, no. 3-4, pp. 355–372, 1991.

[8] P. J. Heaney and D. R. Veblen, "Observations of the α-β phase transition in quartz: a review of imaging and diffraction studies and some new results," *American Mineralogist*, vol. 76, no. 5-6, pp. 1018–1032, 1991.

[9] M. A. Carpenter, E. K. H. Salje, A. Graeme-Barber, B. Wruck, M. T. Dove, and K. S. Knight, "Calibration of excess thermodynamic properties and elastic constant variations associated with the $\alpha \leftrightarrow \beta$ phase transition in quartz," *American Mineralogist*, vol. 83, no. 1-2, pp. 2–22, 1998.

[10] M. Kurt and H. Yurtseven, "Temperature dependence of the Raman frequency shifts and the linewidths in the α phase of quartz," *Balkan Physics Letters*, vol. 19, no. 191042, pp. 362–368, 2011.

[11] M. C. Lider and H. Yurtseven, "Calculation of the Raman linewidths of lattice modes close to the α-β transition in quartz," *High Temperature Materials and Processes*. In press.

[12] E. K. H. Salje, "Hard mode spectroscopy: experimental studies of structural phase transitions," *Phase Transitions*, vol. 37, pp. 83–110, 1992.

[13] P. J. Heaney, D. A. McKeown, and J. E. Post, "Anomalous behavior at the I2/a to Imab phase transition in SiO_2-moganite: an analysis using hard-mode Raman spectroscopy," *American Mineralogist*, vol. 92, no. 4, pp. 631–639, 2007.

[14] P. J. Heaney and J. E. Post, "Evidence for an I2/a to Imab phase transition in the silica polymorph moganite at ~ 570 K," *American Mineralogist*, vol. 86, pp. 1358–1366, 2001.

[15] M. A. Carpenter and E. K. H. Salje, "Elastic anomalies in minerals due to structural phase transitions," *European Journal of Mineralogy*, vol. 10, no. 4, pp. 693–812, 1998.

[16] M. Matsushita, "Anomalous temperature dependence of the frequency and damping constant of phonons near $T\lambda$ in ammonium halides," *The Journal of Chemical Physics*, vol. 65, no. 1, pp. 23–28, 1976.

[17] I. Laulicht and N. Luknar, "Internal-mode line-broadening by proton jumps in KH_2PO_4," *Chemical Physics Letters*, vol. 47, no. 2, pp. 237–240, 1977.

[18] I. Laulicht, "On the drastic temperature broadening of hard mode Raman lines of ferroelectric KDP type crystals near Tc," *Journal of Physics and Chemistry of Solids*, vol. 39, no. 8, pp. 901–906, 1978.

[19] G. Schaack and V. Winderfeldt, "Temperature behaviour of optical phonons near Tc in ferroelectric triglyicine sulphate and triglycine selenate—evidence of non-linear pseudospin-phonon interaction," *Ferroelectrics*, vol. 15, pp. 35–41, 1977.

26

Similarity Solution of Heat and Mass Transfer for Natural Convection over a Moving Vertical Plate with Internal Heat Generation and a Convective Boundary Condition in the Presence of Thermal Radiation, Viscous Dissipation, and Chemical Reaction

S. Mohammed Ibrahim[1] and N. Bhashar Reddy[2]

[1] *Department of Mathematics, Priyadarshini College of Engineering and Technology, Nellore 524004, India*
[2] *Department of Mathematics, Sri Venkateswara University, Tirupat 517501, India*

Correspondence should be addressed to S. Mohammed Ibrahim; ibrahimsvu@gmail.com

Academic Editors: R. R. Burnette, T. M. Inerbaev, and P. Trens

Steady laminar natural convection flow over a semi-infinite moving vertical plate with internal heat generation and convective surface boundary condition in the presence of thermal radiation, viscous dissipation, and chemical reaction is examined in this paper. In the analysis, we assumed that the left surface of the plate is in contact with a hot fluid while the cold fluid on the right surface of the plate contains a heat source that decays exponentially with the classical similarity variable. We utilized similarity variable to transform the governing nonlinear partial differential equations into a system of ordinary differential equations, which are solved numerically by applying shooting iteration technique along fourth-order Runge-Kutta method. The effects of the local Biot number, Prandtl number, buoyancy forces, the internal heat generation, the thermal radiation, Eckert number, viscous dissipation, and chemical reaction on the velocity, temperature, and concentration profiles are illustrated and interpreted in physical terms. A comparison with previously published results on the similar special cases showed an excellent agreement. Finally, numerical values of physical quantities, such as the local skin-friction coefficient, the local Nusselt number, and the local Sherwood number, are presented in tabular form.

1. Introduction

Convective flows with simultaneous heat and mass transfer under the influence of the chemical reaction arise in many transport processes both naturally and artificially in many branches of science and engineering applications. This phenomenon plays an important role in the chemical industry, power and cooling industry for drying, chemical vapour deposition on surfaces, cooling of nuclear reactors, and petroleum industries.

Natural convection flow occurs frequently in nature. It occurs due to temperature differences, as well as due to concentration differences or the combination of these two; for example, in atmospheric flows, there exist differences in water concentration, and hence the flow is influenced by such concentration difference.

Changes in fluid density gradients may be caused by nonreversible chemical reaction in the system as well as by the differences in the molecular weight between values of the reactants and the products. Chemical reactions can be modeled as either homogenous or heterogeneous processes. This depends on whether they occur at an interface or as a single phase value reaction. A homogeneous reaction is one that occurs uniformly throughout a given phase. On the other hand, a heterogeneous reaction takes place in a restricted area or within the boundary of a phase. In most cases of chemical reaction, the reaction rate depends on the concentration of the species itself. A reaction is said to be of first order if the rate of reaction is directly proportional to the concentration itself (Cussler [1]). For example, the formation of smog is a first-order homogeneous reaction. Consider the emission of nitrogen dioxide from automobiles and other smog stacks. This nitrogen dioxide reacts chemically in the atmosphere with unburned hydrocarbons (aided by sunlight) and produces peroxyacetylnitrate, which forms an envelope which is turned photochemical smog. The study of heat and mass transfer with chemical reaction is of great practical importance in many branches of science and engineering. Das et al. [2] studied the effects of mass transfer flow past an impulsively started infinite vertical plate with constant heat flux and chemical reaction. Anjalidevi and Kandasamy [3] found the effects of chemical reaction, heat, and mass transfer on laminar flow along a semi-infinite horizontal plate. More recently, intensive studies have been carried out to investigate the effects of chemical reaction and different flow types (see Seddeek et al. [4]. Salem and Abd El-Aziz [5], Mohamed [6], and Ibrahim et al. [7]).

The study of heat generation or absorption in moving fluids is important in problems dealing with chemical reactions and those concerned with dissociating fluids. Heat generation effects may alter the temperature distribution, and these in turn can affect the particle deposition rate in nuclear reactors, electronic chips, and semiconductor's wafers. Although exact modeling of internal heat generation or absorption is quite difficult, some simple mathematical models can be used to express its general behavior for most physical situations. Heat generation or absorption can be assumed to be constant, space dependent, or temperature dependent. Crepeau and Clarksean [8] have used a space-dependent exponentially decaying heat generation or absorption in their study on flow and heat transfer from vertical plate. Several interesting computational studies of reactive MHD boundary layer flows with heat and mass transfer in the presence of heat generation or absorption have appeared in recent years (see Patil and Kulkarni [9], Salem and Abd El-Aziz [5], Mohamed [6], and Mahdy [10]).

Convective heat transfer studies are very important in processes involving high temperatures, such as gas turbines, nuclear plants, and thermal energy storage. Ishak [11] examined the similarity solutions for flow and heat transfer over a permeable surface with convective boundary condition. Moreover, Aziz [12] studied a similarity solution for laminar thermal boundary layer over a flat plate with a convective surface boundary condition and also studied hydrodynamic and thermal slip flow boundary layers over a flat plate with a constant heat flux boundary condition. Makinde and Olanrewaju [13] investigated the buoyancy effects on a thermal boundary layer over a vertical plate with a convective surface boundary condition. More recently, Makinde [14] studied similarity solution for natural convection from a moving vertical plate with internal heat generation and a convective boundary condition. Olanrewaju et al. [15] examined the effects of internal heat generation, thermal radiation, and buoyancy force on a boundary layer over a vertical plate with a convective surface boundary condition. Makinde and Olanrewaju [16] investigated the combined effects of internal heat generation and buoyancy force on boundary layer flow over a vertical plate with a convective surface boundary condition.

Viscous dissipation changes the temperature distributions by playing a role like an energy source, which leads to the affected heat transfer rates. The merit of the effect of viscous dissipation depends on whether the plate is being cooled or heated. Heat transfer analysis over porous surface is of much practical interest due to its abundant applications. To be more specific, heat-treated materials traveling between a feed roll and wind-up roll or materials manufactured by extrusion, glass-fiber and paper production, cooling of metallic sheets or electronic chips, and crystal growing, just to name a few. In these cases, the final product of desired characteristics depends on the rate of cooling in the process and the process of stretching. The work of Sonth et al. [17] deals with the effect of the viscous dissipation term along with temperature-dependent heat source/sink on momentum and heat and mass transfer in a viscoelastic fluid flow over an accelerating surface. Chen [18] examined the effect of combined heat and mass transfer on MHD-free convection from a vertical surface with ohmic heating and viscous dissipation. The effect of viscous dissipation and Joule heating on MHD-free convection flow past a semi-infinite vertical flat plate in the presence of the combined effect of Hall and nonslip currents for the case of power-law variation of the wall temperature is analyzed by Abo-Eldahab and El Aziz [19].

In many new engineering areas processes such as fossil fuel combustion energy processes, solar power technology, astrophysical flows, gas turbines, and the various propulsion devices for aircraft, missiles, satellites, and space vehicle reentry occur at high temperatures so knowledge of radiation heat transfer beside the convective heat transfer plays a very important role and hence its effect cannot be neglected. Also thermal radiation is of major importance in many processes in engineering areas which occur at a very high temperature for the design of many advanced energy conversion systems and pertinent equipment. The Rosseland approximation is used to describe the radiative heat flux in the energy equation. Pal and Mondal [20] investigate the unsteady two-dimensional MHD non-Darcian mixed convection heat and mass transfer past a semi-infinite vertical permeable plate embedded in a porous medium by taking into account Soret and Dufour effects in the presence of suction or injection, thermal radiation, and first-order chemical reaction. Uwanta [21] studied the effects of chemical reaction and radiation on heat and mass transfer past a semi-infinite vertical porous plate with constant mass flux and dissipation. Olanrewaju et al. [22] found the effects of internal heat generation, thermal radiation, and buoyancy force on a boundary layer over a vertical plate with a convective surface boundary condition.

The objective of this paper was to explore the effects of thermal radiation, heat generation, viscous dissipation, and chemical reaction on the similarity solution for natural convection from a moving vertical plate under a convective boundary condition, which is an extension of Makinde [14] with the addition of thermal radiation, viscous dissipation, and chemical reaction parameter for more physical implications. Using a similarity approach, the governing equations are transformed into nonlinear ordinary equations and solved numerically using shooting iteration technique together with fourth-order Runge-Kutta integration scheme. The pertinent results are displayed graphically and discussed quantitatively.

2. Mathematical Formulation

We consider the steady laminar incompressible natural convection boundary layer flows over the right surface of a vertical flat plate moving with uniform velocity U_0 in contact with a quiescence cold fluid at temperature T_∞ and concentration C_∞. The cold fluid on the right surface of the plate generates heat internally at the volumetric rate $\dot{q}$. The left surface of the plate is heated by convection from a hot fluid at temperature T_f which provides a heat transfer coefficient h_f as shown in Figure 1. Under the Boussinesq for fluid density variation, the continuity, momentum, energy equation, and mass diffusion equations describing the flow can be written as

$$\frac{\partial u}{\partial x} + \frac{\partial v}{\partial y} = 0, \tag{1}$$

$$u\frac{\partial u}{\partial x} + v\frac{\partial u}{\partial y} = \nu\frac{\partial^2 u}{\partial y^2} + g\beta\left(T - T_\infty\right) + g\beta^*\left(C - C_\infty\right), \tag{2}$$

$$\rho C_p\left(u\frac{\partial u}{\partial x} + v\frac{\partial u}{\partial y}\right) = k\frac{\partial^2 T}{\partial y^2} + \dot{q} - \frac{\partial q_r}{\partial y} + \mu\left(\frac{\partial u}{\partial y}\right)^2, \tag{3}$$

$$u\frac{\partial C}{\partial x} + v\frac{\partial C}{\partial y} = D\frac{\partial^2 C}{\partial y^2} - K_r'\left(C - C_\infty\right), \tag{4}$$

where u and v are the x (along the plate) and y (normal to the plate) components of the velocities, respectively, T is the temperature, C is the concentration, μ is the fluid viscosity, ν is the kinematics viscosity of the fluid, k is the thermal conductivity of the fluid, β is the thermal expansion coefficient, β^* is concentration expansion coefficient, $\dot{q}$ is the internally generated heat at volumetric rate, g is the gravitational acceleration, q_r is the radiative heat flux, D is the diffusion coefficient, and K_r' is the chemical reaction parameter.

The boundary conditions at the plate surface and for the cold fluid may be written as

$$u(x,0) = U_0, \qquad v(x,0) = 0,$$
$$-K\frac{\partial T}{\partial y}(x,0) = h_f\left[T_f - T(x,0)\right],$$
$$C_f(x,0) = Ax^\lambda + C_\infty,$$
$$u(x,\infty) = 0, \qquad T(x,\infty) = T_\infty, \qquad C(x,\infty) = C_\infty. \tag{5}$$

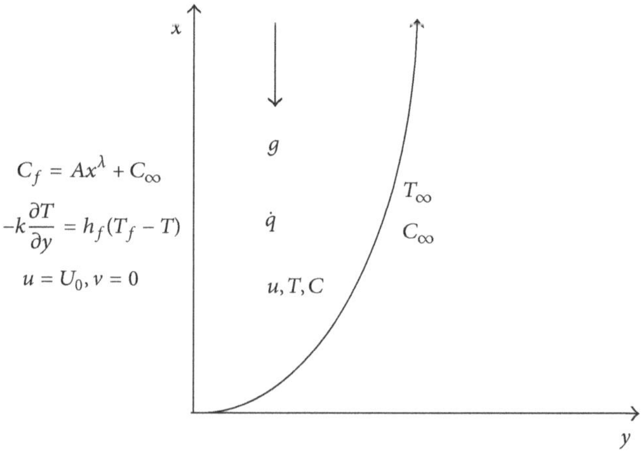

Figure 1: Flow configuration and coordinate system.

By using the Rosseland diffusion approximation Hossain et al. [23] and following Raptis [24] among other researchers the radiative heat flux, q_r is given by

$$q_r = -\frac{4\sigma^*}{3K_s}\frac{\partial T^4}{\partial y}, \tag{6}$$

where σ^* and K_s are the Stefan-Boltzmann constant and the mean absorption coefficient, respectively. Following Chamkha [25], we assume that the temperature differences within the flow are sufficiently small so that T^4 can be expressed as a linear function after using the Taylor series to expand T^4 about the free stream temperature T_∞ and neglecting higher-order terms. This result is the following approximation:

$$T^4 \approx 4T_\infty^3 T - 3T_\infty^4. \tag{7}$$

Using (6) and (7) in (3), we obtain

$$\frac{\partial q_r}{\partial y} = -\frac{16\sigma^*}{3K}\frac{\partial^2 T^4}{\partial y^2}. \tag{8}$$

Introducing a similarity variable η and a dimensionless stream function $f(\eta)$ temperature $\theta(\eta)$ and concentration $\phi(\eta)$ as

$$\eta = y\sqrt{\frac{U_0}{\nu x}} = \frac{y}{x}\sqrt{\mathrm{Re}_x},$$
$$\frac{u}{U_0} = f', \qquad v = \frac{1}{2x}\sqrt{\mathrm{Re}_x}\left(\eta f' - f\right), \tag{9}$$
$$\theta(\eta) = \frac{T - T_\infty}{T_f - T_\infty}, \qquad \phi(\eta) = \frac{C - C_\infty}{C_f - C_\infty},$$

where prime symbol denotes differentiation with respect to η and $\mathrm{Re}_x = U_0 x/\nu$ is the local Reynolds number. These nonlinear partial differential equations are then transformed by similarity transformation into a system of ordinary differential equations given as

$$f''' + \frac{1}{2} f f' + \mathrm{Gr}\,\theta + \mathrm{Gc}\,\phi = 0,$$

$$\theta'' \left[1 + \frac{4}{3}R\right] + \frac{1}{2}\mathrm{Pr}\, f\theta' + \mathrm{Pr}\, Q e^{-\eta} + \mathrm{Ec}\,\mathrm{Pr}\left(f''\right)^2 = 0,$$

$$\phi'' + \frac{1}{2}\mathrm{Sc}\, f\phi' - \mathrm{Kr}\,\mathrm{Sc}\,\phi = 0, \tag{10}$$

$$f(0) = 0, \qquad f'(0) = 1,$$

$$\theta'(0) = -\mathrm{Bi}\left[1 - \theta(0)\right], \qquad \phi(0) = 1,$$

$$f'(\infty) = 1, \qquad \theta(\infty) = 0, \qquad \phi(\infty) = 0,$$

where

$$\mathrm{Bi} = \frac{h_f}{k}\sqrt{\frac{\nu x}{U_0}}, \qquad \mathrm{Pr} = \frac{\nu}{\alpha},$$

$$\mathrm{Gr} = \frac{xg\beta\left(T_f - T_\infty\right)}{U_0^2}, \qquad \mathrm{Gc} = \frac{xg\beta^*\left(C_f - C_\infty\right)}{U_0^2},$$

$$\mathrm{Ra} = \frac{4\alpha^* \sigma T_\infty^3}{kK}, \qquad Q = \frac{x^2 \dot{q} e^{\eta}}{k\mathrm{Re}_x\left(T_f - T_\infty\right)}, \tag{11}$$

$$\mathrm{Ec} = \frac{U_0^2}{k\left(T_f - T_\infty\right)}, \qquad \mathrm{Sc} = \frac{\nu}{D}, \qquad \mathrm{Kr} = \frac{\mathrm{Kr}' x}{U_0}.$$

Bi is the local Biot number, Pr is the Prandtl number, Gr is local Grashof number, Gc is modified local Grashof number, Ra is the radiation parameter, Q is the internal heat generation parameter, Ec is the Eckert number, Sc is the Schmidt number, and Kr is the chemical reaction parameter.

For the momentum and energy equations to have a similarity solution, the parameters Gr, Gc, Q, and Bi_x must be constants and not functions of x as in (11). This condition can be met if the heat transfer coefficient h_f is proportional to $x^{-1/2}$, the thermal expansion coefficient β is proportional to x^{-1}, and the internal generation $\dot{q}$ is proportional to x^{-1}. We therefore assume

$$h_f = cx^{-1/2}, \qquad \beta = mx^{-1} q = lx^{-1}e^{-\eta}, \tag{12}$$

where c, m, and l are constants. Substituting (12) into (13), we have

$$\mathrm{Bi} = \frac{c}{k}\sqrt{\frac{\nu}{U_0}}, \qquad \mathrm{Gr} = \frac{mg\left(T_f - T_\infty\right)}{U_0^2},$$

$$\mathrm{Gc} = \frac{mg\left(C_f - C_\infty\right)}{U_0^2}, \qquad Q = \frac{l\nu}{kU_0\left(T_f - T_\infty\right)}. \tag{13}$$

With Bi, Q, and Gr, Gc is defined by (13). The solutions of (10) yield the similarity solutions. However, the solutions generated are the local similarity solutions whenever i, Q, Gr, and Gc are defined as in (13).

The coupled nonlinear boundary value problems represented by (10) have been solved numerically using the shooting techniques with the fourth-order Runge-Kutta method. From the numerical computations, the plate surface temperature, local skin-friction coefficient, the local Nusselt number, and the local Sherwood number which are, respectively, proportional to $\theta(0)$, $f''(0)$, $-\theta'(0)$, and $-\phi'(0)$ are worked out, and their numerical values are presented in a tabular form.

3. Results and Discussion

To analyze the results, numerical computation has been carried out using the method described in the previous paragraph for various governing parameters, namely, thermal Grashof number Gr, modified Grashof number Gc, Prandtl number Pr, thermal radiation parameter R, heat generation parameter Q, Eckert number Ec, Schmidt number Sc, chemical reaction parameter Kr, and convective parameter Bi. In the present study the following default parameter values are adopted for computations: Gr = 1.0, Gc = 1.0, Pr = 0.72, R = 0.5, Q = 0.5, Ec = 0.5, Sc = 0.6, Kr = 0, and Bi = 0.1. All graphs therefore correspond to these values unless specifically indicated on the appropriate graph.

Table 1 shows the comparison of Makinde [14] work with the present work for Ec = R = Sc = Kr = Gc = 0, and it is noteworthy that there is a perfect agreement. Table 2, shows the values of the skin-friction coefficient, the Nusselt number, the surface temperature, and the Sherwood number in terms of $f''(0)$, $\theta'(0)$, $\theta(0)$, and $\phi'(0)$, respectively, for various values embedded flow parameters. From Table 2, it is understood that the skin friction, the rate of heat transfer, wall surface temperature at the plate surface, and the rate of mass transfer increase with an increase in local Biot number. An increase in buoyancy forces, thermal radiation, internal heat generation, Eckert number there is an increase in skin-friction, surface temperature and the Sherwood number, but decrease in the Nusselt number. An increase in the Prandtl number there is decrease in skin friction, surface temperature and the Sherwood number, but increases the Nusselt number. However, an increase in the Schmidt number and chemical reaction parameter causes a decrease in the skin friction, the Nusselt number, and surface temperature and increase in the surface mass transfer rate, that is, the Sherwood number.

3.1. Velocity Profiles. Figures 2–10 depict the effects of various thermophysical parameters on the fluid velocity profile. It was observed that, generally, the fluid velocity increases gradually away from the plate attain its peak value within the boundary layer and the decreases to the free stream zero value satisfying the boundary conditions. From Figures 2 and 3, we observed that the velocity boundary layer thickness increases with an increase in the values of local Grashof number (Gr) and modified local Grashof number (Gc) due to buoyancy effect. In Figure 4, the influence of Prandtl number on the fluid velocity was displayed, and it is interesting to

TABLE 1: Computations showing comparison with Makinde [14] results for Gc = 0, Ec = 0, $R = 0$, Sc = 0, and $K_r = 0$.

Bi	Gr	Pr	Q	$f''(0)$ Makinde [14]	$\theta'(0)$ Makinde [14]	$\theta(0)$ Makinde [14]	$f''(0)$ present	$\theta'(0)$ present	$\theta(0)$ present
0.1	0.1	0.72	1.0	−0.2000518	0.076578477	1.76578477	−0.253226	0.0353022	1.35302
1.0	0.1	0.72	1.0	−0.2459676	0.281651449	1.28165144	−0.279242	0.128217	1.12822
10	0.1	0.72	1.0	−0.2695171	0.382952717	1.03829527	−0.280211	0.173623	1.01736
0.1	0.5	0.72	1.0	0.4221216	0.048257030	1.48257030	0.250851	0.016076	1.16076
0.1	1.0	0.72	1.0	0.9895493	0.034011263	1.34011263	0.717892	0.00569792	1.05698
0.1	0.1	3.0	1.0	−0.3748695	−0.023814576	0.76185423	−0.26024	0.100968	2.00969
0.1	0.1	7.10	1.0	−0.4138825	−0.057164001	0.42835998	−0.258586	0.153863	2.53863
0.1	0.1	0.72	5	0.3741286	0.576670381	6.76670381	0.198617	0.41344	5.13444
0.1	0.1	0.72	10	0.9010790	1.106605802	12.0660580	0.620946	0.818126	9.18126

TABLE 2: Computation showing $f''(0)$, $\theta'(0)$, $\theta(0)$, and $\phi'(0)$ for different embedded flow parameter values.

Bi	Gr	Gc	Pr	R	Q	Ec	Sc	K_r	$f''(0)$	$\theta'(0)$	$\theta(0)$	$\phi'(0)$
0.1	1.0	1.0	1.0	0.5	0.1	0.1	0.6	0.5	0.617935	0.0690068	0.309932	0.690904
1.0	1.0	1.0	1.0	0.5	0.1	0.1	0.6	0.5	0.995463	0.24145	0.75855	0.705335
10.0	1.0	1.0	1.0	0.5	0.1	0.1	0.6	0.5	1.16294	0.325667	0.967433	0.71137
0.1	2.0	1.0	1.0	0.5	0.1	0.1	0.6	0.5	0.914867	0.0688259	0.311741	0.703017
0.1	3.0	1.0	1.0	0.5	0.1	0.1	0.6	0.5	1.21966	0.0680544	0.319456	0.714747
0.1	1.0	2.0	1.0	0.5	0.1	0.1	0.6	0.5	1.27296	0.067557	0.32443	0.711755
0.1	1.0	3.0	1.0	0.5	0.1	0.1	0.6	0.5	1.8969	0.064555	0.35445	0.729823
0.1	1.0	1.0	1.0	0.5	0.1	0.1	0.6	0.5	0.598557	0.0696409	0.303591	0.689679
0.1	1.0	1.0	3.0	0.5	0.1	0.1	0.6	0.5	0.537444	0.070888	0.29112	0.685736
0.1	1.0	1.0	1.0	1.0	0.1	0.1	0.6	0.5	0.635229	0.0684205	0.315795	0.691995
0.1	1.0	1.0	1.0	1.5	0.1	0.1	0.6	0.5	0.645997	0.0680532	0.319468	0.692676
0.1	1.0	1.0	1.0	0.5	0.2	0.1	0.6	0.5	0.694807	0.0621171	0.378829	0.694271
0.1	1.0	1.0	1.0	0.5	0.3	0.1	0.6	0.5	0.769222	0.0553443	0.446557	0.697471
0.1	1.0	1.0	1.0	0.5	0.1	0.2	0.6	0.5	0.654059	0.0662619	0.337381	0.692653
0.1	1.0	1.0	1.0	0.5	0.1	0.3	0.6	0.5	0.693151	0.0632311	0.367689	0.694521
0.1	1.0	1.0	1.0	0.5	0.1	0.1	0.78	0.5	0.562746	0.0687271	0.312729	0.786616
0.1	1.0	1.0	1.0	0.5	0.1	0.1	1.0	0.5	0.511741	0.068638	0.311362	0.889928
0.1	1.0	1.0	1.0	0.5	0.1	0.1	0.6	1.0	0.514718	0.0688634	0.311366	0.877011
0.1	1.0	1.0	1.0	0.5	0.1	0.1	0.6	1.5	0.488687	0.0687467	0.312533	1.03195

note that velocity boundary layer thickness decreases with an increase in the Prandtl number. Figure 5 depicts the influence of thermal radiation on the fluid velocity, and it is interesting to note that increases the radiation parameter thickness the velocity boundary layer thickness away from the plate surface. Figure 6 depicts the effects of local internal heat generation parameter on the fluid velocity. An increase in the exponentially decaying internal heat generation causes a further increase in the velocity boundary layer thickness. Figure 7 represents the curve of fluid velocity against span-wise co-ordinate η for various values of Eckert number which shows that the increase in Eckert number leads to a sudden increase in the fluid velocity immediately away from the wall plate before satisfying the boundary conditions. It is interesting to note that it thickens the velocity boundary layer thickness close to the wall plate when the velocity profile attains its maximum value point. Figure 8 and Figure 9 show the variation of the boundary layer velocity with the Schmidt number and chemical reaction parameter. From this Figures we observed a slight decrease in the fluid velocity with an increase in Schmidt number and chemical reaction parameter. Figure 10 depicts the effects of the variation of the boundary layer velocity with the intensity of local Biot number (Bi). From this it was observed that the velocity boundary layer thicknesses slightly increase with an increase in the local Biot number due to convective heat transfer the plate surface.

3.2. Temperature Profiles. Figures 11, 12, 13, 14, and 15 illustrate the fluid temperature profiles within the boundary layer. Generally, the fluid temperature is maximum at the plate surface and decreases exponentially to zero value far away from the plate satisfying the boundary conditions. From these figures, it is noteworthy that the thermal boundary layer thickness increases with an increase in the exponentially decaying internal heat generation, thermal radiation, Eckert number, and local Biot number and decreases with an increase in the values of Prandtl number. At high Prandtl

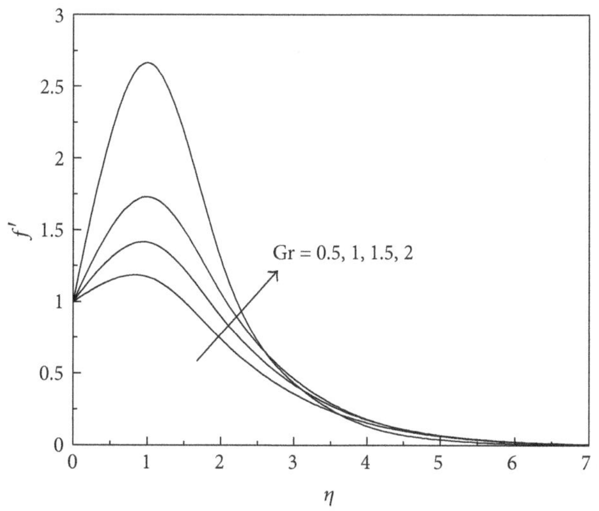

FIGURE 2: Effects of local Grashof number on velocity profile.

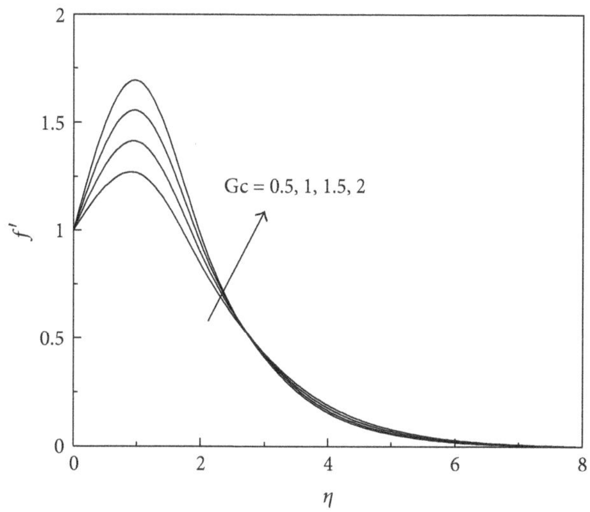

FIGURE 3: Effects of local modified Grashof number on velocity profile.

number has low velocity, which in turn also implies that at lower fluid velocity the species diffusion is comparatively lower and hence higher species concentration is observed at high Prandtl number.

3.3. Concentration Profiles. Figures 16 and 17 depict chemical species concentration profiles against spanwise coordinate η for varying values of physical parameters in the boundary layer. The species concentration is highest at the plate surface and decreases to zero far away from the plate satisfying the boundary condition. From these figures, it is noteworthy that the concentration boundary layer thickness decreases with an increase in the Schmidt number and chemical reaction parameter.

Table 2 shows the values of the skin-friction coefficient, the Nusselt number, surface temperature, and the Sherwood number in terms of $f''(0)$, $\theta'(0)$, $\theta(0)$, and $\phi'(0)$, respectively, for various values of embedded flow parameters. It is understood that the skin friction, the rate of heat transfer, wall surface temperature at the plate surface, and the rate of mass transfer increase with an increase in the local Biot number. An increase in buoyancy forces (Gr, Gc), thermal radiation, internal heat generation, Eckert number there is an increase in skin-friction, surface temperature and Sherwood number and decrease in the Nusselt number. An increase in the Prandtl number there is decrease in skin-friction, surface temperature and Sherwood number, but increases the Nusselt number. However, an increase in the Schmidt number and chemical reaction parameter causes a decrease in the skin friction, the Nusselt number, and surface temperature and increases the surface mass transfer rate.

4. Conclusions

The similarity solution for natural convection from a moving vertical plate with internal heat generation and a convective boundary condition in the presence of thermal radiation, viscous dissipation, and chemical reaction is studied.

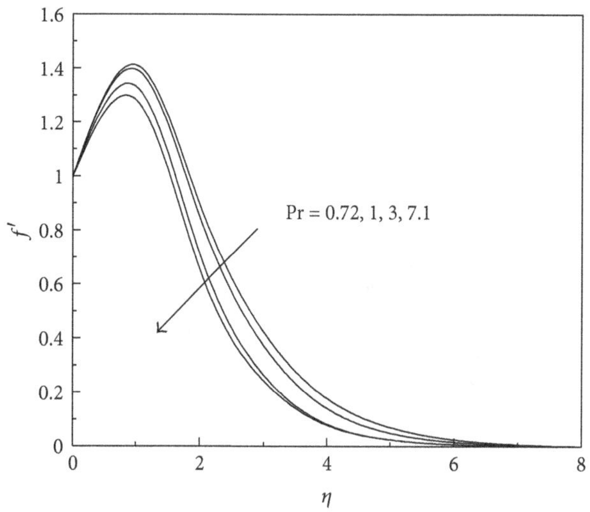

FIGURE 4: Effects of Prandtl number on velocity profile.

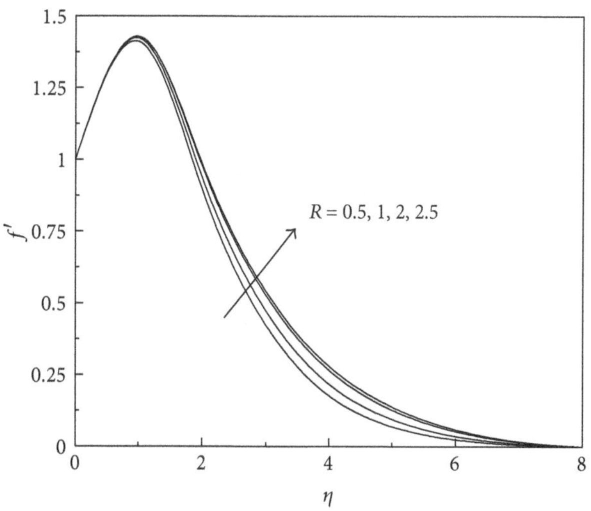

FIGURE 5: Effects of radiation parameter on velocity profile.

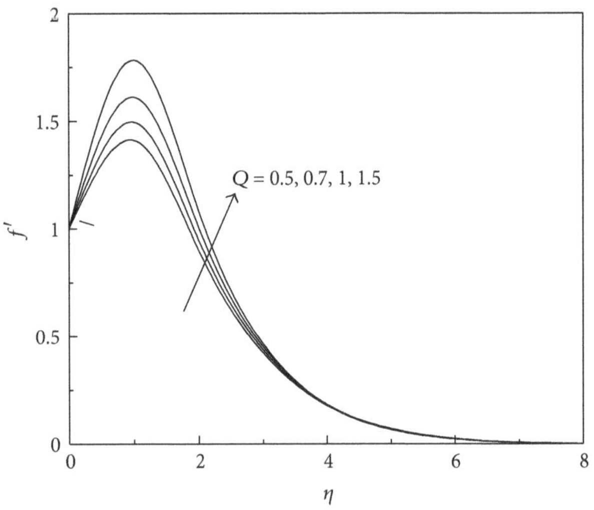

FIGURE 6: Effects of internal heat generation on velocity profile.

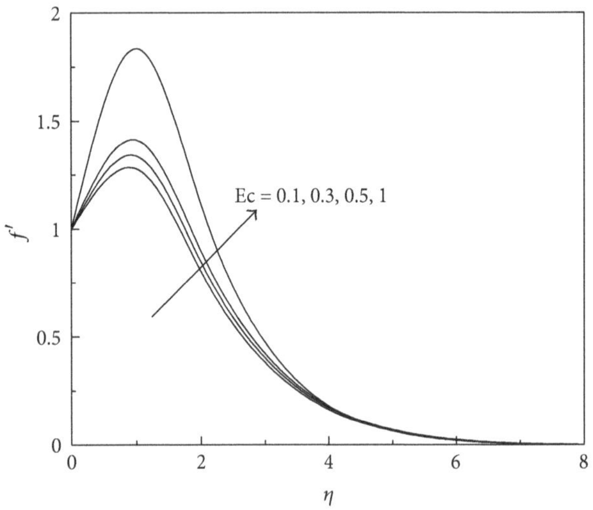

FIGURE 7: Effects of Eckert number on velocity profile.

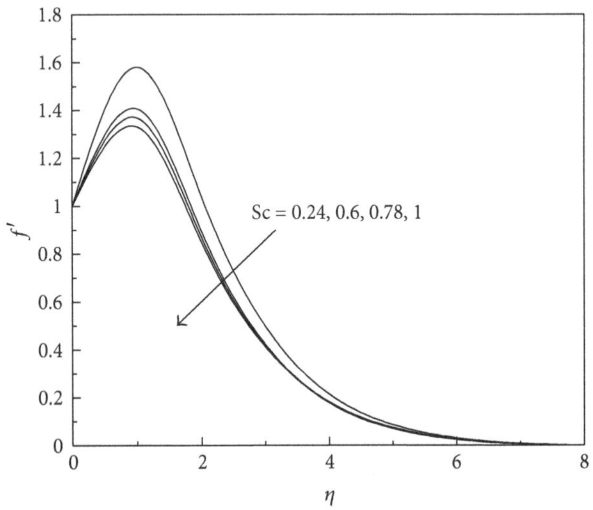

FIGURE 8: Effects of Schmidt number on velocity profile.

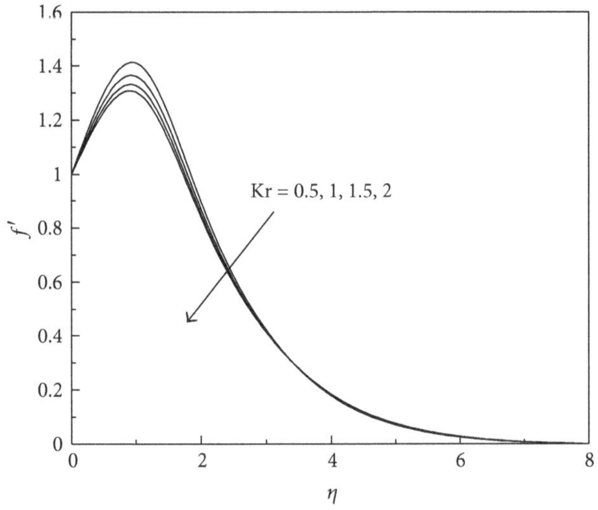

FIGURE 9: Effects of chemical reaction parameter on velocity profile.

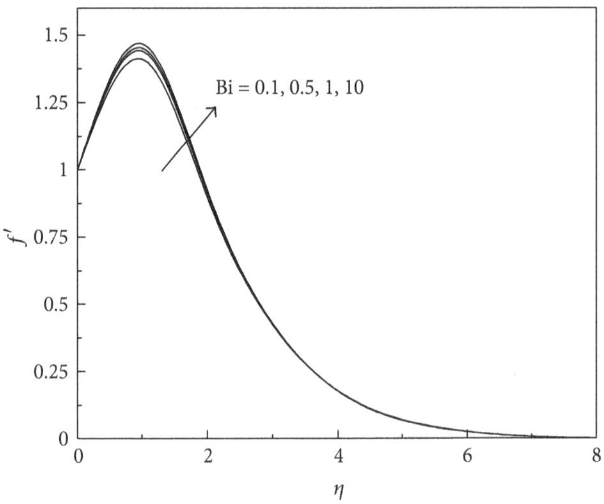

FIGURE 10: : Effects of local Biot number on velocity profile.

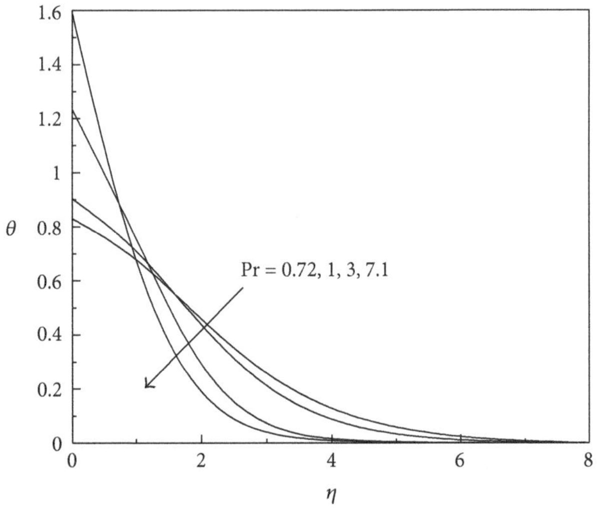

FIGURE 11: Effects of Prandtl number on temperature profile.

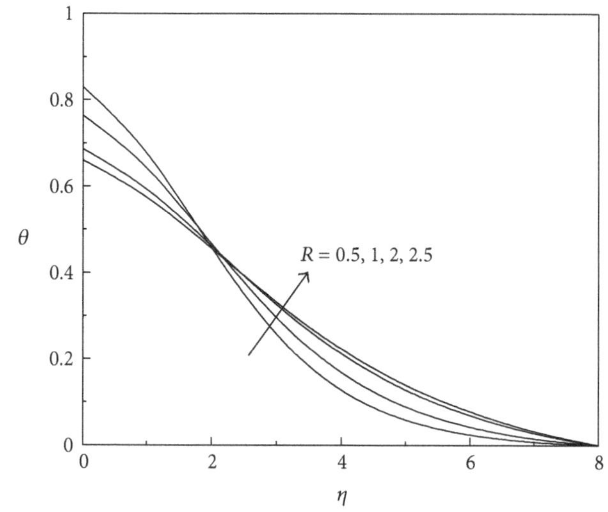

FIGURE 12: Effects of radiation parameter on temperature profile.

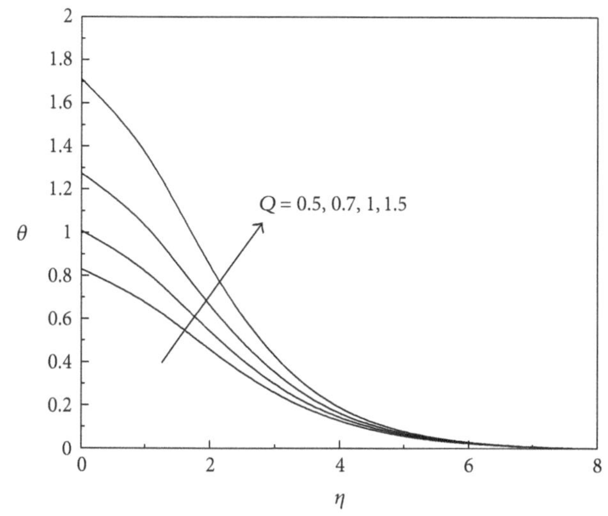

FIGURE 13: Effects of internal heat generation parameter on temperature profile.

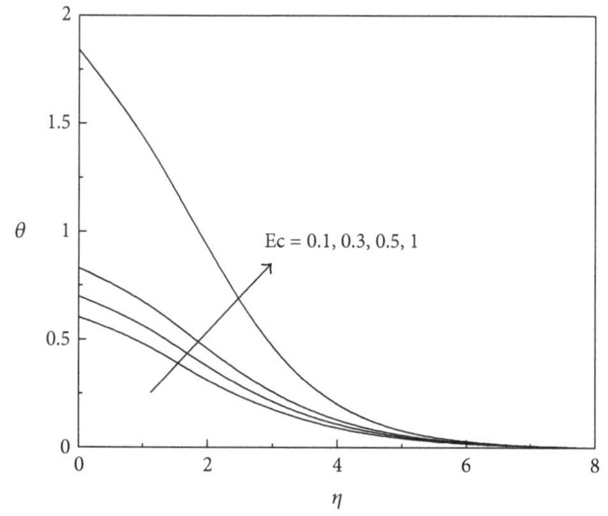

FIGURE 14: Effects of Eckert number on temperature profile.

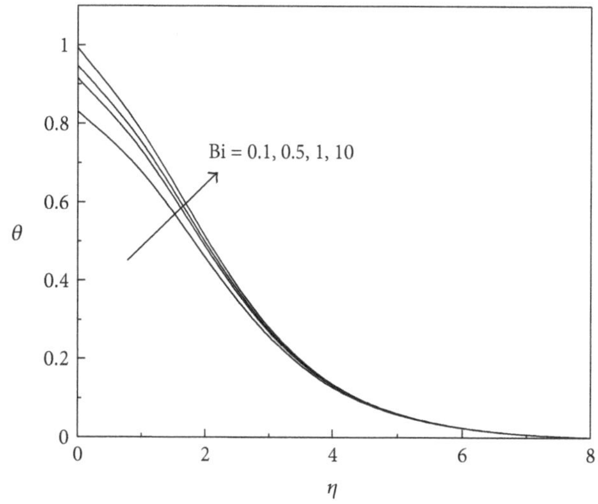

FIGURE 15: Effects of local Biot number on temperature profile.

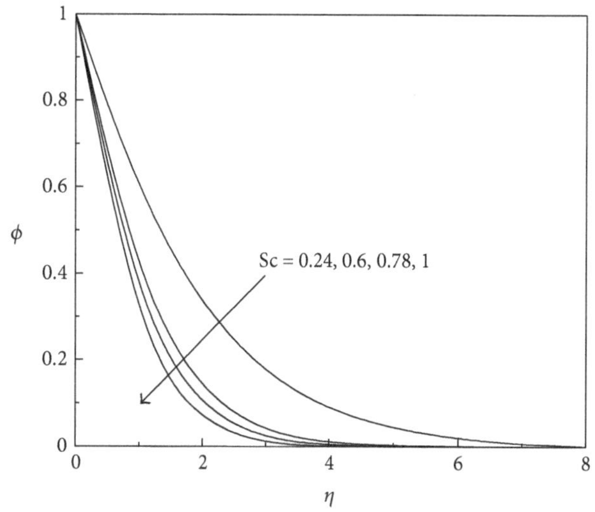

FIGURE 16: Effects of Schmidt number on concentration profile.

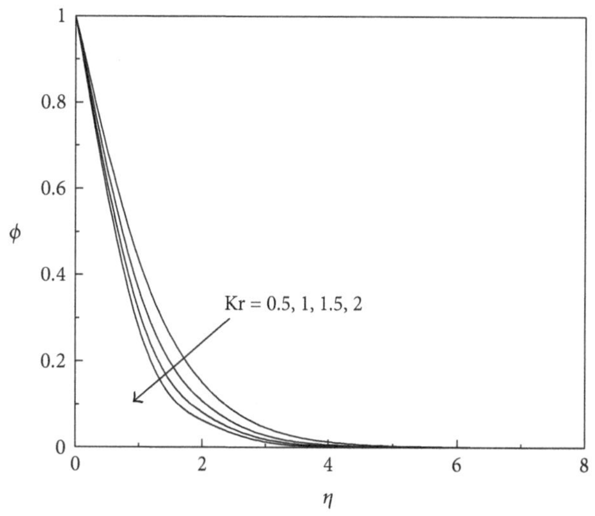

FIGURE 17: : Effects of chemical reaction parameter on concentration profile.

A set of non-linear coupled differential equations governing the fluid velocity, temperature, and concentration is solved numerically for various material parameters. A comprehensive set of graphical results for the velocity, temperature, and concentration is presented and discussed. Our results reveal, among others, that the internal heat generation, thermal radiation, and the Eckert number prevent the flow of heat from the left surface to the right surface of the plate unless the local Grashof number is strong enough to convert away both the internally generated heat in the fluid. Generally, the fluid velocity increases gradually away from the plate, attains its peak value within the boundary layer, and decreases to the free stream zero value satisfying the boundary conditions. It is interesting to note that the fluid velocity within the boundary layer increases with increasing values of the exponentially decaying internal heat generation, thermal radiation, and the Eckert number little away from the wall plate and attains its peak before obeying the boundary conditions. The velocity and concentration both decrease with an in increase in the Schmidt number and the chemical reaction parameter.

References

[1] E. L. Cussler, *Diffusion Mass Transfer in Fluid Systems*, Cambridge University Press, London, UK, 1998.

[2] U. N. Das, R. Deka, and V. M. Soundalgekar, "Effects of mass transfer on flow past an impulsively started infinite vertical plate with constant heat flux and chemical reaction," *Forschung im Ingenieurwesen/Engineering Research*, vol. 60, no. 10, pp. 284–287, 1994.

[3] S. P. Anjalidevi and R. Kandasamy, "Effects of chemical reaction, heat and mass transfer on laminar flow along a semi infinite horizontal plate," *Heat and Mass Transfer*, vol. 35, no. 6, pp. 465–467, 1999.

[4] M. A. Seddeek, A. A. Darwish, and M. S. Abdelmeguid, "Effects of chemical reaction and variable viscosity on hydromagnetic mixed convection heat and mass transfer for Hiemenz flow through porous media with radiation," *Communications in Nonlinear Science and Numerical Simulation*, vol. 12, no. 2, pp. 195–213, 2007.

[5] A. M. Salem and M. Abd El-Aziz, "Effect of Hall currents and chemical reaction on hydromagnetic flow of a stretching vertical surface with internal heat generation/absorption," *Applied Mathematical Modelling*, vol. 32, no. 7, pp. 1236–1254, 2008.

[6] R. A. Mohamed, "Double-diffusive convection-radiation interaction on unsteady MHD flow over a vertical moving porous plate with heat generation and Soret effects," *Applied Mathematical Sciences*, vol. 3, no. 13-16, pp. 629–651, 2009.

[7] F. S. Ibrahim, A. M. Elaiw, and A. A. Bakr, "Effect of the chemical reaction and radiation absorption on the unsteady MHD free convection flow past a semi infinite vertical permeable moving plate with heat source and suction," *Communications in Nonlinear Science and Numerical Simulation*, vol. 13, no. 6, pp. 1056–1066, 2008.

[8] J. C. Crepeau and R. Clarksean, "Similarity solutions of natural convection with internal heat generation," *Journal of Heat Transfer*, vol. 119, no. 1, pp. 183–185, 1997.

[9] P. M. Patil and P. S. Kulkarni, "Effects of chemical reaction on free convective flow of a polar fluid through a porous medium in the presence of internal heat generation," *International Journal of Thermal Sciences*, vol. 47, no. 8, pp. 1043–1054, 2008.

[10] A. Mahdy, "Effect of chemical reaction and heat generation or absorption on double-diffusive convection from a vertical truncated cone in porous media with variable viscosity," *International Communications in Heat and Mass Transfer*, vol. 37, no. 5, pp. 548–554, 2010.

[11] A. Ishak, "Similarity solutions for flow and heat transfer over a permeable surface with convective boundary condition," *Applied Mathematics and Computation*, vol. 217, no. 2, pp. 837–842, 2010.

[12] A. Aziz, "A similarity solution for laminar thermal boundary layer over a flat plate with a convective surface boundary condition," *Communications in Nonlinear Science and Numerical Simulation*, vol. 14, no. 4, pp. 1064–1068, 2009.

[13] O. D. Makinde and P. O. Olanrewaju, "Buoyancy effects on thermal boundary layer over a vertical plate with a convective surface boundary condition," *Journal of Fluids Engineering*, vol. 132, no. 4, Article ID 044502, 4 pages, 2010.

[14] O. D. Makinde, "Similarity solution for natural convection from a moving vertical plate with internal heat generation and a convective boundary condition," *Thermal Science*, vol. 15, pp. S137–S143, 2011.

[15] P. O. Olanrewaju, J. A. Gbadeyan, T. Hayat, and A. A. Hendi, "Effects of internal heat generation, thermal radiation and buoyancy force on a boundary layer over a vertical plate with a convective surface boundary condition," *South African Journal of Science*, vol. 107, no. 9-10, article 476, 6 pages, 2011.

[16] O. D. Makinde and P. O. Olanrewaju, "Combined effects of internal heat generation and buoyancy force on boundary layer flow over a vertical plate with a convective surface boundary condition," *The Canadian Journal of Chemical Engineering*, vol. 90, no. 5, pp. 1289–1294, 2011.

[17] R. M. Sonth, S. K. Khan, M. S. Abel, and K. V. Prasad, "Heat and mass transfer in a visco-elastic fluid flow over an accelerating surface with heat source/sink and viscous dissipation," *Heat and Mass Transfer*, vol. 38, no. 3, pp. 213–220, 2002.

[18] C.-H. Chen, "Combined heat and mass transfer in MHD free convection from a vertical surface with Ohmic heating and viscous dissipation," *International Journal of Engineering Science*, vol. 42, no. 7, pp. 699–713, 2004.

[19] E. M. Abo-Eldahab and M. A. El Aziz, "Viscous dissipation and Joule heating effects on MHD-free convection from a vertical plate with power-law variation in surface temperature in the presence of Hall and ion-slip currents," *Applied Mathematical Modelling*, vol. 29, no. 6, pp. 579–595, 2005.

[20] D. Pal and H. Mondal, "Effects of Soret Dufour, chemical reaction and thermal radiation on MHD non-Darcy unsteady mixed convective heat and mass transfer over a stretching sheet," *Communications in Nonlinear Science and Numerical Simulation*, vol. 16, no. 4, pp. 1942–1958, 2011.

[21] I. J. Uwanta, "Effects of chemical reaction and radiation on heat and mass transfer past a semi-infinite vertical porous plate with constant mass flux and dissipation," *European Journal of Scientific Research*, vol. 87, no. 2, pp. 190–200, 2012.

[22] P. O. Olanrewaju, J. A. Gbadeyan, T. Hayat, and A. A. Hendi, "Effects of internal heat generation, thermal radiation and buoyancy force on a boundary layer over a vertical plate with a convective surface boundary condition," *South African Journal of Science*, vol. 107, no. 9-10, pp. 1–6, 2011.

[23] M. A. Hossain, M. A. Alim, and D. A. S. Rees, "The effect of radiation on free convection from a porous vertical plate," *International Journal of Heat and Mass Transfer*, vol. 42, no. 1, pp. 181–191, 1998.

[24] A. Raptis, "Flow of a micropolar fluid past a continuously moving plate by the presence of radiation," *International Journal of Heat and Mass Transfer*, vol. 41, no. 18, pp. 2865–2866, 1998.

[25] A. J. Chamkha, "Hydromagnetic natural convection from an isothermal inclined surface adjacent to a thermally stratified porous medium," *International Journal of Engineering Science*, vol. 35, no. 10-11, pp. 975–986, 1997.

27

Wavelength Width Dependence of Cavity Temperature Distribution in Semiconductor Diode Laser

S. P. Abbasi and A. Alimorady

Iranian National Center for Laser Science and Technology, No. 71-20th North Kargar, P.O. Box 33665-576, Tehran, Iran

Correspondence should be addressed to A. Alimorady; abolalimo@gmail.com

Academic Editors: J. K. Brennan and D. E. Khoshtariya

The study of heat distribution in laser diode shows that there is nonuniform temperature distribution in cavity length of laser diode. In this paper, we investigate the temperature difference in laser diode cavity length and its effect on laser bar output wavelength width that mounted on usual CS model. In this survey at the first, laser was simulated then the simulations result was compared with experimental test result. The result shows that for each emitter there is difference, about 2.5 degree between the beginning and end of cavity.

1. Introduction

High-power infrared diode laser arrays are effective sources for pumping solid-state lasers [1–3]. These laser arrays are composed of one or more laser bars; each laser bar consists of numerous individual laser emitters formed on a single piece of semiconductor [4, 5]. Various characteristics like quantum efficiency, output power, and their dependence on cavity length and composition have been discussed. There are a number of factors that limit the output power and reliability of diode lasers, for example, catastrophic optical damage and overheating. Cavity length increase was used for increasing output power [4].

In this paper at first four laser diode heat sources were considered and this distribution in the cavity was studied and was simulated. Then the temperature distribution was simulated in the single emitter in the laser diode bar that packaged on the CS mount model and then measured the temperature difference in laser diode points in cavity length. Temperature distribution effect on the wavelength width and the wavelength peak shift and other hand simulation results were compared with experimental results.

2. Laser Diode Structure and Simulation

Our simulation is based on GaAs, 20 W, CW modes, 808 μm CS laser diode. The laser diode has 19 emitters with 100 μm width stripe and 20% fill factor that was produced in INLC (Figure 1). The laser operation specification is listed in Table 1.

The mounting of laser bar on the package and the heat removing direction was shown in Figure 2 and layer structure specification is listed in Table 2 and its arrangement was shown in Figure 3. The bar dimensions, thickness, bar width (cavity length), bar length, are 117, 1000, and 9800 μm, respectively.

Laser diode thermal structure simulated in COMSOL 3.5 Multiphysics software. Geometric symmetry in laser bar can help for simplifying the geometry and then single emitter was simulated. The 150313 triangular meshes were used in this simulation for laser diode, Indium paste, and Cu heat sink body (Figure 4).

The laser was simulated in the temperature condition 27°C, current operation 25 A, and optical output power 20 W. The heat value that must be removed from laser bar equals 24.5 W.

3. Thermal Model

Temperature 3D profiles are found in the laser structure using the thermal conduction equation:

$$\begin{aligned} \operatorname{div}\{k[T(x,y,z),x,y,z]\cdot \operatorname{grad}[T(x,y,z)]\} \\ = -g_T[T(x,y,z),x,y,z], \end{aligned} \quad (1)$$

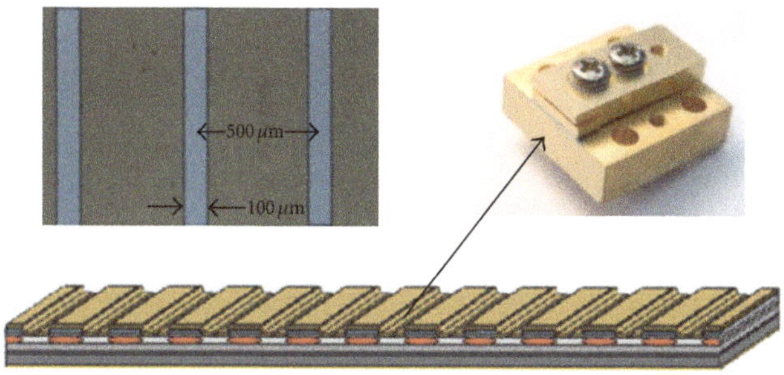

Figure 1: Schematic of laser bar, bar stripe geometry dimension, and CS model packaging laser diode.

Table 1: Laser operation Characteristics.

	Parameters	Value	Unit
1	Output power	20	W
2	Threshold current	0.25	A
3	Operation current	25	A
4	Operation voltage	1.8	V
5	Serial resistance	0.007	Ω
6	Wave length peak	808 ± 3	nm

where the temperature- and position-dependent thermal conductivity k, g_T stand for the 3D distribution of heat generation (in Wm^{-3}). Heat sink is the copper ($5 \times 5\,mm^2$) radiator that is properly taken into account assuming its much larger dimensions than those of the laser chip, so its external walls are assumed to remain at room temperature of the ambient. Top and sidewalls of the laser crystal are assumed to be thermally isolated because of negligible effect of thermal radiation and thermal diffusion of air particles [1].

In this simulation four heat sources were considered:

(1) nonradiative recombination,

(2) reabsorption of radiation,

(3) Joule heating,

(4) mirror absorption.

The produced heat of mirrors absorption is very smaller than the other heat sources but its effect was observed in the results.

The reflectivity of back mirror is 96–98% and for front mirror 7–10% was considered. Nonradiative recombination is proportional to the $1-\eta_i$ that η_i is internal quantum efficiency and relates the waveguide material and doping level (carriers) [4–7]. Reabsorption of radiation occurs in resonator and this process is spatially homogeneous. Diode lasers Joule heating distribution depends on spreading of injection current [4].

Current spreading and the nonuniformity effect of the injection have been studied and simulated in COMSOL 3.5 Multiphysics software in steady state analysis. The electrical model is composed of the Laplace equation:

$$\operatorname{div}\{\sigma(x,y,z)\operatorname{grad}[V(x,y,z)]\} = 0. \quad (2)$$

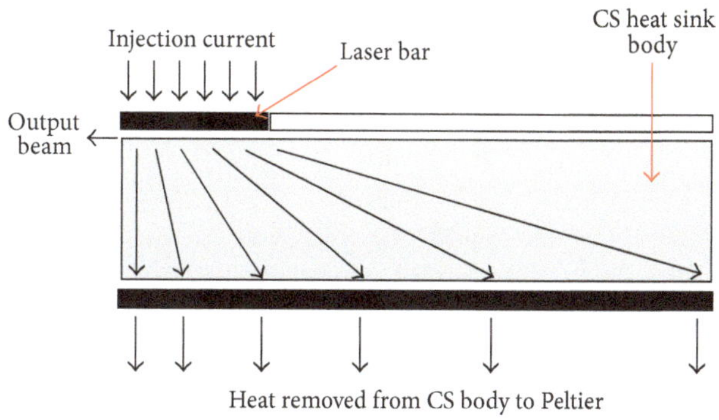

Figure 2: Bar packaging position on CS heat sink and heat removed path from the bar to TEC.

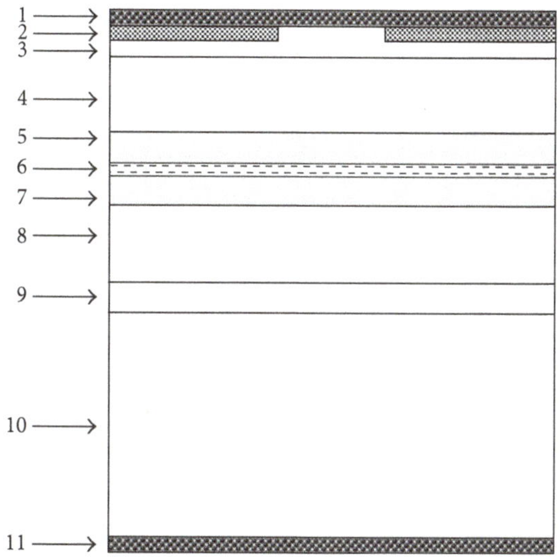

Figure 3: Laser bar structure and epitaxial layer arrangement.

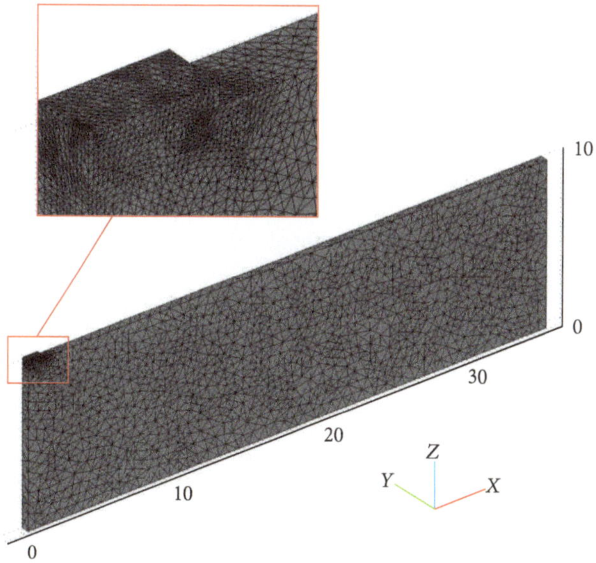

Figure 4: Laser diode and heat sink meshing in COMSOL software.

TABLE 2: Laser bar structure layers specification.

Number	Layers	Function	Thickness (A)	Doping (cm^{-3})
1	P contact Au/Ti/Pt	Contact	3000	—
2	SiO_2	Dielectric	1300	—
3	p-GaAs	Contacting	2500	$(3–15) \times 10^{18}$
4	p-$Al_{0.15}Ga_{0.35}In_{0.5}P$	Cladding	12000	2×10^{18}
5	$Ga_{0.17}In_{0.83}P$	SCH	4500	5×10^{17}
6	$GaAs_{0.81}P_{0.19}$	QW	160	
7	$Ga_{0.17}In_{0.83}P$	SCH	4500	5×10^{17}
8	n-$Al_{0.15}Ga_{0.35}In_{0.5}P$	Cladding	12000	2×10^{18}
9	n-GaInP	Buffer	4000	2×10^{18}
10	n-GaAs	Substrate	110000	2.2×10^{18}
11	N contact Au/Ge/Ni	Contact	4000	—

TABLE 3: Thermal conductivity, electrical resistivity, and electron mobility of material (300 K).

Metals and oxide materials	Thermal conductivity (W/mK)	Electrical resistivity (Ωcm)
Cu	398	1.7×10^{-6}
In	87	8.4×10^{-6}
SiO_2	1.38	10^{18}
Au	318	2.27×10^{-6}
Pt	73	108×10^{-6}
Ti	22	39×10^{-6}
Ni	92	7.2×10^{-6}
GeAu	150	6.5×10^{-3}
Semiconductor materials	Thermal conductivity (W/mK)	Mobility ($cm^2 V^{-1} s^{-1}$)
GaAs	44	8500
$Al_{0.15}Ga_{0.35}In_{0.5}P$	114	2700
$Ga_{0.17}In_{0.83}P$	73	5500
$GaAs_{0.81}P_{0.19}$	61	7850

And the diffusion equation within the active region

$$D(T)\frac{d^2 n_A(y)}{dy^2} - \left[An_A(y) + B(n_A, T)\, n_A^2(y) + Cn_A^3(y)\right] + \frac{j_{pn}(y)}{ed_A} = 0. \quad (3)$$

In the above equations, σ stands for the 3D electrical conductivity profile, V is the 3D potential distribution, D is the temperature-dependent ambipolar diffusion constant, n_A is the active-region carrier-concentration distribution, A, B, and C are the monomolecular, the bimolecular (mostly radiative), and the Auger, respectively, recombination coefficients, $j_{pn}(y)$ stands for the p-n junction current-density distribution, e is the electron charge, and d_A is the cumulative active-region thickness. The Laplace equation, instead of the Poisson one, is used because noncompensated electric charges are confined only to the active-region area, which is treated separately [4]. Current-density profiles are calculated from the potential distribution using the Ohm's law:

$$j(x, y, z) = \sigma(x, y, z)\, \text{grad}\left[V(x, y, z)\right]. \quad (4)$$

The room temperature electrical resistivity of material and layer that was used in simulation was listed in Table 3 [1, 8, 9].

Results show that increasing the current density cannot change the Joule heating distribution in the laser diode and the main part of Joule heating is related to stripe position and only less than 6% in outside of stripe part. Figure 5 shows the current spread in laser diode in a different current.

For simulating the Joule heating, COMSOL 3.5 Multiphysic software was used in steady state analysis in the electrothermal interaction.

For ternary GaInP and GaAsP compounds and the quaternary AlGaInP at the values of the room temperature, thermal conductivities are found in [4, 8] and their relative temperature dependencies are giving finally in ($Wm^{-1}K^{-1}$):

$$k_{In_xGa_{1-x}P} = \frac{100}{1.30 + 20.07x - 19.9x^2},$$
$$k_{GaAs_xP_{1-x}} = \frac{100}{1.30 + 22.57x - 21.6x^2}, \quad (5)$$
$$k_{Al_xGa_yIn_{1-x-y}P} = 147 - 57x - 70y.$$

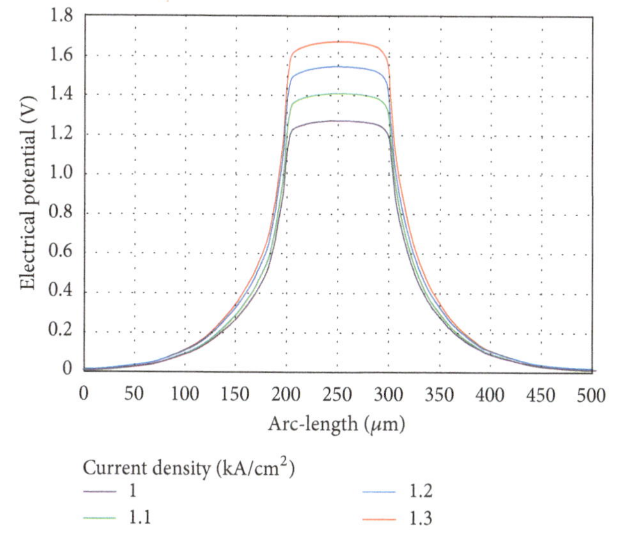

FIGURE 5: The current spread in laser diode in different currents. Stripe width and chip width are 100 μm and 500 μm, respectively.

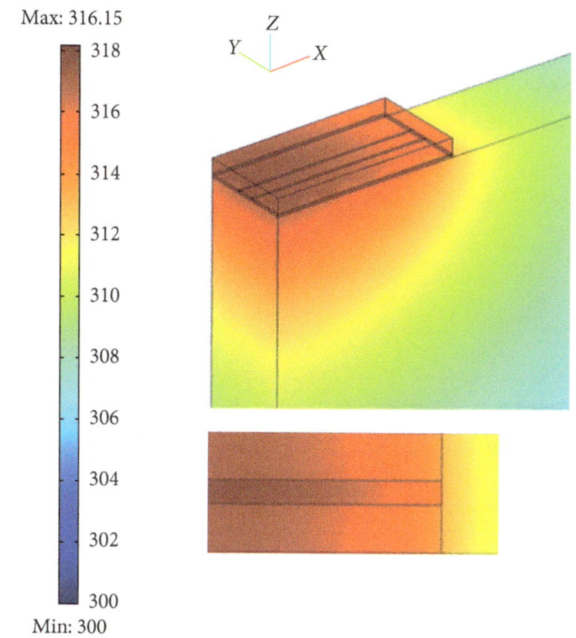

FIGURE 6: Temperature profile.

The values of the room temperature thermal conductivities relative temperature dependencies are giving finally in ($Wm^{-1}K^{-1}$):

$$k_{GaAs} = 44\left(\frac{T}{300}\right)^{-1.25},$$
$$k_{GaP} = 77\left(\frac{T}{300}\right)^{-1.36}, \quad (6)$$
$$k_{InP} = \frac{100}{[1.47 + (t - 300)/111]}.$$

To the best of the authors' knowledge, there are no published measurements of thermal conductivities relative temperature dependencies in Quaternary AlGaInP compounds. Effective thermal conductivity of a two-layer contact is calculated using the relation

$$k_{AB} = \frac{k_A k_B (d_A + d_B)}{d_A k_B + d_B k_A}, \quad (7)$$

where k_A, k_B and d_A, d_B are the thermal conductivities and the thicknesses, respectively, of both A and B layers. For a three-layer contact, this approach should be repeated [2]. The values of thermal conductivities of contact materials that were used were shown in Table 3.

Thermal conductivity of material at room temperature used in simulation was listed in Table 3. This conductivity calculated from the related equations [1].

4. Simulation Results

Figure 6 shows the temperature profile of emitter and heat sink. Figure 6 shows the top view temperature profile of the chip and the temperature difference of regions in the cavity length. There is a temperature difference between 2 regions along the cavity near the front and back mirrors. The reason of this difference is for nonsymmetric position on the heat sink in the straight line of cavity (Figure 2).

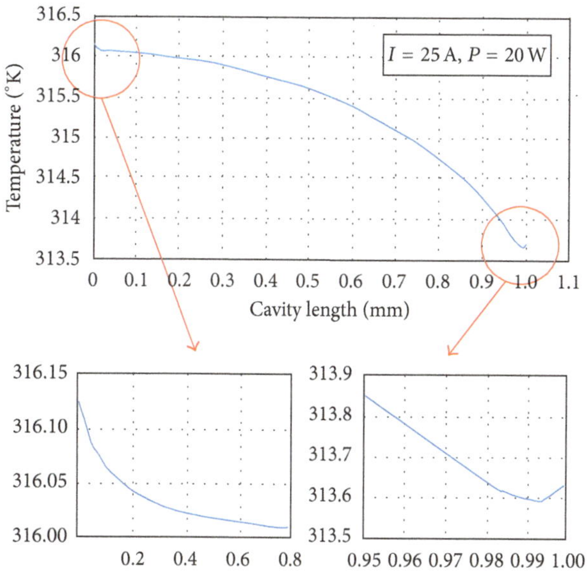

FIGURE 7: The temperature difference in cavity length and mirror effect.

The temperature difference measuring in the cavity length was shown in Figure 7.

There are nonlinear differences near mirrors because of mirror absorption and on other hand the mirror material Al_2O_3 thermal conduction that is less than cavity material GaAs thermal conduction.

The temperature difference in cavity length in different operation currents was shown in Figure 8. The result shows

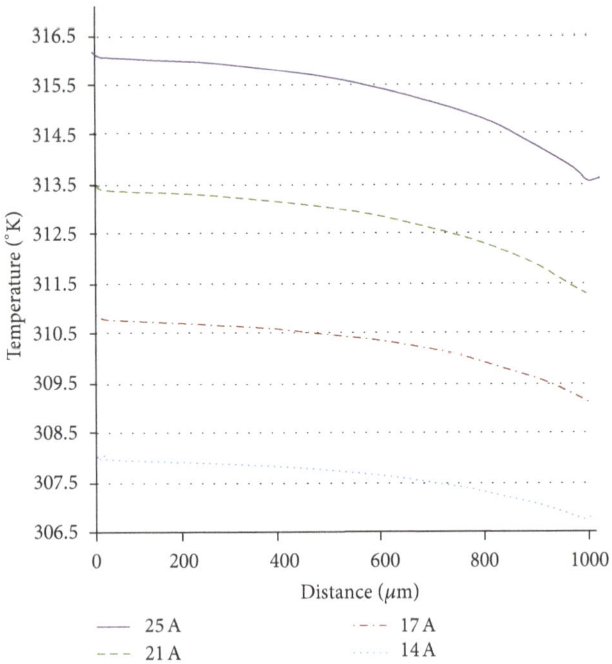

FIGURE 8: The temperature difference in cavity length in different operation currents.

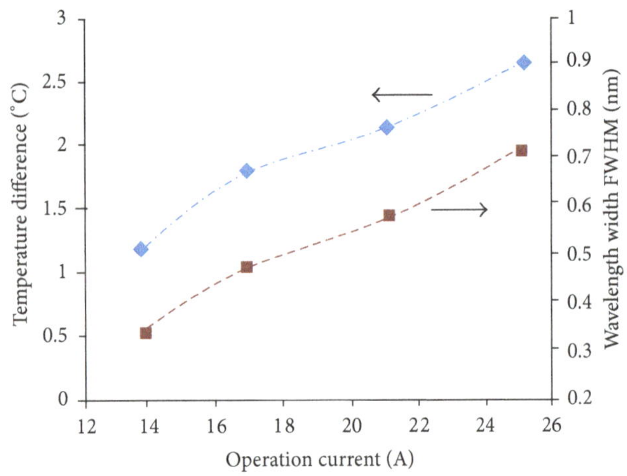

FIGURE 9: The wavelength width increasing in different operation currents affected by temperature difference in cavity length.

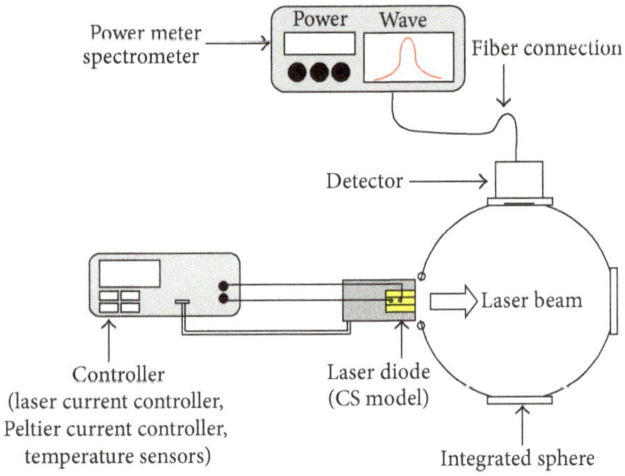

FIGURE 10: Experimental arrangement.

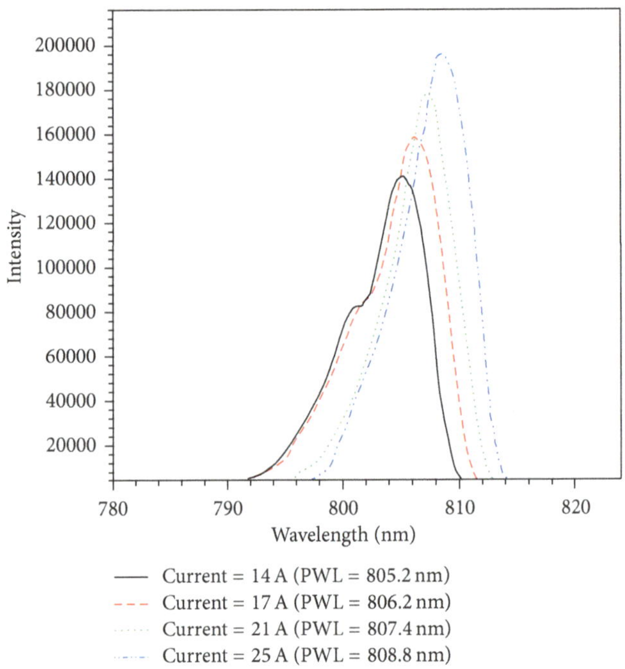

FIGURE 11: The experimental spectra in different operation currents.

the linear increase in this difference with increase of the operation current (Figure 8).

Simulation results for temperature difference in the cavity and the wavelength width variation for this temperature difference was shown in Figure 9. The wavelength shift value in single the cavity in simulation is 0.28 μm/°C that has agreement with experimental results which show that this value is 0.26 μm/°C.

5. Experimental Results

The experiment was arranged according to Figure 10 and the peak wavelength shift and wavelength width were measured in laser diode in different operation currents. The suitable accurate drivers and sensors control the laser current for stable power and TEC (Peltier) current for heat removing.

The spectral result was shown in Figure 11. There are differences in spectral wavelength width that was shown in Figure 12. The peak wavelength shift value is 0.26 μm/°C.

6. Result and Conclusion

Laser diode peak wavelength was shifted by temperature increase. Our survey about the heat distribution in laser diode shows that there is nonuniform temperature distribution in cavity length of laser diode. This temperature difference increases the spectral wavelength width.

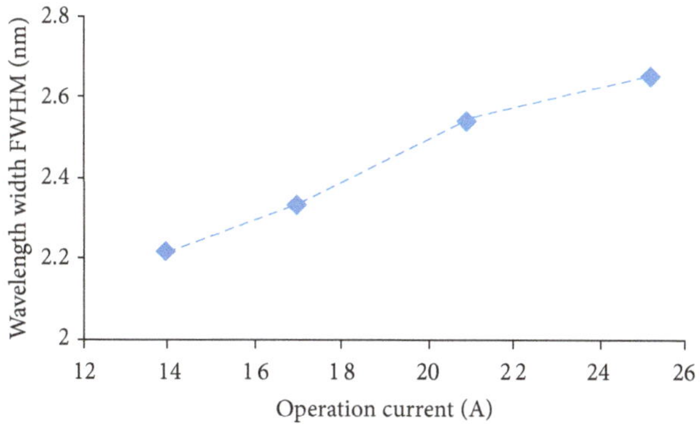

FIGURE 12: The experimental wavelength width increasing in different operation currents.

In this investigation the laser diode CS model was simulated. The result shows that there is 2.5°C difference along cavity length. This difference was increased by increasing operation current. For example, when the operation current was increased from 14 A to 25 A the temperature difference along cavity length was increased from 1°C to 2.7°C; this process increases the wavelength width from 2.2 μm to 2.7 μm. This result was confirmed with experimental results.

References

[1] B. Laikhtman, A. Gourevitch, D. Donetsky, D. Westerfeld, and G. Belenky, "Current spread and overheating of high power laser bars," *Journal of Applied Physics*, vol. 95, no. 8, pp. 3880–3889, 2004.

[2] A. Tomczyk, R. P. Sarzała, T. Czyszanowski, M. Wasiak, and W. Nakwaski, "Fully self-consistent three-dimensional model of edge-emitting nitride diode lasers," *Opto-Electronics Review*, vol. 11, no. 1, pp. 65–75, 2003.

[3] A. Gourevitch, B. Laikhtman, D. Westerfeld et al., "Transient thermal analysis of InGaAsP-InP high-power diode laser arrays with different fill factors," *Journal of Applied Physics*, vol. 97, no. 8, Article ID 084503, 6 pages, 2005.

[4] B. Mroziewicz, M. Bugajski, and W. Nakwaski, *Physics of Semiconductor Lasers*, North-Holland Science Publishers, 1991.

[5] F. Bachmann, P. Loosen, and R. Poprawe, *High Power Diode Lasers Technology and Applications*, Springer, New York, NY, USA, 2007.

[6] D. S. Patil, *Semiconductor Laser Diode Technology and Applications*, InTech, Rijeka, Croatia, 2012.

[7] S. Adachi, *Properties of Semiconductor Alloys: Group-IV, III–V and II–VI Semiconductors*, John Wiley & Sons, New York, NY, USA, 2009.

[8] A. Yariv and P. Yeh, *Photonics Optical Electronics in Modern Communications*, Oxford University Press, New York, NY, USA, 2007.

[9] D. R. Lide, *Handbook of Chemistry and Physics*, CRC Press, New York, NY, USA, 2004.

28

Influence of Variable Permeability and Radiation Absorption on Heat and Mass Transfer in MHD Micropolar Flow over a Vertical Moving Porous Plate

D. Harish Babu[1] and P. V. Satya Narayana[2]

[1] *Department of Humanities and Basic Science, SVPCET, Puttur 517 583, India*
[2] *Department of Mathematics, VIT University, Vellore 632 014, India*

Correspondence should be addressed to P. V. Satya Narayana; pvsatya8@yahoo.co.in

Academic Editors: C. D. Daub, H. Hirao, D. E. Khoshtariya, and Z. Xu

An analysis of unsteady free convection with heat and mass transfer flow for a micropolar fluid through a porous medium with a variable permeability bounded by a semi-infinite vertical plate in the presence of heat generation, thermal radiation, first-order chemical reaction, and the radiation absorption is reported. The plate is assumed to move with a constant velocity in the direction of fluid flow. A uniform magnetic field acts perpendicular to the porous surface which absorbs micropolar fluid with a suction velocity varying with time. The dimensionless governing equations for this investigation are solved analytically using two-term harmonic and nonharmonic functions. To observe physical insight and interesting aspects of the problem, the velocity, angular velocity, temperature, and concentration field are numerically studied and displayed graphically for pertinent parameters.

1. Introduction

The study of heat and mass transfer for an electrically conducting micropolar fluid past a porous plate under the influence of a magnetic field has attracted the interest of many investigators in view of its applications in many engineering problems such as magnetohydrodynamic (MHD) generators, plasma studies, nuclear reactors, oil exploration, geothermal energy extractions, and the boundary layer control in the field of aerodynamics. The theory of microfluids that was developed by Eringen [1], which includes the effect of local rotary inertia, the couple stresses, and inertial spin, satisfactorily provides a model for the non-Newtonian behavior observed in polymers, paints, lubricants, suspended fluids and blood, and so forth. He [2] extended the theory of thermomicropolar fluids and derived the constitutive laws for fluids with microstructure. An excellent review of micropolar fluids and their applications was given by Ariman et al. [3]. Gorla [4] discussed the steady-state heat transfer in a micropolar fluid flow over a semi-infinite plate, and the analysis is based on similarity variables. Kim [5] studied unsteady free convection flow of micropolar fluids through a porous medium bounded by an infinite vertical plate. Charya [6] analyzed the flow of micropolar fluid through a constricted channel. Rees and Pop [7] discussed the free convection boundary layer flow of a micropolar fluid from a vertical flat plate, While Sharma and Gupta [8] studied the effects of medium permeability on thermal convection in micropolar fluids. Kumar et al. [9] have studied the problem of fully developed free convective flow of micropolar and viscous fluids in a vertical channel. Muthu et al. [10] studied peristaltic motion of micropolar fluid in circular cylindrical tubes. Srinivasacharya et al. [11] analyzed the unsteady stokes flow of micropolar fluid between two parallel porous plates. Recently, Muthuraj and Srinivas [12] investigated fully developed MHD flow of micropolar and viscous fluids in a vertical porous space using HAM.

The combined heat and mass transfer problems with chemical reactions are of importance in many processes and, therefore, have received a considerable amount of attention in recent years. In processes, such as drying, evaporation at the surface of a water body, energy transfer in a wet cooling tower, and the flow in a desert cooler, the heat and mass transfer occurs simultaneously. Das et al. [13] considered the effects of a first-order chemical reaction on the flow past an impulsively started infinite vertical plate with constant heat flux and mass transfer. The effects of the chemical reaction and mass transfer on MHD unsteady free convection flow past an infinite/semi-infinite vertical plate were analyzed by [14–17] with constant suction and heat sink. Reddy et al. [18] studied peristaltic transport of a conducting fluid in an inclined asymmetric channel. Recently, Pal and Talukdar [19] studied the unsteady magnetohydrodynamic convective heat and mass transfer in a boundary layer slip flow past a vertical permeable plate with thermal radiation and chemical reaction.

Convection problems associated with heat sources within fluid-saturated porous media are of great practical significance, for that there are a number of practical applications in geophysics and energy-related problems, such as recovery of petroleum resources, geophysical flows, cooling of underground electric cables, storage of nuclear waste materials ground water pollution, fiber and granular insulations, solidification of costing, chemical catalytic reactors, and environmental impact of buried heat generating waste. Abo-Eldahab and El Aziz [20] investigated the problem of steady, laminar, free convection boundary layer flow of a micropolar fluid from a vertical stretching surface embedded in a nondaring porous medium in the presence of a uniform magnetic field, heat generation/absorption, and free stream velocity. Gorla et al. [21] have presented an analysis for the unsteady natural convection from a heated vertical surface placed in a micropolar fluid in the presence of internal heat generation or absorption. Raptis and Perdikis [22] analyzed the effect of a chemical reaction of an electrically conducting viscous fluid on the flow over a nonlinearly (quadratic) semi-infinite stretching sheet in the presence of a constant magnetic field which is normal to the sheet. Kesavaiah et al. [23] investigated the effects of the chemical reaction and radiation absorption on an unsteady MHD convective heat and mass transfer flow past a semi-infinite vertical permeable moving plate. Recently, Vajravelu et al. [24] studied the influence of heat transfer on peristaltic transport of a Jeffrey fluid in a vertical porous stratum.

The role of thermal radiation in the flow heat transfer process is of major importance in the design of many advanced energy conversion systems operating at higher temperatures. Thermal radiation within these systems is usually the result of emission by hot walls and the working fluid. Bakier and Gorla [25] studied thermal radiation effect on mixed convection from horizontal surfaces in porous media. Bakier [26] reported the effect of radiation on the mixed convection flow on an isothermal vertical surface in a saturated porous medium and has obtained self-similar solution. Hossain and Takhar [27] analyzed the effect of radiation on mixed convection along a vertical plate with uniform surface temperature.

In many investigations, while investigating either velocity and temperature distribution or stability characteristics of the flow usually the permeability and porosity are taken as constants. All the above-mentioned studies treat permeability and thermal conductivity of the medium as uniform. It has been observed that when porosity and permeability are taken as constants, the velocity profiles are flat with a thin boundary near the walls. However, the experimental investigations of Schwartz and Smith [28] and Schertz and Bischoff [29] show that the velocity profiles exhibit channeling of velocity near the walls. This hydrodynamic channeling is attributed to the variation of porosity of the medium confined between two boundaries. Further, the measurements of porosity distribution by Benenati and Brosilow [30] in randomly packed beds established a graphical relationship between spatial dependence of porosity and radial position of the container from which we find that the porosity has a maximum value at the wall and minimum value at the center of the bed. Thus, it can be concluded that the porosity and in turn permeability are not constant in a given medium. However, there is no exact mathematical relationship between the porosity variation and the distance from the wall. Chandrasekhara and Namboodiri [31] incorporate the variable permeability concept. They indicated considerable effect on the flow velocity and temperature distributions. Murthy and Feyen [32] studied the influence of variable permeability on the dispersion of a chemically reacting solute in porous media. From the technological point of view, flow arising from temperature and material difference is applied in chemical engineering, geothermal reservoirs, aeronautics, and astrophysics. In some applications, magnetic forces are present and at other times the flow is further complicated by the presence of radiation absorption, a good example of this is in the planetary atmosphere where there is radiation absorption from nearby stars. The influence of a magnetic field on the flow of an electrically conducting viscous fluid with mass transfer and radiation absorption is also useful in planetary atmosphere research [33]. The effect of chemical reaction and radiation absorption on the unsteady MHD free convection flow past a semi-infinite vertical permeable moving plate with heat source and suction was analyzed by [34, 35].

To the best of our knowledge, the problem of magnetohydrodynamics unsteady free convection flow of a micropolar fluid through a porous medium of variable permeability with radiation absorption over a vertical moving porous flat plate has remained unexplored. So the objective of present paper is to study the effects of a first-order homogeneous chemical reaction, variable permeability, and thermal radiation on the heat and mass transfer in MHD micropolar fluid flow over a vertical moving porous flat plate through a porous medium.

2. Formulation

We consider the two-dimensional unsteady flow of a laminar, incompressible micropolar fluid past a semi-infinite vertical porous moving plate embedded in a porous medium with variable permeability. The x^*-axis is measured along the porous plate in the upward direction, and a magnetic field of uniform strength B_0 is applied in the y^*-direction which is normal to the flow direction. Since the motion is two-dimensional and length of the plate is very large, so all the physical variables are independent of x^* and they are functions of normal distance y^* and t^* only. It is assumed that transversely applied magnetic field and magnetic Reynolds number are very small so that the induced magnetic field and the Hall effect are negligible. The motion of the plate as well as permeability of the porous medium and suction velocity increases exponentially with time. The wall is maintained at constant temperature T_w^* and concentration C_w^*, higher than the ambient temperature T_∞^* and concentration C_∞^*, respectively. The permeability of the porous medium is assumed to be of the form $K^* = K_0(1 + B\varepsilon e^{\delta^* t^*})$ and suction velocity is assumed to be $v^* = -V_0(1 + \varepsilon A e^{\delta^* t^*})$, where A, B are real positive constants and ε and εA are small values less than unity, and V_0 is the scale of suction velocity which is nonzero positive constant. The negative sign indicates that the suction is towards the plate. The fluid properties are assumed to be constants except that the influence of density variation with temperature and concentration has been considered in the body-force term. The concentration of diffusing species is very small in comparison to other chemical species; the concentration of species far from the wall C_∞^* is infinitesimally small and hence the Soret and Dufour effects are neglected. The chemical reactions are taking place in the flow and all thermophysical properties are assumed to be constant.

Under these conditions, the governing equations can be written in a Cartesian frame of reference as follows:

$$\frac{\partial v^*}{\partial y^*} = 0, \tag{1}$$

$$\begin{aligned}\frac{\partial u^*}{\partial t^*} + v^* \frac{\partial u^*}{\partial y^*} &= -\frac{1}{\rho}\frac{\partial P^*}{\partial x^*} \\ &\quad + (v + v_r)\frac{\partial^2 u^*}{\partial y^{*2}} + g\beta_f (T^* - T_\infty^*) \\ &\quad + g\beta_c (C^* - C_\infty^*) \\ &\quad - v\frac{u^*}{K_0 (1 + B\varepsilon e^{\delta^* t^*})} - \frac{\sigma}{\rho}B_0^2 u^* + 2v_r \frac{\partial \omega^*}{\partial y^*},\end{aligned} \tag{2}$$

$$\rho j^* \left(\frac{\partial \omega^*}{\partial t^*} + v^* \frac{\partial \omega^*}{\partial y^*}\right) = \gamma \frac{\partial^2 \omega^*}{\partial y^{*2}}, \tag{3}$$

$$\begin{aligned}\frac{\partial T^*}{\partial t^*} + v^* \frac{\partial T^*}{\partial y^*} &= \frac{K}{\rho c_p}\frac{\partial^2 T^*}{\partial y^{*2}} + \frac{Q_0}{\rho c_p}(T^* - T_\infty^*) \\ &\quad + Q_l^* (C^* - C_\infty^*) - \frac{1}{\rho c_p}\frac{\partial q_r^*}{\partial y^*},\end{aligned} \tag{4}$$

$$\frac{\partial c^*}{\partial t^*} + v^* \frac{\partial c^*}{\partial y^*} = D^* \frac{\partial^2 c^*}{\partial y^{*2}} - R^* (C^* - C_\infty^*). \tag{5}$$

The third and fourth terms on the RHS of the momentum equation (2) denote the thermal and concentration buoyancy effects, respectively. The second, third, and fourth terms on RHS of (3) denote the heat absorption, radiation absorption, and thermal radiation effects, respectively.

Under these assumptions, the appropriate boundary conditions for the velocity, angular velocity, temperature, and concentration fields are

$$\begin{gathered} u^* = u_p^*, \qquad T^* = T_w^* + \varepsilon (T_w^* - T_\infty^*) e^{\delta^* t^*}, \\ \omega^* = -n\frac{\partial u^*}{\partial y^*}, \\ C^* = C_w^* + \varepsilon (C_w^* - C_\infty^*) e^{\delta^* t^*} \quad \text{at } y \longrightarrow 0, \\ u^* \longrightarrow U_\infty^* = U_0 \left(1 + \varepsilon e^{\delta^* t^*}\right), \\ T^* \longrightarrow T_\infty^*, \quad C^* \longrightarrow C_\infty^*, \\ \omega^* \longrightarrow 0 \quad \text{at } y \longrightarrow \infty, \end{gathered} \tag{6}$$

where the variables and related quantities are defined in the nomenclature.

Outside the boundary layer, (2) gives

$$-\frac{1}{\rho}\frac{\partial P^*}{\partial x^*} = \frac{dU_\infty^*}{dt^*} + \frac{v}{K^*}U_\infty^* + \frac{\sigma}{\rho}B_0^2 U_\infty^*. \tag{7}$$

The radiative heat flux term by using the Rosseland approximations is given by

$$q_r^* = \frac{4\sigma^*}{3K_1^*}\frac{\partial T^{*4}}{\partial y^*}, \tag{8}$$

where $T^{*4} = 4T_\infty^{*3} - 3T_\infty^{*4}$.

In order to write the governing equations and the boundary conditions in dimensionless form, the following nondimensional quantities are introduced:

$$\begin{gathered} u^* = U_0 u, \qquad v^* = V_0 v, \qquad u_p^* = U_0 U_p, \\ y^* = \frac{vy}{V_0}, \qquad U_\infty^* = U_0 U_\infty, \\ \omega^* = \frac{\omega U_0 V_0}{V_0}, \qquad j^* = \frac{jv^2}{V_0^2}, \qquad t^* = \frac{tv}{V_0^2}, \\ T^* = T_\infty^* + \theta (T_w^* - T_\infty^*), \\ c^* = c_\infty^* + c(c_w^* - c_\infty^*), \qquad \delta^* = \frac{\delta v_0^*}{v}, \\ K^* = \frac{Kv_0^2}{V_0^2}, \qquad s_c = \frac{v}{D^*}, \qquad H = \frac{vQ_0}{\rho v c_p}, \\ \zeta = \frac{R^* v}{V_0^2}, \qquad Q_l = \frac{Q_l^* C (C_w^* - C_\infty^*)}{(T_w^* - T_\infty^*) V_0^2}, \end{gathered}$$

$M = \dfrac{v\sigma B_0^2}{\rho V_0^2}$ is the magnetic field parameter,

$\text{Gr} = \left(\dfrac{vgB_f}{U_0 V_0^2}\right)(T_w^* - T_\infty^*)$ is the thermal Grashof number,

$$\mathrm{Gc} = \frac{\upsilon g B_c}{U_0 V_0^2}\left(c_w^* - c_\infty^*\right) \text{ is the solutal Grashof number,}$$

$$\mathrm{Pr} = \frac{\upsilon \rho c_p}{K} = \frac{\upsilon}{\alpha} \text{ is the Prandtl number,}$$

$$R = \frac{4T_\infty^{*3}\sigma^*}{KK_1^*} \text{ is the radiation parameter,}$$

$$\alpha = \frac{K}{\rho c_p} \text{ is the effective fluid thermal diffusivity.} \tag{9}$$

Furthermore, the spin-gradient viscosity, which gives the same relationship between the coefficients of viscosity and microinertia, is defined as

$$\gamma = \left(\mu + \frac{\Gamma}{2}\right) j^* = \mu j^* \left(1 + \frac{\beta}{2}\right), \tag{10}$$

where $\beta = \Gamma/\mu$.

In view of (7)–(10), the governing equations (2)–(5) reduce to the following dimensionless form:

$$\frac{\partial u}{\partial t} - \left(1 + \varepsilon A e^{\delta t}\right)\frac{\partial u}{\partial y} = \frac{du_\infty}{dt} + (1+\beta)\frac{\partial^2 u}{\partial y^2} + \mathrm{Gr}\theta + \mathrm{Gc}C + \frac{(U_\infty - u)}{K(1 + B\varepsilon e^{\delta t})} + M(U_\infty - u) + 2\beta\frac{\partial \omega}{\partial y}, \tag{11}$$

$$\frac{\partial \omega}{\partial t} - \left(1 + \varepsilon A e^{\delta t}\right)\frac{\partial \omega}{\partial y} = \frac{1}{\eta}\frac{\partial^2 \omega}{\partial y^2}, \tag{12}$$

$$\frac{\partial \theta}{\partial t} - \left(1 + \varepsilon A e^{\delta t}\right)\frac{\partial \theta}{\partial y} = \frac{1}{\mathrm{Pr}}\left(1 + \frac{4R}{3}\right)\frac{\partial^2 \theta}{\partial y^2} + H\theta + Q_l c, \tag{13}$$

$$\frac{\partial c}{\partial t} - \left(1 + \varepsilon A e^{\delta t}\right)\frac{\partial c}{\partial y} = \frac{1}{\mathrm{Sc}}\frac{\partial^2 c}{\partial y^2} - \zeta c, \tag{14}$$

where $\eta = 2/2 + \beta = \mu j^*/\gamma$.

The boundary conditions of (6) are given by the following dimensionless form:

$$\begin{gathered} u = U_p, \quad \theta = 1 + \varepsilon e^{\delta t}, \quad \omega = -n\frac{\partial u}{\partial y}, \\ c = 1 + \varepsilon e^{\delta t} \quad \text{at } y \longrightarrow 0, \\ u = U_\infty \longrightarrow \left(1 + \varepsilon e^{\delta t}\right), \quad \theta \longrightarrow 0, \\ \omega \longrightarrow 0, \quad c \longrightarrow 0 \quad \text{at } y \longrightarrow \infty. \end{gathered} \tag{15}$$

3. Solution of the Problem

Equations (11)–(14) are coupled, nonlinear partial differential equations and these cannot be solved in closed form. However, these equations can be reduced to ordinary differential equations, which can be solved analytically. For these, we may represent the linear and angular velocities, temperature, and concentration as follows:

$$\begin{aligned} u &= u_0(y) + \varepsilon e^{\delta t} u_1(y) + o(\varepsilon^2), \\ \omega &= \omega_0(y) + \varepsilon e^{\delta t} \omega_1(y) + o(\varepsilon^2), \\ \theta &= \theta_0(y) + \varepsilon e^{\delta t} \theta_1(y) + o(\varepsilon^2), \\ c &= c_0(y) + \varepsilon e^{\delta t} c_1(y) + o(\varepsilon^2). \end{aligned} \tag{16}$$

By substituting the above equations (16) into (11)–(14), equating the harmonic and nonharmonic terms, and neglecting the higher-order terms of $O(\varepsilon^2)$, we obtain the following pairs of equations for $(u_0, \omega_0, \theta_0, C_0)$ and $(u_1, \omega_1, \theta_1, C_1)$:

$$\begin{aligned} &(1+\beta)u_0'' + u_0' - Nu_0 \\ &\quad = -N - \mathrm{Gr}\theta_0 - \mathrm{Gc}c_0 - 2\beta\omega_0', \\ &(1+\beta)u_1'' + u_1' - (N+\delta)u_1 \\ &\quad = -Lu_0 + L - \mathrm{Gr}\theta_1 - \mathrm{Gc}c_1 - 2\beta\omega_1' - Au_0', \\ &\omega_0'' + \eta\omega_0 = 0, \\ &\omega_1'' + \eta\omega_1' - \delta\eta\omega_1 = -A\eta\omega_0', \\ &(3+4R)\theta_0'' + 3\mathrm{Pr}\left(\theta_0' + H\theta_0\right) = -3\mathrm{Pr}Q_l c_0, \\ &(3+4R)\theta_1'' + 3\mathrm{Pr}\theta_1' - 3\mathrm{Pr}(\delta - H)\theta_1 \\ &\quad = -3A\mathrm{Pr}\theta_0' - 3\mathrm{Pr}Q_l c_1, \\ &c_0'' + \mathrm{Sc}c_0' - \zeta\mathrm{Sc}c_0 = 0, \\ &c_1'' + \mathrm{Sc}c_1' - (\delta+\zeta)\mathrm{Sc}c_1 = -A\mathrm{Sc}c_0', \end{aligned} \tag{17}$$

where $L = B/K$, $N = (M + 1/K)$ and the primes denote differentiation with respect to y.

The corresponding boundary conditions can be written as

$$\begin{gathered} u_0 = U_p, \quad u_1 = 0, \quad \omega_0 = -nu_0', \quad \omega_1 = -nu_1', \\ \theta_0 = 1, \quad \theta_1 = 1, \quad c_0 = 1, \quad c_1 = 1 \quad \text{at } y \longrightarrow 0, \\ u_0 = 1, \quad u_1 = 1, \quad \omega_0 = 0, \quad \omega_1 = 0, \\ \theta_0 = 0, \quad \theta_1 = 0, \quad c_0 = 0, \quad c_1 = 0 \quad \text{at } y \longrightarrow \infty. \end{gathered} \tag{18}$$

TABLE 1: Effect of Pr on C_f, Nu and Sh for $R = 0.1$, $n = 0.5$, $K = 2$, $H = 0.1$, $\zeta = 0.2$, $\beta = 0.5$, Sc = 0.2, Pr = 1, Gr = 2, $U_p = 0.5$, $Q_l = 0.01$, $B = 0.01$, and Gc = 2.

Pr	Kim [5] $H = 0, Q_l = 0, R = 0$, and $B = 0$			Mohamed and Abo-Dahab [36] $Q_l = 0$ and $B = 0$			Present Results		
	C_f	Nu	Sh	C_f	Nu	Sh	C_f	Nu	Sh
2	5.6985	20.6907	0.3276	5.6964	20.6897	0.3276	5.6933	20.6899	0.3276
3	5.6316	31.9168	0.3276	5.6304	33.9140	0.3276	5.6300	31.9155	0.3276
4	5.5972	43.1064	0.3276	5.5967	43.1001	0.3276	5.5943	43.1034	0.3276
5	5.5763	54.3059	0.3276	5.5752	54.3000	0.3276	5.5751	54.3038	0.3276

TABLE 2: Effect of M on C_f and Sh for $R = 0.1$, $n = 0.5$, $K = 2$, $H = 0.1$, $\zeta = 0.2$, $\beta = 0.5$, Sc = 0.2, Pr = 1, Gr = 2, $U_p = 0.5$, $Q_l = 0.01$, $B = 0.01$, and Gc = 2.

M	Kim [5] $H = 0, Q_l = 0, R = 0$, and $B = 0$		Mohamed and Abo-Dahab [36] $Q_l = 0$ and $B = 0$		Present results	
	C_f	Sh	C_f	Sh	C_f	Sh
1	6.4837	0.3276	6.4892	0.3276	6.4893	0.3276
3	5.6678	0.3276	5.6685	0.3276	5.6686	0.3276
4	5.5941	0.3276	5.5986	0.3276	5.5987	0.3276
5	5.5894	0.3276	5.5800	0.3276	5.5804	0.3276

Solving (17) under the boundary conditions (18), we obtain the linear and angular velocities, temperature, and concentration profiles as follows:

$$\begin{aligned} u(y,t) &= \left(1+a_1 e^{-R_1 y} + a_2 e^{-h_3 y} + a_6 e^{-h_1 y} + a_5 e^{-\eta y}\right) \\ &\quad + \varepsilon e^{\delta t}\left(1+b_1 e^{-R_3 y}+b_2 e^{-h_4 y}+d_1 e^{-h_3 y}+d_2 e^{-h_2 y}\right. \\ &\quad \left.+d_3 e^{-h_1 y}+ b_8 e^{-h_6 y}+b_{10} e^{-R_1 y}+d_4 e^{-\eta y}\right), \omega(y,t) \\ &= \left(c_{11} e^{-\eta y}\right) + \varepsilon e^{\delta t}\left(c_{13} e^{-h_6 y} - \frac{A\eta c_{11}}{\delta} e^{-\eta y}\right), \theta(y,t) \\ &= \left(\left(1-z_3\right) e^{-h_3 y} + z_3 e^{-h_1 y}\right) \\ &\quad + \varepsilon e^{\delta t}\left(c_9 e^{-h_4 y}+ z_4 e^{-h_3 y}+ z_6 e^{-h_2 y} + z_8 e^{-h_1 y}\right), c(y,t) \\ &= e^{-h_1 y} + \varepsilon e^{\delta t}\left(\left(1-z_3\right) e^{-h_2 y} + z_2 e^{-h_1 y}\right), \end{aligned} \tag{19}$$

where the expressions for the constants are given in the appendix.

The skin friction, Nusselt number, and Sherwood number are important physical parameters for this type of boundary layer.

3.1. Skin-Friction. Knowing the velocity field in the boundary layer, we can now calculate the skin-friction at the plate which is given by

$$\tau_w^* = \left(\mu + \Gamma\right) \frac{\partial u^*}{\partial y^*} + \Gamma \omega^* \bigg|_{y^*=0}, \tag{20}$$

and in the dimensionless form, we obtain

$$c_f = \frac{2\tau_w^*}{\rho V_0 U_0} = 2\left[1 + (1-n)\beta\right] u'(0). \tag{21}$$

The couple stress coefficient (C_m) at the plate is written as

$$M_w = \gamma \frac{\partial \omega^*}{\partial y^*}\bigg|_{y^*=0}, \tag{22}$$

and in the dimensionless from, we obtain

$$c_m = \frac{M_w}{\mu j U_0} = \left(1 + \frac{\beta}{2}\right) \omega'(0). \tag{23}$$

3.2. Nusselt Number. Knowing the temperature field, it is interesting to study the effect of the free convection and thermal radiation on the rate of heat transfer q_w^*, and this is given by

$$q_w^* = -k\left(\frac{\partial T^*}{\partial y^*}\right) - \frac{4\sigma^*}{3K_1^*}\left(\frac{\partial T^{*4}}{\partial y^*}\right)\bigg|_{y^*=0}. \tag{24}$$

In view of (8), the above equation becomes

$$q_w^* = \left(-k + \frac{16\sigma^* T^{*3}}{3K_1^*}\right) \frac{\partial T^*}{\partial y^*}\bigg|_{y^*=0}. \tag{25}$$

This is written in dimensionless from as

$$q_w^* = \left(\frac{-k\left(T_w^* - T_\infty^*\right) v_0}{v}\right)\left(1 + \frac{4R}{3}\right) \frac{\partial \theta}{\partial y}\bigg|_{y=0}. \tag{26}$$

The dimensionless local surface heat flux (i.e., Nusselt number) is obtained as

$$\mathrm{Nu}_x = \frac{x q_w^*}{k\left(T_w^* - T_\infty^*\right)}, \quad (27)$$

where

$$\mathrm{Nu}_x \mathrm{Re}_x^{-1} = -\left(1 + \frac{4R}{3}\right)\frac{\partial \theta}{\partial y}\bigg|_{y=0},$$
$$\mathrm{Re}_x = \frac{v_0 x}{v}. \quad (28)$$

3.3. Sherwood Number. The definition of the local mass flux and the local Sherwood number are, respectively, given by

$$j_w = -D^*\left(\frac{\partial c^*}{\partial y^*}\right)\bigg|_{y^*=0},$$
$$\mathrm{Sh}_x = \frac{x j_w}{D^*\left(c_w^* - c_w^*\right)} \quad (29)$$

and with the help of these equations, one can write

$$\mathrm{Sh}_x \mathrm{Re}_x^{-1} = -\left(\frac{\partial c}{\partial y}\right)\bigg|_{y=0}. \quad (30)$$

4. Results and Discussions

The formulation of the problem that accounts for the effects of chemical reaction and thermal radiation on MHD convective flow and mass transfer of an incompressible, micropolar fluid along a semi-infinite vertical porous moving plate in a porous medium with variable permeability in the presence of heat generation has been performed in the preceding sections. The governing equations of the flow field were solved analytically, using a perturbation method, and the expressions for the velocity, microrotation, temperature, concentration, skin-friction, Nusselt number, and Sherwood number were obtained. In order to get a physical insight of the problem, the above physical quantities are computed numerically and presented graphically for different values of the governing parameters: thermal Grashof number Gr, solutal Grashof number Gm, magnetic field parameter M, Schmidt number Sc, heat absorption parameter H, chemical reaction parameter ζ, and thermal radiation parameter R, while the values of some of the physical parameters are taken as constants such as $t = 1, \delta = 0.01, \varepsilon = 0.01, B = 0.1$ and $A = 0.1$, in all the tables and figures. In the present study, the boundary conditions for $y \rightarrow \infty$ are replaced by identical ones at $y_{\max}$ which is a sufficiently large value of y where the velocity profile u approaches the relevant free stream velocity. We choose $y_{\max} = 8$ and a step size $\Delta y = 0.001$.

In order to verify the accuracy of the present work, particular results are compared with those available in the literature through Table 1. The numerical values of skin-friction, Nusselt number, and Sherwood number obtained in the present case are compared with those of Kim [5] and Mohamed and Abo-Dahab [36] for different values of Prandtl number (Pr) and fixed values of remaining parameters. It is clearly seen that there is an excellent agreement between the respective research of our study and those of Kim [5] and Mohamed and Abo-Dahab [36]. Further, it is found from Table 2 that skin-friction decreases with the increase in magnetic parameter M. Also no effect of M is seen on Sherwood number Sh as noted in [36].

For different values of the magnetic field parameter M, the translational velocity and microrotation profiles are plotted in Figure 1. It is obvious that the effect of increasing values of the parameter M results in a decreasing velocity distribution across the boundary layer. The effect of magnetic field is more prominent at the point of peak value; that is, the peak value drastically decreases with the increases in the value of magnetic field because the presence of magnetic field in an electrically conducting fluid introduces a force called the Lorentz force, which acts against the flow if the magnetic field is applied in the normal direction, as in the present problem. This type of resisting force slows down the fluid velocity as shown in this figure. The results also show that the magnitude of the microrotation on the porous plate increases as M increases.

Figure 2 illustrates the variation of velocity and microrotation distribution across the boundary layer for several values of plate moving velocity U_p in the direction of the fluid flow, respectively. Although we have different initial plate moving velocities, the velocity decreases to the constant values for given material parameters. The results also show that the magnitude of microrotation on porous plate increases as U_p increases.

For various values of the permeability K, the profiles of translational velocity and the microrotation across the boundary layer are shown in Figure 3. It is clear that as K increase, the translational velocity profiles across the boundary layer increases, but the magnitude of microrotation decreases with an increase of K-parameter.

Figure 4 presents the velocity and microrotation profiles for different values of the Prandtl number (Pr). The results show that the effect of increasing values of Pr results in a decrease of the translational velocity, while the reverse effect is observed for the microrotation profiles.

For various values of the Schmidt number (Sc), the profiles of translational velocity and the microrotation across the boundary layer are shown in Figure 5. It is clear that as Sc increase, the translational velocity profiles across the boundary layer decreases. The results also show that the magnitude of microrotation on porous plate increases as Sc increases.

For various values of Q_l, the profiles of translational velocity and the microrotation across the boundary layer are shown in Figure 6. It is clear from the graph that the translational velocity increases with an increase of Q_l which is due to the fact that when heat is absorbed the buoyancy force which accelerated the flow rate. The results also show that the magnitude of microrotation on porous plate decreases as Q_l increases.

For various values of radiation parameter R, the translational velocity and microrotation profiles are plotted in

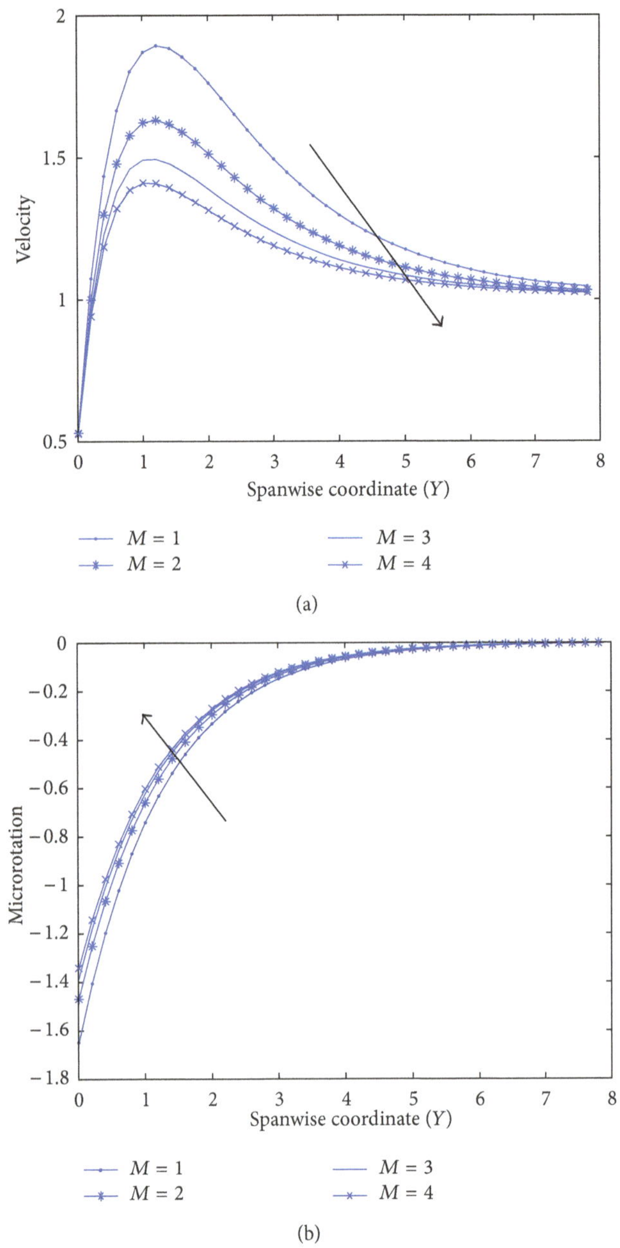

FIGURE 1: Velocity and microrotation distribution.

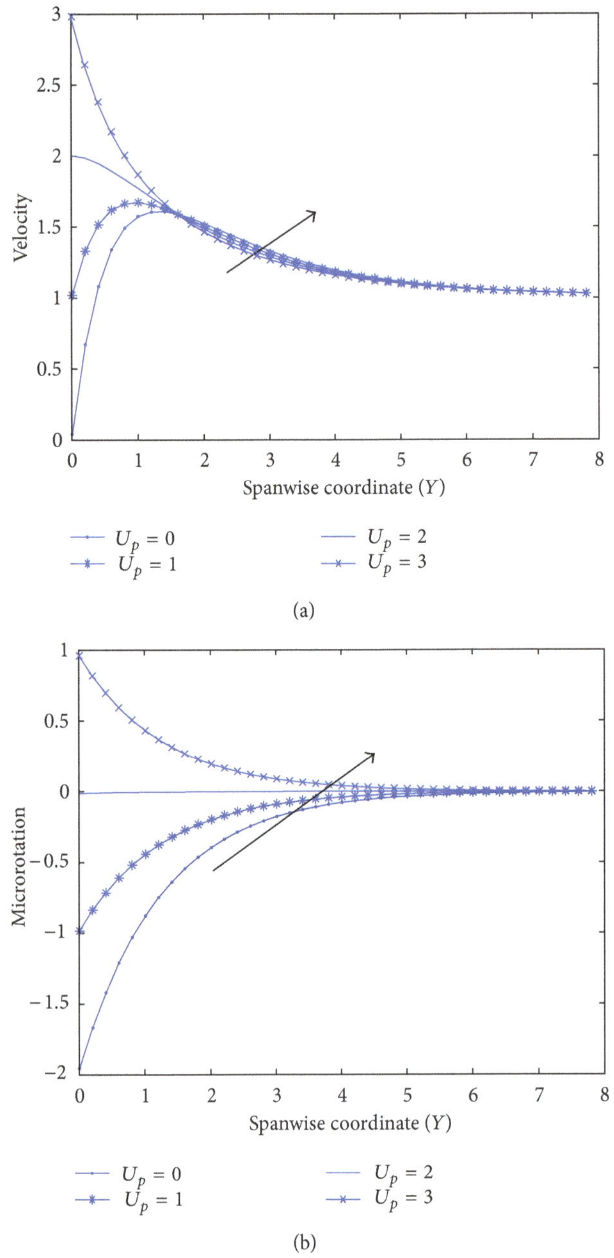

FIGURE 2: Velocity and microrotation distribution.

Figure 7. It is obvious that with increasing values of R-parameter, the velocity distribution across the boundary layer increases. The results also show that the magnitude of microrotation on the porous plate is decreased as R-parameter increases.

The effects of heat generation H on the translational velocity and microrotation profiles across the boundary layer are displayed in Figure 8. It is shown that the translational velocity across the boundary layer increases with an increasing of H. The results also show that the magnitude of microrotation decreases with an increasing of H.

Figure 9 shows the translational velocity and the microrotation profiles across the boundary layer for various values of the chemical reaction parameter ζ. The results display that with an increase of ζ, the translational velocity decreases, but the magnitude of microrotation increases as ζ decreases.

The effects of viscosity ratio β on the translational velocity and the microrotation profiles across the boundary layer are presented in Figure 10. It is clear that the velocity distribution is greater for a Newtonian fluid ($\beta = 0$) with the fixed flow and material parameters, as compared with micropolar fluids. The results display that with an increase of viscosity ratio β, the translational velocity decreases, but the magnitude of microrotation increases as viscosity ratio β increases. The translational velocity and the microrotation profiles against spanwise coordinate y for different values of thermal Grashof number (Gr) are displayed in Figure 11. It is observed that an

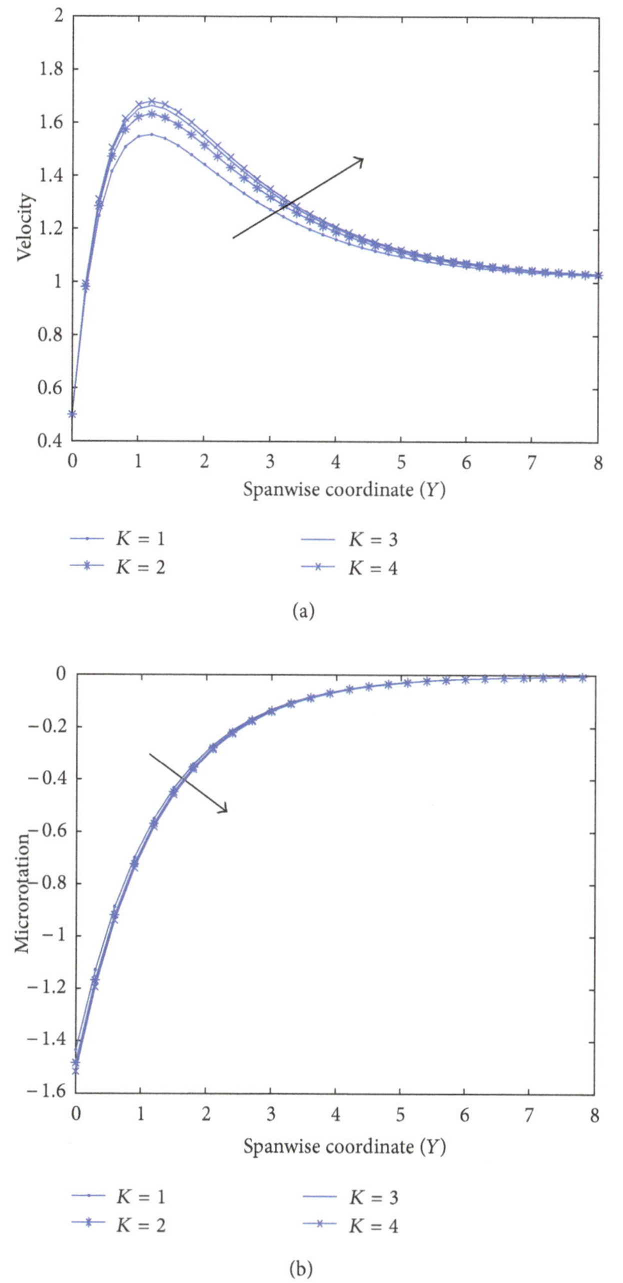

FIGURE 3: Velocity and microrotation distribution.

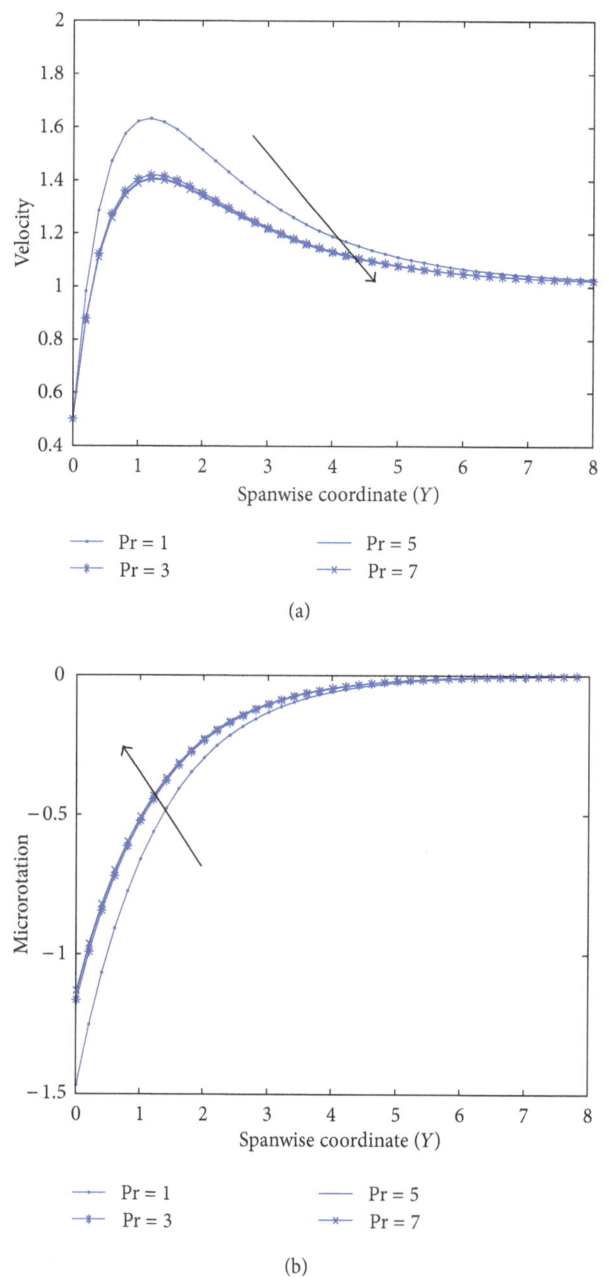

FIGURE 4: Velocity and microrotation distribution.

increase in Gr leads to an increase in the values of velocity and a decrease in the magnitude of microrotation. The translational velocity and the microrotation profiles against spanwise coordinate y for different values of solutal Grashof number (Gc) are displayed in Figure 12. It is observed that an increase in Gc leads to an increase in the values of velocity and a decrease in the magnitude of microrotation.

Figure 13 shows the effect of n-parameter, which is related to microrotation vector and shear stress, on the translational velocity and the microrotation profiles. It is observed that the magnitude of velocity and microrotation increases with an increase of n-parameter, but the magnitude of microrotation decreases with an increase of n-parameter.

Figures 14(a) and 14(b) represent the temperature profiles for different values of thermal radiation parameter R and prandtl number (Pr) in the boundary layer. These figures indicate that the effect of thermal radiation is to enhance heat transfer because of the fact that thermal boundary layer thickness increases with an increase in the thermal radiation. Thus, it is pointed out that the radiation should be minimized to have the cooling process at a faster rate. But the temperature decreases with increasing the values of Prandtl number (Pr) in the boundary layer. From this plot, it is evident that the temperature in the boundary layer falls

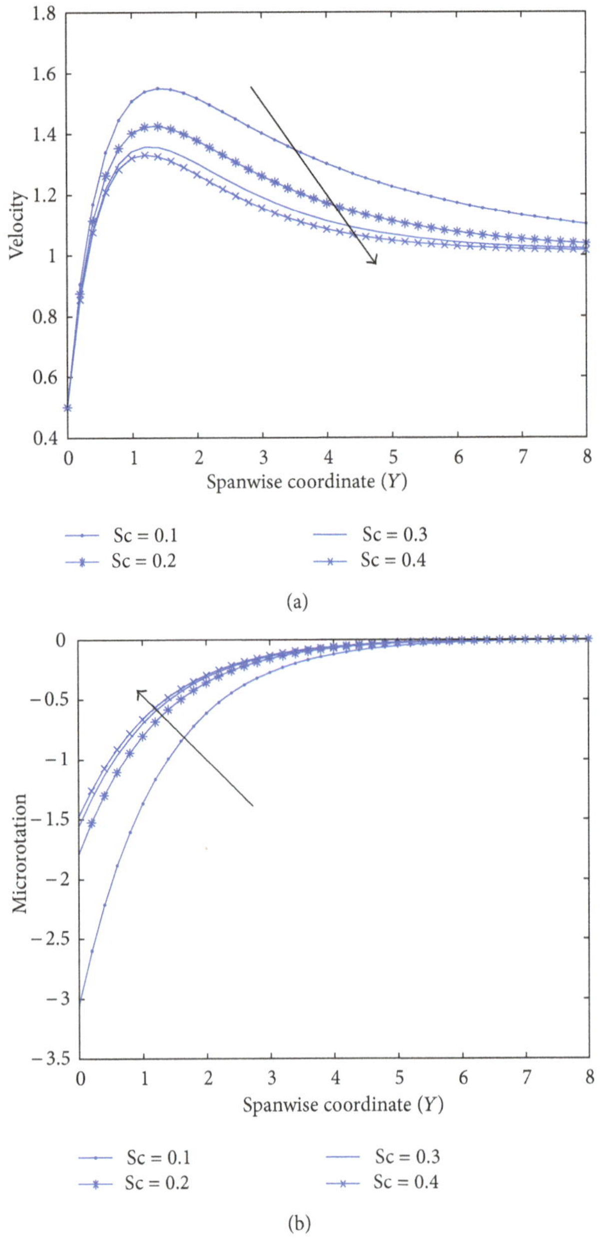

FIGURE 5: Velocity and microrotation distribution.

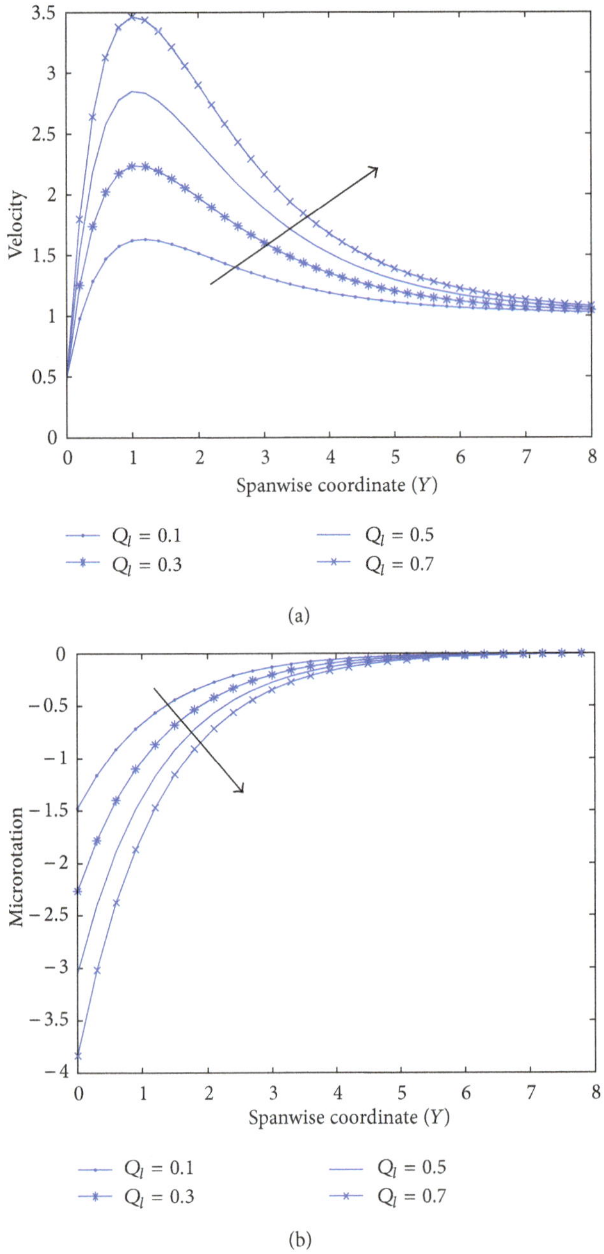

FIGURE 6: Velocity and microrotation distribution.

very quickly for large value of the Prandtl number because of the fact that thickness of the boundary layer decreases with an increase in the value of the Prandtl number.

Typical variations of the temperature profiles along the spanwise coordinate y are shown in Figure 14(c) for various values of Q_l. The results show that the temperature profiles decrease with increasing Q_l. The temperature profiles with respect to spanwise coordinate y for various values of ζ are shown in Figure 14(d). The results show that as the chemical reaction parameter ζ increase, the thermal boundary layer increases.

The effect of the reaction rate parameter ζ on the species concentration profiles for generative chemical reaction is shown in Figure 15(a). It is noticed from the graph that there is a marked effect with increasing the value of the chemical reaction rate parameter ζ on the concentration distribution in the boundary layer. It is clearly observed from this figure that the concentration of species value of 1.0 at vertical plate decreases till it attains the minimum value of zero at the end of the boundary layer, and this trend is seen for all the values of reaction rate parameter. Further, it is observed that increasing the value of the chemical reaction decreases the concentration of species in the boundary layer; this is due to the fact that boundary layer decreases with increase in the value of ζ chemical reaction in this system result in the consumption of the chemical and hence result in decreasing concentration profile. Similar facts are seen in the case when

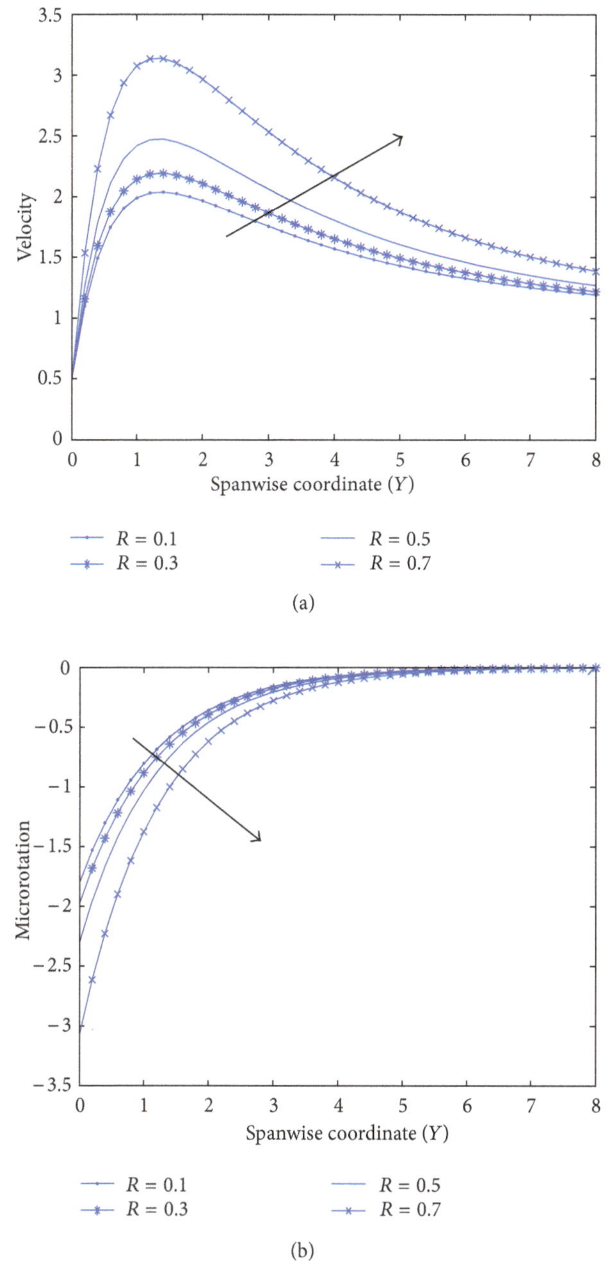

FIGURE 7: Velocity and microrotation distribution.

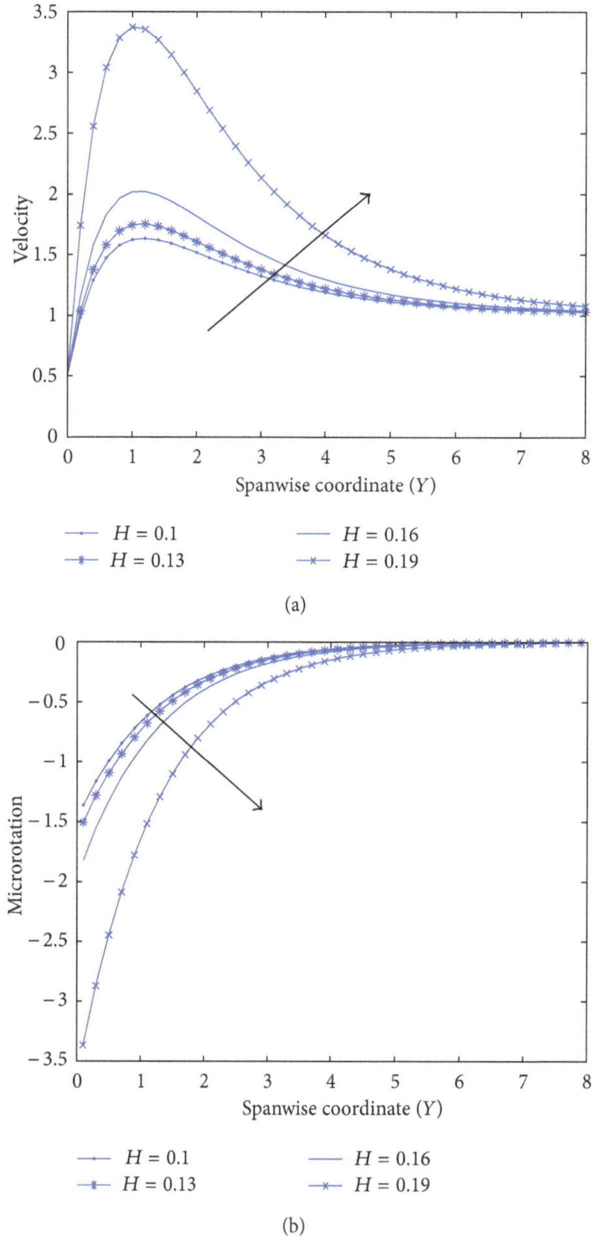

FIGURE 8: Velocity and Microrotation distribution.

Schmidt number is increased as noted in Figure 15(b). It may also be observed from this figure that the effect of Schmidt number (Sc) on concentration distribution decreases slowly for higher values of Sc. It is expected that the mass transfer rate increases as ζ increases with all other parameters fixed; that is, an increase in the Schmidt number (Sc) produces a decrease in the concentration boundary layer thickness, associated with the reduction in the concentration profiles. Physically, the increase in the value of Sc means the decrease of molecular diffusion D. Hence, the concentration of the species is higher for smaller values of Sc and lower for larger values of Sc.

5. Conclusions

The following conclusions are drawn from the present study.

(1) The effect of increasing magnetic parameter M is to retard the transient velocity of the flow field at all points, while a growing Grashof number (Gr) for heat transfer or heat absorption parameter accelerates the transient velocity of the flow field at all points.

(2) A growing magnetic parameter M or Prandtl number (Pr) decelerates the transient temperature of the flow field at all points, while a growing permeability parameter K or heat source parameter S reverses the effect.

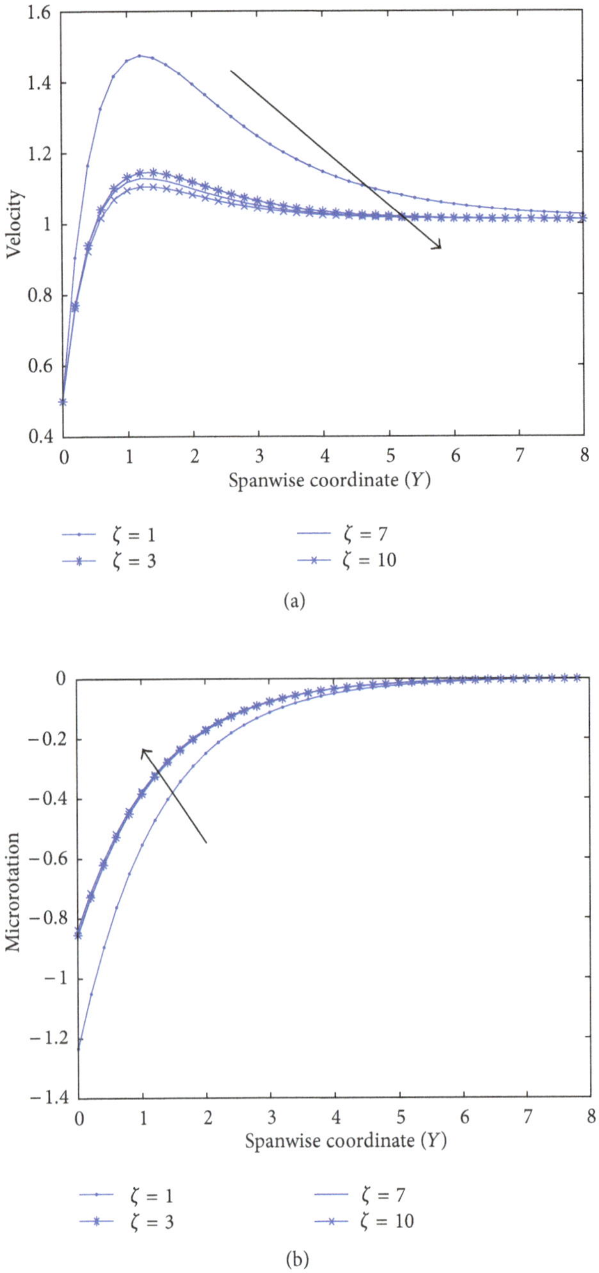

FIGURE 9: Velocity and microrotation distribution.

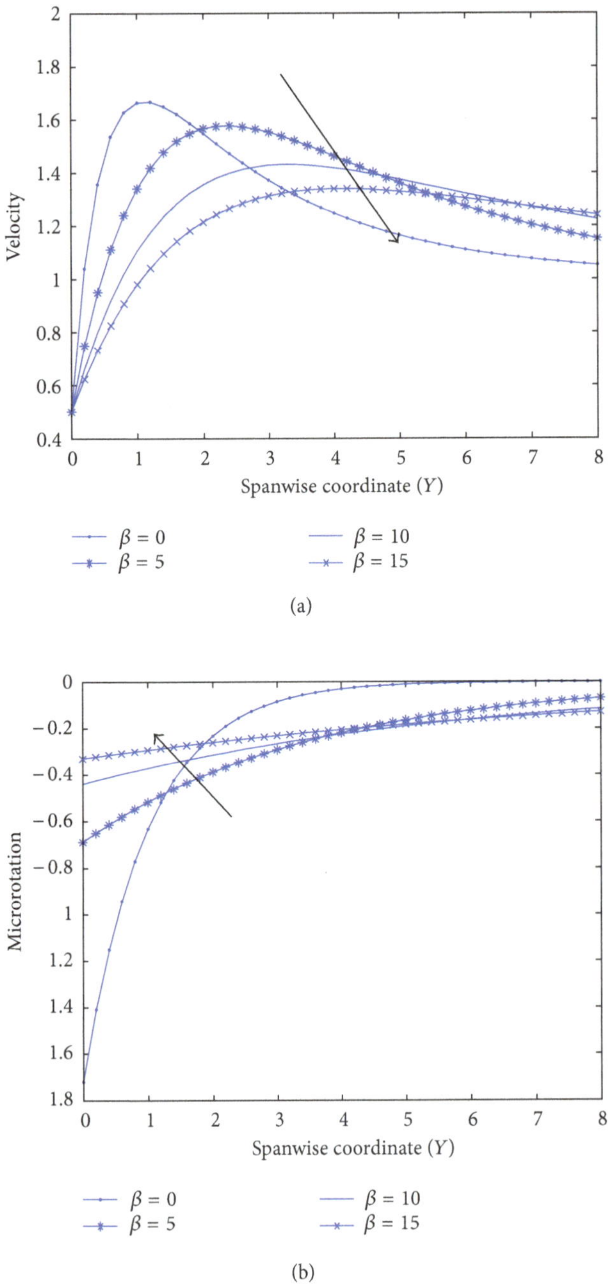

FIGURE 10: Velocity and microrotation distribution.

(3) The effect of increasing magnetic parameter M is to reduce the skin-friction at the wall, while a growing permeability parameter K or heat source parameter S reverses the effect.

(4) A growing Prandtl number (Pr) or magnetic parameter M increases the magnitude of the rate of heat transfer at the wall. On the other hand, a growing heat source S parameter reverses the effect.

(5) Skin-friction decreases when, magnetic field parameter is increased.

Appendix

Consider

$$h_1 = \frac{s_c}{2}\left[1 + \sqrt{1 + \frac{4\zeta}{s_c}}\right],$$

$$h_2 = \frac{s_c}{2}\left[1 + \sqrt{1 + \frac{4(\delta + \zeta)}{s_c}}\right],$$

$$h_3 = \frac{3\mathrm{Pr}}{2(3 + 4R)}\left[1 + \sqrt{1 - \frac{4H(3 + 4R)}{3\mathrm{Pr}}}\right],$$

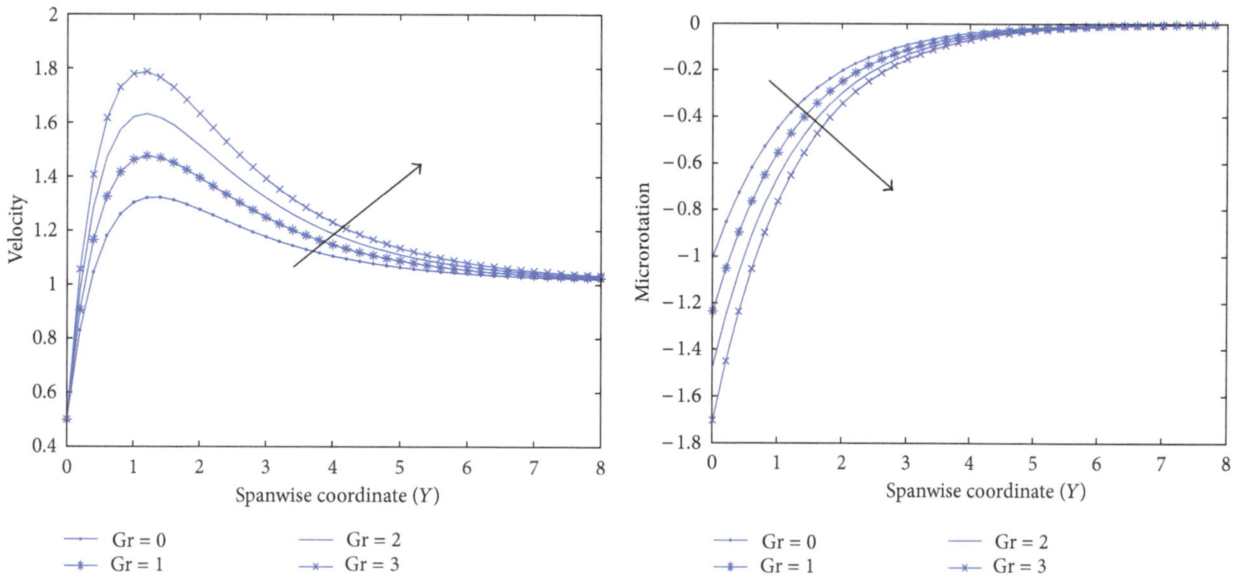

(a) (b)

FIGURE 11: Velocity and microrotation distribution.

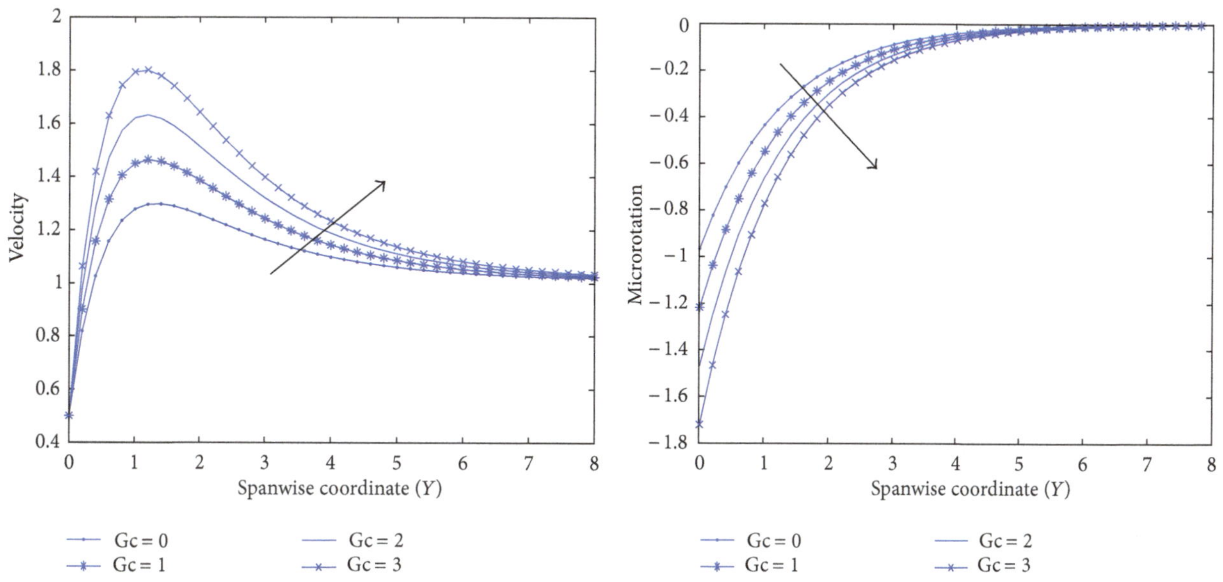

(a) (b)

FIGURE 12: Velocity and microrotation distribution.

$$h_4 = \frac{3\mathrm{Pr}}{2(3+4R)}\left[1+\sqrt{1+\frac{4(\delta-H)(3+4R)}{3\mathrm{Pr}}}\right],$$

$$h_6 = \frac{\eta}{2}\left[1+\sqrt{1+\frac{(4\delta)}{\eta}}\right],$$

$$R_1 = \frac{1}{2(1+\beta)}\left[1+\sqrt{1+(N+\delta)(1+\beta)}\right],$$

$$R_3 = \frac{1}{2(1+\beta)}\left[1+\sqrt{1+4(N+\delta)(1+\beta)}\right],$$

$$z_2 = \frac{As_c h_1}{h_1^2 - h_1 s_c - (\delta+\zeta)s_c},$$

$$z_3 = \frac{-3\mathrm{Pr}Q_l}{(3+4R)h_1^2 - 3\mathrm{Pr}h_1 + 3\mathrm{Pr}H},$$

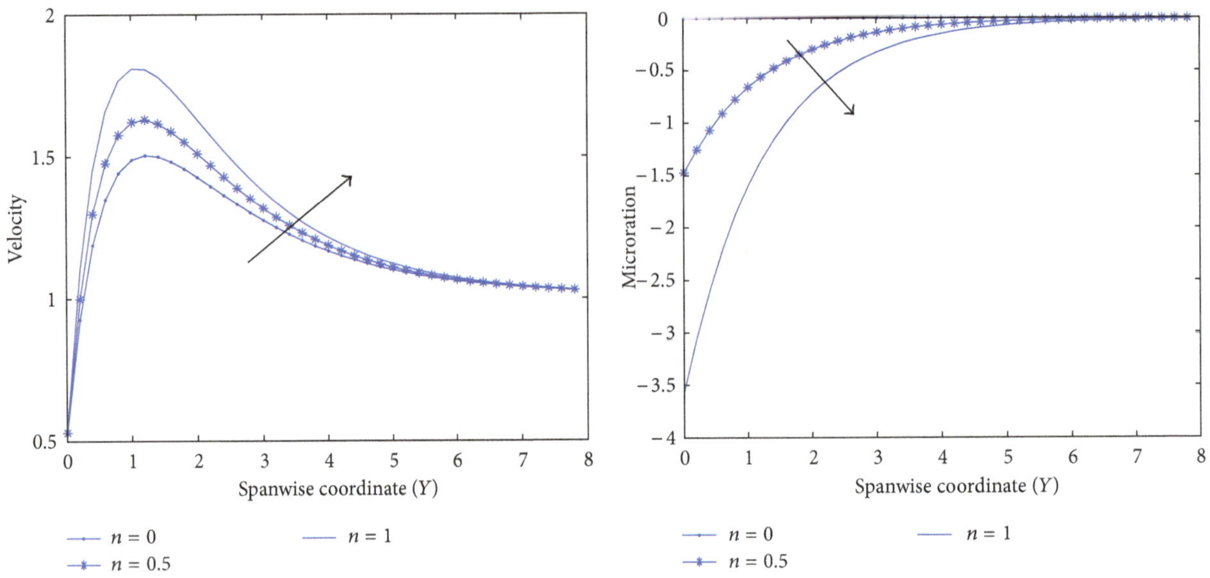

(a) (b)

FIGURE 13: Velocity and microrotation distribution.

$$z_4 = \frac{3\mathrm{Pr}h_3(1 - z_3)}{(3 + 4R)h_3^2 - 3\mathrm{Pr}h_3 - 3\mathrm{Pr}(\delta - H)},$$

$$z_5 = \frac{3\mathrm{Pr}h_1 z_3}{(3 + 4R)h_1^2 - 3\mathrm{Pr}h_1 - 3\mathrm{Pr}(\delta - H)},$$

$$z_6 = \frac{-3\mathrm{Pr}Q_l(1 - z_2)}{(3 + 4R)h_2^2 - 3\mathrm{Pr}h_2 - 3\mathrm{Pr}(\delta - H)},$$

$$z_7 = \frac{-3\mathrm{Pr}Q_l z_2}{(3 + 4R)h_1^2 - 3\mathrm{Pr}h_1 - 3\mathrm{Pr}(\delta - H)},$$

$$a_2 = \frac{-\mathrm{Gr}(1 - z_3)}{(1 + \beta)h_3^2 - h_3 - N},$$

$$a_3 = \frac{-\mathrm{Gr}z_3}{(1 + \beta)h_1^2 - h_1 - N},$$

$$a_4 = \frac{-\mathrm{Gc}}{(1 + \beta)h_1^2 - h_1 - N},$$

$$P = \frac{2\beta\eta}{(1 + \beta)\eta^2 - \eta - N},$$

$$a_6 = a_3 + a_4,$$

$$D = \frac{n}{1 + n(R_1 - \eta)P},$$

$$c_{11} = D\left[\left(U_p - 1\right)R_1 + a_2(h_3 - R_1) + a_6(h_1 - R_1)\right],$$

$$a_5 = P^* c_{11}; \qquad a_1 = U_p - 1 - a_2 - a_6 - a_5,$$

$$z_8 = z_5 + z_7,$$

$$c_9 = 1 - z_4 - z_6 - z_8,$$

$$b_2 = \frac{-\mathrm{Gr}c_9}{(1 + \beta)h_4^2 - h_4 - (N + \delta)},$$

$$b_3 = \frac{-\mathrm{Gr}z_4}{(1 + \beta)h_3^2 - h_3 - (N + \delta)},$$

$$b_4 = \frac{-\mathrm{Gr}z_6}{(1 + \beta)h_2^2 - h_2 - (N + \delta)},$$

$$b_5 = \frac{-\mathrm{Gr}z_8}{(1 + \beta)h_1^2 - h_1 - (N + \delta)},$$

$$b_6 = \frac{-\mathrm{Gc}(1 - z_2)}{(1 + \beta)h_2^2 - h_2 - (N + \delta)},$$

$$b_7 = \frac{-\mathrm{Gc}z_2}{(1 + \beta)h_1^2 - h_1 - (N + \delta)},$$

$$D_1 = \frac{-2\beta A\eta^2 c_{11}}{\delta},$$

$$b_9 = \frac{D_1}{(1 + \beta)\eta^2 - \eta - (N - \delta)},$$

$$b_{10} = \frac{AR_1 a_1}{(1 + \beta)R_1^2 - R_1 - (N + \delta)},$$

$$b_{11} = \frac{Ah_3 a_2}{(1 + \beta)h_3^2 - h_3 - (N + \delta)},$$

$$b_{12} = \frac{Ah_1 a_6}{(1 + \beta)h_1^2 - h_1 - (N + \delta)},$$

$$b_{13} = \frac{A\eta a_5}{(1 + \beta)\eta^2 - \eta - (N + \delta)},$$

$$d_1 = b_3 + b_{11},$$

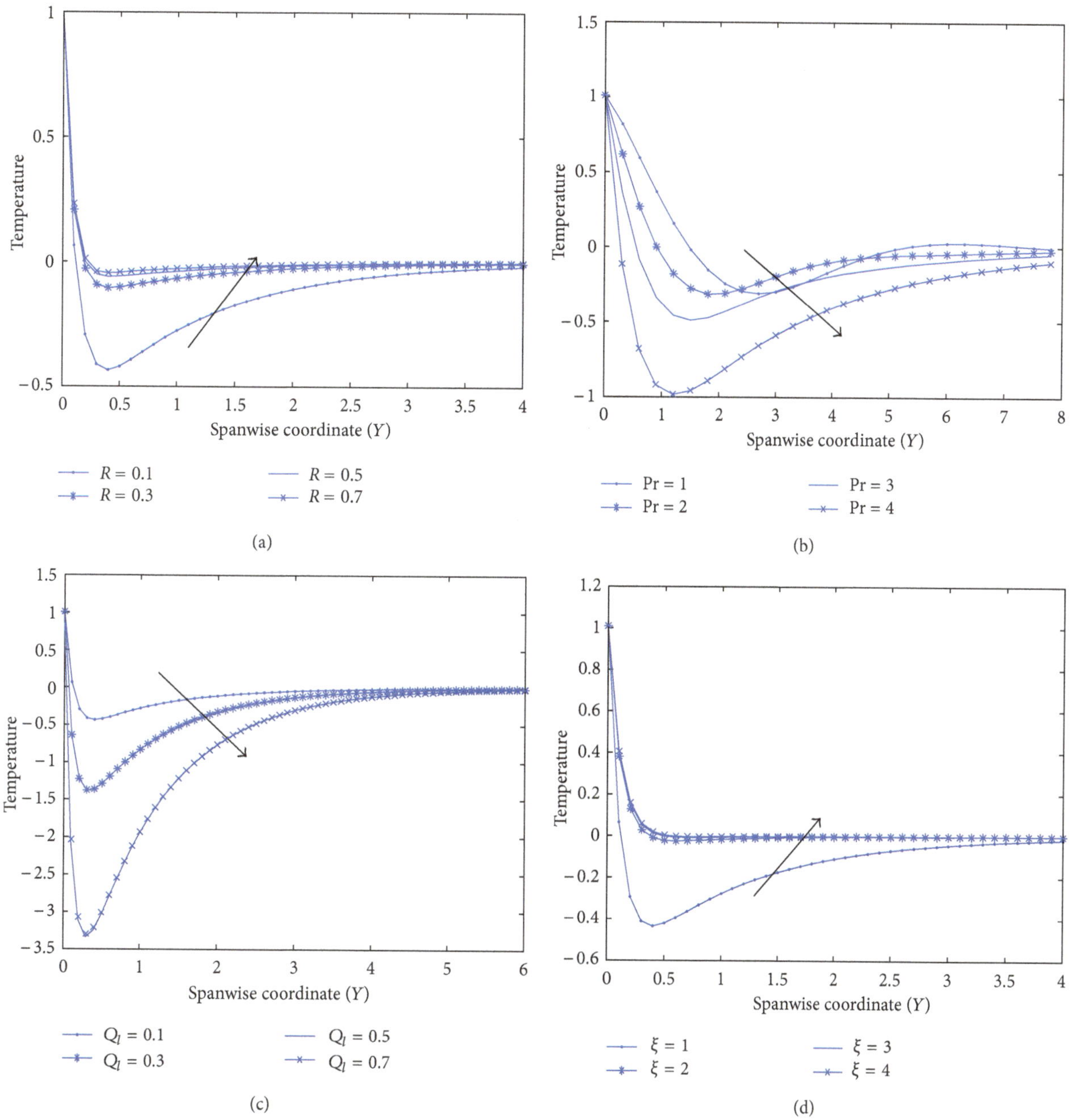

FIGURE 14: Temperature distribution.

$$d_2 = b_4 + b_6,$$

$$d_3 = b_5 + b_7 + b_{12},$$

$$d_4 = b_9 + b_{13},$$

$$Q = \frac{2\beta h_6}{(1+\beta) h_6^2 - h_6 - (N-\delta)},$$

$$k_1 = \frac{A\eta c_{11}}{\delta} + n\left[b_2(b_4 - R_3) + d_1(h_3 - R_3) + d_2(h_2 - R_3) + d_3(h_1 - R_3) + d_4(\eta - R_3) + b_{10}(R_1 - R_3) - R_3\right],$$

$$c_{13} = \frac{1}{1 + n(R_3 - h_6)Q} * \kappa_1,$$

$$b_8 = Qc_{13},$$

$$b_1 = -\left(1 + b_2 + d_1 + d_2 + d_3 + d_4 + b_8 + b_{10}\right). \tag{A.1}$$

Nomenclature

M: The local magnetic field parameter
n: The parameter related to microrotation vector and shear stress
N: The model parameter

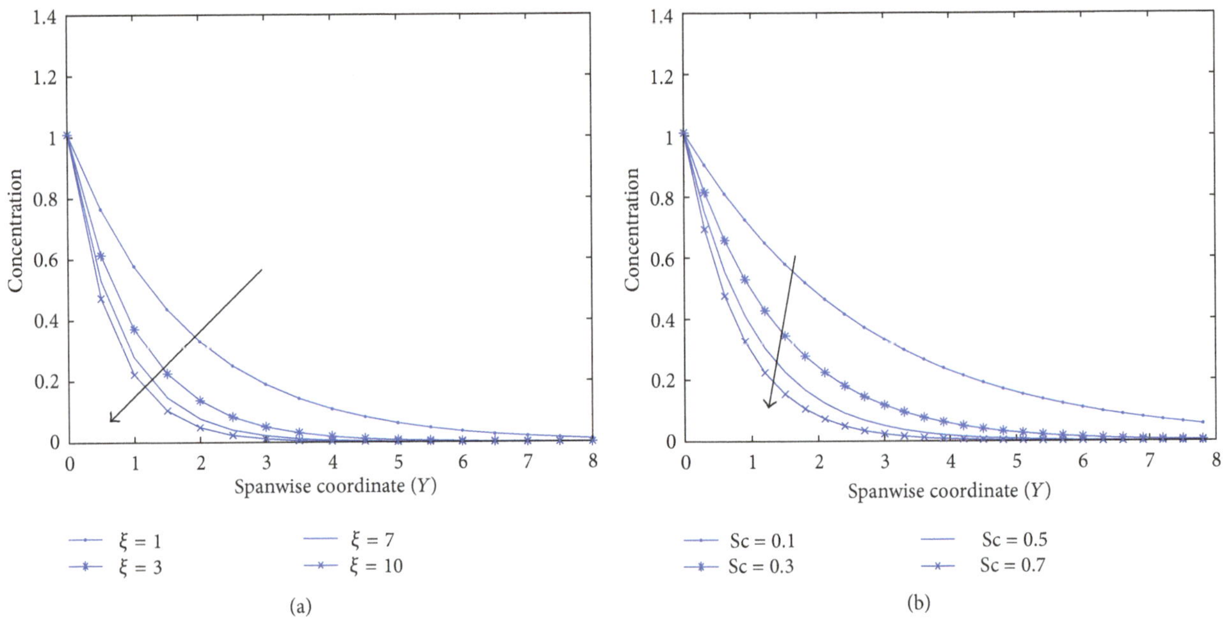

FIGURE 15: Concentration distribution.

Nu: Nusselt number
P^*: The pressure, kg m^{-1} s^{-2}
Pr: The fluid Prandtl number
Q_0: The heat generation coefficient, W m^{-3} K^{-1}
q_r^*: The radiative heat flux, W m^{-2}
R^*: The reaction rate constant, J
R: The radiation parameter
Re: The Reynolds number
Sc: The generalized Schmidt number
Sh: Sherwood number
Q_l: The radiation absorption number
t^*: The time, s
t: The dimensionless time
T^*: The temperature in the boundary layer, K
T_∞^*: The temperature far away from the plate, K
T_w^*: The temperature at the wall, K
u: The dimensionless velocity
u^*, v^*: The velocities along and perpendicular to the plate, m s^{-1}
U_0: The scale of free stream velocity
U_∞^*: The free stream velocity, m s^{-1}
u_p^*: The wall dimensional velocity, m s^{-1}
U_p: The wall dimensionless velocity
V_0: The scale of suction velocity
x^* and y^*: The distances along and perpendicular to the plate, respectively, m
y: The dimensionless spanwise coordinate normal to the plate
γ: The spingradient velocity, kg m s^{-1}
δ: The scalar constant
ε: The scalar constant
θ: The dimensionless temperature
ζ: The chemical reaction parameter
η: The scalar constant
μ: The fluid dynamic viscosity, kg m^{-1} s^{-1}
v: The fluid kinematic rotational viscosity, m^2 s^{-1}
v_r: The kinematic rotational viscosity, m^2 s^{-1}
ρ: Fluid density, kg m^{-3}
σ: The electrical conductivity of the fluid, S m^{-1}
σ^*: The Stefan-Boltzmann constant, W m^{-2} K^{-4}
ω: Angular velocity vector, m s^{-1}
Γ: The coefficient of gyro-viscosity (or vortex viscosity), kg m^{-1} s^{-1}
B_0: The magnetic induction, tesla
C^*: The species concentration, mol m^{-3}
C_w^*: The surface concentration, mol m^{-3}
C_∞^*: The species concentration far from the surface, mol m^{-3}
C_f: The skin-friction coefficient
C_m: The couple stress coefficient
C_p: The specific heat at constant pressure, J kg^{-1} K^{-1}
D^*: The chemical molecular diffusivity, m^2 s^{-1}
Gm: The solutal Grashof number
Gr: Grashof number
g: The acceleration due to the gravity, m s^{-2}
H: The dimensionless heat generation/absorption coefficient
j^*: The microinertia per unit mass, m^2
K: The permeability of the porous media, m^2
K_1^*: The mean absorption coefficient, m^{-1}
K: The dimensionless permeability
k: The thermal conductivity of the fluid, W m^{-1} K^{-1}
α: The fluid thermal diffusivity, m^2 s^{-1}
β: The ratio of vortex viscosity and dynamic viscosity

β_c: Volumetric coefficient of thermal expansion, K^{-1}
β_f: Volumetric coefficient of expansion with concentration, K^{-1}
B: The scalar constant.

References

[1] A. C. Eringen, "Simple microfluids," *International Journal of Engineering Science*, vol. 2, pp. 205–217, 1964.

[2] A. C. Eringen, "Theory of termomicrofluids," *Journal of Mathematical Analysis and Applications*, vol. 38, pp. 480–496, 1972.

[3] T. Ariman, M. A. Turk, and N. D. Sylvester, "Microcontinuum fluid mechanics: a review," *International Journal of Engineering Science*, vol. 11, no. 8, pp. 905–930, 1973.

[4] R. S. R. Gorla, "Mixed convection in a micropolar fluid from a vertical surface with uniform heat flux," *International Journal of Engineering Science*, vol. 30, no. 3, pp. 349–358, 1992.

[5] Y. J. Kim, "Unsteady convetion flow of micropolar fluids past a vertical porous plate embedded in a porous medium," *Acta Mechanica*, vol. 148, no. 1–4, pp. 105–116, 2001.

[6] G. R. Charya, "Flow of micropolar fluid through a constricted channel," *International Journal of Engineering Science*, vol. 15, no. 12, pp. 719–725, 1977.

[7] D. A. S. Rees and I. Pop, "Free convection boundary-layer flow of a micropolar fluid from a vertical flat plate," *IMA Journal of Applied Mathematics*, vol. 61, no. 2, pp. 179–197, 1998.

[8] R. C. Sharma and U. Gupta, "Thermal convection in micropolar fluids in porous medium," *International Journal of Engineering Science*, vol. 33, no. 13, pp. 1887–1892, 1995.

[9] J. P. Kumar, J. C. Umavathi, A. J. Chamkha, and I. Pop, "Fully-developed free-convective flow of micropolar and viscous fluids in a vertical channel," *Applied Mathematical Modelling*, vol. 34, no. 5, pp. 1175–1186, 2010.

[10] P. Muthu, B. V. Rathish Kumar, and P. Chandra, "Peristaltic motion of micropolar fluid in circular cylindrical tubes: effect of wall properties," *Applied Mathematical Modelling*, vol. 32, no. 10, pp. 2019–2033, 2008.

[11] D. Srinivasacharya, J. V. Ramana Murthy, and D. Venugopalam, "Unsteady stokes flow of micropolar fluid between two parallel porous plates," *International Journal of Engineering Science*, vol. 39, no. 14, pp. 1557–1563, 2001.

[12] R. Muthuraj and S. Srinivas, "Fully developed mhd flow of a micropolar and viscous fluids in a vertical porous space using HAM," *Journal of Applied Mathematics and Mechanics*, vol. 6, no. 11, pp. 55–78, 2010.

[13] U. N. Das, R. Deka, and V. M. Soundalgekar, "Effects of mass transfer on flow past an impulsively started infinite vertical plate with constant heat flux and chemical reaction," *Forschung im Ingenieurwesen Engineering Research*, vol. 60, no. 10, pp. 284–287, 1994.

[14] R. Muthucumaraswamy and P. Ganesan, "First-order chemical reaction on flow past an impulsively started vertical plate with uniform heat and mass flux," *Acta Mechanica*, vol. 147, no. 1–4, pp. 45–57, 2001.

[15] P. V. Satya Narayana, D. Ch. Kesavaiah, and S. Venkataramana, "Viscous dissipation and thermal radiation effects on an unsteady MHD convection flow past a semi-infinite vertical permeable moving porous plate," *International Journal of Mathematical Archive*, vol. 2, no. 4, pp. 476–487, 2011.

[16] R. Kandasamy, K. Periasamy, and K. K. S. Prabhu, "Effects of chemical reaction, heat and mass transfer along a wedge with heat source and concentration in the presence of suction or injection," *International Journal of Heat and Mass Transfer*, vol. 48, no. 7, pp. 1388–1394, 2005.

[17] J. Gireesh Kumar, P. V. Satya Narayana, and S. Ramakrishna, "Effects of the chemical reaction and mass transfer on MHD unsteady free convection flow past an infinite vertical plate with constant suction and heat sink," *Ultra Science*, vol. 21, no. 3, pp. 639–650, 2009.

[18] G. R. Reddy, P. V. S. Narayana, and S. Venkataramana, "Peristaltic transport of a conducting fluid in an inclined asymmetric channel," *Applied Mathematical Sciences*, vol. 4, no. 35, pp. 1729–1741, 2010.

[19] D. Pal and B. Talukdar, "Perturbation analysis of unsteady magnetohydrodynamic convective heat and mass transfer in a boundary layer slip flow past a vertical permeable plate with thermal radiation and chemical reaction," *Communications in Nonlinear Science and Numerical Simulation*, vol. 15, no. 7, pp. 1813–1830, 2010.

[20] E. M. Abo-Eldahab and M. A. El Aziz, "Flow and heat transfer in a micropolar fluid past a stretching surface embedded in a non-Darcian porous medium with uniform free stream," *Applied Mathematics and Computation*, vol. 162, no. 2, pp. 881–899, 2005.

[21] R. S. R. Gorla, A. A. Mohammedien, M. A. Mansour, and I. A. Hassanien, "Unsteady natural convection from a heated vertical plate in micropolar fluid," *Numerical Heat Transfer A*, vol. 28, no. 2, pp. 253–262, 1995.

[22] A. Raptis and C. Perdikis, "Viscous flow over a non-linearly stretching sheet in the presence of a chemical reaction and magnetic field," *International Journal of Non-Linear Mechanics*, vol. 41, no. 4, pp. 527–529, 2006.

[23] D. Ch. Kesavaiah, P. V. Satya Narayana, and S. Venkataramana, "Effects of the chemical reaction and radiation absorption on an unsteady MHD convective heat and mass transfer flow past a semi-infinite vertical permeable moving plate embedded in a porous medium with heat source and suction," *Journal of Applied Mathematics and Mechanics*, vol. 7, no. 1, pp. 52–69, 2011.

[24] K. Vajravelu, S. Sreenadh, and P. Lakshminarayana, "The influence of heat transfer on peristaltic transport of a Jeffrey fluid in a vertical porous stratum," *Communications in Nonlinear Science and Numerical Simulation*, vol. 16, no. 8, pp. 3107–3125, 2011.

[25] A. Y. Bakier and R. S. R. Gorla, "Thermal radiation effect on mixed convection from horizontal surfaces in saturated porous media," *Transport in Porous Media*, vol. 23, no. 3, pp. 357–363, 1996.

[26] A. Y. Bakier, "Thermal radiation effect on mixed convection from vertical surfaces in saturated porous media," *International Communications in Heat and Mass Transfer*, vol. 28, no. 1, pp. 119–126, 2001.

[27] M. A. Hossain and H. S. Takhar, "Radiation effect on mixed convection along a vertical plate with uniform surface temperature," *Heat and Mass Transfer*, vol. 46, pp. 1751–1758, 1996.

[28] C. E. Schwartz and J. M. Smith, "Flow distribution in packed beds," *Industrial and Engineering Chemistry*, vol. 45, no. 6, pp. 1209–1218, 1969.

[29] W. M. Schertz and K. B. Bischoff, "Thermal and material transport in nonisothermal packed beds," *AIChE Journal*, vol. 15, no. 4, pp. 597–604, 1969.

[30] R. F. Benenati and C. R. Brosilow, "Void fraction distribution in beds of spheres," *AIChE Journal*, vol. 8, no. 3, pp. 359–361, 1962.

[31] B. C. Chandrasekhara and P. M. S. Namboodiri, "Influence of variable permeability on combined free and forced convection about inclined surfaces in porous media," *International Journal of Heat and Mass Transfer*, vol. 28, no. 1, pp. 199–206, 1985.

[32] S. N. Murthy and J. Feyen, "Influence of variable permeability on the dispersion of a chemically reacting solute in porous media," *International Journal of Engineering Science*, vol. 27, no. 12, pp. 1661–1671, 1989.

[33] J. A. Shercliff, *A Text Book of Magnetohydrodynamics*, Pergamon Press, New York, NY, USA, 1965.

[34] F. S. Ibrahim, A. M. Elaiw, and A. A. Bakr, "Effect of the chemical reaction and radiation absorption on the unsteady MHD free convection flow past a semi infinite vertical permeable moving plate with heat source and suction," *Communications in Nonlinear Science and Numerical Simulation*, vol. 13, no. 6, pp. 1056–1066, 2008.

[35] M. Sudheer Babu and P. V. Satya Narayana, "Effects of the chemical reaction and radiation absorption on free convection flow through porous medium with variable suction in the presence of uniform magnetic field," *JP Journal of Heat and Mass Transfer*, vol. 3, pp. 219–234, 2009.

[36] R. A. Mohamed and S. M. Abo-Dahab, "Influence of chemical reaction and thermal radiation on the heat and mass transfer in MHD micropolar flow over a vertical moving porous plate in a porous medium with heat generation," *International Journal of Thermal Sciences*, vol. 48, no. 9, pp. 1800–1813, 2009.

29

Perturbation Solution for Radiating Viscoelastic Fluid Flow and Heat Transfer with Convective Boundary Condition in Nonuniform Channel with Hall Current and Chemical Reaction

B. J. Gireesha and B. Mahanthesh

Department of Studies and Research in Mathematics, Kuvempu University, Shankaraghatta, Shimoga, Karnataka, 577 451, India

Correspondence should be addressed to B. J. Gireesha; bjgireesu@rediffmail.com

Academic Editors: G. L. Aranovich, C. D. Daub, and A. Ghoufi

A mathematical analysis has been performed for heat and mass transfer of a time-dependent MHD flow of an electrically conducting viscoelastic fluid in nonuniform vertical channel with convective boundary condition. The fluid flow is considered between a vertical long wavy wall and a parallel flat wall saturated with the porous medium. The effects of thermal radiation, heat absorption, chemical reaction, and Hall current are taken into account. The prevailing nonlinear partial differential equations are derived by considering Boussinesq approximation, and the same equations are solved analytically using perturbation technique. Further the expressions for skin friction, Nusselt number, and Sherwood number are presented. The effects of various pertinent parameters on different flow fields are analyzed graphically and tabularly. It is found that effects of Hall parameter and Biot number are unfavorable on velocity profiles, but this trend is reverse for the effect of thermal and solutal Grashof numbers. The expressions of different flow fields satisfy the imposed boundary conditions, which is shown in all graphs; this implies accuracy of the solution.

1. Introduction

The study of viscoelastic fluid has become important in the last few years. Qualitative analysis of these studies has significant bearing on several industrial applications such as polymer sheet extrusion from a dye and drawing of plastic firms. When manufacturing processes at high temperature need cooling, the flow may need viscoelastic fluid to produce a good effect or reduce the temperature. On the other hand, the flow and heat transfer of a viscoelastic fluid between parallel plates have significant role in many engineering fields such as petroleum production, chemical catalytic reactors, and solar power collectors. Boundary layer treatment for an idealized viscoelastic fluid was introduced by Beard and Walters [1]. There has been a continued interest in the investigation of natural convection heat transfer of non-Newtonian fluid, which exhibits the viscoelasticity. Recently, Rajagopal and Na [2] have studied the heat transfer analysis in the forced convection flow of a visco-elastic fluid by considering the Walters model. The problem of MHD flow and heat transfer has wide range of applications in emerging fields due to an electro-magnetic field are relevant to many practical applications in geophysical and astrophysical situations, the metallurgy industry, and cooling of continuous strips and filaments drawn through a quiescent fluid. Sarpkaya [3] was the first who had studied the MHD flows in non-Newtonian fluids. Thereafter, several authors [4–10] have shown their interest in MHD flows under different physical conditions.

Convective flow with simultaneous heat and mass transfer under the influence of a magnetic field and chemical reaction has attracted considerable attention of many researchers because of its applications in various branches of science and technology. On the other hand, radiative heat transfer has many applications in nuclear power plants, in gas turbines, and in the various propulsion devices for space vehicles, missiles, and aircrafts. Motivated by these applications very recently, Pal and Talukdar [11] analyzed the heat and mass transfer in boundary layer slip flow past a vertical permeable plate with thermal radiation and chemical reaction. They reported analytical solution and found that the effect of thermal radiation and chemical reaction decreases the velocity and concentration profiles. Chamkha [12] obtained the numerical solution for the problem of heat and mass transfer by the steady flow of an electrically conducting fluid over a uniformly moving vertical surface in the presence of first-order chemical reaction. Later, the heat and mass transfer of a viscoelastic fluid in a fixed plane with radiation and chemical reaction effect was investigated by Uwanta and Omokhuale [13], and they found that temperature decreases with the increase in Prandtl number and concentration decreases with the increase in chemical reaction parameter. Other examples dealing with radiation and chemical reaction effects can be found in [14, 15].

The study of fluid flow with subject to convective boundary condition plays an important role in several engineering and industrial processes like transpiration cooling process and material drying. Therefore several authors [16–19] have shown their interest in obtaining solutions subject to convective boundary condition. Interestingly, in all of the above investigations the Hall term has been ignored in applying Ohms law as it has no marked effect for small and moderate values of the magnetic field. However, the current trend for the application of magnetohydrodynamics is towards a strong magnetic field, so that the influence of electromagnetic force is noticeable [20]. Under these conditions, the Hall current is important and it has a marked effect on the magnitude and direction of the current density and magnetic force term. With this understanding, very recently, Hayat and Nawaz [21] have studied the effect of Hall current on the mixed convection flow of a second grade fluid in the presence of Soret and Dufour effects.

Previous studies of the flow with heat and mass transfer have focused mainly on a flat wall or a regular channel. It is necessary to study the flow, heat and mass transfer in an irregular channel because such flows find applications in different areas such as transpiration cooling of reentry vehicles and rocket boosters, crosshatching on ablative surfaces, and film vaporization in combustion chambers. In view of these applications, Das [22] analyzed the viscous incompressible fluid confined between a long vertical wavy wall and a parallel flat wall. The effects of time-dependent pulsatile flow and heat transfer of a Newtonian fluid (blood) through a vertical annulus with mild constriction on the outer wall while the inner wall represents the catheter tube were studied by Abd elmaboud and Mekheimer [23]. Many authors like in [24–27] have analyzed the fluid flow in wavy-walled channels.

Keeping all these facts in mind, in this work the effect of radiation on unsteady laminar flow with heat and mass transfer of an electrically conducting, chemically reactive viscoelastic fluid in irregular channel with subject to convective boundary condition has been investigated. The perturbation technique is employed to solve governing coupled nonlinear partial differential equations.

2. Walters B Liquid Model

The constitutive equations for the rheological equation of the state for the visco-elastic fluid (Walters liquid B) are

$$p_{ik} = -pg_{ik} + p_{ik}^{*}, \tag{1}$$

$$p_{ik}^{*} = 2\int_{-\infty}^{t} \psi\left(t-t^{*}\right) e_{ik}^{(1)}\left(t^{*}\right) dt^{*}, \tag{2}$$

where

$$\psi\left(t-t^{*}\right) = \int_{0}^{\infty} \frac{N(\tau)}{\tau} e^{(-(t-t^{*})/\tau)} d\tau. \tag{3}$$

In the above equations p_{ik} is the stress tensor, p is an arbitrary isotropic pressure, g_{ik} is the metric tensor of a fixed coordinate system x^i, and $e_{ik}^{(1)}$ is the rate of strain tensor. $N(\tau)$ is the distribution function of relaxation time (τ). Equation (2) can be put in the following generalized form which is valid for all types of motion and stress; it was shown by Walters [28]:

$$p_{ik}^{*}(x,t) = 2\int_{-\infty}^{t} \psi\left(t-t^{*}\right) \frac{\partial x^{i}}{\partial x_{m}^{*}} \frac{\partial x^{k}}{\partial x_{r}^{*}} e_{mr}^{(1)}\left(x^{*}t^{*}\right) dt^{*}, \tag{4}$$

where $x^{*i} = x^{*i}(x,t,t^{*})$ is the position at the time t^{*} of the element which is instantaneously at the position x^i at the time t. The fluid with equations of state (1) and (4) has been designated as liquid "B." In the case of liquids with short memories, that is, short relaxation times, the above equation of state can be written in the following simplified form:

$$p_{ik}^{*}(x,t) = 2\eta_0 e_{ik}^{(1)} - 2k_0 \frac{\partial e_{ik}^{(1)}}{\partial t}, \tag{5}$$

where $\eta_0 = \int_0^{\infty} N(\tau)\, d\tau$ is the limiting viscosity at small rates of shear. $k_0 = \int_0^{\infty} \tau N(\tau)\, d\tau$ is the Walters-B viscoelasticity parameter, and $\partial/\partial t$ denotes the convected time derivative. This rheological model is very versatile and robust and provides a relatively simple mathematical formulation which is easily incorporated into boundary layer theory for engineering applications.

3. Mathematical Formulation and Solution

Consider an unsteady flow, heat and mass transfer of an electrically conducting, incompressible, chemically reacting viscoelastic fluid between a vertical long wavy wall under the influence of uniform transverse magnetic field of strength B_0 and exponentially decreasing pressure gradient. A parallel flat wall saturated with porous medium is placed in the plane $y = 0$ of a coordinate system. The x-axis is taken in the direction along the channel which is set in motion, and the y-axis is taken perpendicular to it. The inclusion of the Hall currents give's rise to the Lorentz force in z-direction, which induces a crossflow in that direction. The z-axis is assumed to be normal to the x-y plane which is as shown in Figure 1.

The flow field is exposed to the influence of thermal and mass buoyancy effects, Hall effect, thermal radiation, heat absorption, and first order chemically reactive species. The

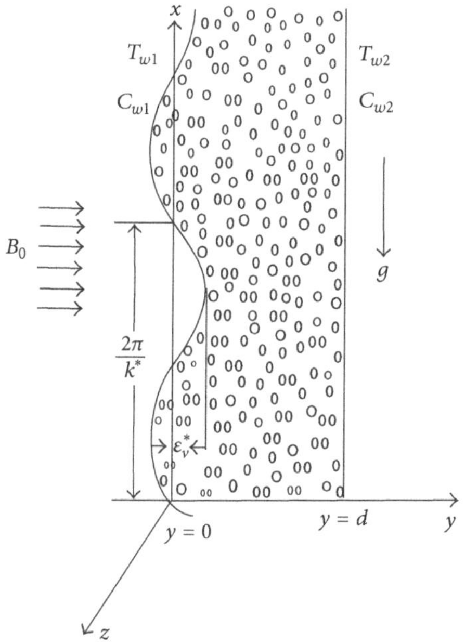

FIGURE 1: The geometry of the problem.

magnetic Reynolds number is assumed to be small enough so that the induced magnetic field is negligible. The wavy wall ($y = \varepsilon_v^* \cos(\lambda^* x)$) maintains a temperature T_{w1} which represents the convective boundary condition $-k(\partial T/\partial y) = h_f(T_1 - T + (T_2 - T_1)\varepsilon e^{-nt})$ and concentration $C_{w1} = C_1 + (C_2 - C_1)\varepsilon e^{-nt}$, respectively. It is assumed that the convective heat exchange with the wall temperature at the channel surface follows Newton's law of cooling.

The equations governing the flow are in vector form.

Momentum Equation

$$\frac{\partial \vec{q}}{\partial t} + (\vec{q}\cdot\nabla)\vec{q} = -\frac{1}{\rho}\nabla P + \nu\nabla^2\vec{q} + \nabla\cdot p_{ik} + \frac{1}{\rho}(\vec{J}\times\vec{B}) - \frac{\nu}{k(t)}\vec{q} + g\beta_T(T - T_1) + g\beta_C(C - C_1). \quad (6)$$

Continuity Equation

$$\nabla\cdot\vec{q} = 0. \quad (7)$$

Energy Equation

$$\frac{\partial T}{\partial t} + (\vec{q}\cdot\nabla)T = \frac{k}{\rho C_p}\nabla^2 T - \frac{1}{\rho C_p}\nabla q_r - \frac{Q_T}{\rho C_p}(T - T_1). \quad (8)$$

Mass Diffusion Equation

$$\frac{\partial C}{\partial t} + (\vec{q}\cdot\nabla)C = D\nabla^2 C - K_R(C - C_1), \quad (9)$$

where ρ is the density of the fluid, ν is the kinematic viscosity of the fluid, ν_1 is the kinematic viscoelasticity of the fluid, $\vec{q}(u, v, w)$ is the velocity vector, u, v, and w are the velocity components along x, y, and z directions, respectively, $\vec{B}(0, B_0, 0)$ is the magnetic induction, $\vec{J}(J_x, J_y, J_z)$ is the current density vector, P is the pressure, p_{ik} is the stress tensor, g is the acceleration due to gravity, β_T is the thermal expansion coefficient, β_C is the concentration expansion coefficient, T is the temperature of the fluid, C is the concentration of the fluid, k is the thermal conductivity, C_p is the specific heat of the fluid at constant pressure, q_r is the radiative heat flux, Q_T is the heat absorption coefficient, K_R is the chemical reaction parameter, T_1 and T_2 are wall temperatures, C_1 and C_2 are wall concentrations, and t is the time. If the Hall term is retained, the current density $\vec{J}$ from the generalized Ohm's law is given by

$$\vec{J} + \frac{\omega_e\tau_e}{B_0}(\vec{J}\times\vec{B}) = \sigma\left(\vec{E} + \vec{q}\times\vec{B} + \frac{\nabla P_e}{e\eta_e}\right), \quad (10)$$

where ω_e is the electron frequency, τ_e is the electron collision time, e is the electron charge, η_e is the number density of electron, P_e is the electron pressure, σ is the electric conductivity, and E is the electric field. In (10) ion slip, thermoelectric effect, and polarization effects are neglected. Solving (10) for $\vec{J}$, it gives

$$\vec{J}\times\vec{B} = \frac{\sigma B_0^2}{(1+m^2)}\left[-(mw + u)\hat{i} + (mu - w)\hat{k}\right], \quad (11)$$

where $m = \omega_e\tau_e$ is the Hall parameter. Under the above stated assumptions and usual Boussinesq approximation, the governing equations in their component form are reduced to:

$$\begin{aligned}
\frac{\partial u}{\partial t} &= -\frac{1}{\rho}\frac{\partial p}{\partial x} + \nu\frac{\partial^2 u}{\partial y^2} - \nu_1\frac{\partial^3 u}{\partial y^2\partial t} - \frac{\sigma B_0^2}{\rho(1+m^2)}(mw + u) \\
&\quad - \frac{\nu}{k(t)}u + g\beta_T(T - T_1) + g\beta_C(C - C_1), \\
\frac{\partial w}{\partial t} &= \nu\frac{\partial^2 w}{\partial y^2} - \nu_1\frac{\partial^3 w}{\partial y^2\partial t} + \frac{\sigma B_0^2}{\rho(1+m^2)}(mu - w) - \frac{\nu}{k(t)}w, \\
\rho C_p\frac{\partial T}{\partial t} &= k\frac{\partial^2 T}{\partial y^2} - \frac{\partial q_r}{\partial y} - Q_T(T - T_1), \\
\frac{\partial C}{\partial t} &= D\frac{\partial^2 C}{\partial y^2} - K_R(C - C_1).
\end{aligned} \quad (12)$$

The suitable initial and boundary conditions for the present problem are

$$t = 0: \quad u = 0, \quad w = 0, \quad T = T_1, \quad C = C_1,$$
$$\text{for } y \in (\varepsilon_v\cos(\lambda^* x), d),$$

$$t>0: \ u=0, \quad w=0,$$

$$-k\frac{\partial T}{\partial y}=h_f\left(T_1-T+\varepsilon\left(T_2-T_1\right)e^{-nt}\right),$$

$$C=C_1+\varepsilon\left(C_2-C_1\right)e^{-nt} \quad \text{at } y=\varepsilon_v\cos(\lambda x),$$

$$: \ u=0, \quad w=0,$$

$$-k\frac{\partial T}{\partial y}=h_f\left(T_2-T+\varepsilon\left(T_2-T_1\right)e^{-nt}\right),$$

$$C=C_2+\varepsilon\left(C_2-C_1\right)e^{-nt} \quad \text{at } y=d, \tag{13}$$

where n is the positive real constant, ε_v^* is the amplitude parameter, h_f is the heat transfer coefficient, d is the dimensional width of the channel, and ε is a small positive constant ($\varepsilon \ll 1$).

Using the relation in [29], the radiative heat flux (q_r) for the optically thin nongray gas near equilibrium is given by

$$\frac{\partial q_r}{\partial y}=4I'\left(T-T_1\right), \quad I'=\int_0^{\infty}K_{\lambda_1 w}\frac{\partial e_{b\lambda_1}}{\partial T}\,d\lambda_1, \tag{14}$$

where q_r is the radiative heat flux, $K_{\lambda_1 w}$ is the radiation absorption coefficient at the wall, and $e_{b\lambda_1}$ is Plank's constant.

The permeability of the porous medium is assumed to be of the following form:

$$k(t)=k_0^+\left(1+\varepsilon e^{-nt}\right), \tag{15}$$

where k_0 is the constant permeability of the medium.

Now we introduce the following non dimensional quantities:

$$(x^*,y^*)=\frac{(x,y)}{d}, \quad u^*=\frac{u}{u_0}, \quad w^*=\frac{w}{u_0},$$

$$t^*=\frac{\nu t}{d^2}, \quad n^*=\frac{nd^2}{\nu}, \quad \varepsilon_v^*=\frac{\varepsilon_v}{d}, \tag{16}$$

$$p^*=\frac{pd}{\nu u_0}, \quad T^*=\frac{T-T_1}{T_2-T_1}, \quad C^*=\frac{C-C_1}{C_2-C_1}.$$

Using (14), (15), and (16) and then introducing complex velocity $\psi^*=u^*+iw^*$ into (12) one can get (dropping the asterisk *)

$$\frac{\partial\psi}{\partial t}=-\frac{1}{\rho}\frac{\partial p}{\partial x}+\frac{\partial^2\psi}{\partial y^2}-E\frac{\partial^3\psi}{\partial y^2\partial t}-\frac{M^2(1-im)}{(1+m^2)}\psi$$

$$-\frac{1}{k_0\left(1+\varepsilon e^{-nt}\right)}\psi+\mathrm{Gr}T+\mathrm{Gc}C,$$

$$\mathrm{Pr}\frac{\partial T}{\partial t}=\frac{\partial^2 T}{\partial y^2}-RT-\alpha_T T, \tag{17}$$

$$\mathrm{Sc}\frac{\partial C}{\partial t}=\frac{\partial^2 C}{\partial y^2}-\mathrm{KrSc}C,$$

where $E=\nu_1/d^2$ is the viscoelastic parameter, $M=B_0 d\sqrt{\sigma_e/\nu\rho}$ is the magnetic parameter, $\mathrm{Gr}=g\beta_T(T_2-T_1)d^2/\nu u_0$ is the thermal Grashof number, $\mathrm{Gc}=g\beta_C(C_2-C_1)d^2/\nu u_0$ is the solutal Grashof number, $\mathrm{Pr}=\mu C_p/k$ is the Prandtl number, $R=4I'd^2/k$ is the radiation parameter, $k_0=k_0^+/d^2$ is the permeability parameter, $\alpha_T=Q_T d^2/k$ is the heat absorption parameter, $\mathrm{Sc}=\nu/d$ is the Schmidt number, and $\mathrm{Kr}=h^2K_R/\nu$ is the chemical reaction parameter.

The corresponding initial and boundary conditions (16) in dimensionless form are

$$t=0: \ \psi=0, \quad T=0, \quad C=0, \quad \text{for } y\in(h,1),$$

$$t>0: \ \psi=0, \quad \frac{\partial T}{\partial y}=B_i\left(T-\varepsilon e^{-nt}\right), \quad C=\varepsilon e^{-nt}$$

$$\text{at } y=h, \tag{18}$$

$$: \ \psi=0, \quad \frac{\partial T}{\partial y}=B_i\left(T-1-\varepsilon e^{-nt}\right),$$

$$C=1+\varepsilon e^{-nt} \quad \text{at } y=1,$$

where $B_i=h_f d/k$ is the Biot number, $h=\varepsilon_v\cos(\lambda x)$, and $\lambda=\lambda^* d$ is the frequency parameter of the wavy wall.

Equations (17) represent the set of partial differential equations that cannot be solved in the closed form. However, these equations can be solved analytically after being reduced to set of ordinary differential equations in dimensionless form. Thus, to this end we can represent the velocity (ψ), temperature (T), and concentration (C) in terms of power of ε ($\varepsilon \ll 1$) and as follows:

$$\psi(y,t)=\psi_0(y)+\varepsilon\psi_1(y,t)e^{-nt}+O\left(\varepsilon^2\right),$$

$$T(y,t)=T_0(y)+\varepsilon T_1(y,t)e^{-nt}+O\left(\varepsilon^2\right), \tag{19}$$

$$C(y,t)=C_0(y)+\varepsilon C_1(y,t)e^{-nt}+O\left(\varepsilon^2\right).$$

Let us assume the pressure gradient is of the form $-(1/\rho)(\partial p/\partial x)=A+\varepsilon Be^{-nt}$, where A and B are constants. Now by substituting (19) into (17), then equating harmonic and nonharmonic terms and neglecting the higher-order terms of ε, then we get the following set of equations:

$$\psi_0''(y)-\left(\frac{M^2(1-im)}{1+m^2}+\frac{1}{k_0}\right)\psi_0(y)$$

$$=-A-\mathrm{Gr}T_0-\mathrm{Gc}C_0,$$

$$(1+nE)\psi_1''(y,t)-\left[\frac{M^2(1-im)}{1+m^2}+\frac{1}{k_0}-n\right]\psi_1(y,t)$$

$$=-B-\mathrm{Gr}T_1-\mathrm{Gc}C_1,$$

$$T_0''(y)-(R+\alpha_T)T_0(y)=0,$$

$$T_1''(y,t)-(R+\alpha_T-n\mathrm{Pr})T_1(y,t)=0,$$

$$C_0''(y) - \mathrm{KrSc}C_0(y) = 0,$$
$$C_1''(y,t) - \mathrm{Sc}(\mathrm{Kr}-n)C_1(y,t) = 0. \tag{20}$$

The corresponding boundary conditions will become

$$\psi_0 = 0, \quad T_0' = B_i T_0, \quad C_0 = 0,$$
$$\psi_1 = 0, \quad T_1' = B_i(T_1 - 1), \quad C_1 = 1 \quad \text{at } y = h,$$
$$\psi_0 = 0, \quad T_0' = B_i(T_0 - 1), \quad C_0 = 1, \tag{21}$$
$$\psi_1 = 0, \quad T_1' = B_i(T_1 - 1), \quad C_1 = 1 \quad \text{at } y = +1,$$

where prime ($'$) denotes the differentiation with respect to y.

The solutions of (20) subject to the boundary conditions (21) are

$$C(y,t) = A_1 e^{\beta_1 y} + A_2 e^{\beta_2 y} + \varepsilon\left(A_3 e^{\beta_3 y} + A_4 e^{\beta_4 y}\right)e^{-nt}, \tag{22}$$

$$T(y,t) = A_5 e^{\beta_5 y} + A_6 e^{\beta_6 y} + \varepsilon\left(A_7 e^{\beta_7 y} + A_8 e^{\beta_8 y}\right)e^{-nt}, \tag{23}$$

$$\begin{aligned}\psi(y,t) &= A_9 e^{\beta_1 y} + A_{10} e^{\beta_2 y} + A_{11} e^{\beta_5 y} + A_{12} e^{\beta_6 y} \\ &\quad + A_{13} + A_{14} e^{\beta_9 y} + A_{15} e^{\beta_{10} y} \\ &\quad + \varepsilon\left(A_{16} e^{\beta_3 y} + A_{17} e^{\beta_4 y} + A_{18} e^{\beta_7 y} + A_{19} e^{\beta_8 y}\right. \\ &\quad \left. + A_{20} + A_{21} e^{\beta_{11} y} + A_{22} e^{\beta_{12} y}\right)e^{-nt}.\end{aligned} \tag{24}$$

On separating real and imaginary parts in (24), we get the axial velocity (u) and transverse velocity (w) profiles.

Skin Friction. The skin friction ($\tau = \tau_1 + i\tau_2$) expression for viscoelastic fluid at a wavy wall $y = h$ is given by

$$\begin{aligned}\tau &= \left(\frac{\partial \psi}{\partial y}\right)_{y=h} \\ &= \beta_1 A_9 e^{\beta_1 h} + \beta_2 A_{10} e^{\beta_2 h} + \beta_5 A_{11} e^{\beta_5 h} \\ &\quad + \beta_6 A_{12} e^{\beta_6 h} + \beta_9 A_{14} e^{\beta_9 h} + \beta_{10} A_{15} e^{\beta_{10} h} \\ &\quad + \varepsilon\left(\beta_3 A_{16} e^{\beta_3 h} + \beta_4 A_{17} e^{\beta_4 h} + \beta_7 A_{18} e^{\beta_7 h} + \beta_8 A_{19} e^{\beta_8}\right. \\ &\quad \left. + \beta_{11} A_{21} e^{\beta_{11} h} + \beta_{12} A_{22} e^{\beta_{12} h}\right)e^{-nt},\end{aligned} \tag{25}$$

where τ_1 and τ_2 are the axial and transverse components of the skin friction, respectively.

Nusselt Number. The Nusselt number (Nu) expression for visco-elastic fluid at a wavy wall $y = h$ is given by

$$\begin{aligned}\mathrm{Nu} = \left(\frac{\partial T}{\partial y}\right)_{y=h} &= \beta_5 A_5 e^{\beta_5 h} + \beta_6 A_6 e^{\beta_6 h} \\ &\quad + \varepsilon\left(\beta_7 A_7 e^{\beta_7 h} + \beta_8 A_8 e^{\beta_8 h}\right)e^{-nt}.\end{aligned} \tag{26}$$

Sherwood Number. The Sherwood number (Sh) expression for visco-elastic fluid at a wavy wall $y = h$ is given by

$$\begin{aligned}\mathrm{Sh} = \left(\frac{\partial C}{\partial y}\right)_{y=h} &= \beta_1 A_1 e^{\beta_1 h} + \beta_2 A_2 e^{\beta_2 h} \\ &\quad + \varepsilon\left(\beta_3 A_3 e^{\beta_3 h} + \beta_4 A_4 e^{\beta_4 h}\right)e^{-nt},\end{aligned} \tag{27}$$

where

$$\beta_1 = -\beta_2 = \sqrt{\mathrm{KrSc}}, \qquad \beta_3 = -\beta_4 = \sqrt{\mathrm{Sc}(\mathrm{Kr}-n)},$$

$$\beta_5 = -\beta_6 = \sqrt{R + \alpha_T},$$

$$\beta_{11} = -\beta_{12} = P = \sqrt{\frac{1}{(1+nE)}\left[\frac{M^2(1-im)}{1+m^2} + \frac{1}{k_0} - n\right]},$$

$$\beta_9 = -\beta_{10} = N = \sqrt{\left(\frac{M^2(1-im)}{1+m^2} + \frac{1}{k_0}\right)},$$

$$\beta_7 = -\beta_8 = \sqrt{R + \alpha_T - nPr},$$

$$A_1 = \frac{e^{\beta_2 h}}{e^{\beta_1} e^{\beta_2 h} - e^{\beta_1 h} e^{\beta_2}}, \qquad A_2 = \frac{e^{\beta_1 h}}{e^{\beta_2} e^{\beta_1 h} - e^{\beta_2 h} e^{\beta_1}},$$

$$A_3 = \frac{e^{\beta_4 h} - e^{\beta_4}}{e^{\beta_3} e^{\beta_4 h} - e^{\beta_3 h} e^{\beta_4}}, \qquad A_4 = \frac{e^{\beta_3 h} - e^{\beta_3}}{e^{\beta_4} e^{\beta_3 h} - e^{\beta_4 h} e^{\beta_3}},$$

$$A_5 = \frac{B_i e^{\beta_6 h}}{(\beta_5 - B_i)\left(e^{\beta_6} e^{\beta_5 h} - e^{\beta_6 h} e^{\beta_5}\right)},$$

$$A_6 = \frac{B_i e^{\beta_5 h}}{(\beta_6 - B_i)\left(e^{\beta_5} e^{\beta_6 h} - e^{\beta_5 h} e^{\beta_6}\right)},$$

$$A_7 = \frac{B_i\left(e^{\beta_8 h} - e^{\beta_8}\right)}{(\beta_7 - B_i)\left(e^{\beta_8} e^{\beta_7 h} - e^{\beta_8 h} e^{\beta_7}\right)},$$

$$A_8 = \frac{B_i\left(e^{\beta_7 h} - e^{\beta_7}\right)}{(\beta_8 - B_i)\left(e^{\beta_7} e^{\beta_8 h} - e^{\beta_7 h} e^{\beta_8}\right)},$$

$$A_9 = \frac{\mathrm{Gc}A_1}{N^2 - \beta_1^2}, \qquad A_{10} = \frac{\mathrm{Gc}A_2}{N^2 - \beta_2^2},$$

$$A_{11} = \frac{\mathrm{Gr}A_5}{N^2 - \beta_5^2}, \qquad A_{12} = \frac{\mathrm{Gr}A_6}{N^2 - \beta_6^2}, \qquad A_{13} = \frac{A}{N^2},$$

$$\begin{aligned}A_{14}^* &= A_9\left(e^{\beta_1} e^{\beta_{10} h} - e^{\beta_1 h} e^{\beta_{10}}\right) + A_{10}\left(e^{\beta_2} e^{\beta_{10} h} - e^{\beta_2 h} e^{\beta_{10}}\right) \\ &\quad + A_{11}\left(e^{\beta_5} e^{\beta_{10} h} - e^{\beta_5 h} e^{\beta_{10}}\right) \\ &\quad + A_{12}\left(e^{\beta_6} e^{\beta_{10} h} - e^{\beta_6 h} e^{\beta_{10}}\right) + A_{13}\left(e^{\beta_{10} h} - e^{\beta_{10}}\right),\end{aligned}$$

$$\begin{aligned}A_{15}^* &= A_9\left(e^{\beta_1} e^{\beta_9 h} - e^{\beta_1 h} e^{\beta_9}\right) + A_{10}\left(e^{\beta_2} e^{\beta_9 h} - e^{\beta_2 h} e^{\beta_9}\right) \\ &\quad + A_{11}\left(e^{\beta_5} e^{\beta_9 h} - e^{\beta_5 h} e^{\beta_9}\right) \\ &\quad + A_{12}\left(e^{\beta_6} e^{\beta_9 h} - e^{\beta_6 h} e^{\beta_9}\right) + A_{13}\left(e^{\beta_9 h} - e^{\beta_9}\right),\end{aligned}$$

$$A_{14} = \frac{A_{14}^*}{e^{\beta_{10}}e^{\beta_9 h} - e^{\beta_{10} h}e^{\beta_9}}, \qquad A_{15} = \frac{A_{15}^*}{e^{\beta_{10} h}e^{\beta_9} - e^{\beta_{10}}e^{\beta_9 h}},$$

$$A_{16} = \frac{\mathrm{Gc}A_3}{(P^2 - \beta_3^2)(1 + nE)}, \qquad A_{17} = \frac{\mathrm{Gc}A_4}{(P^2 - \beta_4^2)(1 + nE)},$$

$$A_{18} = \frac{\mathrm{Gr}A_7}{(P^2 - \beta_7^2)(1 + nE)},$$

$$A_{19} = \frac{\mathrm{Gr}A_8}{(P^2 - \beta_8^2)(1 + nE)}, \qquad A_{20} = \frac{B}{P^2(1 + nE)},$$

$$\begin{aligned} A_{21}^* &= A_{16}\left(e^{\beta_{12}}e^{\beta_3 h} - e^{\beta_{12} h}e^{\beta_3}\right) + A_{17}\left(e^{\beta_{12}}e^{\beta_4 h} - e^{\beta_{12} h}e^{\beta_4}\right) \\ &\quad + A_{18}\left(e^{\beta_{12}}e^{\beta_7 h} - e^{\beta_{12} h}e^{\beta_7}\right) \\ &\quad + A_{19}\left(e^{\beta_{12}}e^{\beta_8 h} - e^{\beta_{12} h}e^{\beta_8}\right) + A_{20}\left(e^{\beta_{12}} - e^{\beta_{12} h}\right), \\ A_{22}^* &= A_{16}\left(e^{\beta_{11}}e^{\beta_3 h} - e^{\beta_{11} h}e^{\beta_3}\right) + A_{17}\left(e^{\beta_{11}}e^{\beta_4 h} - e^{\beta_{11} h}e^{\beta_4}\right) \\ &\quad + A_{18}\left(e^{\beta_{11}}e^{\beta_7 h} - e^{\beta_{11} h}e^{\beta_7}\right) \\ &\quad + A_{19}\left(e^{\beta_{11}}e^{\beta_8 h} - e^{\beta_{11} h}e^{\beta_8}\right) + A_{20}\left(e^{\beta_{11}} - e^{\beta_{11} h}\right), \end{aligned}$$

$$A_{21} = \frac{A_{21}^*}{e^{\beta_{11}}e^{\beta_{12} h} - e^{\beta_{11} h}e^{\beta_{12}}}, \qquad A_{22} = \frac{A_{22}^*}{e^{\beta_{11} h}e^{\beta_{12}} - e^{\beta_{11}}e^{\beta_{12} h}}. \tag{28}$$

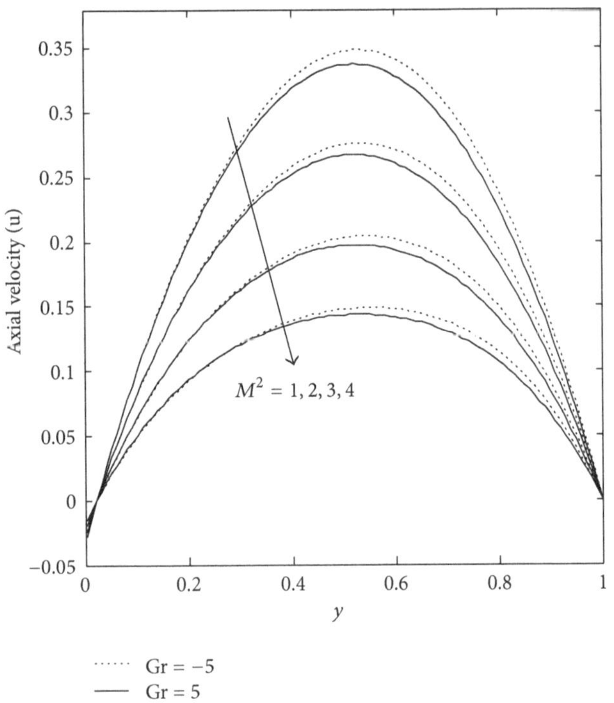

FIGURE 2: Variation of axial velocity profile for the different values of M^2.

4. Results and Discussion

A hydromagnetic flow, heat and mass transfer of a chemically reacting viscoelastic fluid in an irregular channel with convective boundary condition are investigated in the presence of heat absorption, thermal radiation, and Hall effect. The governing partial differential equations are solved analytically using perturbation technique. In order to get a physical insight of the problem, we have written a MATLAB program to perform a parametric study on different flow fields versus y for the different values of flow parameters. The parametric study is carried out for both the cases corresponding to cooling (Gr > 0) and heating (Gr < 0) of the plate by free convection currents. The results of this parametric study are presented graphically and are shown in Figures 2–23. During the numerical computations of velocity and temperature fields, the value of Prandtl number is taken as 3 which physically corresponds to Freon. Freon represents several different chlorofluorocarbons (CFCs) which are used in engineering and industrial applications.

Figures 2 and 3 indicate the effect of magnetic parameter, in both Gr < 0 and Gr > 0 cases on axial velocity and transverse velocity profile, respectively. We can see that in Figure 2 with increasing magnetic parameter, the axial velocity profile decreases in both the cases (Gr < 0 and Gr > 0). Physically, it is justified because transverse magnetic field yields to create a drag-like force called Lorentz force to resist the flow so magnetic field parameter slows down the flow and causes the axial velocity to decrease. But this trend is opposite for transverse velocity profiles in both the cases (Gr < 0 and Gr > 0).

In Figure 4, it is shown that with the increase in visco-elastic parameter, the axial velocity profile descends in both the cases. Because viscoelasticity introduces tensile stress the fluid motion contracts transversely, and hence axial velocity retards. It is also observed that the velocity of Newtonian fluid ($E = 0$) is higher than that of non-Newtonian fluid ($E \neq 0$) in both cases. The secondary velocity field is enhanced due to increase in visco-elastic parameter in both cases, which is depicted in Figure 5.

Figures 6 to 9 elucidate the inclusion of the buoyancy effects. The presence of the buoyancy effects complicates the problem, by coupling of the flow problem with thermal and mass problem. Figures 6 and 7 have been plotted to show the effect of the thermal Grashof number on primary and secondary velocity profiles, respectively. The values of Grashof number Gr $= -8, -6, -4, -2, 2, 4, 6, 8$ are chosen. Here the negative and positive values of Grashof number physically represent heating and cooling of the plate, respectively. Hence, it is observed from Figures 6 and 7 that an increase in the thermal Grashof number leads to increase in both the axial and transverse velocity profiles due to the enhancement of buoyancy forces. Actually, the thermal Grashof number signifies the relative importance of buoyancy force to the viscous hydrodynamic force. The increase in Grashof number indicates small viscous effects in momentum equation and consequently causes increase in the velocity profiles. With the increase in the solutal Grashof number, the axial and transverse velocity profiles increase in both cases (Gr < 0 and Gr > 0) as shown in Figures 8 and 9, respectively. The solutal Grashof number defines the ratio of species buoyancy force

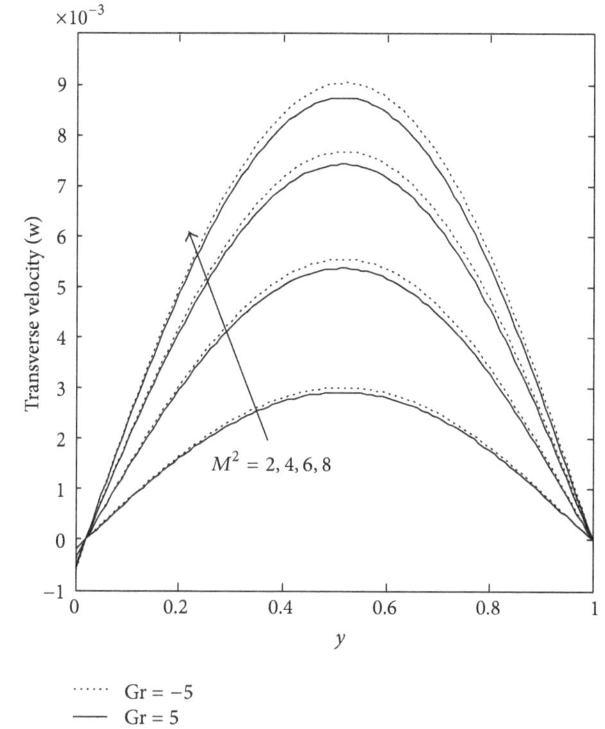

FIGURE 3: Variation of transverse velocity profile for the different values of M^2.

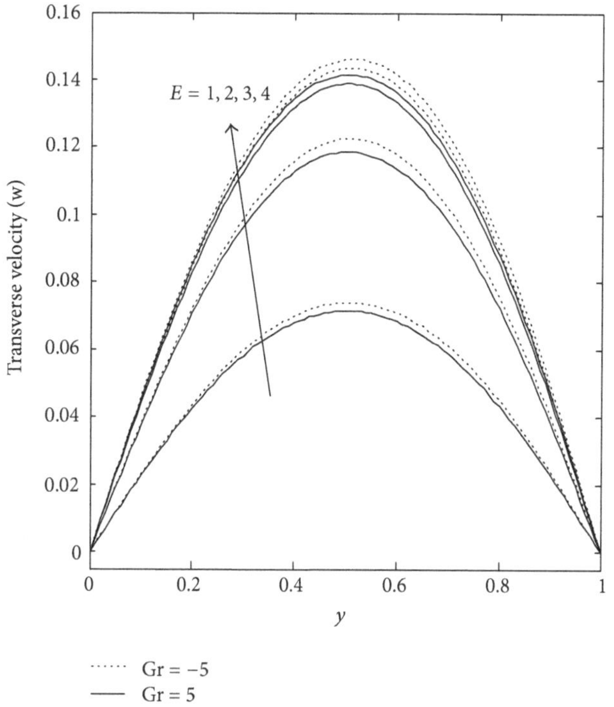

FIGURE 5: Variation of transverse velocity profile for the different values of E.

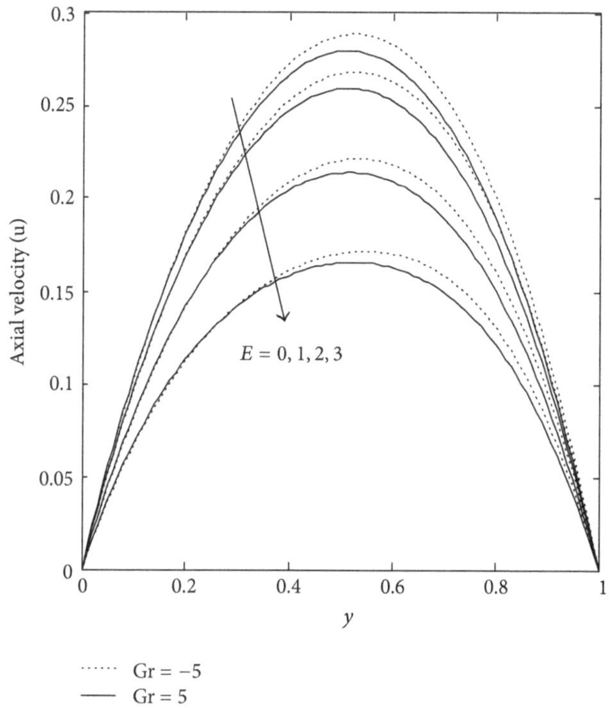

FIGURE 4: Variation of axial velocity profile for the different values of E.

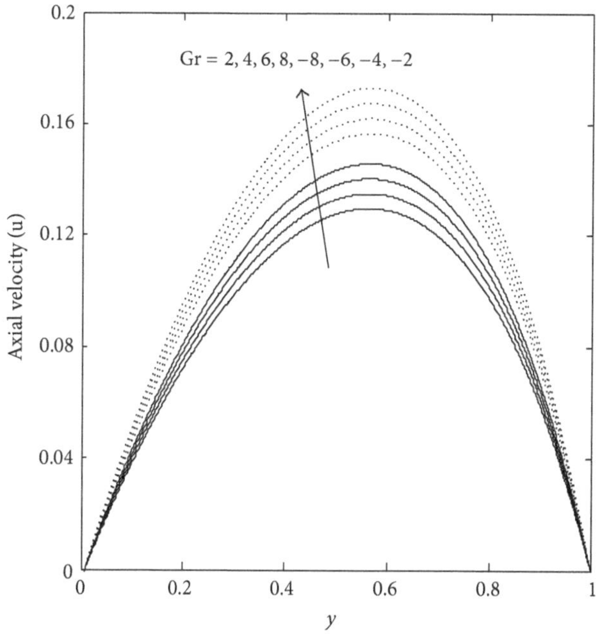

FIGURE 6: Variation of axial velocity profile for the different values of Gr.

to the viscous hydrodynamic force, so as to expect the fluid velocity increases.

One can see that in Figures 10 and 11 with increasing Hall parameter, the axial and transverse velocity profiles decrease in both cases ($\mathrm{Gr} < 0$ and $\mathrm{Gr} > 0$). The effect of chemical reaction parameter is just like the effect of Hall parameter, which are depicted in Figures 12 and 13. Figure 14 elucidates that the axial velocity profile increases with the increase in

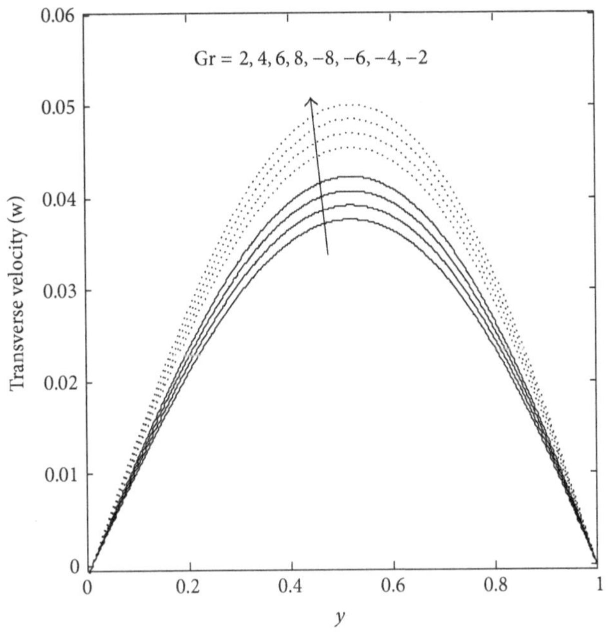

FIGURE 7: Variation of transverse velocity profile for the different values of Gr.

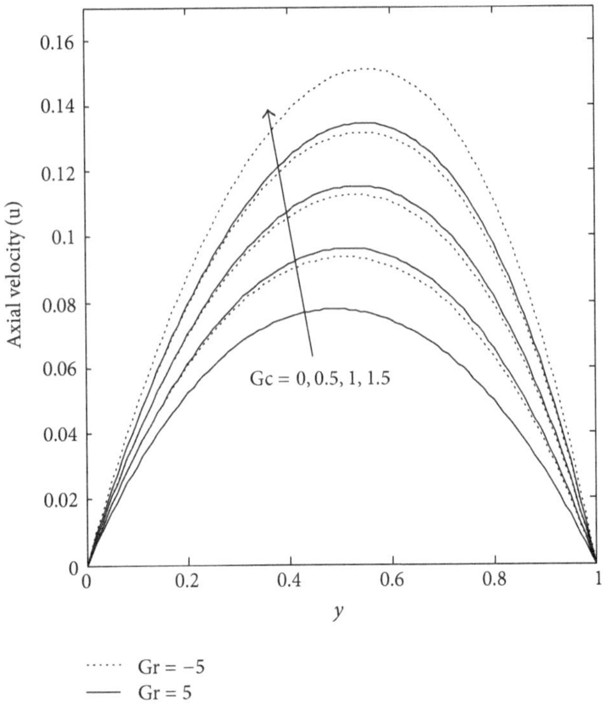

FIGURE 8: Variation of axial velocity profile for the different values of Gc.

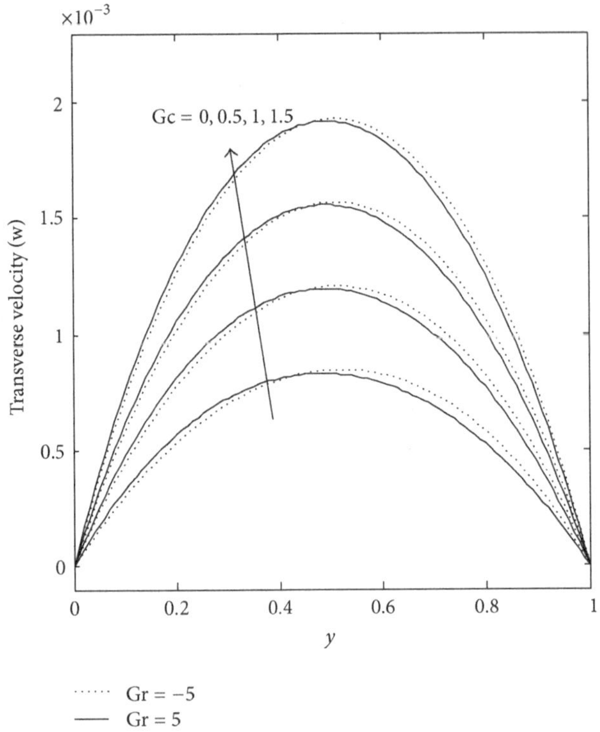

FIGURE 9: Variation of transverse velocity profile for the different values of Gc.

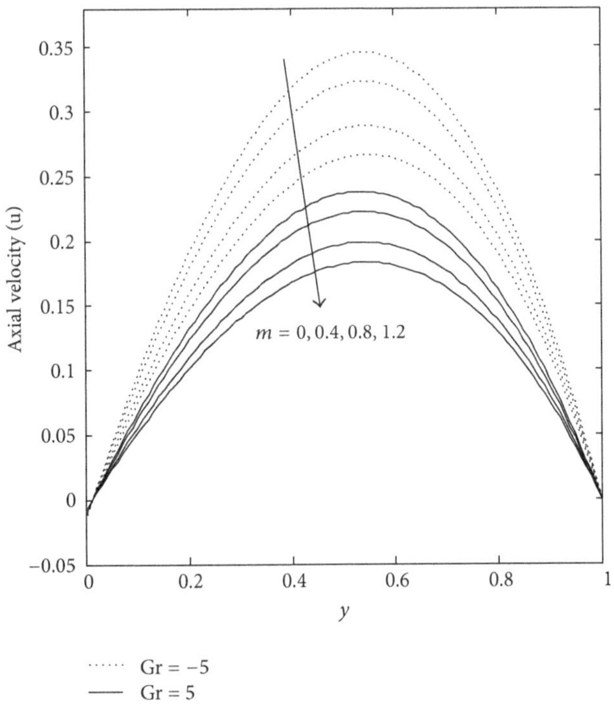

FIGURE 10: Variation of axial velocity profile for the different values of m.

the Biot number in the heating case ($Gr < 0$), and this manner is opposite in cooling case ($Gr > 0$). Further, the effect of Biot number on transverse velocity profile behaves in similar manner as we noticed in the effect of Biot number on axial velocity field, which is shown in Figure 15. Finally from Figures 2–15, it is interesting to note that both the axial and transverse velocity fields are higher in heating case than cooling case.

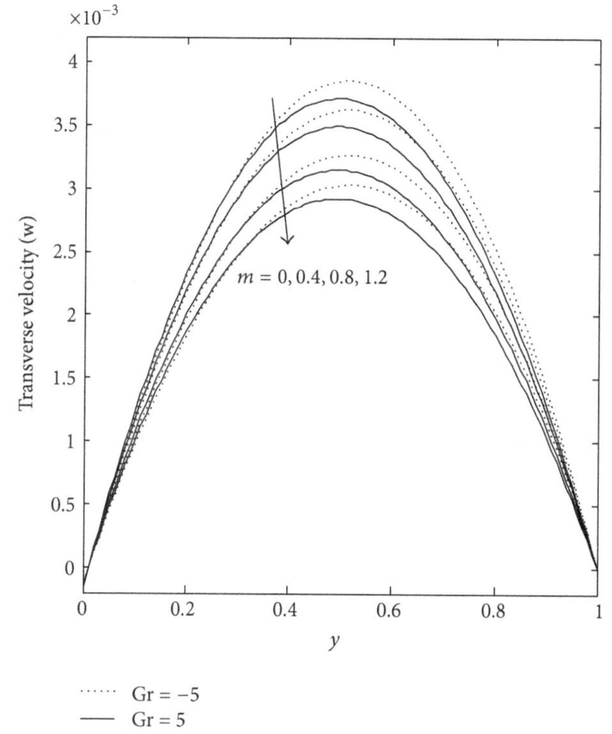

FIGURE 11: Variation of transverse velocity profile for the different values of m.

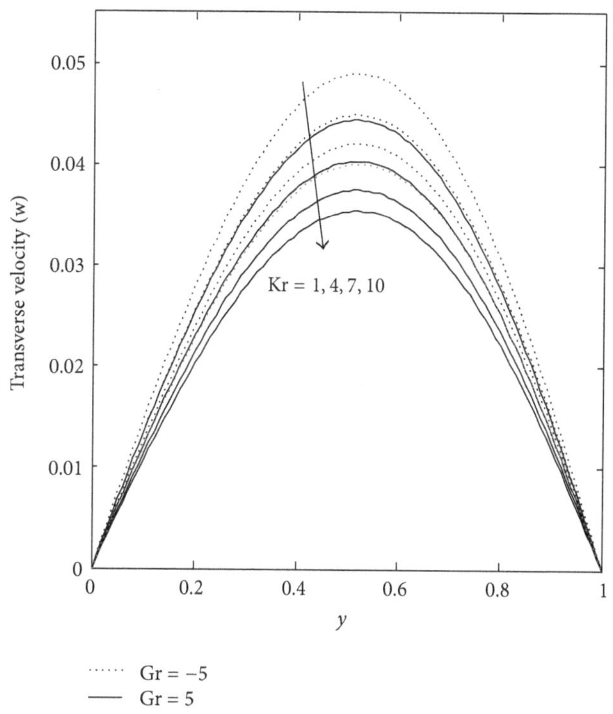

FIGURE 13: Variation of transverse velocity profile for the different values of Kr.

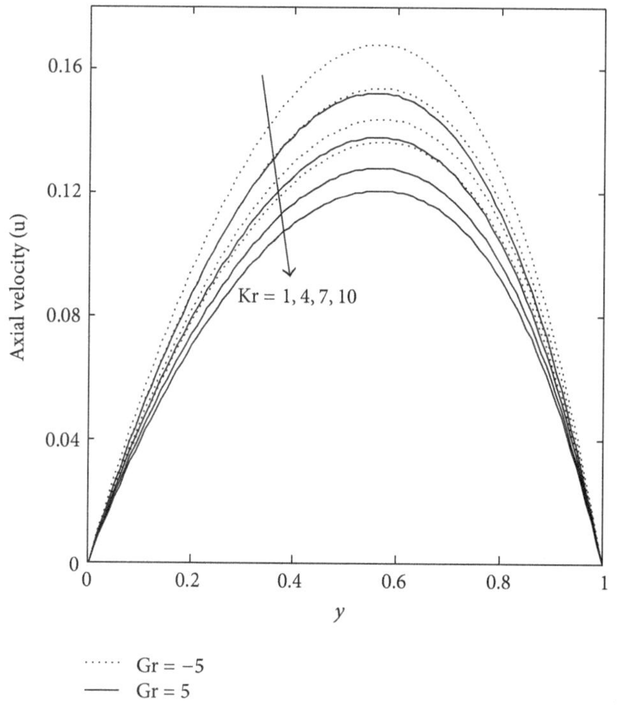

FIGURE 12: Variation of axial velocity profile for the different values of Kr.

Figures 16–19 give the spatial variation of dimensionless temperature distributions for different values α_T, R, Bi, and Pr, respectively. Figure 16 disclosed that the temperature profile decreases with increasing values of the heat absorption parameter α_T with Bi $= 20$ because the heat absorption parameter has propensity to drop off the thermal buoyancy effect, which decreases the heat transfer. Figure 17 reveals an interesting phenomenon that the temperature profile decreases with the increase in the value of radiation parameter with Bi $= 20$. This result qualitatively agrees with expectations since the effect of radiation is to decrease the rate of energy transport to the fluid, thereby decreasing the temperature of the fluid. The effect of Biot number on temperature profiles is figured in Figure 18. The Biot number is the ratio of the internal thermal resistance of the channel to the boundary layer thermal resistance. When Bi $= 0$ inside of the channel with hot fluid is totally insulated, the internal thermal resistance of the channel is extremely high and no convective heat transfer to the cold fluid on the outside of the channel takes place. Moreover, it is noted that the fluid temperature decreases for the increasing values of Biot number since as Bi increases, the thermal resistance of the channel decreases and convective heat transfer of the fluid increases. It is expected that as Biot number goes to infinity, the convective boundary conditions will become the prescribed wall temperature case [18].

Figure 19 has been plotted to depict the variation of temperature profiles against y for different values of Prandtl number by fixing other physical parameters (Bi $= 2$). Four different values of Prandtl number, namely, Pr $= 0.71, 1, 3,$ and 7 corresponding to air, electrolyte, Freon, and water, respectively, are chosen. From this graph we have observed that temperature profile decreases with the increase in the Prandtl

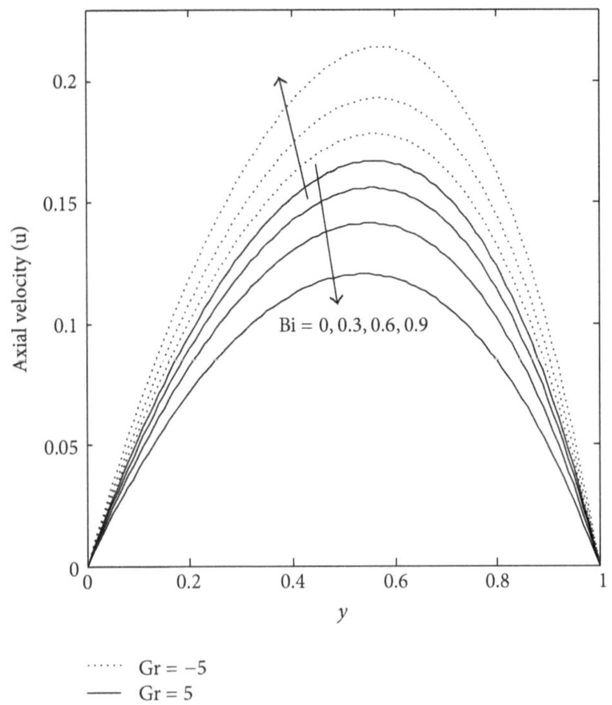

FIGURE 14: Variation of axial velocity profile for the different values of Bi.

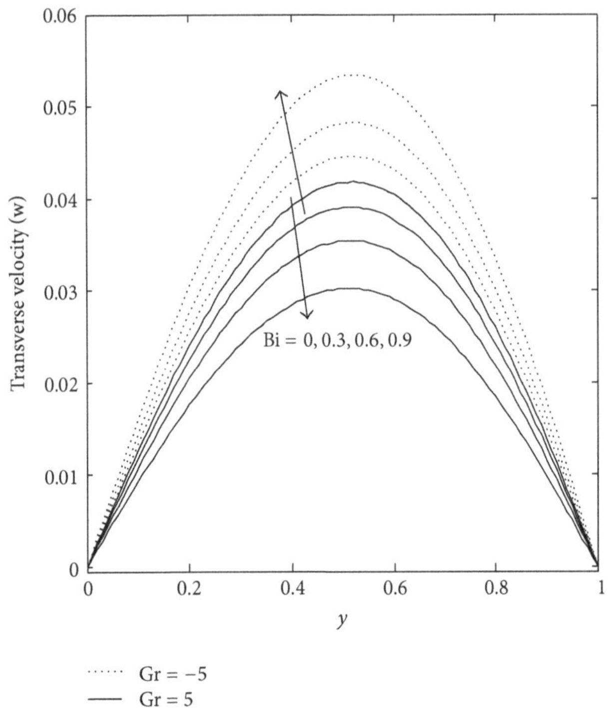

FIGURE 15: Variation of transverse velocity profile for the different values of Bi.

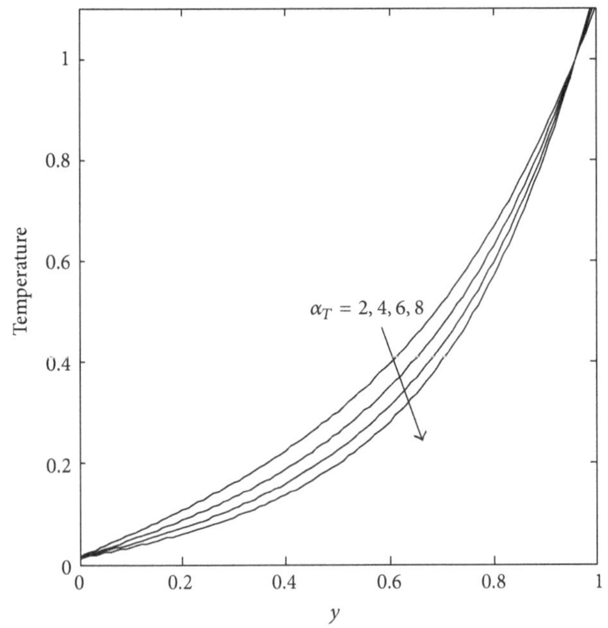

FIGURE 16: Variation of temperature profile for the different values of α_T.

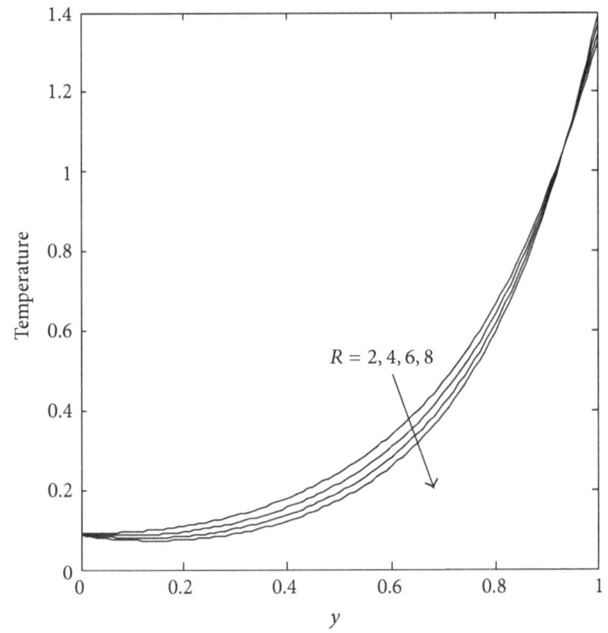

FIGURE 17: Variation of temperature profile for the different values of R.

number. This is because fluids with large Pr have low thermal diffusivity which cause low heat penetration resulting in reducing the temperature distributions.

The effect of the chemical reaction parameter on the species concentration profile is shown in Figure 20. It is clearly observed that the concentration profile decreases for increasing values of chemical reaction parameter. This is due to the fact that destructive chemical reaction reduces the solutal boundary layer thickness and decreases the mass transfer. Figure 21 displays concentration profiles versus y for various gases like hydrogen (Sc = 0.22), helium (Sc = 0.30), water vapour (Sc = 0.60), and ammonia (Sc = 0.78). It

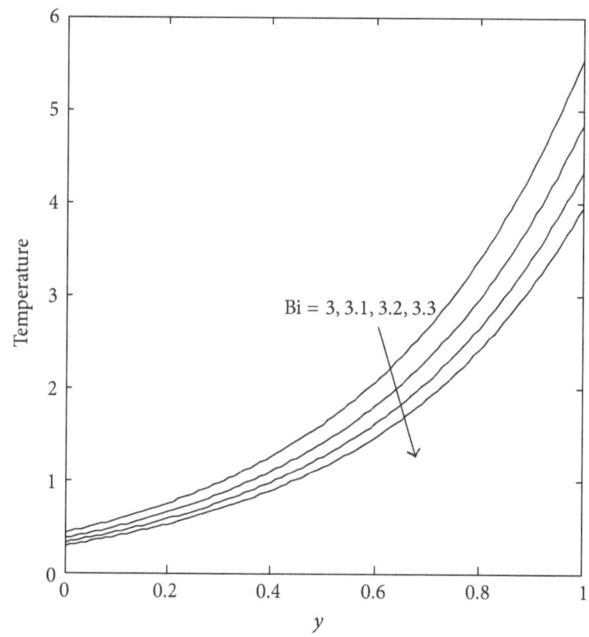

FIGURE 18: Variation of temperature profile for the different values of Bi.

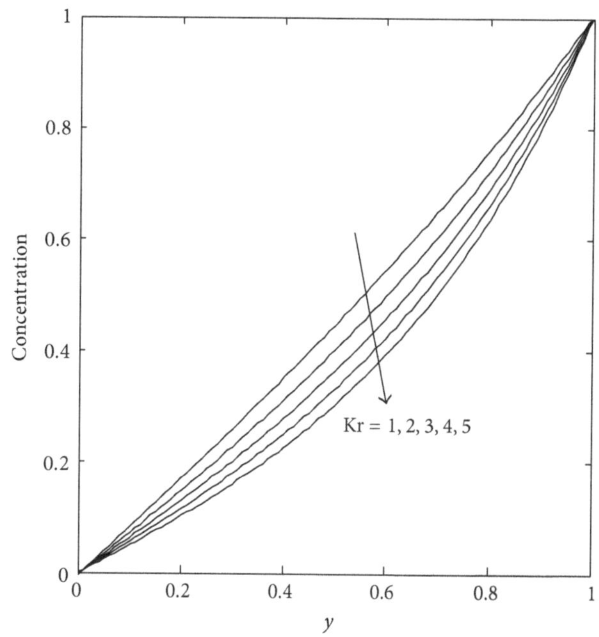

FIGURE 20: Variation of concentration profile for the different values of Kr.

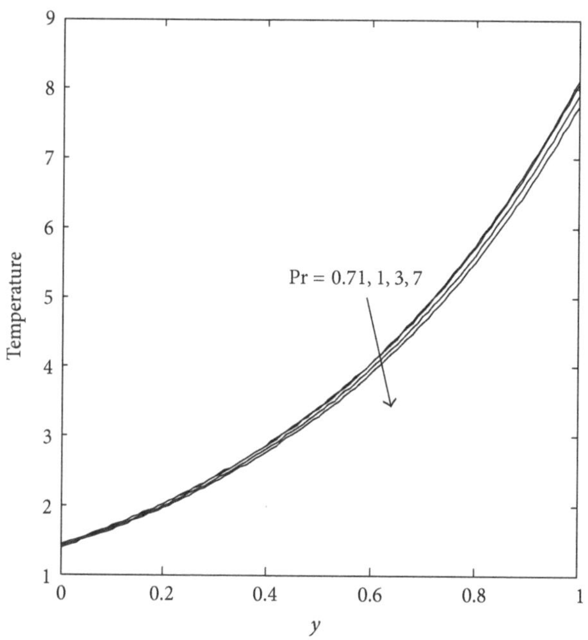

FIGURE 19: Variation of temperature profile for the different values of Pr.

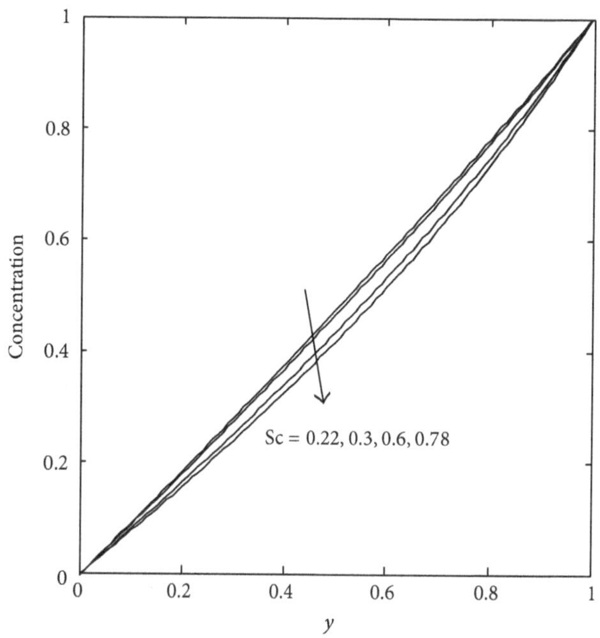

FIGURE 21: Variation of concentration profile for the different values of Sc.

is elucidated that an increase in Schmidt number increases the conduction, which decreases the mass transfer because the increase of Sc means a decrease of molecular diffusivity that results in decrease of concentration profile. Hence the concentration of species is higher for smaller value of Sc and vice versa.

Figures 22 and 23 are plotted to exemplify the effect of magnetic parameter and Hall parameter on skin friction components (τ_1, τ_2) at the wavy wall $(y = h)$ in both cases $(Gr < 0$ and $Gr > 0)$, respectively. Figure 22 illustrates that, the axial skin friction (τ_1) coefficient remarkably decreases for increasing values of magnetic parameter in both cases, while this trend is opposite for transverse skin friction (τ_2) coefficient in both cases. The axial skin friction (τ_1) profile enhances increasing values of Hall parameter in both the

TABLE 1: Effects of E, t, Bi, Kr, and Gc on skin friction distributions at the wavy wall $y = h$ for both cooling (Gr > 0) and heating (Gr < 0) cases.

E	t	Bi	Kr	Gc	τ_1 Gr < 0	τ_2 Gr < 0	τ_1 Gr > 0	τ_2 Gr > 0
2.0	0.1	0.1	2.0	2.0	0.5139	0.1411	0.4591	0.1244
4.0					0.4200	0.2202	0.3767	0.1941
6.0					0.3289	0.2437	0.2967	0.2153
1.0	0.0	0.1	2.0	2.0	0.5690	0.0780	0.4471	0.0596
	2.0				0.5699	0.0787	0.4476	0.0601
	4.0				0.5735	0.0818	0.4515	0.0631
1.0	0.1	0.1	2.0	2.0	0.5470	0.0784	0.4882	0.0694
		0.3			0.6121	0.0885	0.4231	0.0593
		0.5			0.6888	0.1007	0.3464	0.0472
1.0	0.1	0.1	1.0	2.0	0.5558	0.0776	0.4970	0.0687
			5.0		0.5200	0.0766	0.4612	0.0676
			9.0		0.5011	0.0697	0.4423	0.0607
1.0	0.1	0.1	2.0	1.0	0.5731	0.0815	0.4510	0.0628
				2.0	0.6792	0.1038	0.5571	0.0851
				3.0	0.7853	0.1262	0.6633	0.1075

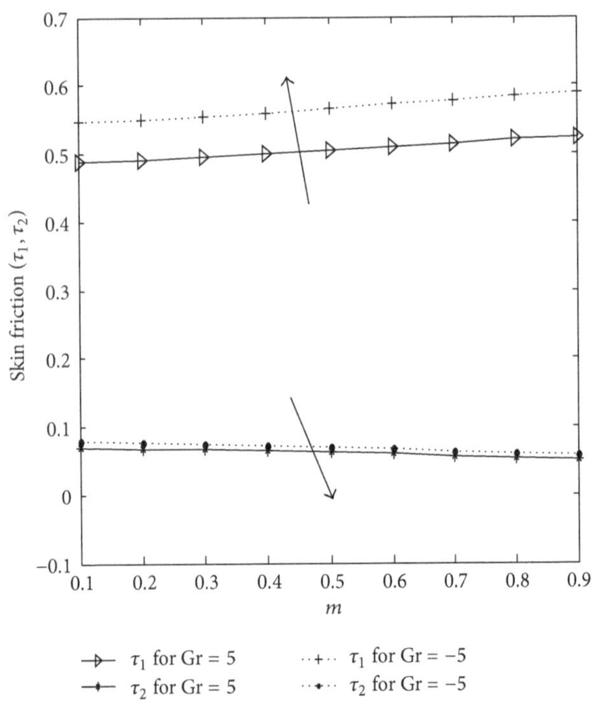

FIGURE 22: Variation of skin friction profile for the different values of M^2.

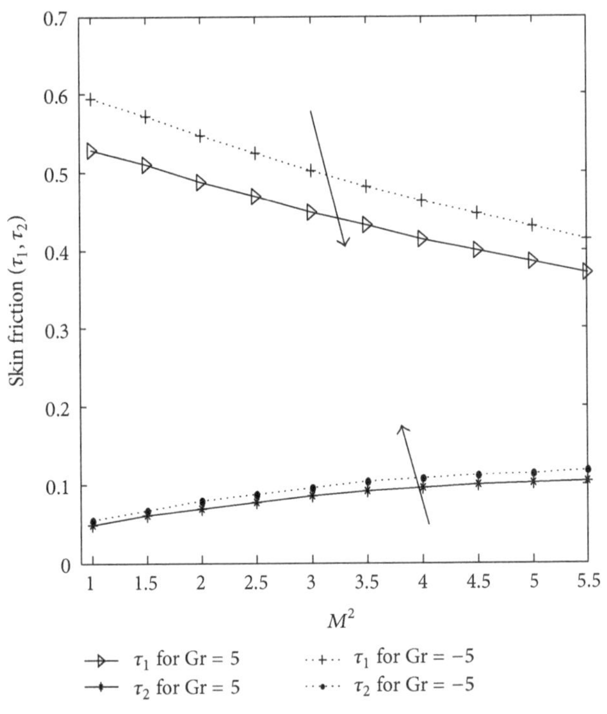

FIGURE 23: Variation of skin friction profile for the different values of m.

cases, but this manner is reverse for transverse skin friction coefficient, which is depicted from Figure 23.

Table 1 illustrates the variation of skin friction components at the wavy wall ($y = h$) for different values of E, t, Bi, Kr, and Gc. Further Tables 2 and 3 illustrate the variation of the Nusselt number and Sherwood number at the wavy wall $y = h$, for different values of Pr, t, and R and Kr, t, and Sc, respectively. Throughout the computations the values of physical parameters Kr $= 2$, $t = 0.1$, $E = 1$, $k_0 = 2$, Pr $= 3$, Gc $= 2$, Sc $= 0.78$, $\alpha_T = 1$, Bi $= 0.1$, $R = 4$, $M^2 = 2$, $A = B = 1$, $n = 0.1$, $\epsilon = 0.02$, $m = 0.1$, and $\lambda = 2\pi$ are used, unless otherwise stated.

5. Conclusions

The present investigation dealt with the combined effect of thermal radiation and Hall current on unsteady MHD

TABLE 2: Effects of Pr, t, and R on Nusselt number distributions at the wavy wall $y = h$.

Parameter	Values	Nu
Pr	0.0	2.6676
	0.72	2.5993
	3.0	2.5613
t	0.5	2.6033
	1.5	2.5767
	2.5	2.5647
R	0.0	1.1447
	1.0	1.4854
	2.0	2.5993

TABLE 3: Effects of Kr, t, and Sc on Sherwood number distributions at the wavy wall $y = h$.

Parameter	Values	Sh
Kr	1.0	1.8162
	3.0	1.4499
	5.0	1.1138
t	0.0	0.6650
	1.0	0.7933
	2.0	0.8405
Sc	0.22	0.9990
	0.30	0.9599
	0.78	0.7409

visco-elastic fluid flow in an irregular channel subject to convective boundary condition. The study is also concerned with the free convective flow and mass transfer with heat absorption and chemical reaction effects. The perturbation method is used to solve the problem, and the results are evaluated numerically and displayed graphically using MATLAB package. In the light of the present investigation, the following conclusions can be summarized.

(i) The velocity profiles are parabolic in nature, and the velocity field is higher in heating ($Gr < 0$) case than that in cooling ($Gr > 0$) case.

(ii) The Hall effect plays an important role in controlling momentum of the fluid.

(iii) The strength of the applied magnetic field should be as low as possible to realize.

(iv) The temperature profile decreases for increasing values of heat absorption parameter.

(v) The effect of increasing values of Prandtl number decreases temperature distributions.

(vi) An increase in the dimensionless time enhances the skin friction profile at the wavy wall, and this manner is opposite for increasing values of chemical reaction parameter.

(vii) The rate of heat transfer decreases for increasing values of heat absorption parameter at the wavy wall.

(viii) Combined effect of chemical reaction parameter and Schmidt number reduces the rate of mass transfer at the wavy wall, but this trend is reverse for the increasing in dimensionless time.

Thus the present study will serve as a good scientific tool for understanding more complex flow problems concerning the various physical parameters.

Acknowledgments

The authors wish to express their deep sense of gratitude to the editor and referee for their encouraging suggestions that significantly improved the paper.

References

[1] D. W. Beard and K. Walters, "Elastico-viscous boundary layer flows I. Two dimensional flow near a stagnation point," *Mathematical Proceedings of the Cambridge Philosophical Society*, vol. 60, pp. 667–674, 1964.

[2] K. R. Rajagopal and T. Y. Na, "On Stokes' problem for a non-Newtonian fluid," *Acta Mechanica*, vol. 48, no. 3-4, pp. 233–239, 1983.

[3] T. Sarpkaya, "Flow of non-Newtonian fluids in a magnetic field," *AIChE Journal*, vol. 7, pp. 324–328, 1961.

[4] C. C. Chang and T. S. Lundgren, "Duct flow in magnetohydrodynamics," *Zeitschrift für angewandte Mathematik und Physik*, vol. 12, no. 2, pp. 100–114, 1961.

[5] H. A. Attia and K. M. Ewis, "Unsteady MHD couette flow with heat transfer of a viscoelastic fluid under exponential decaying pressure gradient," *Tamkang Journal of Science and Engineering*, vol. 13, no. 4, pp. 359–364, 2010.

[6] M. E. Sayed-Ahmed and H. A. Attia, "MHD flow and heat transfer in a rectangular duct with temperature dependent viscosity and Hall effect," *International Communications in Heat and Mass Transfer*, vol. 27, no. 8, pp. 1177–1187, 2000.

[7] H. A. Attia and M. A. M. Abdeen, "Unsteady hartmann flow with heat transfer of a viscoelastic fluid under exponential decaying pressure gradient," *Engineering MECHANICS*, vol. 19, no. 5, pp. 37–44, 2012.

[8] H. A. Attia, "Hall effect on couette flow with heat transfer of a dusty conducting fluid between parallel porous plates under exponential decaying pressure gradient," *Journal of Mechanical Science and Technology*, vol. 20, no. 4, pp. 569–579, 2006.

[9] T. Hayat and O. U. Mehmood, "Slip effects on MHD flow of third order fluid in a planar channel," *Communications in Nonlinear Science and Numerical Simulation*, vol. 16, no. 3, pp. 1363–1377, 2011.

[10] Kh. S. Mekheimer and M. A. El Kot, "Influence of magnetic field and Hall currents on blood flow through a stenotic artery," *Applied Mathematics and Mechanics*, vol. 29, no. 8, pp. 1093–1104, 2008.

[11] D. Pal and B. Talukdar, "Perturbation analysis of unsteady magnetohydrodynamic convective heat and mass transfer in a boundary layer slip flow past a vertical permeable plate with thermal radiation and chemical reaction," *Communications in Nonlinear Science and Numerical Simulation*, vol. 15, no. 7, pp. 1813–1830, 2010.

[12] A. J. Chamkha, "MHD flow of a uniformly streched vertical permeable surface in the presence of heat generation/absorption

and a chemical reaction," *International Communications in Heat and Mass Transfer*, vol. 30, no. 3, pp. 413–422, 2003.

[13] I. J. Uwanta and E. Omokhuale, "Viscoelastic fluid flow in a fixed plane with heat and mass transfer," *Research Journal of Mathematics and Statistics*, vol. 4, no. 3, pp. 63–69, 2012.

[14] S. S. Saxena and G. K. Dubey, "MHD free convection heat and mass transfer flow of viscoelastic fluid embedded in a porous medium of variable permeability with radiation effect and heat source in slip flow regime," *Advances in Applied Science Research*, vol. 2, no. 5, pp. 115–129, 2011.

[15] M. Massoudi and A. K. Uguz, "Chemically-reacting fluids with variable transport properties," *Applied Mathematics and Computation*, vol. 219, pp. 1761–1775, 2012.

[16] R. C. Bataller, "Radiation effects for the Blasius and Sakiadis flows with a convective surface boundary condition," *Applied Mathematics and Computation*, vol. 206, no. 2, pp. 832–840, 2008.

[17] O. D. Makinde and T. Chinyoka, "Numerical study of unsteady hydromagnetic Generalized Couette flow of a reactive third-grade fluid with asymmetric convective cooling," *Computers and Mathematics with Applications*, vol. 61, no. 4, pp. 1167–1179, 2011.

[18] S. Yao, T. Fang, and Y. Zhong, "Heat transfer of a generalized stretching/shrinking wall problem with convective boundary conditions," *Communications in Nonlinear Science and Numerical Simulation*, vol. 16, no. 2, pp. 752–760, 2011.

[19] O. D. Makinde and A. Aziz, "MHD mixed convection from a vertical plate embedded in a porous medium with a convective boundary condition," *International Journal of Thermal Sciences*, vol. 49, no. 9, pp. 1813–1820, 2010.

[20] G. W. Sutton and A. Sherman, *Engineering Magnetohydrodynamics*, McGraw-Hill, New York, NY, USA, 1965.

[21] T. Hayat and M. Nawaz, "Soret and Dufour effects on the mixed convection flow of a second grade fluid subject to Hall and ion-slip currents," *International Journal for Numerical Methods in Fluids*, vol. 67, no. 9, pp. 1073–1099, 2011.

[22] U. N. Das, "Free convective MHD flow and heat transfer in a viscous incompressible fluid confined between a long vertical wavy wall and a parallel flat wall," *Indian Journal of Pure and Applied Mathematics*, vol. 23, pp. 295–304, 1992.

[23] Y. Abd elmaboud and Kh. S. Mekheimer, "Unsteady pulsatile flow through a vertical constricted annulus with heat transfer," *Zeitschrift für Naturforschung A*, vol. 67, pp. 185–194, 2012.

[24] B. Ničeno and E. Nobile, "Numerical analysis of fluid flow and heat transfer in periodic wavy channels," *International Journal of Heat and Fluid Flow*, vol. 22, no. 2, pp. 156–167, 2001.

[25] Kh. S. Mekheimer, S. Z. A. Husseny, and Y. Abd Elmaboud, "Effects of heat transfer and space porosity on peristaltic flow in a vertical asymmetric channel," *Numerical Methods for Partial Differential Equations*, vol. 26, no. 4, pp. 747–770, 2010.

[26] R. Sivaraj and B. Rushi Kumar, "Unsteady MHD dusty viscoelastic fluid Couette flow in an irregular channel with varying mass diffusion," *International Journal of Heat and Mass Transfer*, vol. 55, no. 11-12, pp. 3076–3089, 2012.

[27] R. Taneja and N. C. Jain, "MHD flow with slip effects and temperature-dependent heat source in a viscous incompressible fluid confined between a long vertical wavy wall and a parallel flat wall," *Defence Science Journal*, vol. 54, no. 1, pp. 21–29, 2004.

[28] K. Walters, *On Second-Order Effect in Elasticity,Plasticity and Fluid Mechanics*, IUTAM Int. Symp., Pergamon Press, New York, NY, USA, 1964.

[29] A. C. Cogley, S. E. Giles, and W. G. Vincent, "Differential approximation to radiative heat transfer in a nongrey gas near equilibrium," *AIAA Journal*, vol. 6, no. 3, pp. 551–553, 1968.

30

Phase Behavior at High Pressure of the Ternary System: CO_2, Ionic Liquid and Disperse Dye

Helen R. Mazzer,[1] José C. O. Santos,[1] Vladimir F. Cabral,[1] Claudio Dariva,[2] Marcos H. Kunita,[3] Adley F. Rubira,[3] Martin Aznar,[4] and Lucio Cardozo-Filho[1]

[1] *Department of Chemical Engineering, State University of Maringá, Avenue Colombo 5790, 87020-900 Maringá PR, Brazil*
[2] *Institute for Research and Technology, ITP, PEP/UNIT, Avenue Murilo Dantas 300, 49032-490 Aracaju SE, Brazil*
[3] *Department of Chemistry, State University of Maringá, Avenue Colombo 5790, 87020-900 Maringá PR, Brazil*
[4] *Department of Chemical Process, School of Chemical Engineering, State University of Campinas, Avenue Albert Einstein 500, 13081-970 Campinas SP, Brazil*

Correspondence should be addressed to Lucio Cardozo-Filho, cardozodequem@yahoo.com.br

Academic Editor: Ramesh Gardas

High pressure phase behavior experimental data have been measured for the systems carbon dioxide (CO_2) + 1-butyl-3-methylimidazolium hexafluorophosphate ([bmim] [PF_6]) and carbon dioxide (CO_2) + 1-butyl-3-methylimidazolium hexafluorophosphate ([bmim] [PF_6]) + 1-amino-2-phenoxy-4-hydroxyanthraquinone (C.I. Disperse Red 60). Measurements were performed in the pressure up to 18 MPa and at the temperature (323 to 353 K). As reported in the literature, at higher concentrations of carbon dioxide the phase transition pressure increased very steeply. The experimental data for the binary and ternary systems were correlated with good agreement using the Peng-Robinson equation of state. The amount of water in phase behavior of the systems was evaluated.

1. Introduction

In the past years the development of new technologies to minimize water consumption and wastewater emissions received more attention. In this context, the textile industry has a significant stake in the generation of effluents with highly pollution potential, using large amounts of water with chemical additives [1]. The process of dyeing using supercritical fluids is an alternative method that does not use water and any additive is frequently not required. In addition, such process allows the recovering and the reusing of the dyes. According to van der Kraan et al. [2], in dyeing of textiles consisting of nonpolar materials such as polyester, supercritical CO_2 acts as a blowing agent and increases the rate of diffusion of the dye molecules inside the fiber. Several studies have been reported on this textile-dyeing technology [3–6]. The literature describes applications of supercritical technology in combination with the ionic liquids in both the extraction and the separation processes and as an alternative solvent for organic and inorganic compounds in reaction mediums [7–9]. The ionic liquids are described as "designer solvents," where their characteristics can be adjusted to make the change either of cations or anions, or both. In this way, dyeing process involving association of supercritical technology with ionic liquids (ILs) can be used to improve the solubility of the dyes in the dyeing process of textile fibers and minimize the amount of energy demand.

In this context, the objective of this work is to report new experimental data of CO_2 (1) + 1-butyl-3-methylimidazolium hexafluorophosphate [bmim] [PF_6] + (2) and CO_2 (1) + [bmim] [PF_6] (2) C.I. Disperse Red 60 systems in the temperature range of 323 to 353 K. C.I. Disperse Red 60 is indeed a popular dye and one of the widely used dyes in dyeing process of synthetic textile fibers [2, 10, 11]. [bmim] [PF_6] is one of the most studied ionic liquids (ILs) in the literature with great availability of physical and chemical data as well as phase equilibrium data [12–14]. Keskin et al. [7], Huddleston et al. [15], and Jaitely et al. [16] have reported that the ionic liquid consisting of imidazole as a cation and $[PF_6]^-$ as an anion presents hydrophobic characteristics

TABLE 1: Experimental data of (L → LV) of binary system CO_2 (1) + [bmim] [PF_6] (2).

X_1	Temperature [K]	Pressure [MPa]	% $Water_{after}$	% $Water_{before}$
30.0×10^{-2}	333 ± 0.1	$4{,}53 \pm 0.1$	0.085 ± 0.02	0.26 ± 0.02
40.0×10^{-2}	333 ± 0.1	6.64 ± 0.1	0.085 ± 0.02	0.26 ± 0.02
47.0×10^{-2}	333 ± 0.1	8.06 ± 0.4	0.085 ± 0.02	0.26 ± 0.02
50.0×10^{-2}	333 ± 0.1	9.83 ± 0.3	0.085 ± 0.02	0.26 ± 0.02

compatible to that of disperse dyes with wide uses in dyeing of synthetic fibers. The synthetic static method with variable-volume cell was used in the experimental measurements of phase equilibrium. The amount of water present in ILs was determined by Karl Fischer analysis before and after equilibrium measures once that the ionic liquid [bmim] [PF_6] absorbs the moisture during the synthesis process due to its hygroscopic property [17]. A small amount of water in the ionic liquid can have a dramatic effect on the phase equilibrium behavior [13]. A technique to recover [bmim] [PF_6] from the [bmim] [PF_6] + C.I. Disperse Red 60 mixture was also proposed here once the reuse of [bmim] [PF_6] makes the dyeing process more environmentally benign and economically attractive. Peng-Robinson equation of state with quadratic mixing rules was used to correlate the experimental data.

2. Experimental

2.1. Materials. Commercial CO_2 and nitrogen (99%) were purchased from White Martins (Rio de Janeiro, Brazil). 1-Methylimidazole (99.0% GC) and 1-Bromobutane (96.0% GC) were purchased from Fluka (Sigma-Aldrich Ltda, Brazil). Ethyl acetate (99.5% P.A.), acetone (99.5% P.A.), activated charcoal (P.A.), anhydrous methanol (99.8% HPLC), dichloromethane (99.5% P.A.), and ethyl ether were purchased from Nuclear (CAQ-Chemical of Home Ind. Com. Ltda, Brazil). Potassium hexafluorophosphate (98% GC) was obtained from Aldrich (Sigma-Aldrich Ltda, Brazil). Silica gel mesh 0.035–0.070, 60 Å 99.5% was obtained from Acros (Sigma-Aldrich Ltda, Brazil). The Karl Fisher reagent (99.99% HPLC) was purchased from Merck (Merck S.A-Brazil). All water concentrations were measured through Karl Fischer (Analyzer, KF-1000) analysis. The 1-amino-2phenoxy-4-hydroxyanthaquinone (disperse Red 60) was supplied by DyStar (Brazil).

The ionic liquid [bmim] [PF_6] was synthesized from 1-methylimidazol, bromobutane and potassium hexafluorophosphate (KPF_6) by the process described in the literature [18]. [bmim] [PF_6] was dried under vacuum at 333 K and stored under dried nitrogen. It was purified in a chromatography column with silica gel Acros (Sigma-Aldrich Ltda, Brazil). (silica gel 60; mesh 70–230) and dichloromethane. Hydrogen and Carbon Nuclear Magnetic Resonance (^1H and ^{13}C NMR) analyses of the [bmim] [PF_6] obtained were δH (300 MHz; $CDCl_3$) 8.34 (sl, 3 N–CH_3–N), 7.31 (*t*, CH_2=CH), 7.28 (*t*, CH=CH_2), 4.09 (*t*, N–CH_2), 3.83 (*s*, N–CH_3), 1.78 (quintet, CH_2–CH_2), 1.27 (sextet, CH_2–CH_3), 0.85 (*t*, CH_2–CH_3) for δC (75 MHz, $CDCl_3$) 135.8 (N–CH_2–N), 123.8 (CH_2=CH), 122.5 (CH=CH_2), 49.8 (N–CH_2), 36.1 (N–CH_3), 31.8 (CH_2–CH_2), 19.3 (CH_2–CH_3), 13.3 (CH_2–CH_3). The analyses using NMR were performed in a Varian spectrometer, model Mercury Plus BB using $CDCl_3$ as a solvent.

2.2. Equipment and Procedure. Phase equilibrium experiments (cloud points) were performed through the synthetic static method [19] in a high-pressure variable-volume view cell. The experimental apparatus and experimental procedure are similar to those used in previous studies of phase equilibrium carried out by our group [20–22]. Briefly, the experiment begins with the introduction of a predetermined amount of compounds (CO_2, IL, and dye) in the equilibrium cell so that the global composition of the mixture is known. In the sequence, the system was stirred for 12 h at room temperature for complete solubilization of CO_2 in [bmim] [PF_6]. The procedure used here follows the work of Aki et al. [23]. After the initial procedure, pressure and temperature are adjusted to form a homogeneous solution. After reaching the thermal equilibrium of the system, pressure is slowly decreased until the appearance of a second phase is visually detected.

After the measurements of the phase transitions, the IL + Dye mixture was submitted to a process to recover the IL. The process of IL recovery consisted of the addition of a dichloromethane solution and active coal in the mass ration of 1 : 3 into the mixture of IL + Dye. Afterwards, the solution was stirred at room temperature for 48 h. Then, the mixture was filtrated using qualitative filter paper in a chromatographic column with silica gel (silica gel 60 Å; mesh 0.035–0.70 mm). The purified solution was evaporated under vacuum at 343 K for 10 h.

The values of water content refer to values obtained before and after the phase equilibrium measurements. For the binary system CO_2 + [bmim] [PF_6] the water content was 0.10 ± 0.02% (after phase equilibrium) and 0.23 ± 0.02 (before measurement). For the ternary system CO_2 + [bmim] [PF_6] + C.I. disperse Red 60 the water content corresponded to 0.43 ± 0.04 and 0.26 ± 0.02, after and before measurements, respectively. For the binary system with IL recovered the measurement were 0.19 ± 0.01 and 0.66 ± 0.02, after and before, respectively.

2.3. Thermodynamic Modeling. Peng-Robinson equation [24] of state was used to model the vapor-liquid equilibrium data:

$$P = \frac{RT}{v-b} - \frac{a}{v(v+b)+b(v-b)}. \quad (1)$$

The parameters a and by were calculated through the following mixing rules:

$$a = \sum_{i=1}^{nc}\sum_{j=1}^{nc} y_i y_j \sqrt{a_i a_j}\left(1-k_{ij}\right)$$
$$b = \sum_{j=1}^{nc} y_j b_j, \quad (2)$$

TABLE 2: Experimental data of liquid-vapor ternary system CO_2 (1) + [bmim] [PF6] (2) + Red Dye 60 (3).

X_1	X_2	X_3	Temperature [K]	Pressure [MPa]	% $Water_{after}$	% $Water_{before}$
			323 ± 0.1	4.41 ± 0.2	0.32 ± 0.01	0.83 ± 0.02
30.0×10^{-2}	69.9×10^{-2}	0.1×10^{-2}	333 ± 0.1	5.05 ± 0.1	0.32 ± 0.01	0.83 ± 0.02
			343 ± 0.1	5.69 ± 0.1	0.32 ± 0.01	0.83 ± 0.02
			353 ± 0.1	6.38 ± 0.1	0.32 ± 0.01	0.83 ± 0.02
			323 ± 0.1	5.41 ± 0.1	0.25 ± 0.02	0.34 ± 0.04
33.0×10^{-2}	66.9×10^{-2}	0.1×10^{-2}	333 ± 0.1	6.05 ± 0.1	0.25 ± 0.02	0.34 ± 0.04
			343 ± 0.1	6.76 ± 0.1	0.25 ± 0.02	0.34 ± 0.04
			353 ± 0.1	7.44 ± 0.1	0.25 ± 0.02	0.34 ± 0.04
			323 ± 0.1	5.84 ± 0.1	0.14 ± 0.03	0.24 ± 0.05
35.0×10^{-2}	64.9×10^{-2}	0.1×10^{-2}	333 ± 0.1	6.91 ± 0.1	0.14 ± 0.03	0.24 ± 0.05
			343 ± 0.1	8.10 ± 0.1	0.14 ± 0.03	0.24 ± 0.05
			353 ± 0.1	9.39 ± 0.1	0.14 ± 0.03	0.24 ± 0.05
			323 ± 0.1	6.25 ± 0.1	0.32 ± 0.01	0.83 ± 0.02
40.0×10^{-2}	59.9×10^{-2}	0.1×10^{-2}	333 ± 0.1	7.21 ± 0.1	0.32 ± 0.01	0.83 ± 0.02
			343 ± 0.1	8.25 ± 0.1	0.32 ± 0.01	0.83 ± 0.02
			353 ± 0.1	9.37 ± 0.1	0.32 ± 0.01	0.83 ± 0.02
			323 ± 0.1	7.05 ± 0.1	0.14 ± 0.03	0.24 ± 0.05
43.0×10^{-2}	56.9×10^{-2}	0.1×10^{-2}	333 ± 0.1	8.45 ± 0.1	0.14 ± 0.03	0.24 ± 0.05
			343 ± 0.1	10.01 ± 0.1	0.14 ± 0.03	0.24 ± 0.05
			353 ± 0.1	11.62 ± 0.1	0.14 ± 0.03	0.24 ± 0.05
			323 ± 0.1	8.15 ± 0.1	0.25 ± 0.02	0.34 ± 0.04
46.0×10^{-2}	53.9×10^{-2}	0.1×10^{-2}	333 ± 0.1	9.31 ± 0.1	0.25 ± 0.02	0.34 ± 0.04
			343 ± 0.1	10.54 ± 0.1	0.25 ± 0.02	0.34 ± 0.04
			353 ± 0.1	11.99 ± 0.1	0.25 ± 0.02	0.34 ± 0.04
			323 ± 0.1	8.42 ± 0.1	0.14 ± 0.03	0.24 ± 0.05
50.0×10^{-2}	49.9×10^{-2}	0.1×10^{-2}	333 ± 0.1	10.28 ± 0.1	0.14 ± 0.03	0.24 ± 0.05
			343 ± 0.1	12.41 ± 0.1	0.14 ± 0.03	0.24 ± 0.05
			353 ± 0.1	14.72 ± 0.1	0.14 ± 0.03	0.24 ± 0.05
			323 ± 0.1	10.87 ± 0.1	0.25 ± 0.02	0.34 ± 0.04
53.0×10^{-2}	46.9×10^{-2}	0.1×10^{-2}	333 ± 0.1	13.01 ± 0.1	0.25 ± 0.02	0.34 ± 0.04
			343 ± 0.1	15.26 ± 0.1	0.25 ± 0.02	0.34 ± 0.04
			353 ± 0.1	17.68 ± 0.1	0.25 ± 0.02	0.34 ± 0.04

where y_i is the molar fraction of the component i and a_i and b_i are the Peng-Robinson parameters for the pure component i. The binary interaction parameters, k_{ij}, were determined through the correlation of experimental data obtained in this work. The optimization procedure employed for fitting k_{ij} was based on the Nelder and Mead simplex method [25].

3. Results and Discussion

Experimental data of phase transitions at high pressures of the CO_2 (1) + [bmim] [PF_6] (2) binary system at 333 K are presented in Figure 1 and Table 1. All phase transitions observed were vapor-liquid type. The CO_2 (1) solubility in [bmim] [PF_6] (2) presented in Figure 1 is compared with experimental data reported by Blanchard and Brennecke [26] and Fu et al. [17]. Solubility obtained here is in agreement with Fu et al. [17] data. However, experimental data reported by Blanchard and Brennecke [26] show a significant divergence compared to the results obtained in this work. Fu et al. [17] state that the discrepancy in the solubility measurements can be attributed to different experimental techniques, uncertainties in the experimental procedure, impurity in the IL, among other factors. From own our point of view, the differences verified in this figure are mainly due to the water content in the ionic liquids used in the experiments. Blanchard and Brennecke [26] reported 0.15% water content in IL prior to experimental measures. However, water content data after the experiments have not yet been reported.

It is possible to notice that CO_2 solubility in the IL increases continually with pressure, reaching 0.50 at 333 K and 9.83 MPa. Such behavior is also reported by Bermejo et al. [27], Kamps et al. [14], and Shariati and Peters [12].

Figure 2 and Table 2 show the experimental phase behavior for the ternary system CO_2 (1) + [bmim] [PF_6] (2) +

TABLE 3: Experimental data of liquid-vapor binary system carbon dioxide + [bmim] $[PF6]_{recovered}$.

X_1	Temperature [K]	Pressure [MPa]	% $Water_{after}$	% $Water_{before}$
30.0×10^{-2}	333 ± 0.1	5.70 ± 0.1	0.190 ± 0.010	0.66 ± 0.02
40.0×10^{-2}	333 ± 0.1	7.40 ± 0.1	0.190 ± 0.010	0.66 ± 0.02
45.0×10^{-2}	333 ± 0.1	8.45 ± 0.4	0.190 ± 0.010	0.66 ± 0.02
50.0×10^{-2}	333 ± 0.1	10.61 ± 0.3	0.190 ± 0.010	0.66 ± 0.02

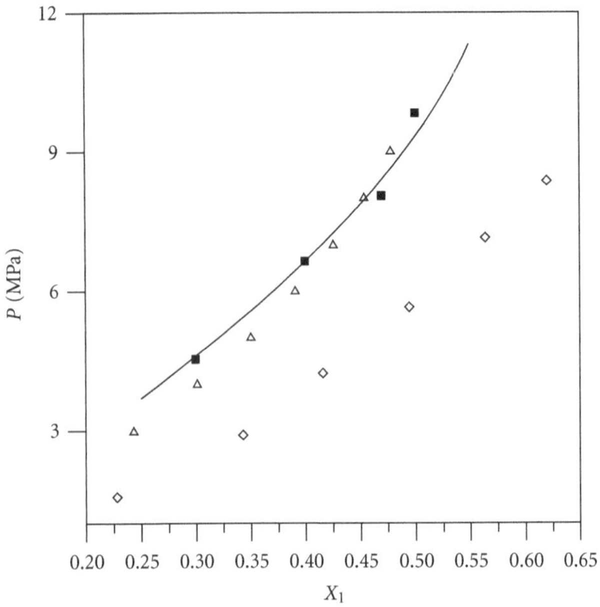

FIGURE 1: Experimental data obtained in this work (■) and reported by Blanchard et al. [13] (◇), Fu et al. [17] (△) for the $CO_2(1)$ + [bmim] $[PF_6]$ (2) binary system at 333 K isotherm. Thermodynamic modeling (—).

TABLE 4: Critical properties and acentric factor of carbon dioxide, ionic liquid and disperse Red 60.

Components	Mw	T_c (K)	P_c (MPa)	ω
CO_2	44.01	304.2	73.8	0.239
[bmim] $[PF_6]$	284.18	708.9	17.3	0.755
Disperse red 60	331.32	1361	30.2	1.370

disperse dye Red 60 (3). In this case, it is possible to notice the increase of CO_2 solubility with increasing pressure, an analogue behavior to that of the binary system. Moreover, CO_2 solubility decreases as temperature increases. The IL addition in supercritical environment induces a significant solubilization of the dispersed dye Red 60, which is almost insoluble in supercritical CO_2 in the temperature and pressure range investigated. The solubility of the dye Red 60 in CO_2 is around 10^{-6} in the investigated range [28]. On the other hand, in this range the solubility of the dye Red 60 in CO_2 and IL has an average value of 10^{-4}. For all experimental conditions (Figures 1 and 2), a good agreement of the Peng-Robinson equation with the experimental data is observed.

Figure 3 shows a comparison of the phase diagram for the systems CO_2 (1) + [bmim] $[PF_6]$ (2) and CO_2 (1) + [bmim] $[PF_6]$ (2) + disperse dye Red 60 (3) at 333 K. It is possible to observe that CO_2 shows lower solubility in ionic liquid for the ternary system when compared to the binary system.

Figure 4 shows the phase behavior for the binary systems CO_2 (1) + [bmim] $[PF_6]$ (2) and CO_2 (1) + [bmim] $[PF_6]_{recovered}$ (2). Such results are summarized in Table 3. It can be observed that the phase behavior of both systems (CO_2 (1) + [bmim] $[PF_6]$ (2) and CO_2 (1) + [bmim] $[PF_6]_{recovered}$ (2)) are similar. Solubility differences between these two systems can be attributed to the water content present in each situation. Furthermore, the RMN spectra of IL after removal of the disperse dye Red 60 were consistent with values found for pure LI produced in our laboratory and with those reported in literature [29].

TABLE 5: PR-EoS binary interaction parameters for the binary and ternary systems.

System	T/K	K_{ij}	AAD%
[bmim] $[PF_6]_{system\ binary}$	333	0.1441	1.34
[bmim] $[PF_6]_{recovered}$	333	0.1577	3.45
	323	0.1536	4.65
[bmim] $[PF_6]_{system\ ternary}$	333	0.1616	3.87
	343	0.1709	4.58
	353	0.1787	4.15

The PR-EoS with the quadratic mixing rules was used to modeling the phase transition experimental data for all systems. The binary interaction parameters in the mixing rules were optimized using the P, x data for each system. The critical properties and acentric factors of the IL were obtained of the Valderrama and Robles [30], and the critical properties, and acentric factors of the disperse dye Red 60 were obtained of the Mishima et al. [31]. The critical properties are presented in Table 4 and Table 5 shows the interaction parameters for each system.

4. Conclusions

It was verified that CO_2 solubility in the ionic liquid decreased by the presence of dye. Such phenomenon results in higher vapor-liquid pressure transitions for the ternary system when compared to the binary system. Water content (impurities) in the systems studied considerably change in the pressure values of phase transition, corroborating literature observations [19, 31, 32]. Therefore, water content in IL synthesis should be carefully controlled. IL's purification protocol for dye removal presented good results without IL physical chemical changes, which enables IL to be reused.

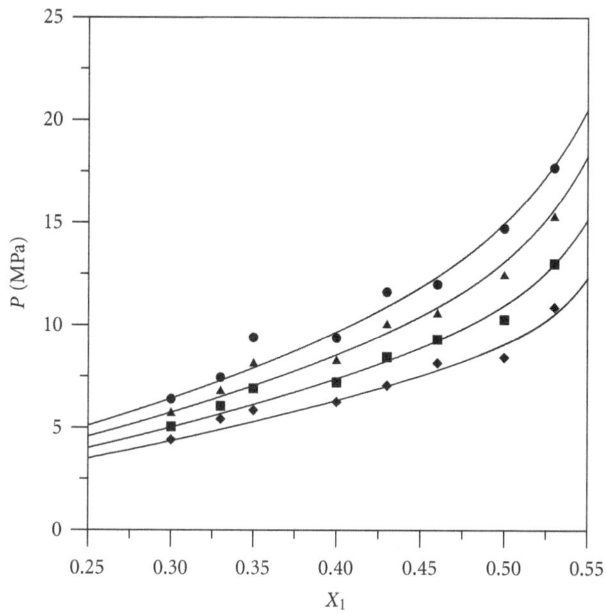

FIGURE 2: Experimental data of bubble-type phase transition at (♦) 323 K, (■) 333 K, (▲) 343 K, and (•) 353 K for the ternary system $CO_2(1)$ + [bmim] [PF_6] (2) + disperse dye Red 60 (3). Thermodynamic modeling using the Peng-Robinson equation of state (—).

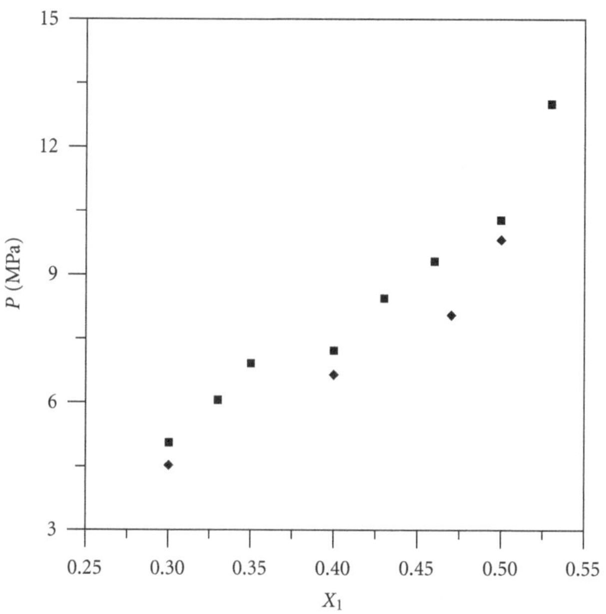

FIGURE 3: Comparison of Px diagram for the binary system $CO_2(1)$ + [bmim] [PF_6] (2) (♦) and ternary system $CO_2(1)$ + [bmim] [PF_6] (2) + disperse dye Red 60 (3) (■) at 333 K.

Peng-Robinson equation of state presented a good performance in describing the experimental data observed. The results obtained might be useful in the development of dyeing processes using supercritical fluids in association with ionic liquids. Finally, the results of partition of the undisclosed dye composition between $SCCO_2$ and an ionic liquid demonstrated that it is possible to increase significantly the

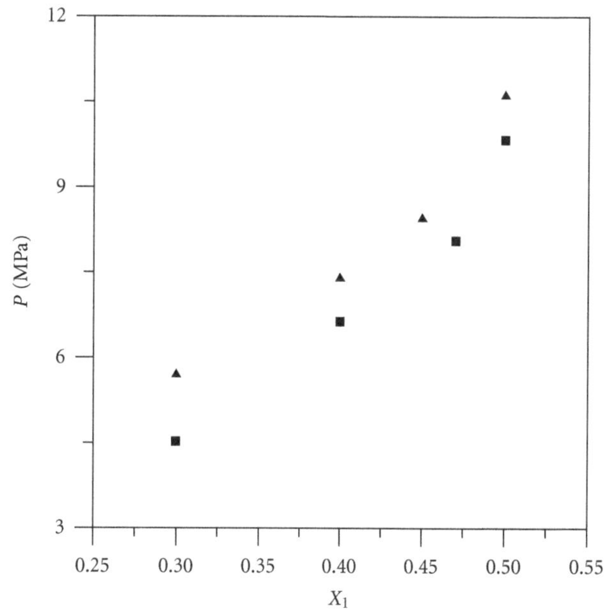

FIGURE 4: Phase transition Px diagram for the binary system $CO_2(1)$ + [bmim] [PF_6] (2) (■) and $CO_2(1)$ + [bmim] $[PF_6]_{recovered}$ (2) (▲) at 333 K.

disperse dye solubility in $SCCO_2$ using ionic liquids as an addictive. This effect can be used to improve the color aspects in the textiles, one of the most desirable characteristics in the dyeing of textiles.

Acknowledgments

The authors would like to thank CNPq for the financial support. This paper is in memory of Proffessor D. S. Octávio Augusto Ceva Antunes with thanks for the orientation, friendship and dedication during the investigation.

References

[1] S. Papić, N. Koprivanac, A. Lončarić Božić, and A. Meteš, "Removal of some reactive dyes from synthetic wastewater by combined Al(III) coagulation/carbon adsorption process," *Dyes and Pigments*, vol. 62, no. 3, pp. 291–298, 2004.

[2] M. van der Kraan, M. V. F. Cid, G. F. Woerlee, W. J. T. Veugelers, and G. J. Witkamp, "Equilibrium study on the disperse dyeing of polyester textile in supercritical carbon dioxide," *Textile Research Journal*, vol. 77, no. 8, pp. 550–558, 2007.

[3] E. Bach, E. Cleve, and E. Schollmeyer, "Past, present and future of supercritical fluid dyeing technology—an overview," *Review of Progress in Coloration and Related Topics*, vol. 32, pp. 88–102, 2002.

[4] A. Schmidt, E. Bach, and E. Schollmeyer, "Damage to natural and synthetic fibers treated in supercritical carbon dioxide at 300 bar and temperatures up to 160°C," *Textile Research Journal*, vol. 72, no. 11, pp. 1023–1032, 2002.

[5] A. S. Özcan, A. A. Clifford, and K. D. Bartle, "Solubility of disperse dyes in supercritical carbon dioxide," *Journal of Chemical and Engineering Data*, vol. 42, no. 3, pp. 590–592, 1997.

[6] T. Shinoda and K. Tamura, "Solubilities of C.I. Disperse orange 25 and C.I. Disperse blue 354 in supercritical carbon dioxide," *Journal of Chemical and Engineering Data*, vol. 48, no. 4, pp. 869–873, 2003.

[7] S. Keskin, D. Kayrak-Talay, U. Akman, and Ö. Hortaçsu, "A review of ionic liquids towards supercritical fluid applications," *Journal of Supercritical Fluids*, vol. 43, no. 1, pp. 150–180, 2007.

[8] L. A. Blanchard, D. Hancu, E. J. Beckman, and J. F. Brennecke, "Green processing using ionic liquids and CO_2," *Nature*, vol. 398, no. 6731, pp. 28–29, 1999.

[9] J. Dupont, R. F. De Souza, and P. A. Z. Suarez, "Ionic liquid (molten salt) phase organometallic catalysis," *Chemical Reviews*, vol. 102, no. 10, pp. 3667–3692, 2002.

[10] A. Ferri, M. Banchero, L. Manna, and S. Sicardi, "Dye uptake and partition ratio of disperse dyes between a PET yarn and supercritical carbon dioxide," *Journal of Supercritical Fluids*, vol. 37, no. 1, pp. 107–114, 2006.

[11] P. Bao and J. Dai, "Relationships between the solubility of C. I. disperse red 60 and uptake on PET in supercritical CO_2," *Journal of Chemical and Engineering Data*, vol. 50, no. 3, pp. 838–842, 2005.

[12] A. Shariati and C. J. Peters, "High-pressure phase equilibria of systems with ionic liquids," *Journal of Supercritical Fluids*, vol. 34, no. 2, pp. 171–176, 2005.

[13] L. A. Blanchard, Z. Gu, and J. F. Brennecke, "High-pressure phase behavior of ionic liquid/CO_2 systems," *Journal of Physical Chemistry B*, vol. 105, no. 12, pp. 2437–2444, 2001.

[14] A. P. S. Kamps, D. Tuma, J. Xia, and G. Maurer, "Solubility of CO_2 in the ionic liquid [bmim][PF_6]," *Journal of Chemical and Engineering Data*, vol. 48, no. 3, pp. 746–749, 2003.

[15] J. G. Huddleston, A. E. Visser, W. M. Reichert, H. D. Willauer, G. A. Broker, and R. D. Rogers, "Characterization and comparison of hydrophilic and hydrophobic room temperature ionic liquids incorporating the imidazolium cation," *Green Chemistry*, vol. 3, no. 4, pp. 156–164, 2001.

[16] V. Jaitely, A. Karatas, and A. T. Florence, "Water-immiscible room temperature ionic liquids (RTILs) as drug reservoirs for controlled release," *International Journal of Pharmaceutics*, vol. 354, no. 1-2, pp. 168–173, 2008.

[17] D. Fu, X. Sun, J. Pu, and S. Zhao, "Effect of water content on the solubility of CO_2 in the ionic liquid [bmim][PF_6]," *Journal of Chemical and Engineering Data*, vol. 51, no. 2, pp. 371–375, 2006.

[18] J. Dupont, P. A. Z. Suarez, C. S. Consorti et al., "Preparation of 1-butyl-3-methyl imidazolium-based room temperature ionic liquids," *Organic Syntheses*, vol. 79, pp. 236–243, 2002.

[19] M. A. McHugh and V. J. Krukonis, *Supercritical Fluid Extraction: Principles and Practice*, Butterworth-Heinemann, Boston, Mass, USA, 1994.

[20] R. Favareto, P. H. Fregadolli, V. F. Cabral, O. A. Antunes, and L. Cardozo-Filho, "Phase equilibria of acrylonitrile and p-bromobenzaldehyde in carbon dioxide," *Journal of Chemical and Engineering Data*, vol. 53, no. 5, pp. 1080–1084, 2008.

[21] A. T. Souza, M. L. Corazza, L. Cardozo-Filho, R. Guirardello, and M. A. A. Meireles, "Phase equilibrium measurements for the system clove (Eugenia caryophyllus) Oil + CO_2," *Journal of Chemical and Engineering Data*, vol. 49, no. 2, pp. 352–356, 2004.

[22] M. L. Corazza, L. C. Filho, O. A. C. Antunes, and C. Dariva, "High pressure phase equilibria of the related substances in the limonene oxidation in supercritical CO_2," *Journal of Chemical and Engineering Data*, vol. 48, no. 2, pp. 354–358, 2003.

[23] S. N. V. K. Aki, A. M. Scurto, and J. F. Brennecke, "Ternary phase behavior of ionic liquid (IL)-organic-CO_2 systems," *Industrial and Engineering Chemistry Research*, vol. 45, no. 16, pp. 5574–5585, 2006.

[24] D. Y. Peng and D. B. Robinson, "A new two-constant equation of state," *Industrial and Engineering Chemistry Fundamentals*, vol. 15, no. 1, pp. 59–64, 1976.

[25] J. A. Nelder and R. Mead, "A Simplex-method for function minimization," *Computer Journal*, vol. 7, pp. 308–313, 1965.

[26] L. A. Blanchard and J. F. Brennecke, "Recovery of organic products from ionic liquids using supercritical carbon dioxide," *Industrial and Engineering Chemistry Research*, vol. 40, no. 1, pp. 287–292, 2001.

[27] M. D. Bermejo, M. Montero, E. Saez et al., "Liquid-vapor equilibrium of the systems butylmethylimidazolium nitrate -CO_2 and hydroxypropylmethylimidazolium nitrate -CO_2 at high pressure: influence of water on the phase behavior," *Journal of Physical Chemistry B*, vol. 112, no. 43, pp. 13532–13541, 2008.

[28] S. N. V. K. Aki, B. R. Mellein, E. M. Saurer, and J. F. Brennecke, "High-pressure phase behavior of carbon dioxide with imidazolium-based ionic liquids," *Journal of Physical Chemistry B*, vol. 108, no. 52, pp. 20355–20365, 2004.

[29] R. B. Gupta and J. J. Shin, *Solubility in Supercritical Carbon Dioxide*, CRC Press, New York, NY, USA, 2007.

[30] J. O. Valderrama and P. A. Robles, "Critical properties, normal boiling temperatures, and acentric factors of fifty ionic liquids," *Industrial and Engineering Chemistry Research*, vol. 46, no. 4, pp. 1338–1344, 2007.

[31] K. Mishima, K. Matsuyama, H. Ishikawa, K. I. Hayashi, and S. Maeda, "Measurement and correlation of solubilities of azo dyes and anthraquinone in supercritical carbon dioxide," *Fluid Phase Equilibria*, vol. 194–197, pp. 895–904, 2002.

[32] J. L. Anthony, E. J. Maginn, and J. F. Brennecke, "Solution thermodynamics of imidazolium-based ionic liquids and water," *Journal of Physical Chemistry B*, vol. 105, no. 44, pp. 10942–10949, 2001.

Permissions

All chapters in this book were first published by Hindawi Publishing Corporation; hereby published with permission under the Creative Commons Attribution License or equivalent. Every chapter published in this book has been scrutinized by our experts. Their significance has been extensively debated. The topics covered herein carry significant findings which will fuel the growth of the discipline. They may even be implemented as practical applications or may be referred to as a beginning point for another development.

The contributors of this book come from diverse backgrounds, making this book a truly international effort. This book will bring forth new frontiers with its revolutionizing research information and detailed analysis of the nascent developments around the world.

We would like to thank all the contributing authors for lending their expertise to make the book truly unique. They have played a crucial role in the development of this book. Without their invaluable contributions this book wouldn't have been possible. They have made vital efforts to compile up to date information on the varied aspects of this subject to make this book a valuable addition to the collection of many professionals and students.

This book was conceptualized with the vision of imparting up-to-date information and advanced data in this field. To ensure the same, a matchless editorial board was set up. Every individual on the board went through rigorous rounds of assessment to prove their worth. After which they invested a large part of their time researching and compiling the most relevant data for our readers.

The editorial board has been involved in producing this book since its inception. They have spent rigorous hours researching and exploring the diverse topics which have resulted in the successful publishing of this book. They have passed on their knowledge of decades through this book. To expedite this challenging task, the publisher supported the team at every step. A small team of assistant editors was also appointed to further simplify the editing procedure and attain best results for the readers.

Apart from the editorial board, the designing team has also invested a significant amount of their time in understanding the subject and creating the most relevant covers. They scrutinized every image to scout for the most suitable representation of the subject and create an appropriate cover for the book.

The publishing team has been an ardent support to the editorial, designing and production team. Their endless efforts to recruit the best for this project, has resulted in the accomplishment of this book. They are a veteran in the field of academics and their pool of knowledge is as vast as their experience in printing. Their expertise and guidance has proved useful at every step. Their uncompromising quality standards have made this book an exceptional effort. Their encouragement from time to time has been an inspiration for everyone.

The publisher and the editorial board hope that this book will prove to be a valuable piece of knowledge for researchers, students, practitioners and scholars across the globe.

List of Contributors

A. S. Madhusudhan Rao
Department of Physics, Varadha Reddy College of Engineering, Warangal 506009, India

K. Narender
Kakatiya University, Warangal 506009, India

Muhamed BijediT and Sabina BegiT
Faculty of Technology, University of Tuzla, 75000 Tuzla, Bosnia and Herzegovina

Dileep Singh Chauhan and Rashmi Agrawal
Department of Mathematics, University of Rajasthan, Jaipur 302004, India

Jonas Johansson
Solid State Physics, Lund University, P.O. Box 118, 221 00 Lund, Sweden

Harmandeep Singh Gill and V. K. Rattan
Dr. SSB University Institute of Chemical Engineering and Technology, Panjab University, Chandigarh 160014, India

S. Mohammed Ibrahim
Department of Mathematics, Priyadarshini College of Engineering and Technology, Nellore 524004, Andhra Pradesh, India

T. Sankar Reddy
Deparment of Mathematics, Annamacharya College of Engineering and Technology, Cuddapa 516001, Andhra Pradesh, India

N. Bhaskar Reddy
Department of Mathematics, Sri Venkateswara University, Tirupathi 517502, Andhra Pradesh, India

M. V. Rathnam and Devappa R. Ambavadekar
Physical Chemistry Research Laboratory, B.N. Bandodkar College of Science, Thane 400601, India

M. Nandini
Department of Chemistry, Dr. P.R. Ghogrey Science College, Deopur, Dhule 424005, India

Mulu Berhe Desta
Department of Chemistry, College of Natural and Computational Sciences, Mekelle University, P.O. Box 231, Mekelle, Ethiopia

Daming Gao, Hong Sun, Hui Zhang, Dechun Zhu and Jianjun Shi
Department of Chemistry and Materials Engineering, Hefei University, Anhui, Hefei 230022, China
Sino-German Research Center for Process Engineering and Energy Technology, Anhui, Hefei 230022, China

Peter Lücking
Sino-German Research Center for Process Engineering and Energy Technology, Anhui, Hefei 230022, China
Department of Engineering, Jade University of Applied Science, 26389 Wilhelmshaven, Germany

Jingyu Si and Hong Chen
Department of Chemistry and Materials Engineering, Hefei University, Anhui, Hefei 230022, China

Arvind R. Mahajan and Sunil R. Mirgane
P. G. Department of Chemistry, Jalna Education Society's R. G. Bagdia Arts, R. Benzonji Science College, S. B. Lakhotia Commerce, Jalna, Maharashtra 431 203, India

Kh. Abdul Maleque
American International University-Bangladesh, House 23, 17, Kamal Ataturk Avenue, Banani, Dhaka-1213, Bangladesh

A. Salhoumi
Faculté des Sciences Juridiques Economiques et Sociales Ain Sebaa, Université Hassan II-Mohammedia, Beausite, BP 2634, Ain sebaa, Casablanca, Morocco
Laboratoire de Physique de la Matiére Condensée (LPMC), Faculté des Sciences Ben M'sik, Université Hassan II-Mohammedia, BP 7955, Casablanca, Morocco

Vladimir N. Pokrovskii
Moscow State University of Economics, Statistics and Informatics, Moscow 119501, Russia

Hamit Yurtseven and Özcan Tilki
Department of Physics, Middle East Technical University, 06531 Ankara, Turkey

Mustafa Kurt
Department of Physics, C¸ anakkale 18 Mart University, 17100 C¸ anakkale, Turkey

Nejib Hidouri, Imen Chermiti and Ammar Ben Brahim
University of Gabes, Engineers National School, Chemical and Process Engineering Department, Applied Thermodynamics Unit, Omar Ibn El Khattab Street, 6029 Gabés, Tunisia

Krishnendu Bhattacharyya
Department of Mathematics, The University of Burdwan, Burdwan, West Bengal 713104, India

Zeinab A. M. Khalel, Ali A. Rabah and Taj Alasfia M. Barakat
Department of Chemical Engineering, Faculty of Engineering, University of Khartoum, P.O. Box 321, Khartoum 11111, Sudan

Olaseni Segun Esan and Christopher Olumuyiwa Aboluwoye
Chemistry Department, Adekunle Ajasin University, Akungba-Akoko 001, Nigeria

Medinat Olubunmi Osundiya
Chemistry Department, Lagos State University, Ojo Apapa, Lagos 0001, Nigeria

Owoyomi Olanrewaju and Jide Ige
Department of Chemistry, Obafemi Awolowo University, Ile Ife 432, Nigeria

Samuel O. Adesanya
Department of Mathematical Sciences, Redeemer's University, Redemption City, PMB 2011, Sagamu, Ogun State, Nigeria

Adnan Saeed Butt and Asif Ali
Department of Mathematics, Quaid-I-Azam University, Islamabad 45320, Pakistan

Boris I. Sedunov
The Information Systems and Computer Technologies Department, Russian New University, Radio St. 22, Moscow 105005, Russia

Arunagiri Appusamy, Prabisha Purushothaman and Kalaichelvi Ponnusamy
Department of Chemical Engineering, National Institute of Technology, Tiruchirappalli, Tamilnadu 620015, India

Anantharaj Ramalingam
Department of Chemical Engineering, Faculty of Engineering, University of Malaya, Kuala Lumpur 50603, Malaysia

Ravinder Kumar and Avdhesh Kr. Sharma
Department of Mechanical Engineering, D.C.R University of Science and Technology, Murthal, Sonepat, India

P. C. Tewari
Department of Mechanical Engineering, N.I.T Kurukshetra, Haryana, India

Sk. Md Nayeem
Department of Physics, KRK Govt. Degree College, Addanki 523 201, India

M. Kondaiah
Department of Physics, NM Govt. Degree College, Jogipet, Medak District, Telangana 502270, India

K. Sreekanth
Department of Physics, PBN College, Guntur District, Nidubrolu 522124, India

D. Krishna Rao
Department of Physics, Acharya Nagarjuna University, Nagarjuna Nagar 522 510, India

Mustafa Cem Lider and Hamit Yurtseven
Department of Physics, Middle East Technical University, 06531 Ankara, Turkey

S. Mohammed Ibrahim
Department of Mathematics, Priyadarshini College of Engineering and Technology, Nellore 524004, India

N. Bhashar Reddy
Department of Mathematics, Sri Venkateswara University, Tirupat 517501, India

S. P. Abbasi and A. Alimorady
Iranian National Center for Laser Science and Technology, No. 71-20th North Kargar, P.O. Box 33665-576, Tehran, Iran

D. Harish Babu
Department of Humanities and Basic Science, SVPCET, Puttur 517 583, India

P. V. Satya Narayana
Department of Mathematics, VIT University, Vellore 632 014, India

B. J. Gireesha and B.Mahanthesh
Department of Studies and Research in Mathematics, Kuvempu University, Shankaraghatta, Shimoga, Karnataka, 577 451, India

Helen R.Mazzer, José C. O. Santos, Vladimir F. Cabral and Lucio Cardozo-Filho
Department of Chemical Engineering, State University of Maringá, Avenue Colombo 5790, 87020-900 Maringá PR, Brazil

Claudio Dariva
Institute for Research and Technology, ITP, PEP/UNIT, Avenue Murilo Dantas 300, 49032-490 Aracaju SE, Brazil

Marcos H. Kunita and Adley F. Rubira
Department of Chemistry, State University of Maringá, Avenue Colombo 5790, 87020-900 Maringá PR, Brazil

Martin Aznar
Department of Chemical Process, School of Chemical Engineering, State University of Campinas, Avenue Albert Einstein 500, 13081-970 Campinas SP, Brazil